Online Panel Research

Online Panel Research

A Data Quality Perspective

Editors

Mario Callegaro
Survey Research Scientist, Quantitative Marketing Team, Google UK

Reg Baker
Senior Consultant, Market Strategies International, USA

Jelke Bethlehem
Senior Advisor, Statistics Netherlands, The Netherlands

Anja S. Göritz
Professor of Occupational and Consumer Psychology
University of Freiburg, Germany

Jon A. Krosnick
Professor of Political Science, Communication and Psychology
Stanford University, USA

Paul J. Lavrakas
Independent Research Psychologist/Research Methodologist, USA

WILEY

Library of Congress Cataloging-in-Publication Data

Online panel research : a data quality perspective / edited by Mario Callegaro, Reg Baker, Jelke Bethlehem, Anja Goritz, Jon A. Krosnick, Paul J. Lavrakas.
       pages cm
    Includes bibliographical references and index.
    ISBN 978-1-119-94177-4 (pbk.)
    1.   Panel analysis.   2.   Internet research.   3.   Marketing research–Methodology   4.   Social sciences–Methodology
I. Callegaro, Mario, editor of compilation.
    H61.26.O55 2014
    001.4'33–dc23

                                                                                              2013048411

A catalogue record for this book is available from the British Library.

ISBN: 978-1-119-94177-4

Set in 10/12pt TimesLTStd by Laserwords Private Limited, Chennai, India

1 2014

Contents

Part V   NONRESPONSE AND MEASUREMENT ERROR   311

Part VII OPERATIONAL ISSUES IN ONLINE PANELS   409

Preface

Web survey data collected via online panels are an important source of information that thousands of researchers use to gain insight and which influence their decisions about the topics under study. Thousands upon thousands of completed surveys can be collected in a few days and for little money, as low as $2 per completed questionnaire (Terhanian & Bremer, 2012). Therefore, with the data user's point of view in mind, it is prudent to ask the questions: how reliable are these data and what risks do we assume when we use them?

Content of the book

This edited volume is one of the first attempts to carefully examine the quality of data from online panels. The book is organized into seven parts and contains 19 chapters. The first two chapters are critical reviews of what is known so far about the development and current status of online panel research and the quality of data obtained using these panels in terms of point estimates, relationships across variables, and reproducibility of results.

Part I discusses coverage error, in which three international case studies from probability-based panels in Germany, Finland, and the United States are reviewed.

Nonresponse is a much-debated topic that is addressed in Part II, in four chapters, examining the issue related to nonresponse in online panels from different points of view. Chapter 6 looks at nonresponse and attrition in the Dutch LISS panel, followed by Chapter 7, a review of determinants of response quantity, analyzing a large dataset of online panel surveys. In Chapter 8, data from Austria are presented to advance understanding of the reasons respondents join an online panel. The final chapter explores the effectiveness of experiments in engaging panel members by providing feedback on their participation.

Part III examines measurement error from two angles: the issue of professional respondents (Chapter 10) and the issue of speeding when completing web surveys (Chapter 11).

In Part IV, the next issue explored is adjustment of data obtained from online panels. Chapter 12 describes an example of how propensity score weighting was used to calibrate a non-probability panel and its effect when comparing the results to official statistics benchmarks. Imputation, which is another way to compensate for nonresponse, is the topic of Chapter 13. Taking advantage of a unique study in which online panel members were invited to join a specific project and thus some variables for the nonrespondents were already known, the chapter evaluates the effectiveness of imputation.

Survey researchers have demonstrated on many occasions (e.g., Groves & Peytcheva, 2008) the importance of examining the relation between nonresponse and measurement error. This relation is evaluated in Part V, in contributions by two teams, one from the United States (Chapter 14) and one from the United Kingdom (Chapter 15), both using multiple datasets. These complementary chapters demonstrate, using advanced analytics, the numerous ways of exploring the relation between nonresponse and measurement error.

Part VI on special domains looks at two recent topics: (1) the impact of smartphones on panel data collection in Chapter 16; and (2) online ratings panels in Chapter 17. In the latter type of panel, the Internet behavior of its members is tracked, providing a constant flow of traffic that can be analyzed.

The volume ends with Part VII, an examination of two important operational issues: (1) the software used to manage online panels in Chapter 18; and (2) how to validate respondents who join a panel in Chapter 19.

References

Groves, R. M., & Peytcheva, E. (2008). The impact of nonresponse rates on nonresponse bias: A meta-analysis. *Public Opinion Quarterly, 72*, 167–189.

Terhanian, G., & Bremer, J. (2012). A smarter way to select respondents for surveys? *International Journal of Market Research, 54*, 751–780.

Acknowledgments

The book's initial idea was suggested by Lars Lyberg to Wiley. We are indebted to Lars for his forward thinking. We thank our Wiley editors Richard Davies, Kathryn Sharples, Heather Kay, and Jo Taylor.

Companion datasets and book format

All authors agreed to make the datasets used in the analysis publicly available. Wiley is providing a section of the book's webpage dedicated to showing readers how to obtain the anonymized datasets. The GESIS data archive kindly provided support for some datasets not archived elsewhere. We hope other similar volumes will do the same in the future. Transparency like this is a key to scientific credibility and progress. By allowing other scholars to analyze and enhance the findings of studies, these chapters will have even more impact and value over time. (Cooper, 2013; Fang, Steen, & Casadevall, 2012).

The book is released in different formats: print, e-book, and pdf version. Single chapters can be purchased or accessed online on the Wiley website where an abstract is also available. Online appendies and instructions on how to access the datasets used in the chapters are available at: www.wiley.com/go/online_panel

References

Cooper, J. (2013). On fraud, deceit and ethics. *Journal of Experimental Social Psychology*, *49*, 314.

Fang, F. C., Steen, R. G., & Casadevall, A. (2012). Misconduct accounts for the majority of retracted scientific publications. *Proceedings of the National Academy of Sciences of the United States of America*, *109*, 17028–17033.

Mario Callegaro
Reg P. Baker
Jelke Bethlehem
Anja S. Göritz
Jon A. Krosnick
Paul J. Lavrakas

About the Editors

Reg Baker is the former President and Chief Operating Officer of Market Strategies International, a full-service research company specializing in healthcare, energy, financial services, telecommunications, and information technology. Prior to joining Market Strategies, he was Vice President for Research Operations at NORC where he oversaw the national field staff, the company's CATI centers, and its technology infrastructure.

Over the course of his almost 40-year career, Reg has focused on the methodological, operational and management challenges of new survey technologies including CATI, CAPI, Web, and now mobile. He writes, presents and consults on these and related issues to diverse national and international audiences and has worked with a wide variety of clients in both the private and public sectors, including substantial research with academic survey methodologists on web survey methods. He was the Chair of the AAPOR Task Force on Online Panels and Co-Chair of the AAPOR Task Force on Non-Probability Sampling. He serves as a consultant to the ESOMAR Professional Standards Committee and has worked on numerous project teams for that association, producing a variety of quality and ethical guidelines including its "28 Questions to Help Buyers of Online Samples" and its "Guideline for Conducting Mobile Marketing Research." He also is an adjunct instructor in the Master of Science in Marketing Research Program at Michigan State University and is a member of the Executive Editorial Board of the *International Journal of Market Research*. He continues to consult with Market Strategies and other private and public organizations, including the U.S. Bureau of the Census.

He blogs off and on as The Survey Geek.

Jelke Bethlehem, PhD, is Senior Advisor in the Methodology Team of the Division of Process Development, IT and Methodology at Statistics Netherlands. He is also Professor of Survey Methodology at Leiden University. He studied mathematical statistics at the University of Amsterdam. His PhD was about nonresponse in surveys. He worked for over 30 years at Statistics Netherlands. His research topics were disclosure control, nonresponse, weighting adjustment, and web surveys. In the 1980s and 1990s he was in charge of the development of Blaise, a software system for computer-assisted survey data collection. He has participated in a number of European research projects financed by the European Union.

Dr. Bethlehem's current research interests include web surveys, computer-assisted survey information collection, and graphical techniques in statistics. He is the author of other books published by Wiley: *Applied Survey Methods, Handbook of Nonresponse in Household Surveys* and *Handbook of Web Surveys*.

Mario Callegaro is Survey Research Scientist in the Quantitative Marketing team at Google UK. He works on web survey design and focuses on measuring customer satisfaction. He also consults with numerous internal teams regarding survey design, sampling, questionnaire design, and online survey programming and implementation.

Mario holds a BA in Sociology from the University of Trento, Italy, and an MS and PhD in Survey Research and Methodology from the University of Nebraska, Lincoln. Prior to joining Google, Mario was working as survey research scientist for Gfk-Knowledge Networks.

Mario has published over 30 peer-reviewed papers and book chapters and made over 100 conference presentations nationally and internationally in the areas of web surveys, telephone and cell phone surveys; question wording, polling and exit polls; event history calendar; longitudinal surveys, and survey quality. He is associate editor of *Survey Research Methods*, a member of the editorial board of the *International Journal of Market Research*, and reviewer for other survey research-oriented journals. His latest book entitled *Web Survey Methodology* (with Katja Lozar Manfreda and Vasja Vehovar) is forthcoming with Sage.

Anja S. Göritz is a Full Professor of Occupational and Consumer Psychology at the University of Freiburg in Germany. Anja holds a graduate degree in Psychology from the University of Leipzig and a PhD in Organizational and Social Psychology from the University of Erlangen-Nürnberg, Germany. Her research focuses on web-based data collection, market psychology and human–computer interaction. She also consults with international clients regarding design, programming and implementation of web surveys.

Anja has taught graduate and post-graduate courses on research methods and web-based data collection. Moreover, she has regularly been an instructor in the Advanced Training Institute "Performing Web-Based Research" of the American Psychological Association. In 2000, she built and has since maintained Germany's first university-based online panel with more than 20000 panelists. Anja programmed a number of open-source tools for web-based data collection and released them into the public domain.

Anja has published over 60 peer-reviewed papers and book chapters and made over 100 presentations at national and international academic conventions. In 2008, she chaired the program of the General Online Research Conference. She is an associate editor of *Social Science Computer Review* and a member of the editorial board of *International Journal of Internet Science*.

Jon A. Krosnick is Frederic O. Glover Professor in Humanities and Social Sciences and Professor of Communication, Political Science, and Psychology at Stanford University. A leading international authority on questionnaire design and survey research methods, Professor Krosnick has taught courses for professionals on survey methods for 25 years around the world, and has served as a methodology consultant to government agencies, commercial firms, and academic scholars. His books include *Introduction to Survey Research, Polling, and Data Analysis* and *The Handbook of Questionnaire Design* (forthcoming, Oxford University Press), which reviews 100 years of research on how different ways of asking questions can yield different answers from survey respondents and on how to design questions to measure most accurately. His recent research has focused on how other aspects of survey methodology (e.g., collecting data by interviewing face-to-face vs. by telephone or on paper questionnaires) can be optimized to maximize accuracy.

Dr. Krosnick is also a world-recognized expert on the psychology of attitudes, especially in the area of politics. He is co-principal investigator of the American National Election Study, the nation's preeminent academic research project exploring voter decision-making and political campaign effects. For 30 years, Dr. Krosnick has studied how the American public's political attitudes are formed, change, and shape thinking and action. His publications explore the causes of people's decisions about whether to vote, for whom to vote, whether to approve of the President's performance, whether to take action to influence government policy-making on a specific issue, and much more.

Dr. Krosnick's scholarship has been recognized with the Phillip Brickman Memorial Prize, the Pi Sigma Alpha Award, the Erik Erikson Early Career Award for Excellence and Creativity, a fellowship at the Center for Advanced Study in the Behavioral Sciences, and membership as a fellow of the American Academy of Arts and Sciences.

As an expert witness in court, he has testified evaluating the quality of surveys presented as evidence by opposing counsel and has conducted original survey research to inform courts in cases involving unreimbursed expenses, uncompensated overtime work, exempt/non-exempt misclassification, patent/trademark violation, health effects of accidents, consequences of being misinformed about the results of standardized academic tests, economic valuation of environmental damage, change of venue motions, and other topics.

At Stanford, Dr. Krosnick directs the Political Psychology Research Group (PPRG). PPRG is a cross-disciplinary team of scholars who conduct empirical studies of the psychology of political behavior and studies seeking to optimize research methodology for studying political psychology. The group's studies employ a wide range of research methods, including surveys, experiments, and content analysis, and the group often conducts collaborative research studies with leading news media organizations, including ABC News, The Associated Press, the *Washington Post*, and *Time Magazine*. Support for the group's work has come from U.S. Government agencies (e.g., the National Science Foundation, the Bureau of Labor Statistics), private foundations (e.g., the Bill and Melinda Gates Foundation, the Robert Wood Johnson Foundation), and Institutes at Stanford (e.g., the Woods Institute for the Environment). Dr. Krosnick also directs the Summer Institute in Political Psychology, an annual event that brings 60 students and professionals from around the world to Stanford for intensive training in political psychology theory and methods.

Paul J. Lavrakas, PhD, a research psychologist, is a methodological research consultant for several private sector and not-for-profit organizations, and also does select volunteer service projects. He also is a Visiting Scholar, teaching research method courses, at Northern Arizona University.

He was a Nielsen Vice President and their chief methodologist (2000–2007); Professor of Journalism/Communications (Northwestern, 1978–1996; Ohio State, 1996–2000); and founding faculty director of the NU Survey Lab (1982–1996) and the OSU Center for Survey Research (1996–2000). Born in Cambridge, MA, educated in the public schools of Birmingham, MI, he earned a BA from Michigan State University (Psychology/ Political Science/History). He taught fifth grade on Chicago's Southside (1968–1972), then earned an MA in experimental social psychology (1975) and a PhD in applied social psychology (1977) from Loyola University. He helped establish the Westinghouse Evaluation Institute and conducted two decades of federally funded anti-crime evaluation projects (1974–1993).

Dr. Lavrakas has written two editions *of Telephone Survey Methods: Sampling, Selection and Supervision* (Sage; 1987, 1993) and several chapters on telephone survey methods (e.g., in *Handbook of Survey Research*; Elsevier, 2010). He was editor of the 2007 *Public Opinion Quarterly* issue on cell phone surveying and co-editor of *Advances in Telephone Survey Methodology* (Wiley, 2008). He is also the editor of the *Encyclopedia of Survey Research*

Methods (Sage, 2008). He was the editor of a special issue of *Public Opinion Quarterly* on Cell Phone Surveying, and organized and chaired the 2007–2008 and 2009–2010 AAPOR task forces on Cell Phone Surveying. He also has edited three books on media election polling and co-authored four editions of *The Voter's Guide to Election Polls* (with M. Traugott). He has served on the election night decision teams of VNS (1996–2002), ABC (2008), and The AP (2010–2012).

An AAPOR member since the 1980s, he has served as Conference Operations chair (1995–1997), Conference Program chair (1998–1999), Counselor at Large (2008–2010); Vice President (2011–2012), President (2012–2013) and Past-President (2013–2014). In addition, he is a MAPOR Fellow (1997) and has received NYAAPOR's Award for Outstanding Achievement (2007).

About the Contributors

Nick Allum is Professor of Sociology at the University of Essex. He earned his PhD from the London School of Economics and his research interests are in survey methodology, social and political trust and on attitudes, beliefs and knowledge about science and technology.

Wolfgang Bandilla is a researcher at the department of Survey Design and Methodology at GESIS – Leibniz Institute for the Social Sciences, Mannheim, Germany. His research interests include web survey methodology and mixed-mode surveys.

Bernad Batinic is Full Professor of Work, Organizational, and Media Psychology at the Institute of Education and Psychology, University of Linz, Austria. Before joining the University of Linz, he was a Deputy Professor at the Department of Work and Organizational Psychology, University of Marburg, Germany, and Assistant Professor at the Department of Psychology, especially Organizational and Social Psychology at the University of Erlangen-Nuremberg, Germany. Batinic received his Diploma in Psychology from the University of Giessen, Germany, and his PhD (Dr.rer.pol.) from the University of Erlangen-Nuremberg. His research spans a wide range of basic and applied topics within (online) market research, unemployment and health, latent deprivation model (Marie Jahoda), opinion leadership, and media psychology.

Annamaria Bianchi is a researcher in Statistics applied in Economics at the University of Bergamo, Italy. She obtained her PhD in Mathematical Statistics in 2007 at the University of Milan. Her main research interests are online panels, panel data analysis and representativeness, performance indicators and M-quantile regression with application to problems of innovation and entrepreneurship.

Silvia Biffignandi is Professor of Economic and Business Statistics and Director of the Centre for Statistical Analyses and Survey Interviewing (CASI) at the University of Bergamo, Italy. She currently focuses her research in the areas of web surveys, online panels, and official statistics.

Marcel Das holds a PhD in Economics from Tilburg University (1998). In 2000, he became the director of CentERdata, a survey research institute specializing in web-based surveys and applied economic research, housed at the campus of Tilburg University. As a director of CentERdata he has managed a large number of national and international research projects. Since February 2009, Das has also been Professor of Econometrics and Data Collection at Tilburg

University. He has published in international peer-reviewed journals in the field of statistical and empirical analysis of survey data and methodological issues in web-based (panel) surveys.

Frank Drewes is Director of Marketing Science at Harris Interactive AG, Germany. He is responsible for the development and implementation of innovative market research methods and tools. His research experience in market research agencies spans more than 12 years. Frank Drewes holds a Diploma in Psychology from the University of Bielefeld, Germany. His major fields of study were Psychological Diagnostics and Psychological Methodology.

Robert Greszki is research associate at the Chair of Political Sociology at the University of Bamberg. His research interests include survey methodology, online surveys, quantitative methods of empirical social research, and political attitudes.

Kimmo Grönlund is a Professor of Political Science and the Director of the Social Science Research Institute at Åbo Akademi University, Finland. He is the Principal Investigator of the Finnish National Election Study, 2012–2015. His major research interests include political behavior in general and electoral behavior in particular, the role of social and institutional trust in democracy, as well as experimental research especially in deliberative democracy. He has published on the topics in journals such as the *European Political Science Review*, *Political Studies*, *Scandinavian Political Studies*, *Electoral Studies* and the *American Review of Public Administration*.

D. **Sunshine Hillygus** is Associate Professor of Political Science at Duke University, NC, USA, and Director of the Duke Initiative on Survey Methodology. Her research specialties include public opinion, campaigns and elections, survey research, and information technology and society. She is co-author of *The Hard Count: The Social and Political Challenges of the 2000 Census* (Russell Sage Foundation, 2006) and *The Persuadable Voter: Wedge Issues in Political Campaigns* (Princeton University Press, 2008). She serves on Census Scientific Advisory Committee, the board of the American National Election Study, and the editorial boards of several leading academic journals. She received her PhD in political science from Stanford University in 2003.

Natalie Jackson is Senior Analyst at the Marist College Institute for Public Opinion (Marist Poll). She has a PhD in political science from the University of Oklahoma with additional training at the ICPSR Summer Program for Quantitative Methods, and was a postdoctoral associate with the Duke Initiative on Survey Methodology prior to joining the Marist Poll staff. She has worked in the survey research field since 2006, and has designed and implemented surveys in a wide variety of contexts. At Marist, she focuses on transparency issues and methodological developments in the field in addition to broader work on American public opinion and voting behavior.

Dinaz Kachhi is Senior Manager, Research Insights, at uSamp. She consults with clients on the use of appropriate methodology for gauging brand awareness, lift, advocacy, campaign, and ad effectiveness. Dinaz also is involved in research in the online space to continuously improve and enhance current measurement practices.

Lars Kaczmirek is team leader and researcher in the department Monitoring Society and Social Change at GESIS – Leibniz Institute for the Social Sciences in Germany. He studied psychology with minors in neurology and computer science and holds a PhD in psychology from the University of Mannheim. He was the project leader of the GESIS Online Panel Pilot.

Lars is editor of *Survey Methods: Insights from the Field* and on the board of the German Society for Online Research (DGOF). He has published on survey research methods, for example, questionnaire design, mixed-mode, survey software tools, and human–survey interaction.

Olena Kaminska is a survey statistician at the Institute for Social and Economic Research at the University of Essex, UK. She has received a Doctoral Degree from the University of Nebraska-Lincoln, USA. Her research interests comprise a wide range of topics within the survey methodology field, including topics on statistical weighting, retention in mixed mode studies, and studying measurement error with a real-world eye-tracker, among others.

Florian Keusch is a Post-doctoral Research Fellow in Survey Methodology at the Institute for Social Research (ISR), University of Michigan. Before joining ISR he was a Research Associate at the Institute for Advertising and Marketing Research at WU, Vienna University of Economics and Business, Austria. He received a PhD in Social and Economic Sciences (Dr.rer.soc.oec.) and an MSc in Business (Mag.rer.soc.oec.) from WU. His main research interests include various aspects of survey methodology with a particular focus on factors influencing response behavior and measurement error in web surveys.

Keith Lange is Senior Technical Manager at uSamp. His work focuses on statistical analysis, survey programming, and software development.

Peter Lugtig is a survey methodologist with a particular interest in modeling survey errors in longitudinal surveys. He holds a PhD in survey methodology, and currently works as an Assistant Professor in the Department of Methods and Statistics at Utrecht University, and as a Senior Research Officer at the Institute for Social and Economic Research at the University of Essex. He has published scientific articles in international journals on attrition in panel surveys, modeling survey errors in mixed-mode surveys, and the effects of questionnaire characteristics on survey data quality.

Tim Macer is founder of Meaning Ltd, a UK-based consulting company specializing in the application of technology to market and opinion research. He is widely published and contributes regularly to ESOMAR's *Research World* and *Quirk's Marketing Research Review* on software and technology. Tim's work on a technology project for the Internet bank Egg won the MRS Research Excellence Best New Thinking award in 2007. In 2008, with Dr. David F. Birks, he edited *Marketing Research: Critical Perspectives*, published by Routledge. In 2012, Tim was appointed Honorary Research Fellow at the University of Winchester, working in the field of digital research.

Neil Malhotra is Associate Professor of Political Economy at the Stanford Graduate School of Business. His research interests include American politics, public opinion, and survey methodology. His research has been published in the *American Political Science Review*, the *American Journal of Political Science*, the *Journal of Politics*, *Public Opinion Quarterly*, and the *Proceedings of the National Academy of Science*, among other outlets.

Wolfgang Mayerhofer is Associate Professor and Deputy Head of the Institute for Marketing and Consumer Research at WU, Vienna University of Economics and Business, Austria. He received a PhD in Social and Economic Sciences (Dr.rer.soc.oec.) from WU. His main research interests include various aspects of marketing and consumer research such as brand image, brand positioning, brand extensions, brand equity, country-of-origin effect, and measuring the effectiveness of advertising, including methods that employ biological and observable procedures such as eye tracking or electrodermal activity.

Allan L. McCutcheon is Donald O. Clifton Chair of Survey Science, Professor of Statistics, founding Director of the Gallup Research Center, and founding Chair of the Survey Research and Methodology graduate program at the University of Nebraska-Lincoln. He is a Senior Scientist with Gallup and Gallup-Europe, and an elected Fellow of the American Statistical Association, the Royal Statistical Society, and the Midwest Association for Public Opinion Research. McCutcheon serves as General Secretary of the World Association for Public Opinion Research. His interests include statistical models for the analysis of categorical data, web surveys and online panels, and cross-national survey research.

Marco Meyer is research associate at the Chair of Political Sociology at the University of Bamberg. His research interests include political attitudes, political behavior, and questions of political psychology.

Chuck Miller is President of Digital Marketing & Measurement, LLC, an agency focused on emerging media. During his 20+ years in marketing research Chuck has built significant capabilities for both provider and end client organizations, including a dozen years as a VP at AOL and Time Warner, leading the Consumer Insights and Business Intelligence departments.

Joanne M. Miller is Associate Professor of Political Science at the University of Minnesota. Her first research program explores the direct and indirect effects of the media on citizens' political attitudes. Her second research program examines the impact of psychological motivations on citizens' decisions to become politically active. She applies theories from political science, psychology, and mass communication and uses multiple research methods to illuminate the processes by which citizens become politically engaged. Her research has been published in the *American Journal of Political Science*, the *Journal of Politics*, and *Public Opinion Quarterly*, among other outlets.

Philip M. Napoli (PhD, Northwestern University) is a Professor of Journalism & Media Studies in the School of Communication and Information at Rutgers University. His research focuses on media institutions, audiences, and policy. He is the author/editor of five books. His most recent book is *Audience Evolution: New Technologies and the Transformation of Media Audiences* (Columbia University Press, 2011). Professor Napoli's research has received awards from organizations such as the National Business & Economics Society, the Association for Education in Journalism & Mass Communication, and the International Communication Association. His research has been funded by organizations such as the Ford Foundation, the Open Society Institute, and the New America Foundation.

Kumar Rao has been Director of the Statistical Center of Innovation at The Nielsen Company and also the co-editor of a survey research and public opinion online journal – *Survey Practice* (www.surveypractice.org). While his recent interest is in web analytics and online surveys, his previous research has focused on mode effects and analysis of panel data for examining panel attrition, recruitment and retention. More information on his research work can be found on his website (www.kumarrao.net). Mr. Rao holds a Master's degree in Survey Methodology and Master's and Bachelor's degree in engineering and management.

Caroline Roberts is Assistant Professor in Survey Research Methods at the University of Lausanne, Switzerland. Her research interests are in the field of data collection methodology and survey error. Before Lausanne, she worked at Stanford University as a member of the team coordinating the 2008–2009 American National Election Studies Internet Panel Survey. From 2004–2008, she was Senior Research Fellow at City University, London, where she coordinated methodological research into mixed mode data collection for the Core Scientific Team of the European Social Survey. She has a PhD in Social Psychology from the London School of Economics.

Ines Schaurer is a researcher at the department of Survey Design and Methodology at GESIS – Leibniz Institute for the Social Sciences in Germany. She studied social sciences with focus on methods of social research, sociology, and social psychology at the University of Mannheim and at the University of Utrecht. Currently, she is working in the project team which is setting up the GESIS Panel, a probability-based mixed-mode access panel for the social sciences. She is writing her PhD about recruitment processes of probability-based online panels.

Annette Scherpenzeel is methodologist and the cohort manager of the Youth and Identity cohort study at the Utrecht University. Prior to joining Utrecht, she was the project leader of the MESS project and LISS panel at CentERdata (a research institute associated with the Tilburg University), the Netherlands. She has more than 20 years of experience in panel surveys and online interviewing. Starting at the Telepanel in Amsterdam, she worked at the University of Amsterdam and at Telecom Netherlands (KPN). Next, she helped build a Swiss social-economical panel and worked for several years as a methodologist for this panel. Her research fields are online interviewing, panel data, structural equation modeling, response effects, well-being, health and satisfaction.

Harald Schoen is Professor of Political Science at the University of Bamberg and holds the Chair of Political Sociology. His research interests include in particular voting behavior and the cognitive foundations of voter choice.

Stephanie Steinmetz is an Assistant Professor of Sociology at the University of Amsterdam, the Netherlands, and a senior researcher in the project "Improving web survey methodology for social and cultural research" at the Erasmus Studio Rotterdam, where she studies methodological issues concerning volunteer web surveys. She obtained her PhD at the University of Mannheim, Germany, in 2009. Her main research interests are quantitative research methods and web survey methodology, social stratification and gender inequalities.

Kim Strandberg holds a PhD in Political Science and is a Senior Researcher and Associate Professor at the Social Science Research Institute in the Department of Politics and Administration at Åbo Akademi University, Finland. His primary areas of research are political communication, citizen deliberation, and political uses of the Internet. He has published on these topics in journals such as *Party Politics*, *New Media & Society*, *Information Polity*, *Scandinavian Political Studies* and *Journal of Information Technology and Politics*.

Bella Struminskaya is a researcher in the Department of Survey Design and Methodology at GESIS – Leibniz Institute for the Social Sciences in Germany. She studied sociology and economics at Novosibirsk State University and at the University of Marburg. She holds a Master's degree in sociology from the University of Mannheim. Bella is writing her doctoral dissertation on data quality in online panel surveys. Currently, she is working in the project team, which is setting up the GESIS Panel, a probability-based mixed-mode access panel for the social sciences. She has published on identifying nonworking cell-phone numbers in the RDD samples in *Survey Practice*.

Patrick Sturgis is Professor of Research Methodology in the Department of Social Statistics at the University of Southampton and Director of the ESRC National Centre for Research Methods. He is Principal Investigator of the Wellcome Trust Monitor study and President of the European Survey Research Association. His main research interests are in the areas of survey and statistical methods, public opinion and political behavior, particularly regarding social cohesion and trust and public attitudes to science and technology.

Kea Tijdens is a Research Coordinator at the Amsterdam Institute of Advanced Labour Studies (AIAS) at the University of Amsterdam, the Netherlands, and a Professor of Women's Work at the Department of Sociology, Erasmus University Rotterdam. She is the scientific coordinator of the continuous WageIndicator web-survey on work and wages, and has analysed the data at large. Her research concentrates on wage setting processes, women's work, occupational segregation, and industrial relations, as well as on web survey methodologies, such as the measurement of occupations and the bias in the volunteer web survey.

Vera Toepoel is Assistant Professor at the Department of Methods & Statistics at Utrecht University, the Netherlands. Vera previously worked for the Dutch Knowledge Centre Tourism, University of Applied Sciences INHOLLAND, CentERdata, and Tilburg University. She recently obtained a VENI grant from the Dutch Scientific Organization and is the Chairwoman of the Dutch and Belgian Platform for Survey Research. She has published among others in: *Public Opinion Quarterly*, *Journal of Official Statistics*, *Field Methods*, *Sociological Research Methods*, *Personality and Individual Differences*, and *Social Science Computer Review*. Vera is currently working on the book, *Doing Surveys Online*, which will be published by Sage (expected in 2014).

Jacob Tucker is a Research Analyst at uSamp where he designs, executes, analyzes, and presents results of research studies to internal and external clients. His current research interest is the use of mobile phones as a research platform.

Ana Villar is Research Fellow at the Centre for Comparative Social Surveys of City University London. She coordinates the mixed-mode research program of the European Social Survey. Her research interests include survey translation, answer scales design, web surveys, cross-cultural surveys and nonresponse error related to break-offs. Dr Villar studied Psychology at the University of Santiago de Compostela, Spain, where she also worked as researcher in the Department of Behavioral Sciences Methodology. She completed her doctorate at the University of Nebraska-Lincoln at the Survey Research and Methodology program, while working as research assistant at the Gallup Research Center.

Justin Wedeking is Associate Professor of Political Science at the University of Kentucky. His research and teaching interests focus on American politics, law and courts, judicial politics, public opinion, and political psychology. His research has been published in the *American Journal of Political Science*, the *Journal of Politics*, and *Law & Society Review*, among other outlets.

Lisa Wilding-Brown is Vice President, Global Panel & Sampling Operations at uSamp, a premier provider of technology and online samples. Lisa has managed the development and ongoing operation of a wide range of online research applications at both uSamp and Harris Interactive.

David Yeager is an Assistant Professor of Developmental Psychology at the University of Texas at Austin. His research focuses on social cognitive development during adolescence and on the psychology of resilience, particularly during difficult school transitions such as the transition to high school or college. He is also a research methodologist, with an emphasis on the psychology of asking and answering questions and on the role of representative sampling in social scientific research. He holds an MA in psychology and a PhD in education from Stanford University, and a BA in the Program of Liberal Studies from the University of Notre Dame.

McKenzie Young is currently an Associate at Global Strategy Group in the Research Team. She received her Master's in Political Science from Duke University in 2013 where she studied survey methodology, minority coalitions, and southern voting patterns in the United States. She received her BA from Elon University in 2011.

Weiyu Zhang is Assistant Professor in the Department of Communications and New Media, National University of Singapore. She holds a PhD in Communication from Annenberg School for Communication, University of Pennsylvania. Her research focuses on civic engagement and ICTs, with an emphasis on Asia. Her published works have appeared in *Communication Research*, *Information, Communication, & Society*, *International Communication Gazette*, *Asian Journal of Communication*, *Chinese Journal of Communication*, *Computers in Human Behavior*, *First Monday*, *Journal of Media Psychology*, *Javnost: the Public*, and many others. Her recent project was a cross-nation study on youth, new media, and civic engagement in five Asian countries.

1

Online panel research

History, concepts, applications and a look at the future

Mario Callegaro[a], Reg Baker[b], Jelke Bethlehem[c], Anja S. Göritz[d], Jon A. Krosnick[e], and Paul J. Lavrakas[f]

[a]*Google UK*

[b]*Market Strategies International, USA*

[c]*Statistics Netherlands, The Netherlands*

[d]*University of Freiburg, Germany*

[e]*Stanford University, USA*

[f]*Independent Research Psychologist/Research Methodologist, USA*

1.1 Introduction

Online panels have become a prominent way to collect survey data. They are used in many fields including market research (Comley, 2007; Göritz, 2010; Postoaca, 2006), social research (Tortora, 2008), psychological research (Göritz, 2007), election studies (Clarke, Sanders, Stewart, & Whiteley, 2008), and medical research (Couper, 2007).

Panel-based online survey research has grown steadily over the last decade. ESOMAR has estimated that global expenditures on online research as a percentage of total expenditures on quantitative research grew from 19% in 2006 to 35% in 2012.[1] Inside Research (2012), using

---

[1] Authors' computation of ESOMAR Global market research reports, 2006–2013.

Online Panel Research: A Data Quality Perspective, First Edition.
Edited by Mario Callegaro, Reg Baker, Jelke Bethlehem, Anja S. Göritz, Jon A. Krosnick and Paul J. Lavrakas.
© 2014 John Wiley & Sons, Ltd. Published 2014 by John Wiley & Sons, Ltd.
Companion website: www.wiley.com/go/online_panel

data from 37 market research companies and their subsidiaries, estimated that clients spent about $1.5 billion on online research in the United States during 2006 versus more than $2 billion during 2012. In Europe, expenditures were $490 million in 2006 and a little more than $1 billion in 2012.

1.2   Internet penetration and online panels

In principle, one might expect the validity of research using online panels is a function of the Internet penetration in the country being studied. The higher the household Internet penetration, the greater the chance that a panel may reflect the socio-demographic characteristics of the entire population. When Internet penetration is less than 100%, people who cannot access the Internet are presumably *not missing at random*. It seems likely that such individuals have socio-demographic, attitudinal, and other characteristics that distinguish them from the population of individuals who do have Internet access (Ragnedda & Muschert, 2013). In the United States, Internet access is positively associated with being male, being younger, having a higher income and more education, being Caucasian or Asian, and not being Hispanic (Horrigan, 2010). Similar patterns but with different race and ethnicity features were found in the Netherlands (Bethlehem & Biffignandi, 2012).

In the mid-1990s in the United States, Internet access at the household level was about 20% and increased to 50% by 2001 (U.S. Department of Commerce & National Telecommunications and Information Administration, 2010). By end of 2011, it had reached 75%[2]. As Comley (2007) has pointed out, the excitement over cost savings and fast turnaround in the United States was stronger than were concerns about sample representativeness and the ability to generalize the results to a larger population. In Europe, the average individuals Internet penetration among 28 countries in 2013 was 75% (Seybert and Reinecke 2013).[3] However, increasing Internet penetration does not necessarily mean that coverage bias has been decreasing. As the non-Internet population becomes smaller, it may become more different from the Internet population than when it was larger. So these figures do not necessarily assure increasing accuracy of online surveys. Further, measuring Internet penetration has become more challenging (Horrigan, 2010).[4]

1.3   Definitions and terminology

The meaning of the word "panel" in "online panel" is different from the traditional meaning of that word in the survey research world (Göritz, Reinhold, & Batinic, 2002). According to the traditional definition, "panel surveys measure the same variables with identical individuals at several points in time" (Hansen, 2008, p. 330). The main goal of a panel in this usage is to study change over time in what would be called a "longitudinal panel." In contrast, an online panel is a form of *access panel*, defined in the international standard, ISO 20252 "Market, opinion and social research – Vocabulary and Service Requirements," as "a sample database of potential respondents who declare that they will cooperate for future

---

[2] Our elaboration using the October CPS 2012 supplement in Dataferret.

[3] This is just a straight (unweighted) average of individual who used the internet within the last three months among 28 countries considered in the Eurostat survey.

[4] Measuring Internet penetration has become more challenging as more people use mobile devices to access the Internet. For example, in the United States, a growing portion of the population goes online only or primarily via their smartphone (Horrigan, 2010).

data collection if selected" (International Organization for Standardization, 2012, p. 1). These panels sometimes include a very large number of people (often one million or more) who are sampled on numerous occasions and asked to complete a questionnaire for a myriad of generally unrelated studies. Originally, these panels were called *discontinuous access panels* whose "prescreened respondents report over time on a variety of topics" (Sudman & Wansink, 2002, p. 2).[5] Panel members can be re-sampled (and routinely are) to take part in another study with varying levels of frequency.

1.3.1   Types of online panels

There are a number of different types of online panels. The most important distinction is between probability and nonprobability panels (described below). In the latter type, there is considerable variation in how panels are recruited, how panel members are sampled, how they are interviewed, the types of people on the panel, and the kinds of data typically collected.

Some panel companies "sell" potential respondent samples to researchers but do not host surveys. In these cases, panel members selected for a study receive a survey invitation from the panel company directing them to another website where the survey is hosted. Through the use of links built into the survey questionnaire, the panel provider can track which members started the survey, which were screened out, which aborted the survey, and which completed the survey. In this model, a panelist's experience with a survey is different every time in terms of the questionnaire's look and feel.

In contrast, other panel companies host and program all the questionnaires. Panel members therefore complete questionnaires that are consistent in terms of their layout, look and feel.

Finally, some panel companies use both approaches, depending on the preference of each client.

1.3.2   Panel composition

Panels also differ in terms of the types of members. In terms of membership, there are generally four types of online panels (see also Baker et al., 2010, p. 8):

- general population panel;
- specialty panel;
- proprietary panel;
- election panel.

General population panels are the most common. These panels tend to be very large and are recruited to include the diversity of the general population, sometimes including people in hard-to-reach subpopulations. A general population panel typically is used as a frame, from which samples are drawn based on the requirements of individual studies. Examples of studies using general population panels are found in Chapters 5, 6, 8, 10, and 11 in this volume.

Specialty panels are designed to permit study of subpopulations defined by demographics and/or behavioral characteristics. Examples include Hispanics, car owners, small business owners, and medical professionals. One form of a specialty panel is a *B2B panel*, the goal of which is to include diverse professionals working in specific companies. Individuals are

---

[5] Sudman and Wansink (2002) described the precursors of online panels, which involved regular data collection from individuals via paper questionnaire, face-to-face interviews, or telephone interview.

selected based on their roles in the company and the company's *firmographics*, meaning characteristics such as size, industry, number of employees, and annual revenues (Vaygelt, 2006). Specialty panels typically are built through a combination of recruiting from sources believed to have large numbers of people who fit the panel's specification and by screening a general population panel.

Proprietary panels are a subclass of specialty panels in which the members participate in research for a particular company. These panels are also called *client panels* (Poynter, 2010, p. 9), *community panels*, and, more recently, *insight communities* (Visioncritical University, 2013). They provide the opportunity for a company to establish a long-term relationship with a group of consumers – typically customers of products or services offered by the company – in a setting that allows for a mix of qualitative and quantitative research, of which surveying panels' members is just one method of research.

In *election panels*, people eligible to vote are recruited, and then the panel is subsampled during the months before (and perhaps after) an election to study attitude formation and change (Clarke et al., 2008). These panels resemble more traditional longitudinal panels, because each member is surveyed at each wave before and after the election. An example of an election panel is described in Chapter 4, and election panels are studied in Chapters 14 and 15.

Finally, some online panels rely on passive data collection rather than surveys. *Internet audience ratings panels* (Napoli, 2011) track a panelist's browsing behavior via software installed on the panelist's computer or by using other technologies, such as a router,[6] to record the sites he or she visits, the amount of time spent on each site, and the actions taken on that site. This type of panel is discussed in detail in Chapter 17.

1.4   A brief history of online panels

1.4.1   Early days of online panels

As described above, online panels are essentially access panels moved online. Göritz et al. defined an online panel as a "pool of registered persons who have agreed to take part in online studies on a regular basis" (Göritz, Reinhold, & Batinic, 2002, p. 27). They are the natural evolution of the consumer panels (Delmas & Levy, 1998) in market research, used for decades as a sample source for mail, phone, and face-to-face surveys (Sudman & Wansink, 2002).

The attraction of online panels is threefold: (1) fast data collection; (2) promised lower cost per interview than most other methods; and (3) sampling efficiency due to extensive profiling. Panel companies typically invest substantial resources to recruit and maintain their panels so that the entire data collection process for clients can be streamlined and data can be turned around in a matter of weeks, if not days.

It is difficult to pinpoint exactly when the first online panels were launched. The idea of having respondents participate in computer-assisted self-interviewing goes back to the mid-1980s, when British Videotext and France Telecom Minitel terminals were used to conduct survey interviews (Saris, 1998). In the Netherlands, the probability-based Dutch Telepanel was launched in 1986 with about 1000 households. A PC and modem were placed in each randomly-selected household, and every weekend, a new questionnaire was downloaded to the PC. The system selected the household member to be interviewed, and after the survey was completed, the system dialed the central computer to upload the data (Saris, 1998).

---

[6] A router is a physical device that handles Internet traffic.

The next wave of online panels were implemented in the mid-1990s (Postoaca, 2006), primarily in the United States, but with a few other early adopters in Europe, such as those in the Netherlands (Comley, 2007; Faasse, 2005).

Comley (2007) listed a number of reasons why online panel research took off as quickly as it did, at least in the United States:

1. Many US research buyers already were using panels who mailed in completed paper questionnaires, so the switch to online was easy to accept.

2. The main US online panels were created during the dot com boom, when investments in online businesses of all kinds were at their peak.

3. Research buyers in the US were especially interested in lowering their data collection costs.

4. Response rates for US random-digit dialing (RDD) surveys were declining.

5. The cost and turnaround time of RDD, face-to-face, and mail surveys were increasingly viewed as problematic by many clients.

1.4.2   Consolidation of online panels

The period from the mid-1990s until about 2005 was one of explosive growth of online panels, especially in the United States and Europe. That was followed by a period of consolidation, driven by two complementary forces. The first was the need to build much larger panels that could more effectively service the growing demand for online survey respondents. The second was the internationalization of market research. As US companies increasingly looked to expand into global markets, their research needs shifted to emphasize, first, the European Union (EU) and Japan, followed by emerging markets, especially the so-called BRIC countries (Brazil, Russia, India, and China). Examples include Survey Sampling International's acquisition of Bloomerce Access Panels in 2005, Greenfield Online's acquisition of CIAO that same year, and the subsequent acquisition of Greenfield Online, first by Microsoft in 2008, and then by Toluna in 2009.

1.4.3   River sampling

A competing methodology, called *river sampling*, offered a different attraction to researchers. In 1995, Digital Marketing Insights created Opinion Place, giving researchers access to the roughly 24 million users of America Online (AOL). The goal of Opinion Place was not to build a panel but rather to invite people using the Internet to interrupt their activities and complete one of dozens or more waiting surveys. Rather than relying on previously provided personal data (such as demographics) to assign respondents to a specific survey, respondents were profiled at the time of recruitment and routed to a particular questionnaire.

The argument for river sampling was that it provided researchers access to a much larger and more diverse pool of respondents than that of even a very large online panel (PR Newswire Association, 2000). This broader pool also made it possible to screen a very large number of individuals for membership in low incidence subgroups. One weakness of river sampling was that it was very difficult to predict how long it might take for a study to achieve the desired number of completed interviews.

In 2006, Greenfield Online introduced a similar product called "Real-Time Sampling." Greenfield expanded the potential pool even more by creating relationships with hundreds of high-traffic sites across the Internet which displayed survey invitations (Research Live, 2006).

At the same time, the demand for greater diversity in online samples, increased targeting of lower incidence populations, and client requirements for "category exclusions" drove panel companies to supplement their panels through contracting with competitors. Alternatively, agencies routinely doing research with difficult-to-reach populations developed relationships with multiple suppliers. One immediate complication was the potential for individuals who belonged to more than one panel to be sampled more than once for the same survey, creating the need to develop methods for de-duplicating the sample.

1.5   Development and maintenance of online panels

1.5.1   Recruiting

A key distinction among panel providers is in terms of panel recruitment methodology. In addition to probability-based and nonprobability online panels (see Couper, 2000; Lozar Manfreda & Vehovar, 2008; Sikkel & Hoogendoorn, 2008) are invitation panels, as we explain below.

1.5.2   Nonprobability panels

The recruitment methods for nonprobability panels are numerous and varied, but with virtually all of them, anyone who is aware of an open invitation to join can volunteer to become a panel member. That is, people select themselves into the panel, rather than the researcher selecting specific individuals from a sampling frame that contains all members of a target population. It is impossible to know in advance who will see the invitation or how many times a given individual might encounter it. As a result, it is impossible for the panel recruiter to know the probability of selection of each member of the panel. Because none of these methods are probability-based, researchers have no scientific basis for calculating standard statistics such as confidence intervals. Of course, if a subset of members of a panel were randomly selected from the panel for a particular survey, the researcher would be justified in generalizing to the panel but not to any known population outside of the panel.

Companies that created nonprobability panels tend to be secretive about the specifics of their recruiting methods, perhaps believing that their methods provide them a competitive advantage (Baker et al., 2010). For this reason, there are few published sources to rely on when describing recruitment methods (Baker et al., 2010; Comley, 2007; Postoaca, 2006). The following list is based on the above references plus our personal experiences dealing with online panel providers and reviewing their website.

Online recruitment has been done by using the following methods:

- Placing banner ads on various website, often chosen to target specific types of people. Interested visitors click to go to the panel company's website, where they enroll.

- Invitations to join a panel distributed via newsgroups or mailing lists.

- Search engine advertisements (Nunan & Knox, 2011). Panel vendors bid on keywords (such as "survey money" or "online survey"). When a person uses one of the keywords as a search term, he or she is shown an ad inviting him or her to join a panel.

- Ads on social networking sites.

- Co-registration agreements integrating online panel enrollment with other online services. For example, when registering with a portal, e-commerce site, news site, or other special interest site, a person may also enroll in an online panel.

- Affiliate hubs, also called online panel consolidators, allow panel members to join one or more online panels simultaneously.

- Member-get-a-member campaigns (snowballing), which use current panel members to recruit their friends and family by offering an incentive.

- Recruiting at the end of an online survey, especially when the participants have been recruited via river sampling.

Offline methods also are used to solicit online panel members, though they tend to be more expensive than online methods. They include piggybacking at the end of an existing offline survey (e.g., mail, face-to-face, or phone) or directly recruiting via offline methods such as face-to-face, mail, or telephone.

1.5.3   Probability-based panels

Probability-based online panels recruit panel members using established sampling methodologies such as random-digit dialing (RDD), address-based sampling (ABS), and area probability sampling. Regardless of the specific sampling method used, a key requirement is that all members of the population of interest have a known, non-zero probability of receiving an invitation to join. No one is allowed to join the panel unless he or she is first selected in a probability sample. Panel members cannot refer friends, family, or colleagues to join the panel.

Recruitment begins with the selection of a probability sample. Potential panel members may be initially contacted via a letter or by an interviewer. For example, the LISS panel in the Netherlands first sent an advance letter along with a panel brochure to sampled households. When an address could be matched with a telephone number, an interviewer attempted to contact the household via telephone. When no telephone number was available, a field interviewer made an in-person visit (Scherpenzeel & Das, 2010; Scherpenzeel, 2011). In the United States, GfK Custom Research (formerly Knowledge Networks and before that, InterSurvey) uses an ABS frame (DiSogra, Callegaro, & Hendarwan, 2010) and sends a mail recruitment package to sampled households. The package contains instructions on how to join the panel either via mail, telephone, or online (Knowledge Networks, 2011). EKOS (2011) in Canada and Gallup in the United States used RDD samples to recruit panels (Tortora, 2008). Finally, face-to-face recruitment of an area probability sample was used to create the Face-to-Face Recruited Internet Survey Platform (FFRISP) in the United States (Krosnick et al., 2009).

To assure representativeness, probability-based panels need a mechanism to survey households that do not have convenient Internet access. For example, GfK Custom Research/Knowledge Networks and the LISS panel offered these households a free computer and Internet connection. Other companies, such as EKOS and Gallup, survey their offline members via phone and/or mail (DiSogra & Callegaro, 2010; Rookey, Hanway, & Dillman, 2008). A third strategy is to offer Internet and non-Internet households the same device. For example, FFRISP members were given a laptop, whereas the French panel Elipps[7] provided

---

[7] http://www.elipss.fr/elipss/recruitment/.

a tablet. This strategy standardizes the visual design of the questionnaire because every panel member can answer the survey with the same device.

1.5.4 Invitation-only panels

One set of online panels includes those whose members are invited from lists (invitation-only panel), such as members of airline frequent flier programs or hotel rewards programs. Firms building such panels do not permit volunteers to join; only invited individuals may do so. If the invited individuals are all members of a list or are a random subset, then the obtained panel is a probability sample of the list. If researchers then wish to generalize results to members of the list, there is a scientific justification for doing so. But generalization to other populations would not be justified on theoretical grounds.

1.5.5 Joining the panel

It has become a best practice to use a *double opt-in* process when people join an online panel. As defined by ISO 26362:2009 "Access Panels in Market Opinion and Social Research," double opt-in requires "explicit consent at two separate points to become a panel member." The double opt-in process typically requires that the potential panel member first indicates their intent to join by providing some basic information (e.g., name and email address) on the panel's join page. He or she is then sent an email with a unique link. After clicking on this link, the potential panel member completes an enrollment survey that may also include an extensive profiling questionnaire. During this process, the potential member is often asked to read material describing how the company will use his or her personal information, how cookies are used, and other general information about membership and rewards. Some companies consider this sufficient to meet the double opt-in requirement. Others require still another step, during which the member is profiled.

1.5.6 Profile stage

The profile stage entails answering a series of questions on various topics. The data obtained through profiling are useful at the sampling stage, when the characteristics of members are used to select samples that reduce the amount of screening required in a client's survey. Therefore, panel companies attempt to obtain a completed profile from every new member. Profile data are refreshed regularly, depending on how often specific items in the profile are likely to change. Some panels allow respondents to update their profile data at any time, while others invite panelists to a profile updating survey on a regular basis. Sometimes, the profile survey is "chained" to the end of a survey done on behalf of a client. Thus, profiling can be an on-going process rather than a one-off event.

Among the most important profile data are the demographic characteristics of each member. It has become something of a standard in online research for clients to specify samples that fill common demographic quotas in terms of age, sex, ethnicity, and region. Without full demographic information on its members, a panel company cannot meet that kind of specification. During the demographic profiling stage, many companies also collect information such as the mailing address and phone numbers to be used for respondent verification (see Chapter 19) and manage communications with members. These demographic variables can also be used as benchmarks to adjust for attrition in subsequent surveys. Once a potential panel member has completed the demographic profile survey, he or she is officially a panel member.

Thematic profiling on topics such as shopping habits, political preferences, voting behavior, religious belief, health and disease states, media consumption, and technology use and ownership also is common. Thematic profiling allows the panel company to create targeted samples without having to screen the entire panel to find a specific group of people for a particular survey (Callegaro & DiSogra, 2008).

When a client requests a study of a specified target population (e.g., respondents with a particular medical condition), the provider can draw a sample of respondents based on profile survey data. This strategy is called *targeted sampling* (Sikkel & Hoogendoorn, 2008) or *pre-screening* (Callegaro & DiSogra, 2008).

The ability to reach a specific target of the population was recognized early as a major appeal of online panels. In market research, niche marketing is the result of greater segmentation, so companies want to find out the opinions of every segmented, specific group of customers, users, or potential customers.

The data obtained through profiling can be useful at the analysis stage in helping to understand nonresponse bias within the panel on specific surveys. But there is little reported indication that panel companies or researchers take advantage of this on a regular bias.

1.5.7   Incentives

Incentives are employed in surveys for many reasons, with the two most often cited being increased participation and data quality improvement (Lavrakas et al., 2012). Incentives can be classified along different dimensions (Göritz, 2014; Lavrakas et al., 2012). First, with regard to timing, there are prepaid (noncontingent) and postpaid (contingent) incentives. As their names suggest, prepaid incentives are given to potential respondents before their participation, whereas postpaid (i.e., promised) incentives are delivered only after the respondent has complied with the survey request.

A review of the literature as well as our personal experience suggests that prepaid incentives are rarely used in online panels, perhaps because they are logistically more challenging, and because they are perceived to be more expensive, since everyone sampled is paid the incentive. However, the research literature on survey incentives is very consistent over the past 80 years, continually showing that prepaid incentives are more effective in increasing response rates than postpaid/contingent incentives of similar value. Almost all that research is with "first-time" respondents who have had no prior experiences with the researchers – experiences which may have helped build trust and feelings of reciprocity. Thus, in the case of panel members who have extensive prior experience with the panel company, promised (contingent) incentives may have a greater impact on cooperation than the literature would suggest.

A second dimension is whether everybody or only some respondents get an incentive. Here, we distinguish between per-capita versus lottery incentives (in the United States, usually called "sweepstakes"). With per-capita incentives, every panelist who completes the survey gets the incentive. With a lottery incentive, panel members essentially get a "ticket" for that monthly draw each time they complete a survey, thereby increasing their chance of winning by completing multiple surveys within a given month.

A third dimension is the character of the incentive, most often either monetary (cash, checks, electronic payments, gift cards, etc.) or points that can be accrued and redeemed for goods or services. The literature is consistent in showing that "cash is king," but almost all these findings come from studies with first-time respondents, so this "best practice" may not generalize to online panels.

1.5.8   Panel attrition, maintenance, and the concept of active panel membership

As Callegaro and DiSogra (2008) explained, panel membership changes constantly. Most panels continuously recruit new panel members. At the same time, panels continually suffer from attrition of four kinds:

- voluntary

- passive

- mortality

- panel-induced

Voluntary attrition. Voluntary attrition is the proactive action of panel members to contact the company and ask to be removed from the panel. This occurs for various reasons, including fatigue, growing concerns about privacy, and lack of satisfaction with the rewards earned. This form of attrition is relatively infrequent.

Passive attrition. More frequently, panel members simply stop answering surveys, or they change their email addresses without notifying the company. These members are also referred as "sleepers," as they are not active, but some of them can be "awakened" with specific initiatives (Scherpenzeel & Das, 2010). This form of attrition is relatively common.

Mortality attrition. This occurs when a panel member dies or is no longer physically or mentally capable of answering surveys. This is relatively uncommon.

Panel-induced attrition. Lastly, the panel company can decide to "retire" or force panel members out of the panel. For example, Nielsen calls this "forced turnover."[8] Some panels have a limit on panel tenure. Others have rules that place limits on noncompliance. For example, the Gallup panel has a five-strikes rule: panel members are dropped from the panel if they do not answer five consecutive surveys to which they were invited (McCutcheon, Rao, & Kaminska, Chapter 5 in this volume).

Panel attrition can be measured in terms of size and type. In terms of size, the simple ratio of the number of panelists who have left the panel within a certain period of time, whatever the reason, indicates whether the extent of attrition is small, medium, or large. In terms of the type of attrition, *differential panel attrition* occurs when a nonrandom subset of panelists leave the panel. These two types of attrition are independent of one another. For example, overall panel attrition might be high, but differential attrition low. The opposite also can be the case.

Panel attrition can be measured at two points in time: (1) at a time reference point (e.g., monthly); and (2) at a wave reference point when looking at longitudinal designs (Callegaro & DiSogra, 2008). In the first case, the magnitude of the attrition can be measured by looking at a specific cohort of respondents and following them month after month. The monthly attrition for that cohort will be equal to the number of active panel members at Time t minus the number of active panel members at Time $t + 1$, divided by the number of active panel members at Time t. In the second case, for longitudinal designs, the formula will be the same, but substituting wave with time.

How a panel manages its attrition can affect how well the panel performs. For example, aggressive panel management practices that purge less active members may increase the participation rate of each survey (Callegaro & DiSogra, 2008). But this comes at the price of

---

[8] http://www.agbnielsen.net/glossary/glossaryQ.asp?type=alpha&jump=none#alphaF.

reducing the active panel size and increasing the risk of bias, because more active panel members may respond differently than less active panel members (Miller, 2010).

Attrition is particularly problematic in longitudinal studies. The reduction in available sample size can lower the power of statistical analysis, and differential attrition will lower data quality. Attrition can be tackled by the panel provider with specific initiatives that are intended to reduce it (Kruse et al., 2010). Examples of such initiatives are re-recruitment of members who have left the panel and an incentive program to retain members in the panel for longer periods of time.

As the foregoing suggests, attrition can significantly impact the quality of survey-based estimates, depending on the causes of the attrition. The best case is that attrition is *missing completely at random* (MCAR), that is, unrelated (uncorrelated) to the variables measured in a particular survey. Sample size may be reduced, but that does not necessarily lead to biased estimates. However, when data are simply *missing at random* (MAR), and nonparticipation is related to variables that have been measured in earlier surveys, there is an increased risk of bias. Fortunately, these variables then can be used in weighting adjustment techniques that reduce or remove any resulting bias. Finally, attrition can result in data being *not missing at random* (NMAR), meaning that attrition is correlated with and possibly caused by variables that are only measured in the current survey.

Once a panel is built, maintenance methods are often similar across probability and nonprobability online panel types (Callegaro & DiSogra, 2008). The effort that goes into maintaining an online panel is significant. Knowing how many active panel members are available and their characteristics is a key statistic for an online panel. The definition of what constitutes an *active panel member* varies considerably from company to company. This variability is apparent in the answers that different online panel companies give to Questions 20 and 21 of the ESOMAR "28 Questions to Help Buyers of Online Panels" (2012) online panel companies make decisions about which services are best for their own needs. Section 3.3 of ISO 26362: 2009 (2009) defines an active panel member as a "panel member who has participated in at least one survey if requested, has updated his/her profile data or has registered to join the access panel, within the last 12 months" (p. 2).

1.5.9   Sampling for specific studies

Few publicly-available documents describe how different panels draw and balance their samples. However, our experience suggests two extremes, depending on the specifications provided by the client. At one extreme are targeted samples, for which clients describe the characteristics of sample members in ways that match variables in the panel member profiles constructed by the panel company. At the other end are more general population surveys, for which there is little or no match between the desired sample characteristics and the panel company's profile data. In this case, a very large sample is often needed, so that respondents can be screened to yield the desired distribution of characteristics among people who complete the survey. Most studies probably fall between these two extremes.

Viewed through the lens of traditional sampling techniques, we can distinguish three primary methods:

- *Simple random sample* or *stratified random sample*. This method is similar to, if not the same as, traditional sampling methods that have been used in survey research for the past 60 years. Using a complete list of active panel members and extra variables describing each of them, it is straightforward to draw a simple or stratified random sample.

- *Quota sampling.* Quota sampling is currently the most commonly used method for selecting a sample from nonprobability online panels (Rivers, 2007). It entails setting up quotas or maximum numbers of respondents in key subgroups, usually demographically defined, but sometimes behaviorally defined as well. Quotas are enforced during questionnaire completion, rather than during the sample draw. Once a quota is filled, new respondents who might quality for that cell are screened out and typically are politely informed that their responses are not needed.

- *Sample matching.* A number of panel companies used more complex sampling methods designed to maximize sample representativeness. For example, YouGov (former Polimetrix) developed a sample matching method (Rivers, 2006, 2007) that starts with an enumeration of the target population using pseudo-frames constructed from high quality probability-based surveys (e.g., American Community Survey or Current Population Survey) and commercially-available databases, such as lists of registered voters (when the topic is election polling). A random sample is drawn from the pseudo-frame and matched to panel members who share the same characteristics. Multiple panel members are then selected for each line in the pseudo-frame to increase the likelihood of getting a response. This method is used simultaneously for all open studies sharing the same sample specifications. If a panel member reaches a study and another closest match already completed it, he/she is rerouted in real time to the second best match study, so s/he is not turned away.

The *Propensity Score Select* or *SmartSelect* by Toluna (Terhanian & Bremer, 2012) is another method that relies on sample matching. It starts by conducting two parallel surveys with a set of shared questions related to the survey topic or believed to distinguish people who are online compared to those who are not. The specific set of questions asked is important and must be carefully chosen. One survey uses a probability sample and a traditional method (e.g., telephone, face-to face), and the other is done online using the Toluna panel. The first group is the external target population, and the second group is the "accessible" population. The results are used in a logistic regression to estimate the probability that a Toluna panel member belongs to the target population rather than the accessible population. For future surveys, the online panel members are asked these key set of questions. The process can be used with multiple panel sources and river samples as the propensity is computed in real time. Once the distribution of the propensity scores is determined, it can be used to select respondents for future surveys on the same topic. The methodology is also combined with traditional quotas when inviting panel members to a new survey.

Sample matching and *SmartSelect* methodologies have not been extensively tested, and very few investigations of them are available in the literature. For example, sample matching has been used in pre-election polling (YouGov, 2011). To the best of our knowledge, these methods also have not been tested independently, so it is not known how well they work in surveys on a variety of topics, or single surveys that are used to produce estimates on a broad range of topics.

Some sample matching techniques *vet* or prescreen respondents (Terhanian & Bremer, 2012) prior to sample selection. For example, Global Market Insight's (GMI) *Pinnacle* methodology (Eggers & Drake, 2011; GMI, 2012) is based on profiling respondents using 60 demographic, psychographic, and behavioral questions from the US General Social Survey (GSS). The GSS is a high quality, probability-based, high response rate long-standing survey considered a gold standard for the attitudes and beliefs of the US population. Samples are then drawn from the panel so that the distribution of characteristics matches that of a GSS sample.

Marketing Inc.'s *Grand Mean Project* also uses a vetting approach, in which participating panels profile their members according to buying behavior, socio-graphics, media, and market segments (Gittelman & Trimarchi, 2010; Gittelman, Trimarchi, & Fawson, 2011). The resulting data are used to create a series of segmentation profiles. The *Grand Mean* is an average of the percentage of members per profiles across panels within the same country. Each single online panel can compare their segments to the grand mean and use this information to sample its panel members. Another use of the grand mean is to track changes in panel composition over time.

Other vetting approaches are tailored to exclude people who may complete the same survey more than once, or who show low engagement with the survey through behaviors such as speeding and straight-lining. These approaches work by using multiple sample sources at the same time, respondent verification using third party databases, and digital fingerprinting to identify respondents who take the survey multiple times. Examples of the above strategies are MarketTools' *TrueSample* (MarketTools, 2009) and Imperium's suite of products (Relevant ID, Verity, Capture, etc.) (http://www.imperium.com/).

1.5.10   Adjustments to improve representativeness

Researchers generally conduct surveys so that they can make statistical inferences to a larger population. Within the probability-sampling paradigm, valid inference is only possible if a sample has been selected properly. "Properly selected" means that every person in the target population has a known non-zero probability of being selected. When these conditions are met, the sample can be described as being representative of the target population. In these cases, researchers can compute unbiased estimates and measures of the accuracy of those estimates (Horvitz & Thompson, 1952).

By definition, nonprobability panels do not satisfy these conditions. In particular, when they purport to represent the general population of some geo-political area, they typically suffer from high noncoverage and considerable coverage error. Some of that coverage error is due to the less than 100% household Internet penetration, but more often, it is due to the fact that panels are comprised of volunteers who have self-selected into the panel, rather than being selected from a frame that contains the full population. Probabilities of selection are unknown, so the panel cannot be described as representative. Proponents of nonprobability panels generally have come to accept this proposition but also argue that the bias in samples drawn from these panels can be reduced through the use of auxiliary variables that make the results representative. These adjustments can be made in sample selection, in analysis, or both. The sample matching techniques described above are one such method.

Weighting adjustments are another. They are intended to reduce bias and improve the accuracy of survey estimates by using auxiliary information to make post-survey adjustments. Auxiliary information is defined as a set of variables that have been measured in the survey, and for which information on their population distribution (or complete sample distribution) is available.

By comparing the distribution of a variable from the survey with an auxiliary variable that measures the same characteristic in the target population, researchers can assess whether the sample is representative of the population with respect to that particular variable. If the distributions differ considerably, researchers may conclude that the sample is biased and can attempt to reduce or eliminate the bias(es) through weighting. Estimates of population characteristics are then computed by using the weighted values instead of the unweighted values. Overviews of weighting procedures have been offered by Bethlehem and Biffignandi (2012) and Särndal and Lundström (2005).

The two most common methods to weight online panels are post-stratification and propensity score weighting. A more detailed discussion of these methods can be found in the introduction to Part IV on adjustment techniques in this volume. With post-stratification, the sample typically is divided into a number of strata. All cases within a stratum are assigned the same weight, and this weight is such that the sample percentage of people in a stratum is equal to the population percentage of people in that stratum. In other words, the sample is made to look like the population it is meant to represent on the variables that are used by the researchers to do these adjustments.

Other weighting techniques make use of response propensities. Harris Interactive first introduced propensity weighting for nonprobability panels in the late 1990s (Terhanian, Smith, Bremer, & Thomas, 2001). Other applications of propensity score weighting on nonprobability panels were described by Lee and Valliant (2009). First, the (unknown) response probabilities are estimated. Next, the estimated probabilities (propensities) can be used in several ways. One way is to adapt the original selection probabilities by multiplying them by the estimated response probabilities. Another way is to use the estimated response probabilities as stratification variables and to apply post-stratification. For more information about the use of response probabilities, see Bethlehem, Cobben, and Schouten (2011).

As with probability samples, there is no guarantee that weighting will be successful in reducing or eliminating bias in estimates due to under-coverage, the sampling design used, nonresponse, or self-selection. The bias is only reduced if the weighting model contains the proper auxiliary variables. Such variables should satisfy four conditions:

- They must have been measured in the survey.

- Their population distributions must be known.

- They must be correlated with all the measures of interest.

- They must be correlated with the response probabilities.

Unfortunately, such variables are often unknown to the researchers, or if known, are not available, or there is only a weak correlation. Instead, researchers typically use the "usual suspects," such as demographics like sex, age, race, education, etc., because they are readily available for both the sample and the population, and they correlate significantly (albeit weakly) with the key measures being surveyed and the response behavior.

When relevant auxiliary variables are not available, one might consider conducting a reference survey to create them. This reference survey is based on a probability sample, where data collection takes place in a mode leading to high response rates and little bias. Such a survey can be used to produce accurate estimates of population distributions of auxiliary variables. These estimated distributions can be used as benchmarks in weighting adjustment techniques. The reference survey approach has been applied by several market research organizations (see e.g., Börsch-Supan et al., 2007; Duffy et al., 2005; Terhanian & Bremer, 2012) and discussed in the academic literature by Valliant and Dever (2011).

An interesting aspect of the reference survey approach is that any variable can be used for adjustment weighting as long as it is measured both in the reference survey and in the online panel. For example, some market research organizations use "webographics" or "psychographic" variables that divide the population into mentality groups (see Schonlau et al., 2004; 2007, for more details about the use of such variables). Yet despite the advantages that reference surveys offer, researchers are often ignorant about what are the key auxiliary variables they should be measuring.

When a reference survey is conducted to create relevant auxiliary variables, it should be realized that the reference survey only estimates their population distribution. This introduces an extra source of variation. Therefore, the variance of the weighting adjusted estimates is increased. The increase in variance depends on the sample size of the reference survey: the smaller the sample size, the larger the variance. So using a reference survey can reduce bias, but at the cost of an increased variance. Depending on the derived weights, this approach also can reduce the effective sample size.

1.6   Types of studies for which online panels are used

Online panels allow for *cross-sectional* and *longitudinal* research. In the cross-sectional case, panel members participate in a variety of surveys on different topics, but they are not interviewed on the same topic repeatedly. Cross-sectional surveys can be done once, or the same survey can be conducted multiple times but with different respondents. A classic example is a tracking study designed to collect repeated measures, often related to customer satisfaction, advertising effectiveness, perceptions of a company's brand, or likely voting intentions. The same questionnaire is administered on a regular basis, but with different respondents every time.

Online panels can be also utilized for longitudinal purposes, where the same panelists are interviewed at different points in time on the same topic (Göritz, 2007). This type of design is the closest to the traditional concept of household panels, where the same people are followed over the years and interviewed on the same topics to document change. Every measurement occasion is called a survey *wave*. Re-asking the same question at different points in time can be used to study the reliability of items (test-retest), and this information can be used to increase the overall data quality (Sikkel & Hoogendoorn, 2008). At the same time, it is possible to use a longitudinal design with a cross-sectional component, where specific or thematic questions are asked only once.

As discussed above, it is common for online panels to run thematic surveys on a diversity of topics on the whole panel in a census fashion – a.k.a. profile surveys. In principle, at least, one advantage of profile surveys is that they can eliminate the need to ask the profile questions (e.g., demographics that do not change over time) over and over in each client study, making the entire questionnaire shorter. This also can reduce respondent burden and avoid annoying panel members with the same repeated questions. Previously collected data can also help making questionnaire routing more efficient. Unfortunately, few clients take advantage of these efficiencies, believing that profile data is not sufficiently reliable.

1.7   Industry standards, professional associations' guidelines, and advisory groups

Especially in the last decade, industry and professional associations worldwide have sought to guide their members on the proper and effective use of samples from online panels. We provide a brief overview of these efforts below. We urge the reader to become familiar with these various activities and undertakings, especially as they may differ from country to country.

In 2009, the International Organization for Standardization (ISO) issued ISO 26362 "Access Panels in Market, Opinion, and Social Research" a service standard primarily focused on nonprobability online panels. The goal of this standard was to apply the quality framework of ISO 20252 (originally issued in 2006) to panels. ISO 26362 presents a terminology and a series of requirements and practices that companies should follow when recruiting, managing, and providing samples from an online panel. The more recent version of ISO 20252 (2012) incorporates many of the principles from ISO 26362, especially in Section 4.5.1.4, which focuses on nonprobability samples. Panel companies can be certified under these standards by agreeing to a series of external audits that verify their compliance.

Another global organization, ESOMAR, has produced two guidance documents. Their "Guideline for Online Research" (2011) offers guidance on the full range of online research methods, including online panels. A second document, "28 Questions to Help Research Buyers of Online Samples" (2012), is the third in a series "designed to provide a standard set of questions a buyer can ask to determine whether a sample provider's practices and samples fit with their research objectives" (ESOMAR, 2011).

A number of both industry and professional associations at the country and regional levels endorse global standards, refer to them, or have their own specific quality standard documents that incorporate many of the same principles. For example, EFAMRO, the European Federation of Market, Social and Opinion Research Agency Trade Associations, has endorsed ISO 20632 and ISO 20252. Other national professional associations have specific documents. The Canadian Market Research and Intelligence Association (MRIA) has their "10 Questions to Ask Your Online Survey Provider" (2013). In Italy, Assirm, inspired by ISO 20252, has specific rules for online panels. In Australia, the Association of Market and Social Research Organizations (AMSRO) has developed a certification process for its members called Quality Standards for Online Access Panels (QSOAP) (AMSRO, 2011). This was partly an interim process until ISO 26362 was issued in December 2009. From 2010 on, AMSRO decided not to accept any more applications for the QSOAP but instead to endorse ISO 26362. In the United Kingdom, the Market Research Society has issued a document called "Guidelines for Online Research" (MRS, 2012) where there are specific sections dedicated to online panels. Still in the United Kingdom, the British Standard Institution (BSI) produced the second edition of its study *Quality in Market Research: From Theory to Practice* (Harding & Jackson, 2012). In this book, a very broad approach to quality in market research is taken from every angle: market research as a science, its professionals and members, the clients, the legislation, and the interviewers. Other Chapters discuss ISO 9001, ISO 26362 and ISO 20252.

There also are guidelines written by advisory groups. For example, the Canadian Advisory Panel on Online Public Opinion Survey Quality (2008) has developed standards and guidelines for probability and nonprobability online research that focus on areas such as pre-field planning, preparation and documentation, sampling, data collection, use of multiple panels, success rate (response rates for web surveys), data management and processing, data analysis/reporting, and survey documentation.

The Advertising Research Foundation (ARF) in the United States has established their Research Quality Council (RQC) (http://www.thearf.org/orqc-initiative.php). This group has organized funding and resources for two major research initiatives known as Foundations of Quality (FOQ) 1 and 2 (http://thearf.org/foq2.php). These initiatives fund quality studies and produce reports such as the 17 online panel comparison study (Walker, Pettit, &

Rubinson, 2009) described in Chapter 2. In 2011, the ARF commissioned a report called "Online Survey Research: Findings, Best practices, and Future Research" (ARF, 2011; Vannette, 2011) as a prelude to the launch of FOQ 2. The report is a literature review that "represents what we believe is the most comprehensive and representative aggregation of knowledge about online survey research compiled to date" (Vannette, 2011, p. 4). The ARF has since designed an experimental study aimed at improving our understanding of a broad range of online panel practices. Data collection is now complete, and analysis is ongoing.

The American Association for Public Opinion Research (AAPOR) has issued three documents that specifically address online panels. In the "Final Dispositions of Case Codes and Outcome Rates for Surveys" (2011), methods for computing standard quality metrics for probability and nonprobability online panels are detailed. The second document is the result of a task force on online panels (Baker et al., 2010). The charge of the task force was

> reviewing the current empirical findings related to opt-in online panels utilized for data collection and developing recommendations for AAPOR members ... Provide key information and recommendations about whether and when opt-in panels might be best utilized and how best to judge their quality.
>
> (p. 712)

The most recent effort is the work of another AAPOR task force on nonprobability sampling (Baker et al., 2013). As its name suggests, this report is primarily focused on the range of nonprobability sampling methods used across disciplines, some of which may be especially useful to researchers relying on online panels.

1.8   Data quality issues

Despite the rapid growth of online panels and their emergence as a major methodology worldwide, concerns about data quality persist. A full investigation of this issue is the primary theme of this volume. For example, Chapter 2 in this volume: "A critical review of studies investigating the quality of data obtained with online panels" discusses the major studies to date comparing the quality of online panels with other survey data and benchmarks. It also discusses topics such as multiple panel membership and the "life" of online panel members. The issue of professional respondents is discussed in Chapter 10, "Professional respondents in nonprobability online panels" where a review of the major studies on the topic is highlighted before presenting new original data on professional respondents. Chapter 11: "The impact of speeding on data quality in nonprobability and freshly recruited probability-based online panels" present the controversial issue of "speeders". Finally, the issue of respondents' identity and validation is discussed in Chapter 19, "Validating respondents' identity in online panels."

1.9   Looking ahead to the future of online panels

As we noted at the outset, approximately one-third of quantitative market research is now done using online surveys, the majority of which relies on online panels. Online panels are here to stay, and given the increasing cost of such traditional methods as face-to-face and telephone surveys, the use of online panels is likely to continue to grow.

From a scientific perspective, probability-based online panels generally are preferable to those that rely on nonprobability methods. Best practices for building and maintaining probability-based panels are established and well known (e.g., Scherpenzeel & Das, 2010; Scherpenzeel, 2011). However, they also are expensive to build and maintain, and often too small to support the low incidence and small area geographic studies that are a significant part of the attraction of online panel research. For these and other reasons, they are likely to continue to represent a small proportion of the overall online panel business. As Internet penetration reaches very high coverage and devices to browse the web become financially more accessible to the general population, especially for those in the lowest economic tiers, the cost to build and maintain a probability-based online panel will decrease. For example, in Europe, new probability-based panels are being built or under consideration (Das, 2012).

That said, nonprobability panels continue to face some very serious challenges. First among them is data quality. This volume investigates online panel data quality from a wide range of perspectives, but arguably the biggest challenge that panels face is developing more robust methods for sampling and weighting to improve representativeness, resulting in more accurate estimates, and making reliable statistical inference possible. There are some promising developments, especially in sample selection, but a good deal more work is needed to fully validate these approaches and articulate their assumptions in ways that lead researchers to adopt them with confidence. At present, the advantages of cost, speed, and reach do not always go hand in hand with high data quality. We look forward to more research being done on online panel quality, especially on sampling and weighting methods, and hope this volume can serve as a basis for conceptualizing it.

At the same times, at least in the United States, the traditional panel model is rapidly falling into obsolescence, as the demand for online respondents grows, clients and researchers alike look for greater diversity in their online samples, and interest in studying low incidence populations increases. Using a model called *multi-sourcing* or *dynamic sourcing*, providers of online samples are increasingly relying on a range of sources that expands beyond their proprietary panels to the panels of competitors, social networking sites, and the placement of general survey invitations on a variety of website across the Internet, much like river sampling. These respondents often do not receive invitations to a specific survey on a specific topic. Instead they receive a general invitation to do a survey that directs them to a website where they are screened and then routed to a waiting survey for which they already have qualified, at least partially. The software that controls this process is called a *router*. Its goal is to ensure that anyone willing to do a survey online gets one. As of this writing, there is a good deal of variation in how these router systems are designed, how they operate, and what impacts, if any, they have on the data. And, of course, many of the metrics that we are accustomed to using to evaluate samples become difficult to compute. Nonetheless, online sample providers are moving in this direction quickly. In all probability, it will become the dominant paradigm in the next few years. The impact, if any, on data quality is unknown. If nothing else, it standardizes the sample selection process, at least within a specific provider, and that may be good thing. But whether it ultimately leads to improved data quality in online research remains to be seen.

References

American Association for Public Opinion Research. (2011). *Final dispositions of case codes and outcomes rates for surveys* (7th ed.). Deerfield, IL: AAPOR.

AMSRO. (2011). Background: QSOAP. Retrieved January 1, 2013, from: http://www.amsro.com.au/background-qsoap.

ARF. (2011). Online survey research: Findings, best practices, and future research. Paper presented at the Research Quality Forum, New York. Retrieved January 1, 2013, from: http://my.thearf.org/source/custom/downloads/2011-04-07_ARF_RQ_Presentation.pdf.

Baker, R., Blumberg, S. J., Brick, J. M., Couper, M. P., Courtright, M., Dennis, J. M., Dillman, D. A., et al. (2010). Research synthesis: AAPOR report on online panels. *Public Opinion Quarterly*, *74*, 711–781.

Baker, R., Brick, M. J., Bates, N., Battaglia, M. P., Couper, M. P., Deever, J. A., Gile, K. J., et al. (2013). *Non-probability sampling: AAPOR task force report*. Deerfield, IL: AAPOR.

Bethlehem, J., & Biffignandi, S. (2012). *Handbook of web surveys*. Hoboken, NJ: John Wiley & Sons, Inc.

Bethlehem, J., Cobben, F., & Schouten, B. (2011). *Handbook of nonresponse in household surveys*. Hoboken, NJ: John Wiley & Sons, Inc.

Börsch-Supan, A., Elsner, D., Faßbender, H., Kiefer, R., McFadden, D., & Winter, J. (2007). How to make internet surveys representative: A case study of a two-step weighting procedure. In *CenterData workshop: Measurement and experimentation with Internet panels. The state of the art of Internet interviewing*. Tilburg University: CenterData. Retrieved January 1, 2013, from: http://www.mea.mpisoc.mpg.de/uploads/user_mea_discussionpapers/loil50ozz320r55b_pd1 _040330%20geschuetzt.pdf.

Callegaro, M., & DiSogra, C. (2008). Computing response metrics for online panels. *Public Opinion Quarterly*, *72*, 1008–1032.

Clarke, H. D., Sanders, D., Stewart, M. C., & Whiteley, P. (2008). Internet surveys and national election studies: A symposium. *Journal of Elections, Public Opinion & Parties*, *18*, 327–330.

Comley, P. (2007). Online market research. In M. van Hamersveld & C. de Bont (Eds.), *Market research handbook* (5th ed., pp. 401–419). Chichester: John Wiley & Sons, Ltd.

Couper, M. P. (2000). Web surveys: A review of issues and approaches. *Public Opinion Quarterly*, *64*, 464–494.

Couper, M. P. (2007). Issues of representation in eHealth research (with a focus on web surveys). *American Journal of Preventive Medicine*, *32(5S)*, S83–S89.

Das, M. (2012). Innovation in online data collection for scientific research: The Dutch MESS project. *Methodological Innovations Online*, *7*, 7–24.

Delmas, D., & Levy, D. (1998). Consumer panels. In C. McDonald, & P. Vangelder (Eds.), *ESOMAR handbook of market and opinion research* (4th ed., pp. 273–317). Amsterdam: ESOMAR.

DiSogra, C., & Callegaro, M. (2010). Computing response rates for probability-based online panels. In AMSTAT (Ed.), *Proceedings of the Joint Statistical Meeting, Survey Research Methods Section* (pp. 5309–5320). Alexandria, VA: AMSTAT.

DiSogra, C., Callegaro, M., & Hendarwan, E. (2010). Recruiting probability-based web panel members using an Address-Based Sample frame: Results from a pilot study conducted by Knowledge Networks. In *Proceedings of the Annual Meeting of the American Statistical Association* (pp. 5270–5283). Paper presented at the 64th Annual conference of the American Association for Public Opinion Research, Hollywood, FL: AMSTAT.

Duffy, B., Smith, K. Terhanian, G., & Bremer, J. (2005). Comparing data from online and face-to-face surveys. *International Journal of Market Research*, *47*, 615–639.

Eggers, M., & Drake, E. (2011). Blend, balance, and stabilize respondent sources. Paper presented at the 75th Annual Conference of the Advertising Research Foundation, New York: ARF. Retrieved January 1, 2013, from: http://thearf-org-aux-assets.s3.amazonaws.com/annual/presentations/kif/04_D1_KIF _Eggers_Drake_v04.pdf.

EKOS. (2011). What is Probit? Retrieved from: http://www.probit.ca/?page_id=.7

ESOMAR. (2011). ESOMAR guideline for online research. Retrieved July 1, 2011, from: http://www .esomar.org/uploads/public/knowledge-and-standards/codes-and-guidelines/ESOMAR_Guideline -for-online-research.pdf.

ESOMAR. (2012). 28 Questions to help research buyers of online samples. Retrieved December 12, 2012, from: http://www.esomar.org/knowledge-and-standards/research-resources/28-questions-on -online-sampling.php.

Faasse, J. (2005). *Panel proliferation and quality concerns*. Paper presented at the ESOMAR panel research conference, Budapest.

Gittelman, S., & Trimarchi, E. (2010). Online research … and all that jazz! The practical adaption of old tunes to make new music. *ESOMAr Online Research 2010*. Amsterdam: ESOMAR.

Gittelman, S., Trimarchi, E., & Fawson, B. (2011). A new representative standard for online research: Conquering the challenge of the dirty little "r" word. Presented at the ARF Key Issue Forum, Re:Think conference, New York. Retrieved January 1, 2013, from: http://www.mktginc.com/pdf /Opinionology%20and%20Mktg%20Inc%20%20ARF%202011_March.pdf.

GMI. (2012). GMI Pinnacle. Retrieved January 1, 2013, from: http://www.gmi-mr.com/uploads/file /PDFs/GMI_Pinnacle_10.7.10.pdf.

Göritz, A. S. (2014). Incentive effects. Chapter 28 In U. Engel, B. Jann, P. Lynn, A. Scherpenzeel, & P. Sturgis (Eds.), *Improving survey method*. New York: Taylor & Francis.

Göritz, A. S. (2007). Using online panels in psychological research. In A. N. Joinson, K. Y. A. McKenna, T. Postmes, & U.-D. Reips (Eds.), *The Oxford handbook of Internet psychology* (pp. 473–485). Oxford: Oxford University Press.

Göritz, A. S. (2010). Web panels: replacement technology for market research. In T. L. Tuten (Ed.), *Enterprise 2.0: How technology, eCommerce, and Web 2.0 are transforming business virtually* (Vols. 1, 2, Vol. 1, pp. 221–236). Santa Barbara, CA: ABC-CLIO.

Göritz, A. S., Reinhold, N., & Batinic, B. (2002). Online panels. In B. Batinic, U.-D. Reips, & M. Bosnjak (Eds.), *Online social sciences* (pp. 27–47). Seattle, WA: Hogrefe & Huber.

Hansen, J. (2008). Panel surveys. In M. W. Traugott & W. Donsbach (Eds.), *The Sage handbook of public opinion research* (pp. 330–339). Thousand Oaks, CA: Sage.

Harding, D., & Jackson, P. (2012). *Quality in market research: From theory to practice* (2nd ed.). London: British Standards Institution.

Horrigan, J. B. (2010). *Broadband adoption and use in America* (No. OBI working papers series No 1). Federal Communication Commission. Retrieved from: http://hraunfoss.fcc.gov/edocs_public /attachmatch/DOC-296442A1.pdf.

Horvitz, D. G., & Thompson, D. J. (1952). A generalization of sampling without replacement from a finite universe. *Journal of the American Statistical Association*, *47*, 663–685.

Inside Research. (2012). Worldwide online research spending. *Inside Research*, March, 5.

International Organization for Standardization. (2009). *ISO 26362* Access panels in market, opinion, and social research: Vocabulary and service requirements. Geneva: ISO.

International Organization for Standardization. (2012). ISO 20252 Market, opinion and social research: Vocabulary and service requirements (2nd ed.). Geneva: ISO.

Knowledge Networks. (2011). Knowledge panel design summary. Retrieved August 6, 2011, from: http://www.knowledgenetworks.com/knpanel/KNPanel-Design-Summary.html.

Krosnick, J. A., Ackermann, A., Malka, A., Sakshaug, J., Tourangeau, R., De Bell, M., & Turakhia, C. (2009). Creating the Face-to-Face Recruited Internet Survey Platform (FFRISP). Paper presented at the Third Annual Workshop on Measurement and Experimentation with Internet Panels, Santpoort, the Netherlands.

Kruse, Y., Callegaro, M., Dennis, M. J., DiSogra, C., Subias, T., Lawrence, M., & Tompson, T. (2010). Panel conditioning and attrition in the AP-Yahoo news election panel study. In *Proceedings of the Joint Statistical Meeting, American Association for Public Opinion Research Conference* (pp. 5742–5756). Washington, DC: AMSTAT.

Lavrakas, P. J., Dennis, M. J., Peugh, J., Shand-Lubbers, J., Lee, E., & Charlebois, O. (2012). Experimenting with noncontingent and contingent incentives in a media measurement panel. Paper presented at the 67th Annual conference of the American Association for Public Opinion Research, Orlando, FL.

Lee, S., & Valliant, R. (2009). Estimation for volunteer panel web surveys using propensity score adjustment and calibration adjustment. *Sociological Methods Research*, *37*, 319–343.

Lozar Manfreda, K., & Vehovar, V. (2008). Internet surveys. *International handbook of survey methodology* (pp. 264–284). New York: Lawrence Erlbaum.

MarketTools. (2009). MarketTools TrueSample. Retrieved January 1, 2013, from: http://www.truesample.net/marketing/DataSheetTrueSample.pdf.

Miller, J. (2010). The state of online research in the U.S. Paper presented at the MRIA Net Gain 4.0, Toronto, Ontario.

MRIA. (2013). Ten questions to ask your online survey provider. Retrieved January 1, 2013, from: http://www.mria-arim.ca/STANDARDS/TenQuestions.asp.

MRS. (2012, January). MRS Guidelines for online reseach. Retrieved January 1, 2013, from: http://www.mrs.org.uk/pdf/2012-02-16%20Online%20Research%20Guidelines.pdf.

Napoli, P. M. (2011). *Audience evolution: New technologies and the transformation of media audiences*. New York: Columbia University Press.

National Telecommunication and Information Administration. (2013). *Exploring the digital nation: America's emerging online experience*. Washington, DC: U.S. Department of Commerce. Retrieved from: http://www.ntia.doc.gov/files/ntia/publications/exploring_the_digital_nation_-_americas_emerging_online_experience.pdf.

Nunan, D., & Knox, S. (2011). Can search engine advertising help access rare samples? *International Journal of Market Research*, *53*, 523–540.

Postoaca, A. (2006). *The anonymous elect: Market research through online access panels*. Berlin: Springer.

Poynter, R. (2010). *The handbook of online and social media research: Tools and techniques for market researchers*. Chichester: John Wiley & Sons, Ltd.

PR Newswire Associaton. (2000). DMS/AOL's Opinion Place expands research services to offer broadest online representation available. Retrieved January 1, 2013, from: http://www.thefreelibrary.com/DMS%2FAOL%27s+Opinion+Place+Expands+Research+Services+to+Offer+Broadest ... -a066296354.

Public Works and Government Services Canada. (2008). *The advisory panel on online public opinion survey quality: Final report June 4, 2008*. Ottawa: Public Works and Government Services Canada. Retrieved from: http://www.tpsgc-pwgsc.gc.ca/rop-por/rapports-reports/comiteenligne-panelonline/tdm-toc-eng.html.

Ragnedda, M., & Muschert, G. (Eds.). (2013). *The digital divide: The internet and social inequality in international perspective*. New York: Routledge.

Research Live. (2006, November 29). Greenfield unveils real-time sampling. Retrieved January 1, 2013, from: http://www.research-live.com/news/greenfield-unveils-real-time-sampling/3002563.article.

Rivers, D. (2006). Understanding people: Sample matching. Retrieved January 1, 2013, from: http://psfaculty.ucdavis.edu/bsjjones/rivers.pdf.

Rivers, D. (2007). Sampling for web surveys. *Joint Statistical Meeting, section on Survey Research Methods*. Paper presented at the 2007 Joint Statistical Meeting, Salt Lake City: AMSTAT. Retrieved January 1, 2013, from: http://www.laits.utexas.edu/txp_media/html/poll/files/Rivers_matching.pdf.

Rookey, B. D., Hanway, S., & Dillman, D. A. (2008). Does a probability-based household panel benefit from assignment to postal response as an alternative to Internet-only? *Public Opinion Quarterly*, *72*, 962–984.

Saris, W. E. (1998). Ten years of interviewing without interviewers: The telepanel. In M. P. Couper, R. P. Baker, J. Bethlehem, C. Z. F. Clark, J. Martin, W. L. Nicholls II,, & J. M. O'Reilly (Eds.), *Computer assisted survey information collection* (pp. 409–429). New York: John Wiley & Sons, Inc.

Särndal, C.-E., & Lundström, S. (2005). *Estimation in surveys with nonresponse*. Chichester: John Wiley & Sons, Ltd.

Scherpenzeel, A. (2011). Data collection in a probability-based Internet panel: How the LISS panel was built and how it can be used. *Bulletin of Sociological Methodology/Bullétin de Méthodologie Sociologique, 109*, 56–61.

Scherpenzeel, A. C., & Das, M. (2010). True longitudinal and probability-based Internet panels: Evidence from the Netherlands. *Social and behavioral research and the internet: Advances in applied methods and research strategies* (pp. 77–104). New York: Routledge.

Schonlau, M., van Soest, A., & Kapteyn, A. (2007). Are "Webographic" or attitudinal questions useful for adjusting estimates from Web surveys using propensity scoring? *Survey Research Methods, 1*, 155–163.

Schonlau, M., Zapert, K., Payne Simon, L., Hayness Sanstand, K., Marcus, S. M., Adams, J., Spranka, M., et al. (2004). A comparison between responses from a propensity-weighted web survey and an identical RDD survey. *Social Science Computer Review, 22*, 128–138.

Seybert, H., & Reinecke, P. (2013). Internet use in households and by individual in 2013. Eurostat Statistics in Focus 29/2013. Eurostat. Retrieved from: http://epp.eurostat.ec.europa.eu /statistics_explained/index.php/Internet_use_statistics_-_individuals.

Sikkel, D., & Hoogendoorn, A. (2008). Panel surveys. In E. De Leeuw, J. Hox, & D. A. Dillman (Eds.), *International handbook of survey methodology* (pp. 479–499). New York: Lawrence Erlbaum Associates.

Sudman, S., & Wansink, B. (2002). *Consumer panels* (2nd ed.). Chicago, IL: American Marketing Association.

Terhanian, G., & Bremer, J. (2012). A smarter way to select respondents for surveys? *International Journal of Market Research, 54*, 751–780.

Terhanian, G., Smith, R., Bremer, J., & Thomas, R. K. (2001). Exploiting analytical advances: Minimizing the biases associated with non-random samples of internet users. *Proceedings from the 2001 ESOMAR/ARF Worldwide Measurement Conference* (pp. 247–272). Athens.

Tortora, R. (2008). Recruitment and retention for a consumer panel. In P. Lynn (Ed.), *Methodology of longitudinal surveys* (pp. 235–249). Hoboken, NJ: John Wiley & Sons. Inc.

U.S. Department of Commerce, & National Telecommunications and Information Administration. (2010). *Digital Nation: 21st century America's progress toward universal broadband Internet access.* Washington, DC: National Telecommunications and Information Administration.

Valliant, R., & Dever, J. A. (2011). Estimating propensity adjustments for volunteer web surveys. *Sociological Methods & Research, 40*, 105–137.

Vannette, D. L. (2011). *Online survey research: Findings, best practices, and future research. Report prepared for the Advertising Research Foundation.* New York: Advertising Research Foundation.

Vaygelt, M. (2006). Emerging from the shadow of consumer panels: B2B challenges and best practices. *Panel Research 2006.* ESOMAR.

Visioncritical University. (2013). Insight communities. Retrieved July 16, 2013, from: http://vcu .visioncritical.com/community-panel/.

Walker, R., Pettit, R., & Rubinson, J. (2009). A special report from the Advertising Research Foundation. The foundations of quality initiative: A five-part immersion into the quality of online research. *Journal of Advertising Research, 49*, 464–485.

YouGov. (2011). YouGov's record. Public polling results compared to other pollsters and actual outcomes. Retrieved January 1, 2013, from: http://cdn.yougov.com/today_uk_import/yg-archives-pol -trackers-record2011.pdf

2

A critical review of studies investigating the quality of data obtained with online panels based on probability and nonprobability samples[1]

Mario Callegaro[a], Ana Villar[b], David Yeager[c], and Jon A. Krosnick[d]

[a]*Google UK*

[b]*City University London, UK*

[c]*University of Texas at Austin, USA*

[d]*Stanford University, USA*

2.1 Introduction

Online panels have been used in survey research as data collection tools since the late 1990s (Postoaca, 2006). The potential great cost and time reduction of using these tools have made research companies enthusiastically pursue this new mode of data collection.

---

[1] We would like to thank Reg Baker and Anja Göritz for their useful comments on preliminary versions of this chapter.

Online Panel Research: A Data Quality Perspective, First Edition.
Edited by Mario Callegaro, Reg Baker, Jelke Bethlehem, Anja S. Göritz, Jon A. Krosnick and Paul J. Lavrakas.
© 2014 John Wiley & Sons, Ltd. Published 2014 by John Wiley & Sons, Ltd.
Companion website: www.wiley.com/go/online_panel

The vast majority of these online panels were built by sampling and recruiting respondents through nonprobability methods such as snowball sampling, banner ads, direct enrollment, and other strategies to obtain large samples at a lower cost (see Chapter 1). Only a few companies and research teams chose to build online panels based on probability samples of the general population. During the 1990s, two probability-based online panels were documented: the CentER data Panel in the Netherlands and the Knowledge Networks Panel in the United States. Since then, a few probability panels started in the 2000s, including the Face-to-Face-Recruited-Internet-Platform (FFRISP) and the American Life Panel in the United States, the Longitudinal Internet Studies for the Social Sciences (LISS) in the Netherlands (Callegaro & DiSogra, 2008), and a handful of new panels are being built in European countries, including Germany,[2] France[3] (Das, 2012), Norway, and Sweden (Martinsson, Dahlberg, & Lundmark, 2013).

In the minds of many is the question: how do online panels of nonprobability samples compare in terms of quality to online panels of probability samples? The reasons why many online panels were built using nonprobability sampling and recruitment methods stem from methodological as well as financial reasons and are discussed in Chapter 1. In this chapter, we review a set of studies comparing survey estimates obtained from online panels to estimates from other data collection methods in order to assess the quality of the former, capitalizing on more than a decade's worth of studies and experiments.

We aim to provide data-driven answers to four main research questions:

1. How accurate are point estimates computed from online panels of probability and nonprobability samples?

2. How useful are weighting procedures in improving accuracy of these estimates?

3. How do relationships and predictive relations of data collected from online panels of probability and nonprobability samples compare to benchmark surveys?

4. How do experiments on online panels of probability and nonprobability samples replicate over time and across panels?

2.2   Taxonomy of comparison studies

The existing studies comparing statistics from online panels of nonprobability samples to other sources differ with respect to whether the comparison is made against surveys using probability or nonprobability samples, their mode of data collection, and whether benchmark estimates are available. We found six types of designs in the literature depending on these aspects (Table 2.1). These designs are not mutually exclusive; many studies use a combination of two or more designs, for example, an online panel from a nonprobability sample can be compared against an online panel and a telephone survey both using probabilistic sampling.

Next, each type of design will be described, together with their strengths and weaknesses:

Design 1: Comparison of two online panels with nonprobability samples. Design number 1 has the advantage of keeping the mode of data collection constant (online) and possibly

---

[2] http://reforms.uni-mannheim.de/english/internet_panel/home/index.html; http://www.gesis.org/en/services/data-collection/.

[3] http://www.sciencespo.fr/dime-shs/content/dime-shs-web.

Table 2.1 Possible designs used in studies comparing nonprobability online panels results to other results collected in a different way.

Design	Reference study	Comparison study	Mode	Benchmarks
1	Online panel with a nonprobability sample	Online panel with a nonprobability sample	Self-administered, Online	Yes – No
2	Online panel with a nonprobability sample	Online panel with a probability sample	Self-administered, Online	Yes – No
3	Online panel with a nonprobability sample	Telephone cross-sectional survey with a probability sample	Interviewer, Telephone	Yes – No
4	Online panel with a nonprobability sample	Face-to-face cross-sectional survey with a probability sample	Interviewer, Face-to-Face	Yes – No
5	Online panel with a nonprobability sample	Mail cross-sectional survey with a probability sample	Self-administered	Yes – No
6	Online panel with a nonprobability sample	Same online panel with a nonprobability sample	Self-administered, Online	No

the questionnaire administration constant. Three alternatives for questionnaire administration are possible: (a) each panel redirects their sample to a third party site where the survey is taken; (b) each panel programs and hosts the survey itself; and (c) a survey is centrally located and administered but the look and feel of the questionnaire are specific to each panel provider. In the first case, we have the purest case from an experimental point of view because the visual design of the instrument, the instructions, the prompts and real-time checks are the same for every respondent. However, redirecting panel members to another third party site can introduce nonresponse bias difficult to quantify because some panel members can be reluctant to complete the survey on a site that is not the panel site they belong to. In the second case, the same questionnaire is programmed individually by each panel provider. With this strategy, panel members see the survey on the same site they are familiar with, experiencing the look and feel they are used to. Design 1 allows direct comparison across panels but in order to assess accuracy of each panel, external benchmarks or other forms of data validation need to be available. This is also the case for the other five designs encountered in the literature.

Design 2: Comparison of an online panel with a nonprobability sample to an online panel with a probability sample. Design 2 allows comparison of online panels with different sampling designs, while keeping the mode of data collection constant. This design is similar to design 1, but there are usually a number of restrictions associated with the way probability-based online panels are run: (a) members are typically not allowed to be redirected to other website for survey completion; and (b) in connection with this, surveys are typically programmed in-house. When using design 2, it will be necessary to decide whether or not to include households from the probability-based online panels that did not have Internet at the moment of recruitment and were provided with a device and Internet connection for the study, given that such households would not be, in general, part of the online panels from nonprobability samples.

Design 3 and Design 4: Comparison of an online panel with a nonprobability sample to a face-to-face or a telephone survey with probability sample. These two modes are interviewer-administered and the questions are generally presented to the respondent orally (with the possible addition of show cards to present response options and other materials). As a consequence, any differences could be due to measurement effects as well as coverage, sampling, or differential nonresponse error. Therefore, when comparing results, possible mode effects need to be taken into account.

Design 5: Comparison of an online panel with a nonprobability sample to a mail survey with a probability sample. We found fewer examples of design 5 among the reviewed studies; however, this design has the strength of keeping the mode of administration (self-administered) closer across survey implementations than designs 3 and 4. At the same time, mode effects in mail and web surveys are also possible due to differences in visual design.

Design 6: Replication within panel. Design 6 is very different in nature and has a distinctive goal. Here the same questionnaire is implemented on non-overlapping cross-sectional samples of the same nonprobability-based online panel at different points in time. The questionnaire is generally comprised of questions that are not subject to rapid change and the time across the different administration is usually kept reasonably short (Gittelman & Trimarchi, 2010). The goal of this design is to test if a panel is "deteriorating" in any way. The hypothesis behind it is that if the quality of the panel is good, the results from one wave to the next one should not be too different. Additional quality metrics are generally computed for each wave such as percentage of speeders, straight-liners, inconsistency in the same questionnaire, and failure to follow an instruction.

All these designs can be further compared to benchmark estimates. Benchmarks are typically demographic and behavioral measures (such as health status, race, or number of rooms in the household), and usually come from official government statistics such as the American Community Survey. Attitudinal benchmarks come from high-quality surveys with probability samples such as the National Election Studies, or the General Social Survey. Until now, benchmarks have generally been collected by an interviewer in surveys that achieve extremely high response rates.

If benchmarks are available and usable for some or all questions, then each panel can be compared against the benchmark, and a measure of error can be computed from that comparison. However, in order to compare the results from surveys to benchmark estimates, two requirements should ideally be met:

1. *Question wording should be identical across the compared surveys.* Question wording is something to keep in mind when comparing studies, regardless of design. Small wording changes have shown to sometimes produce large effects on measurement (e.g., Smith, 1995), therefore to avoid confounding effects, the exact same question wording should be used in all surveys. At the same time, this can be difficult to achieve when mode differs across the surveys being compared and question adaptation becomes necessary. Specifically, benchmarks and other probability-based studies are often collected in interviewer-administered formats where questions are delivered orally, therefore questions selected from these surveys to include in the online panels for later comparison will need to be adapted to the self-administered, visual delivery mode.

2. *The populations represented by each survey need to be comparable.* If the benchmark survey includes population members without Internet access, these will have to be excluded from the estimation if the online panel includes only respondents with Internet access, as is usually the case. Problems may emerge if the definition of the Internet population used by the agency providing the benchmarks does not match the population from which the study respondents were recruited. This is further complicated when no question is asked on the benchmark study that identifies Internet users.

In Section 2.3 we provide a review of accuracy metrics that have been used to evaluate the differences in data quality between online panels and other surveys.

2.3   Accuracy metrics

When comparing results from online panels to benchmarks, different accuracy metrics are used in the literature:

1. *Direct comparisons* (panel by panel) to benchmarks of response distributions are the most commonly reported metric (e.g.,Vonk, van Ossenbruggen, & Willems, 2006; Walker, Pettit, & Rubinson, 2009) and look at the variability of estimates from different

sources. Panel names are usually not disclosed, with the exception of a few studies with a smaller number of panels (e.g., Duffy & Smith, 2005; Malhotra & Krosnick, 2007).

2. The *lowest and highest values* provide the reader with a range of possible estimates computed from data from the surveys used in the study (van Ossenbruggen, Vonk, & Willems, 2006).

3. The *average estimates across panels* are compared to a benchmark in the NOPVO (van Ossenbruggen et al., 2006) and the ARF study (Walker et al., 2009). This metric focuses on one estimate at a time and has the disadvantage of masking differences across panels; even if the overall average of an estimate across panels is equal to the benchmark, individual panels might grossly underestimate or overestimate the phenomenon, which would mean that using a single panel to address a research question would most likely result in biased estimates.

4. To solve the previous measurement issue, Yeager, Krosnick, et al. (2011) propose the *average absolute error* as a metric. The average absolute error is the average of the absolute difference between the modal category of the benchmark and the survey estimate for that category. It has the advantage of avoiding differences to cancel out.

5. The *largest absolute error* is used to summarize more than one estimate and it is measured as the error of the variable estimate in which the survey was least accurate (Yeager, Krosnick, et al., 2011).

6. The *number of significant differences from the benchmark* is the percentage of variables considered in the study that are statistically significantly different from the benchmark. It can be reported panel by panel or as the average percentage across panels (Yeager, Krosnick, et al., 2011).

All the above metrics can be reported either weighted or unweighted and, of course, more than one metric can be reported and compared to each other. We treat the issue of weighting later in the chapter.

2.4   Large-scale experiments on point estimates

Among the numerous studies that compare accuracy of estimates from online panels, many focus on comparing one panel to another survey, and a smaller number compare accuracy of several online panels. For space reasons, we focus on the largest comparisons experiments on point estimates that have been conducted since 2006, starting with the pioneering NOPVO project conducted in the Netherlands.

2.4.1   The NOPVO project

The first published large-scale experiment was initially presented at the 2006 ESOMAR panel research conference. Vonk, van Ossenbruggen, and Willems (2006) illustrated the Dutch online panel comparison (NOPVO) project (http://www.nopvo.nl/english/english.htm). The study compared the results of fielding the same survey on samples of approximately 1000 panel members from 19 different online panels of nonprobability samples in the Netherlands, which captured 90% of all Dutch online panel respondents at the time (Van Ossenbruggen et al., 2006). An omnibus questionnaire was administered in each panel during the same week of 2006, and was in field during seven days after the initial invitation. No quota sampling was used in selecting each sample from each panel. In total, 18999 panel members were invited to participate and 9514 completed the survey for a completion rate (Callegaro & DiSogra, 2008) of 50.04%.

To investigate data quality, the data were compared, when possible, to known benchmarks from Statistics Netherlands (CBS). Together with the omnibus questionnaire, panel member historical data were attached to the records and used in the analysis. When compared to known benchmarks, respondents across all 19 panels were more likely to be heavy Internet users (81% reported going online daily compared to the CBS benchmark of 68%), less likely to belong to a minority group and more likely to live in big cities. The average estimate of voter turnout, for example, was 90%, but the actual turnout was 79%. Voters for the Christian Democrats were on average underrepresented in the panels (16% vs. 29%) whereas voters of the Socialist Party were overestimated (14% vs. 6%). Some 23% of online panel members claimed to belong to a religious community as compared to a benchmark of 36%. The percentage of respondents who reported doing paid work for more than 15 hours a week varied across all panels from 53% to 82% (28 percentage point difference), whereas the percentage of respondents surfing the web for more than 10 hours had a range of variation of 29 percentage points across the lowest to the highest panel estimate. Although in the original NOPVO study no data were collected online from probability-based samples, a recent study (Scherpenzeel & Bethlehem, 2011) conducted using the Dutch probability-based online panel Longitudinal Internet Studies for the Social Sciences (LISS) compares the same statistics (collected on the LISS panel in 2006) to the benchmark data used by the NOPVO experiment. The bias from the LISS panel, measured as the difference from the benchmark, was smaller than that of the average NOPVO bias in five of the six benchmarks.

2.4.2 The ARF study

Based on concerns raised by early research on panel data quality, the Advertising Research Foundation (ARF) set up the Online Research Quality Council (ORQC) in August 2007 (Walker et al., 2009). One of the council's plans was to arrange a comparison study (NOPVO style) among 17 US online panel providers (all using nonprobability samples) a telephone sample panel, and a mail sample panel. A two-wave study was conducted in October and November 2008. One version of the questionnaire was fielded at a local market level (selected local markets). The online questionnaire was administered by a third independent party and contained: (1) factual and behavioral questions to be compared against known benchmarks; and (2) other common market research attitudinal questions such as intention to purchase items. Factual and behavioral questions were asked with the same question wording as the benchmarks they would be compared against. Of 1038616 invites, 76310 panel members completed the study for a completion rate of 7.34%. Various findings were obtained from this large study, whose estimated "book value" cost exceeded $1 million. When compared to known benchmarks, the study showed a similar pattern to the NOPVO study, with wide variation across panels in the survey estimates of interest. For instance, most panels overestimated smoking behavior; the estimates ranged from 42% (matching the benchmark value from NHIS) of members admitting having smoked at least 100 cigarettes in their entire life, to 58%, depending on the panel. Cell phone ownership was also overestimated across panels ranging from 85–93%, all above the benchmark value of 79%. Where panels showed the highest variance was in purchase intent and likelihood to recommend questions, typical market research questions. Two products were tested: the intention to purchase a new soup and a new paint. The percentage of panel members who chose the two response options indicating highest likelihood of purchase for the new soup varied from 32%–53% across panels. The authors also found that sample tenure (how long the respondent had belonged to the panel) was negatively related to the intention of purchase. Panel members with self-reported three or more years of membership were less willing (37%) to recommend the new kind of soup than panel members with three months or less of panel tenure (50%). A similar picture emerged for intent to recommend a new type of paint, 48% versus 62%.

The ARF redid the above study in 2012 with a similar design under the umbrella of the Foundation of Quality 2 (FOQ2) taskforce.[4] At the time of writing, there are no publicly available results to report.

2.4.3   The Burke study

The research firm Burke commissioned a study across 20 online panels with nonprobability samples and one online panel with a probability sample (Miller, 2007, 2008). The main purpose of the study was to investigate fraudulent respondents and satisficers. The same questionnaire, which included qualifying (screening) questions, "trap questions," and other standard market research questions was commissioned to the 21 online panels. No quota control in the within-panel sample design was set and the survey length was of about 25 minutes. Completion rates had an extremely large variability, similar to the NOPVO study, going from 3%–91% with an average of 18%. Few of the estimates had the potential to be benchmarked.[5] One of the benchmarked items asked in 11 of the panels was a question about whether the respondent was left-handed or ambidextrous. The absolute average error was of 1.7 percentage points for the proportion of left-handed respondents (ranging from a difference from the benchmark of −2 percentage points to +3 percentage points) and of 4.5 for the proportion of ambidextrous respondents (ranging from a +2 percentage-point to a +6 percentage-point difference from the benchmark). When comparing estimates of usage of blood glucose monitors, the range varies from a minimum of 10% to a maximum of 17% and the incidence of respondents claiming to have pet health insurance from a minimum of 4% to a maximum of 22%.

2.4.4   The MRIA study

A study similar to the ARF study was conducted in 2009 for the Marketing Research and Intelligence Association (MRIA) among 14 Canadian panels, one of which was Probit, an online panel with a probability sample (Chan & Ambrose, 2011). In this study, quotas for age, gender, and income were used to draw the sample. In terms of coverage of the target population, the authors reported that some panels could not deliver enough respondents for Quebec whereas others vastly under represented the French-speaking population. When look-ing at differences across panels for newspaper, magazine and radio consumption, the variation was small across panels. Further research steps were announced in the article but (to our

---

[4] http://thearf.org/foq2.php.
[5] No details are provided in the article about the source used for the benchmark estimates.

knowledge) no publication was available at the time of printing. Despite the fact that each panel was anonymized in the article, there was only one panel with a probability sample (Probit), which was therefore self-identified. At the same annual conference in 2010, Probit (EKOS, 2010) reanalyzed the MRIA study using the average of the panels with nonprobability samples and compared it against the Probit estimates collected in the same experiment. Official benchmarks were also added to the study. The authors found that Probit panel members were less likely to be heavy Internet users, to use coupons when shopping, and to have joined the panel for the money or incentives than members of the online panels of nonprobability samples. When compared to the distribution of income for the Internet population according to official benchmarks, online panels of nonprobability samples recruited more members with lower income than the Probit panel, which yielded estimates of income that were however closer to the benchmark.

2.4.5   The Stanford studies

Finally, Yeager, Kosnick, et al. (2011) compared estimates from an RDD telephone survey to estimates from six online panels of nonprobability samples, one online panel with a probability sample, and one cross-sectional sample recruited via river sampling. The exact same online questionnaire was used in all surveys. Data were collected in the fall of 2004 and early 2005 for a total sample size of 1000 respondents per company (study 1). A second round of data collection was done in 2009 with the same probability sample of 2004 and two nonprobability panels of the previous study (study 2). The questionnaire contained items on basic and secondary demographics such as marital status, people living in the households, and home ownership. Other questions asked were frequency of smoking, passport ownership and health status. The uniqueness of the Stanford study is that *every* question was selected so that known gold standards collected by US federal agencies were available for comparison. The authors were then able to compute and compare the absolute average error of each sample source.

Results indicated that the RDD and the probability-based online panel data were on average closer to the benchmarks than any of the online panels with nonprobability samples. The same findings were found for the more recent data collection of 2009: the average absolute error among the same panel providers was close to that in the 2004/2005 study. The probability sample was also more accurate than the two nonprobability samples.

2.4.6   Summary of the largest-scale experiments

To better summarize the findings from these large-scale experiments we have compiled two tables where data from the above studies are compared with known benchmarks coming from official, high-quality surveys with probability samples. In Table 2.2 we have compiled the comparison with smoking benchmarks across different studies. In order to standardize the comparison across studies the average absolute difference metric described above has been used. We could not use other metrics, such as the largest absolute error and the number of significant differences from the benchmark, because detailed panel-by-panel original estimates are not available for the studies considered, with the exception of the Stanford study.

To shed more light on the variability of smoking estimates across panels, in Table 2.3 we reproduce Figure 1 of Walker et al. (2009, p. 474).

Probability sample panels were always closer to the smoking benchmarks than nonprobability sample panels (see Table 2.3). This is true for studies conducted in different years and countries. Online panels of nonprobability samples in the United States and in Canada tend

Table 2.2 Average absolute error of smoking estimates across different studies.

Study	Variable	Benchmark compared to	Average absolute error	Range min–max
Stanford study 1	Non-smoker	1 RDD sample	2.6	–
Stanford study 1	Non-smoker	1 Probability sample panel	4.2	–
Stanford study 1	Non-smoker	Average of 6 nonprobability sample panels	9.6	5.8–17.8
ARF	Ever smoked	Average of 17 nonprobability sample panels	10.0	–
ARF	Currently smoke	Average of 17 nonprobability sample panels	5.6	0–12
MRIA	Currently smoke	Average of 13 nonprobability sample panels + 1 probability sample panel	10.5	–
MRIA	Currently smoke	1 Probability sample panel	2.1	–

Table 2.3 Comparison of weighted percentages regarding smoking behaviors across the 17 nonprobability sample panels in the ARF study.

Source	Currently smoke	Smoked at least 100 cigarettes in your entire life
NHIS/CDC benchmark	*18*	*42*
Panel A	19	42
Panel B	20	47
Panel C	20	47
Panel D	21	48
Panel E	23	49
Panel F	24	50
Panel G	26	50
Panel H	26	50
Panel I	27	50
Panel L	27	51
Panel M	28	51
Panel N	28	51
Panel O	30	52
Panel P	30	55
Panel Q	31	57
Panel R	32	57
Panel S	33	58

Notes: The data come from two different panels which are organized in order of magnitude so the readers should not assume that the results from the same row come from the same panels. Data shown in order of magnitude.

Table 2.4 Average absolute error of average estimates of different variables across different studies.

Study	Variables	Benchmark compared to	Average absolute error	Range min–max
NOPVO	6 variables	Average of 19 nonprobability sample panels	8.5	Cannot be computed[1]
NOVPO	6 variables	1 probability sample panel	4.0	–
Stanford study 1	13 variables	1 RDD sample	2.9	–
Stanford study 1	13 variables	1 probability sample panel	3.4	–
Stanford study 1	13 variables	Average of 6 nonprobability sample panels	5.2	4.5–6.6
ARF	6 variables	Average of 17 nonprobability sample panels	5.2	0–10
Stanford study 2	13 variables	1 RDD sample	3.8	–
Stanford study 2	13 variables	1 nonprobability sample panel	4.7	–
Stanford study 2	13 variables	1 probability sample panel	2.8	–

Note: [1]Data for each single panel included in the NOVPO experiment are not available so we cannot report the minimum and maximum value.

to estimate a higher proportion of smokers than the proportion of smokers in the population according to the benchmark, even after weighting.

The same finding is replicated using other variables (see Table 2.4). Most of the variables analyzed in this study are behavioral or factual in nature such as work status, number of bedrooms in the house, number of vehicles owned, having a passport, drinking and quality of health, having a landline or cell phone, and party voted for in the last election. Here again, probability sample panels and RDD telephone surveys are closer to the benchmarks than online panels based on nonprobability samples.

Sometimes benchmarks are not available, either because more accurate population estimates are impossible to collect for a given variable or because they are not readily available when analyses are conducted. In these cases it is not possible to use an accuracy metric but it is still possible to study the variability of estimates across panels. This kind of data is still relevant and informative for survey commissioners to appreciate how reliable data from online panels might be.

The NOPVO study addressed this question by studying familiarity with brands ("Have you seen a commercial of the following [brand]?"). The original values were not reported in the study; instead a mean value was computed across panels together with the top three estimates plus the bottom three estimates, providing an indication of variability across estimates from different panels. In comparison to the average brand awareness across panels, estimates varied from −5 to +7 percentage points for Citroën, from −9 to +9 for Mazda, from −6 to +6 for T-mobile and from −11 to +5 for Volkswagen (see Table 2.4).

In the ARF estimates about willingness to buy the new soup and paint,[6] the percentage of respondents who selected the top two answers (definitely and probably would buy) varied from a low range of 34% to a high range of 51% for the soup and from 37% to 62% for the new paint (weighted results). In the same ARF study, the mail sample estimate for the intention to buy the new soup was 32%, and for the phone sample 36%.

2.4.7   The Canadian Newspaper Audience Databank (NADbank) experience

In 2006, the Newspaper Audience Databank (NADbank), the Canadian daily newspaper audience measurement agency, initiated a test to assess the feasibility of collecting newspaper readership data using an online panel rather than the until then traditional data collection protocol based on RDD telephone surveys (Crassweller, D. Williams, & Thompson, 2006). In the experiment, the results from their standard telephone data collection (spring and fall) were compared to results from 1000 respondents from an online panel with a nonprobability sample (same time periods) for the Toronto CMA.[7] The online sample estimates for average number of hours per week of TV and Internet usage, as well as for average number of newspapers read per week, were higher than the estimates from the telephone sample (Crassweller et al., 2006). Most importantly, the key readership metrics by newspaper differed with the different sampling approaches and there was no consistent pattern or relationship in the differences.

Based on these initial results NADbank decided to broaden the scope of the test and include more online panels (Crassweller, Rogers, & Williams, 2008). In 2007, another experiment was conducted in the cities of Toronto, Quebec City, and Halifax. Again, the four nonprobability sample panels received identical instructions for project design, implementation, weighting, and projection and were run in parallel with the telephone RDD sample in those markets. The results from the four panels varied substantially in terms of demographic composition (unweighted and after weighting to census data for age, gender, and household size) and in terms of media habits; panels did not mirror the benchmark in any of the metrics of interest.

Compared to the benchmark, all panel estimates of readership for both print (paper versions) and online newspapers were over estimated to varying degrees. This was true for all newspapers in all markets. No one panel performed better than another. The authors concluded that there was no obvious conversion factor to align panel estimates to RDD estimates and that the panel recruitment approach could not provide a sample that reflected the general population. Without such a sample it would be impossible to gain valid insights regarding the population's newspaper readership behavior. The outcome of the test resulted in NADbank maintaining their current RDD telephone sampling methodology. It was clear that at that time "a web-based panel does not provide a representative sample, and secondly that different panels produce different results" (Crassweller et al., 2008, p. 14).

Four years later, NADbank commissioned another study, this time comparing the results from their RDD sample to Probit, an online panel with a probability sample recruited using landline and cell-phone exchanges with an IVR recruitment protocol (Crassweller, J. Rogers, Graves, Gauthier, & Charlebois, 2011). The findings from the online panel were more accurate than the previous comparisons. In terms of unweighted demographics, Probit was better able to match census benchmarks for age and gender than previous panels. The probability-based panel recruitment approach resulted in closer estimates of print readership but over estimated

---

[6] Assuming this product was available at your local store and sold at an acceptable price, which of the following statements best describes how likely you would be to buy it?

[7] Statistics Canada Census Metropolitan Areas (CMA).

online readership. The authors concluded that this approach was an improvement on previous panel recruitment approaches but still reflected the limitations of the recruitment method (IVR) and the predisposition for mediaphiles to participate in online media surveys. The key strength of the Probit (IVR) approach is that it "has the potential to provide a one-stop shop for online and offline consumers" (p. 6). The authors warned that more work still needed to be done before quantitative research studies can be conducted using online panels of nonprobability samples but concluded that incorporating RDD sampling approaches with the use of online panel measurement provided alternatives for the near future.

2.4.8 Conclusions for the largest comparison studies on point estimates

The main conclusion from this set of studies is that different results will be obtained using different panels or, in other words, that online panels "are not interchangeable". In the NOPVO study Vonk, Ossenbruggen & Willems, (2006, p. 20) advise: "Refrain from changing panel when conducting continuous tracking research". Similar statements are made in the ARF study: "The findings suggest strongly that panels are not interchangeable" (Walker et al., 2009, p. 484), and in the comparison done in Canada by MRIA (Chan & Ambrose, 2011, p. 19) "Are Canadian panels interchangeable? Probably not for repetitive tracking". On a different note, the authors from the Stanford study conclude their paper by saying: "Probability samples, even ones without especially high response rates, yielded quite accurate results. In contrast, nonprobability samples were not as accurate and were sometimes strikingly inaccurate" (Yeager, Krosnick, et al., 2011, p. 737).

2.5   Weighting adjustments

Differences across panels' estimates could potentially disappear after each panel has been weighted. Unfortunately in the reviewed studies that was not the case. The ARF weighting on common demographics made almost no difference in reducing the discrepancy among panels and in comparison to the benchmarks. A second type of weighting was then attempted. In this approach, in addition to post-stratification, duplicates and respondents who belonged to multiple panel were removed. This second approach improved data quality to some extent, but significant differences from the benchmarks still remained (Walker et al., 2009). The ARF study stated: "Sample balancing (weighting) survey data to known census targets, ... removed variance but did not completely eliminate it. Likewise, the test of a pseudodemographic weighting variable (panel tenure) did not eliminate variance" (Walker et al., 2009, p. 473).

In the NADbank report the authors conclude that: "There is no firm basis on which to develop a conversion factor or weight that could bridge telephone and online findings" (Crassweller et al., 2008, p. 14). Finally, in the Stanford study, the authors concluded: "Post-stratification of nonprobability samples did not consistently improve accuracy, whereas post-stratification did increase the accuracy of probability sample surveys" (Yeager, Krosnick, et al., 2011, p. 733).

Finally, Tourangeau, Conrad, and Couper (2013) presented a meta-analysis of the effect of weighting on eight online panels of nonprobability samples in order to reduce bias coming from coverage and selection effects. Among different findings, they concluded that the adjustment removed at most up to three-fifths of the bias, and that a large difference across variables still existed. In other words, after weighting, the bias was reduced for some variables but at the same time it was increased for other variables. The estimates of single variables after weighting would shift up to 20 percentage points in comparison to unweighted estimates.

A promising approach that has been developed during the year is the use of propensity score weighting, as discussed in Chapter 12.

2.6   Predictive relationship studies

Findings reported until now suggest that researchers interested in univariate statistics should avoid using panels from nonprobability samples to obtain these estimates. However, more often than not, researchers are interested in investigating relationships between variables, and some argue that multivariate analyses might not be biased when computed using panels of nonprobability samples.

This section summarizes findings from four studies that have compared estimates of association between variables in probability sample panels against nonprobability sample panels.

2.6.1   The Harris-Interactive, Knowledge Networks study

Parallel studies on global climate change and the Kyoto Protocol were administered to an RDD telephone sample, to two independent samples drawn five months apart on the nonprobability sample Harris Interactive panel (HI), and on the probability sample panel Knowledge Networks (KN) (Berrens, Bohara, Jenkins-Smith, Silva, & Wiemer, 2003). The authors compared the relationships between environmental views and ideology across the four samples. When combining the samples and looking at an ordered probit model predicting environmental threat (on an 11-point scale: 0 = No real threat; 10 = brink of collapse), the model showed that ideology was a strong predictor of perceived threat, where the more conservative respondents were, the least of a threat they saw in global warming. There were, however, large significant interactions of the Internet samples (taking the RDD sample as baseline) where the relationship between ideology and perceived threat was less strong in the two nonprobability samples. When controlling for demographics, the effect of the sample source disappeared. In a logistic regression analysis predicting if respondents would vote for or against (0–1) ratification of the Kyoto Protocol given an increased amount of taxes, the authors found that respondents to all the online panels were less supportive of the Kyoto Protocol that respondents to the telephone survey. However, in all samples "the analyst would make the same policy inference (...) – the probability of voting yes on the referendum is significantly and inversely related to the bid price (or cost) of the policy" (p. 20).

2.6.2   The BES study

Parallel to the British Election Studies (BES) of 2005 (a face-to-face survey where addresses were selected from a postal address file in the United Kingdom with a response rate of over 60%), an Internet sample was selected from the YouGov panel (based on a nonprobability sample) with the goal of comparing the accuracy of estimates from both designs (Sanders, Clarke, Stewart, & Whiteley, 2007). The authors found significant differences between the two samples with respect to point estimates of political choice, voter turnout, party identification, and other questions about political issues, where the probability sample was overall, but not always, more accurate than the nonprobability sample. Models predicting three different variables were tested in each sample.

1. The first model used 16 variables to predict voting turnout and found significant differences across samples in five of the 21 estimated parameters. For two of the parameters

(efficacy/collective benefits and education), the relationship was significantly stronger for the face-to-face probability sample. For two other parameters (personal benefits and Midlands Region), the coefficient was significant in one sample but not in the other. Finally, according to the face-to-face probability sample, females were less likely to have voted than males. The opposite was found in the Internet nonprobability sample.

2. The second model was a regression on party choice in the 2005 election, where significant differences were found in 5 of the 27 estimated parameters. Again, for two parameters (Blair effect and Kennedy effect) the coefficient was larger in the face-to-face probability sample than in the Internet nonprobability sample. Two other parameters (party-issue proximity and Southwest region) were significant in one sample and not in the other, and one parameter (age) was negative in the face-to-face sample (suggesting, as one would expect, that older respondents were less likely to vote for the Labour Party) and positive in the Internet nonprobability sample.

3. In the third set of models, rather than comparing coefficients, different competing models were compared to try to find the one that better explained the performance of the rival party. Both samples led to the same conclusions when inspecting the associated explained variance and other goodness-of-fit statistics.

2.6.3   The ANES study

Around the same time, Malhotra and Krosnick (2007) conducted a study comparing the 2000 and 2004 American National Election Study (ANES), traditionally recruited and interviewed face-to-face, to data collected from nonprobability Internet samples. Response rates in the ANES were above 60%; the 2000 ANES sample was compared to a sample obtained from the Harris Interactive panel survey, and the 2004 ANES sample was compared to a sample from the YouGov panel. The questions asked of each sample were not always identical, but only those questions with similar questions and equal number of response options were used to compare the face-to-face probability samples to their Internet nonprobability counterparts. In contrast to the multivariate regression approach followed by Sanders et al., Malhotra and Krosnick analyzed bivariate logistic regressions that predicted "predicted" vote choice, actual vote choice, and actual turnout.

Results showed that the design of the surveys (which used a different mode and sampling strategy) had an impact on survey estimates of voting intention and behavior as well as on estimates of bivariate relationships. For example, in the 2004 study, 10 out of 16 parameters predicting "predicted" vote choice were significantly different in the two sources of data; in the 2000 study, 19 out of 26 parameters were significantly different. When predicting actual vote choice using data from 2000, 12 out of the 26 parameters were significantly different across samples. Weighting the data did not reduce these differences, and they were not entirely explained by different levels of interest in politics of respondents in both types of sample. As in the BES study, even though the true values of the regression parameters are unknown, we do know that point estimates about vote choice and turnout were more accurate in the face-to-face sample than in the nonprobability Internet sample.

2.6.4   The US Census study

The third study investigating differences in relationships between variables compared a series of RDD telephone surveys collecting data from about 200–250 respondents per day for almost

5 months to online surveys fielded on weekly nonprobability samples from the E-Rewards panel. This resulted in about 900 completes per week for approximately 4.5 months (Pasek & Krosnick, 2010). Using questions that were identical or virtually identical, they first compared the demographic composition of the two sets of data and found that the telephone samples were more representative than the Internet samples. When comparing response distributions for the substantive variables, there were also sizeable differences (often differing by 10 to 15 percentage points) between the two samples.

Pasek and Krosnick (2010) first compared bivariate and multivariate models predicting two different variables tested in each sample. When predicting intent to complete the Census Form, 9 of the 10 substantive variables had similar bivariate associations in the expected direction. For example, in both samples, respondents were more likely to report intent to complete the Census form if they thought the Census could help them, or if they agreed that it is important to count everyone. For the tenth variable the relationship was in the expected direction for the telephone sample, but panel respondents who did not think it was important to count everyone were *more* likely to intend to complete the census form. For eight of the substantive variables where the direction of the relationship was the same in both samples, however, the relationships were stronger for the panel sample than for the telephone sample for five variables and weaker for three variables. Demographic predictors were often significantly different in the two samples, supporting different conclusions. When predicting actual Census form completion, differences were less pronounced but still present, suggesting again that which sample is used to investigate the research questions can have an impact on the conclusions that are ultimately reached.

Pasek and Krosnick also compared all possible correlations among the variables measured in both surveys, finding that correlations were significantly stronger in the panel sample than in the telephone sample. It is worth noting that in both the BES and the US Census study the relationship between age and the predicted variable differed significantly between the nonprobability online panel sample and the alternative probability sample. The relationship was significant for both samples but had opposite signs in each. In the nonprobability online survey, the relationship was the opposite of what was expected from theory. In addition, both the ANES and the US Census studies bivariate relationships tended to be significantly stronger for predictors in the online nonprobability sample than in the alternative sample. This suggests that respondents in the former were systematically different from the alternative method respondents.

Although some authors conclude that researchers would make similar conclusions when using probability or nonprobability panels (Berrens et al., 2003; Sanders et al., 2007) when looking at the signs of the coefficients, they are not always in the same direction (Pasek & Krosnick, 2010) and the strength of relationships varies across samples (Malhotra & Krosnick, 2007; Sanders et al., 2007; Pasek & Krosnick, 2010). We hope more studies will follow up this topic.

2.7   Experiment replicability studies

An important question for market researchers and behavioral scientists involves replicability – in terms of both significance and effect sizes – of random-assignment experimental studies that use as participants respondents from online panels. Indeed, market researchers often seek to understand what influences consumers' behaviors and attitudes. Experiments are an effective method to assess the impact of some change in message or marketing strategy on a person's preference for or likelihood of purchasing a given product. Likewise, public opinion researchers often seek to understand the impact of a candidate's policy on the public's vote. Experiments that present respondents with randomly assigned messages, can allow campaigns to estimate the proportion of the vote that might be won when taking one strategy or another. The estimates of this impact can then be used to calculate the expected gain, in terms of sales or votes that might be found when taking one strategy versus another. This allows for more efficient use of resources. Therefore, it is often of interest to know both *whether* a given change is likely to alter Americans' behaviors or preferences, and also *how much* this change would affect them. Put another way, researchers who conduct experiments using online panels are often interested in both the *significance* of an experimental comparison and the *effect size* of that comparison. What does the research say about replicating experimental results – in terms of both significance and effect sizes – in probability and nonprobability-based samples?

The research literature on this topic is sparse. To date, there has been no published extensive empirical or theoretical analysis of this question. Much research has focused on whether probability sample panels provide more accurate point estimates of the prevalence of various behaviors or characteristics as just discussed in this chapter, while no published study has comprehensively investigated whether probability versus nonprobability sample panels yield similar conclusions about causal relationships as assessed through experiments. However, there are a number of studies that happened to have used both probability and nonprobability samples when testing causal relationships using experiments (e.g., Bryan, Walton, T. Rogers, & Dweck, 2011; Yeager & Krosnick, 2011, 2012; Yeager, Larson, Krosnick, & Tompson, 2011). Furthermore, disciplines such as social psychology have a long history of discussing the potential impact of sample bias on experimental results (Henrich, Heine, & Norenzayan, 2010; Jones, 1986; Sears, 1986). In this section, then, we review: (1) the key theoretical issues to consider regarding the results of experiments in online panels; (2) the emerging empirical evidence and what future research needs to be conducted in order to sufficiently address this question.

2.7.1   Theoretical issues in the replication of experiments across sample types

One important starting point for theory about the replicability of experiments comes from researchers in social, cognitive, and personality psychology. These researchers have a long history of using nonprobability samples to conduct experiments – specifically, samples of undergraduate students who are required to participate in psychology studies to complete course credit. This large body of work has contributed greatly to our understanding of patterns of thinking and social behavior. However, at various times in the field's history it has responded to criticisms of its database. For instance, Sears (1986) proposed that the narrow age range, high educational levels, and other unique characteristics of college students make them different from adults in ways that may limit the generalizability of findings (see also Henry, 2008). Likewise, Wells (1993), a prominent consumer behavior researcher, said that: "students are not typical consumers" because of their restricted age range and educational levels and that ignoring these uniquenesses "place[s] student-based conclusions at *substantial risk*" (pp. 491–492, emphasis added).

Psychologists have responded to these criticisms by arguing that the objective of much academic research is not to produce point estimates but rather to assess the causal relation between two conceptual variables in any segment of the population. For instance, Petty and Cacioppo (1996) stated:

> If the purpose of most psychological or marketing laboratory research on college students were to assess the absolute level of some phenomenon in society (e.g., what percentage of people smoke or drink diet coke?) … then Wells's criticism would be cogent. However, this is not the case. [A laboratory study using college students] examines the viability of some more general hypothesis about the relationship between two (or more) variables and ascertains what might be responsible for this relationship. Once the relationship is validated in the laboratory, its applicability to various specific situations and populations can be ascertained.
>
> (pp. 3–4)

Similarly, Ned Jones (1986) has argued that:

> Experiments in social psychology are informative mainly to the extent that they clarify relationships between theoretically relevant concepts. Experiments are not normally helpful in specifying the frequency of particular behaviors in the population at large.
>
> (p. 234)

Indeed, as noted above, research to assess point estimates is distinct from research to understand relations between variables. However, marketing and political researchers are often not interested in whether a given relationship could exist in *any* segment of the population during any time period, but whether it exists *right now in a population they care about*, that is, consumers and voters. Further, as noted above, the size of an effect is often a substantive question. Understanding not only that something *might* matter under some specified set of conditions is sometimes less important when making decisions about how to invest resources than knowing *how much* something matters. And there is no strong statistical rationale for assuming that the size or significance of results from a small biased sample will be true in the population as a whole. To the contrary, statistical sampling theory suggests that any estimate of a parameter will be more accurate when that parameter is estimated using data from a random sample, compared to a biased (nonrandom) sample.

While there is no statistical basis for assuming homogeneity of effect sizes in a biased versus probability-based sample, the logic of random assignment assumes that whatever characteristic that might affect the outcome variable will be distributed equally across the two conditions (see Morgan & Winship, 2007). Given a large enough sample so that participant characteristics are truly randomly distributed across conditions, sample selection bias would only be expected to bias the size of the treatment effect in the event that the sample is biased in terms of some characteristic that is correlated with a person's responsiveness to the experimental manipulation.

For instance, imagine an experiment to test two framings of a campaign issue. If these two framings are judged as equally different by everyone regardless of their cognitive ability, then a nonprobability sample that underrepresents high-education respondents might not result in different treatment effects. However, if only people who think carefully about the issues will notice the difference between the issue framings – that is, if only highly-educated people were expected to show a treatment effect – then a nonprobability sample that includes too-few college educated respondents might show a smaller or even nonexistent treatment effect. Therefore, one theoretical issue that will likely determine the replicability of an experiment in probability versus nonprobability samples is whether the treatment effect is likely to be different for people with different characteristics, and whether the sampling methods are likely to produce respondents that differ on those characteristics.

A related issue involves research hypotheses that are explicitly designed to test whether a given subgroup of people (for instance, low-education respondents) will show an experimental effect (for instance, whether they will distinguish between the persuasiveness of two advertising campaigns). One assumption might be that *any* sample that includes enough respondents in that sub-group to allow for a test with reasonable power will provide an accurate estimate of the treatment effect for that group. That is, all low-education respondents may be thought to respond identically to the experimental manipulation, whether they were recruited through probability or nonprobability methods. Indeed, this is the perspective of much of psychology, which treats any member of a group (such as "low cognitive ability" vs. "high cognitive ability," (e.g., West, Toplak, & Stanovich, 2008); or "westerners" or "easterners" (Markus & Kitayama, 1991)) as a valid representative of the psychological style of that group. By this logic, it is unimportant whether such a study includes proportions of members of a sub-group that match the population. Instead, the crucial feature is whether the sample has enough people in that group to adequately allow for the estimation of experimental effects.

However, another perspective is that members of subgroups may only be considered to be informative about the thinking styles or behaviors of that subgroup if they were randomly sampled from the population. That is, the "low-education" respondents in a given sample may not resemble low-education respondents more generally in terms of their receptivity to an experimental manipulation. If this is true, then experiments using nonprobability samples to test for effects within a given subgroup may lead researchers astray.

In summary, if researchers are looking for main effects of an experimental manipulation, and if people's responsiveness to that manipulation is uncorrelated with a person's characteristics, then a nonprobability sample would be expected to provide similar estimates of an effect size as a probability-based sample (all other methodological details being equal). However, if responsiveness to the manipulation depends on some characteristic that is over- or under-represented in a nonprobability sample, then experimental effects might vary between that sample and a probability-based sample. Further, if researchers are hoping to assess experimental effects within some subgroup (e.g., low-income respondents, women, Latinos, etc.) and if respondents are not a random sample of people from that subgroup, then it is possible that the subgroup analysis will yield a different result in probability-based and nonprobability-based samples. With these issues in mind, we turn to the limited evidence available, in addition to future studies that are needed to further understand these issues.

2.7.2 Evidence and future research needed on the replication of experiments in probability and nonprobability samples

A large number of studies in psychology and behavioral economics have assessed the different results obtained in experiments with nonprobability samples of college students and nonprobability samples of nonstudent adults. Peterson (2001) meta-analyzed 30 meta-analyses that tested for moderation by sample type and found a great deal of variance in college student versus noncollege student samples. In many cases, findings that were significant and in one direction in one sample were nonsignificant or significant in the opposite direction in the other sample. Similarly, Henrich, Heine, and Norenzayan (2010) compared results from experiments conducted with samples of college students in the United States to results from the same experiments conducted with nonprobability samples of adults in other countries in

Africa or Asia. These authors found many cases of nonreplication or of studies that produced effects in the opposite direction. An obvious limitation in these studies, however, is that both of the samples were recruited using nonprobability methods. It is thus unclear which sample was biased in its estimate of the effect size.

A small number of studies have begun to test for experimental effects using a college-student sample and then have replicated the study using a probability-based sample. One prominent example is a series of experiments conducted by Bryan, Walton, Rogers, and Dweck (2011). These researchers assessed the impact of a brief framing manipulation the day before an election (referring to voting as "being a voter in tomorrow's election" vs. "voting in tomorrow's election") on registered potential voters' actual voting behavior (as assessed by looking for research participants in the validated voter file). In one study conducted with Stanford students, Bryan et al. (2011) found that the framing manipulation increased actual voter turnout by roughly ten percentage points. In a second study conducted with a probability-based sample of voters – members of the GFK Knowledge Panel – the authors replicated the significance of the effect, and the size of the effect was nearly identical. Thus, in at least one case, both significance and effect size were replicated in a probability-based sample.

Two other investigations have conducted randomized experiments to assess the impact of a small change in question wording on the validity of respondents' answers (Yeager & Krosnick, 2011, 2012; Yeager, Larson, et al., 2011). Yeager and Krosnick (2012) examined whether questions types that employ a stem that first tells respondents what "some people" and "other people" think before asking for the respondent's own opinions yields more or less valid data relative to more direct questions. They tested this in both nationwide probability-based samples (the General Social Survey, the FFRISP, and the Knowledge Panel) and in nonprobability-based Internet samples (from Lightspeed Research and Luth Research). These authors found that "some/other" questions yielded less validity, and this was true to an equal extent in both probability and nonprobability-based cases. Furthermore, they reached identical conclusions when they tested the "some/other" format in convenience samples of adolescents (Yeager & Krosnick, 2011). Replicating these overall findings, Yeager, Larson et al. (2011) found that the significance and size of the impact of changes in the "most important problem" question[8] were no different in an RDD telephone survey or in a nonprobability sample of Internet volunteers. Thus, the limited evidence so far does not suggest that there are substantial differences in either replication or size of effects across probability and nonprobability-based samples.

The evidence is not adequate, however, to assess the more general question of whether the two types of samples are always likely to replicate experimental effects. The studies noted above do not have likely *a priori* moderators that could have existed in substantially different proportions across the types of samples. Therefore, it will be important in future research to continue to examine effects that are likely to be different for different people. Furthermore, the studies above were not interested in sub-group analyses. It is an open question whether, for instance, studies assessing the impact of a manipulation for women versus men, or rich versus poor, would yield different conclusions in probability or nonprobability-based samples.

2.8   The special case of pre-election polls

The advantage of pre-election polls is that the main statistics of interest (voter turnout and final election outcome) can be evaluated against a benchmark for all panels. Baker et al.

---

[8] Respondents were asked: "What do you think is the most important problem facing the country today?"

(2010, p. 743) and Yeager, Larson et al. (2011, p. 734) provide a list of studies showing that nonprobability online panels can provide as good and sometimes better accuracy than probability sample panels. In the United States, for example, this goes as far back as the 2000 election (Taylor, Bremer, Overmeyer, Siegel, & Terhanian, 2001) and in the United Kingdom, this goes back to 2001 (YouGov, 2011). In the 2012 US election, nonprobability panels performed as well and sometimes better than traditional probability polling (Silver, 2012).

At the same time pre-election studies differ in several ways from other survey research. Pre-election polls are focused mainly on estimating one variable (the election outcome), which is most of the time (depending on the country) a binary variable. In addition, pre-election studies are often conducted in an environment where during weeks before the election many other studies, generally pre-election telephone polls, are publicly available. In fact, unlike the majority of surveys, in pre-election polls there are continuous sources of information that help guide additional data collection and refine predictive models based on identification of likely voters, question wording, handling of undecided and nonrespondents, and weighting mechanisms. Thus, differences in accuracy do not just reflect differences in accuracy of nonprobability samples panels but also differences on how all these variables are handled. As the recent AAPOR report from the nonprobability samples states: "Although nonprobability samples often have performed well in electoral polling, the evidence of their accuracy is less clear in other domains and in more complex surveys that measure many different phenomena" (Baker et al., 2013, p. 108).

As Humphrey Taylor recognized early on (Taylor, 2007), the secret to generating accurate estimates in online panels is to recognize their biases and properly correct them. In the specific case of election polls, some companies are better than others in doing so. The case of pre-election polls is encouraging and we hope that many more studies are published trying to extend successful bias correction methodologies to other survey topics.

2.9   Completion rates and accuracy

In online panels of nonprobability samples, response rates cannot be really computed because the number of total people invited to sign up (the "initial base") is unknown. Completion rates can still be computed by dividing the number of unique complete survey responses by the number of email invitations sent for a particular survey (Callegaro & DiSogra, 2008).

In the NOPVO study (Vonk et al., 2006), completion rates ranged from 18%–77%. The authors explained the differences as a function of panel management: some companies "clean up" their database from less active members more than others and they found that fresh respondents were more responsive than members who had been panelists a year or longer. Yeager, Krosnick et al. (2011) studied the effect of completion rates on accuracy of the responses finding that in the nonprobability samples, higher completion rates were strongly associated with higher absolute error ($r = .61$). A similar but slightly weaker relationship was found for the response rates of the seven RDD studies ($r = .47$) and for the response rates of the seven samples drawn from the probability-based Knowledge Networks online panel, ($r = .47$). These results add to an increasing body of literature suggesting that efforts to increase response rates do not result in improvements in accuracy, as previously expected.

2.10   Multiple panel membership

Multiple panel membership is an issue that has attracted the attention of the research community since the beginning of online panels. Also called *panel duplication* (Walker et al., 2009),

Table 2.5 Average number of membership per panel member, and percentage of members belong to five or more panels.

Studies	Year	$\overline{X}$ panel member	% belonging to 5+	Country
Multiple panels studies				
Chan & Ambrose	2011		45	CA
Walker et al.	2009	3.7	45	US
Gittleman & Trimarchi	2010	4.4	45	US
Gittleman & Trimarchi	2010		25	FR
Gittleman & Trimarchi	2010		19	ES
Gittleman & Trimarchi	2010		23	IT
Gittleman & Trimarchi	2010		28	DE
Gittleman & Trimarchi	2010		37	UK
Gittleman & Trimarchi	2010		38	AU
Gittleman & Trimarchi	2010		39	JP
Vonk et al.	2006	2.7	23	NL
Fulgoni	2005	8.0		US
Single panel studies				
Casdas et al.	2006		11	AU[1]
De Wulf & Bertellot	2006		29[2]	BE
Comley	2005		31	UK[3]

[1] Measured in one panel only, AMR interactive.
[2] Measured in one panel only, XL Online. Some 29% of members declared they belonged to more than one panel.
[3] Measured in one panel only, UK Opinion Advisors.

or *panel overlap* (Vonk et al., 2006), this is a phenomenon found in as many countries as we could find a study for. In Table 2.5 we list the average number of memberships per panel member and the percentage of members belonging to more than five panels, according to different studies. All these studies were undertaken by comparing online panels of nonprobability samples. At the current stage we could not locate studies of probability-based panels or of panels where membership is restricted by invitation only as described in Chapter 1.

It is not uncommon that members belong to multiple panels with as high as 45% of panel members belonging to five or more panels in the most recent estimates in the United States and Canada. The issue of multiple panel membership is important from two points of view: diversity of panel members, and data quality. The first aspect resonates with the concern that Fulgoni (2005) voiced that a minority of respondents might be responsible for the majority of surveys collected.

In the pioneering NOPVO study (Vonk et al., 2006), the number of multiple panel membership varied by recruitment methods: panels who bought addresses or recruited via link or banners had a higher amount of overlap (average of 4.3 and 3.7 panels per member respectively) than panels who recruited by phone or snowballing (2.0 and 2.3 respectively). Panel offering self-registration had an average overlap of 3.3, while panels recruiting via traditional research had an overlap of 2.4. Interestingly but not surprisingly, respondents with high Internet activity had an average multiple panel membership of 3.5 in comparison to low Internet users: 1.8 (i.e., respondents who checked their email once or twice a week). We will return to the issue of frequency of Internet usage with more up-to-date data later on in this chapter. Casdas, Fine, and Menictas (2006) compared multiple panel member demographics with Australian census data, finding that they were more likely to be younger, less educated, female, working part-time and renting their living space. In the ARF study (Walker et al., 2009) multiple panel membership was again related to the recruitment method: higher multi-panel memberships occurred with unsolicited registrations, affiliate networks and email invitations. Multiple panel membership was also three times higher for African Americans and Latinos.

2.10.1   Effects of multiple panel membership on survey estimates and data quality

Most studies examining the effects of multiple panel membership on data quality have been conducted in the area of traditional market research questions such as shopping attitudes, brand awareness, and intention to purchase. In one of the first multiple panel membership studies, Casdas and colleagues (2006) noted that members belonging to more than two panels were more likely to be more price-driven than brand-driven in comparison to members belonging to one panel only and to a CATI parallel interview. The comparison was done with a multivariate model controlling for demographics characteristics. In terms of brand awareness Vonk et al. (2006) compared multiple panel members' answers to the average awareness results from all the 19 panels in their study. Multiple panel members had above average brand awareness but below average advertisement familiarity. Lastly, in the ARF study (Walker et al., 2009), members belonging to four or five and more panels were more likely to say that they would buy a new soup, or paint (intention to purchase concept test) than panel members belonging to less panels. For example, the percentage of respondents saying that they will definitely buy a new soup was of 12% for members belonging to one panel, 15% for two panels, 16% for three panels, 22% for four panels and 21% for five or more panels.

Vonk, van Ossenbruggen, and Willems (2006) noted a strong correlation ($r = .76$) between being a professional respondent (defined as number of multiple panel memberships + number of surveys completed in the last 12 months) and inattentive respondents (defined as completing the survey in a shorter amount of time and providing shorter answers to open-ended questions).

2.10.2   Effects of number of surveys completed on survey estimates and survey quality

Loyal respondents are desirable from the panel management point of view because they constantly provide answers to the surveys they are invited to. In a context of declining response rates, this can be seen as encouraging. At the same time, we need to explore the possibility that frequent survey-takers provide different answers than less frequent takers and what effect this might have in nonresponse and measurement error. In a Survey Spot panel study, Coen, Lorch, and Piekarski (2005) noted that experienced respondents (who had responded to 4–19 surveys) and very experienced responders (who had responded to 20+ surveys) gave much lower scores than inexperienced respondents (who had completed 1–3 surveys) on questions such as intention to buy, brand awareness, liking, and future purchase frequency. These results were true even after weighting the three groups to make sure they all represented 33% of responses and also after weighting by demographics.

The US bank Washington Mutual (WaMu) switched their market research data collection from telephone surveys to fully nonprobability online panels (the company used more than one). During the gradual switch, researchers at the company noted substantial variations between online panels, across themselves, and in comparison to RDD telephone studies. The bank then started a program of research, pooling together 29 studies across different online panels for a total of 40000 respondents (Gailey, Teal, & Haechrel, 2008). One of the main findings was that respondents who took more surveys in the past three months (11 or more) gave lower demand ratings for products and services than respondents who took fewer surveys (10 or fewer surveys). When controlling for age, the same patterns held true. The second finding was that not only was the number of surveys a predictor of lower demand (for product and service) but also panel tenure. This prompted the bank to ask every online sample vendor to append survey experience auxiliary variables for their project.

In a very recent study, Cavallaro (2013) compared the responses of tenured Survey Spot members with new members on a variety of questions such as concept testing, propensity to buy, early adoption, and newspaper readership. In the study design, the same questions were asked twice to the same respondents a year apart. The data showed that tenured respondents were less enthusiastic about concepts (e.g., a new cereal brand) and more likely to be "early technology adopters" than new panelists. Differences over time for tenured respondents were small, suggesting that the observed differences between tenured respondents and new respondents are not due to changes in answers but rather to changes in panel composition due to attrition.

From the above studies it seems that the respondents who stay longer in a panel have different psychographic attitudes (at least in the topics discussed above) than new panel members. In this context, it is definitely worth mentioning the pioneering work done on the probability-based CentERdata panel in the Netherlands (Felix & Sikkel, 1991; Sikkel & Hoogendoorn, 2008) where panel members were profiled at the early stage with a set of 22 standardized psychological test on traits such as loneliness, social desirability, need for cognition and innovativeness. When looking at all respondents' scores on the 22 traits and correlating them with the length of stay in the panel, the authors barely found any statistically significant correlations. This study strengthens the Cavallaro (2013) hypothesis that the difference between new and tenured panel members is a matter of attrition, and not of panel conditioning at least on psychological traits. We look forward to new research in this area.

The issue of multiple panel membership is also debated in the context of *professional respondents*. We refer the reader to Chapter 10 of this volume for a thorough discussion on professional respondents and their impact on data quality.

2.11   Online panel studies when the offline population is less of a concern

By definition, the offline population is not part of online panels of nonprobability samples. In other words, individuals from the population of interest without Internet access cannot sign up for nonprobability-based online panels. Although it can be argued that weighting can compensate for the absence of the offline population from a survey error point of view, the percentage of people or households that are not online for a specific country contributes to potential noncoverage error. For this reason, probability-based panels so far have provided Internet access to the non-Internet population units or have surveyed them in a different mode such as mail or telephone (Callegaro & DiSogra, 2008).

In the commercial and marketing sector, the issue of representativeness and noncoverage of the offline population sometimes has a different impact than it has for surveys of the general population. As discussed by Bourque and Lafrance (2007), for some topics, customers (that is, the target population) might be mostly online (e.g., wireless phone users) while for other topics (e.g., banking) the offline population is "largely irrelevant from a strategic decision-making standpoint" (p. 18).

However, comparison studies focused on the online population only provide increasing evidence that respondents joining online panels of nonprobability samples are different from the general online population in that they are heavier Internet users, and more interested in technology. For example, in a study comparing two online panels of nonprobability samples with a face-to-face survey of a probability sample, Baim and colleagues (2009) found large differences in Internet usage. According to the face-to-face survey, 37.7% of the adult population in the United States used the Internet five or more times a day, compared to a 55.8% in panel A and a 38.1% in panel B. For Internet usage of about 2–4 times a day, the face-to-face survey estimated that 24.8% of the population fell in this category whereas the nonprobability sample panel A estimated 31.9% and B 39.6%. A more recent study conducted in the United Kingdom compared government surveys to the TNS UK online panel (Williams, 2012). When looking at activities done during free time, the demographically calibrated online panel over estimated using the Internet by 29 percentage points and playing computer games by 14 percentage points. The author concludes: "the huge overestimate of Internet and computer games activity levels illustrates a general truth that access panels will not provide accurate prevalence about the use of technology" (p. 43).

Higher time spent online and heavier technology usage in comparison to benchmarks were also found in the Canadian comparison studies of Crassweller, Rogers, and Williams (2008) – higher number of time spent online "yesterday," and by Duffy and Smith (2005) – higher time spent online and higher usage of technology in the United Kingdom. Therefore, studies who are only interested in the online population might also be affected by differences related to Internet usage between those who belong to their target population and the subgroup that signs up for panels that recruit respondents online.

2.12   Life of an online panel member

As mentioned in Chapter 1 of this volume, online panels do not openly share details about their practices and strategies for fear of giving the competition an advantage. For this reason, it is not easy to know what is requested of online panel members. One way to obtain some information is to sign up in online panel portals that allow to do so, and monitor the activity as panel members. The company Grey Matter Research has used this approach. Staff and other volunteers signed up on different US online panels that allowed those who wanted to become members and monitored the traffic for three months (Grey Matter Research, 2012). At sign-up they did not lie on their demographics, nor did they try to qualify for studies they would not qualify for otherwise. In other words, the study was done with participants being on their "best behavior" – each member attempted to complete each survey to the best of their knowledge and in a reasonable time frame of three days maximum from the moment the email invitation was received. In Table 2.6, we report the results of the 2012 study. A similar study had also been conducted three years before (Grey Matter Research, 2009) with similar results.

Each volunteer monitored the number of invitations per panel. As we can see from Table 2.6, the range is quite wide, where the panel with highest invitation level sent on

Table 2.6   Life of an online panel member.

Panel	Average # of invitations in 30 days	% of surveys closed within 72 hours	Average questionnaire length in minutes
1	42.3	9.5	22.1
2	10.3	19.4	20.2
3	20.0	16.0	17.3
4	9.8	33.7	17.7
5	51.3	27.3	17.5
6	11.3	0.0	16.1
7	6.5	0.0	10.7
8	34.0	42.1	21.2
9	8.5	0.0	18.3
10	7.7	22.1	9.6
11	23.0	29.1	19.6
Average	**20.4**	**18.1**	**17.3**

average 51 invitations within 30 days and the panel with the lowest invitation level sent on average 6.5 invitations in 30 days.

A sizeable number of surveys were already closed when the participants attempted to complete them, with an average of 18.1% and a high of 42.1%. Surveys varied in length but they were on average 17.3 minutes long, with the panel with the shortest questionnaires lasting on average 9.6 minutes and the panel with the longest questionnaires having an average of 22.1 minutes. The above picture highlights likely levels of participation requests and burden on online panels. If we take the mean of the panels, for example, we can estimate that an "average" panel member would spend about seven hours filling out questionnaires in a month with high burden panels topping about 16 hours a month (e.g., panel 1) or low burden panels asking less than 2 hours of commitment a month (e.g., panel 7). These results are per single panel; if a respondent is a member of multiple panels, then the commitment quickly increases.

This rare study, which confirmed the results from the company's previous research conducted in 2009, sheds some light on the kind of data obtained by online panels. Active panel members are requested to participate in numerous surveys for a substantial amount of time each month. The importance of the studies lies in realizing the online panels are victim of their own success. It is hard for companies managing online panels to satisfy every client request. That translates into numerous survey requests per month. There is plenty of room for research to investigate the effects of heavy participation in surveys on their data quality.

2.13   Summary and conclusion

In this chapter we have systematized and brought together the disparate and sometimes hard to find literature on the quality of online survey panels, focusing on the critical review of the largest studies on data quality conducted thus far. This review should provide a starting point for additional studies as well as stimulate the publication of existing and new studies. It was apparent from our review that many of these studies appear on conference presentations, blogs, and few are published in peer-reviewed journals. This creates a problem of transparency because for most studies some of the key survey information, such as the original questionnaire or descriptive statistics, was not available.

The chapter started with the proposal of a taxonomy of different comparison study techniques, together with a review of their strengths and weaknesses. The hope is that the taxonomy can be useful when researchers design future studies on online panel quality. In order to tackle the issue of quality of data obtained from online panels, we looked at three key quality criteria: accuracy of point estimates, relationships across variables, and reproducibility of results. Our recommendation is that researchers and data users/buyers analyzing data coming from online panels should use these criteria to assess the quality of the survey estimates.

The outcome of our review on point estimates, relationships across variables, and reproducibility of results points to quality issues in data obtained from online panels of nonprobability samples. Pre-election online polls are one exception to the general findings, where many web panels of nonprobability samples performed as good as and sometimes better than probability based pre-election polls. Weighting could have the potential to minimize the noncoverage and selection bias observed in online panels, but so far, again with the exception of pre-election polls, this strategy does not seem to be effective.

The final part of the chapter was devoted to common issues debated in the market and survey research arena, specifically the debate on the relationship between completion rates and accuracy, the issue of multiple panel memberships, and studies focusing only on the online population. In the first case, the agreement from the literature is that such relationship does not follow the expected direction. For probability-based panels (though we found only one study: Yeager, Krosnick, et al., 2011), higher completion rates lead to higher bias. For nonprobability panels, what makes a large difference in completion rates seems to be how the company manages the panel in terms of invitation and "panel member deletions" with different, mostly undocumented rules. Multiple panel membership was noticed early on, at the latest since the first study conducted by the NOPVO consortium in the Netherlands in 2006. Given the self-selected nature of panels of nonprobability samples, it is not uncommon for a panel member to sign up for multiple panels. Our review of the limited evidence highlights some issues of data quality for particular questions (e.g., purchase intent or product recommendation) and the fact that members who are more active (in terms of survey completion) than the average tend to have a different psychographic profile from members less active in the panel.

The reader might think that nonprobability sample panels are better suited to study the online population only. However, panel members tend to be heavy Internet users and heavy technology consumers, thus are less representative of the online population overall than sometimes is presumed.

We concluded the chapter by presenting an image of the life of an online panel member. The two studies we reviewed suggest that some panel members spend a high number of hours completing surveys each month. This burden is a new phenomenon, where the population has a higher chance than ever of being selected and receive survey requests, compared to the pre-Internet era when cross-sectional surveys where the norm and even in panel studies frequency of invitation tends to be considerably lower than that of online panels. Protecting the population from overload of survey requests might be important to maintain future cooperation from respondents. Hence, further research is needed investigating the optimal frequency of survey requests (and their length) is for online panel members.

We agree with Farrell and Petersen (2010) that Internet research should not be stigmatized. At the same time, it is worth noting that research conducted using online panels of nonprobability samples has still numerous quality issues that have not been fully resolved. We hope this review can serve as a baseline for a more transparent research agenda on the quality of data obtained from online panels. However, we lament the fact that the existing commercial studies (NOPVO, ARF, and MRIA) produced insufficient documentation and did not share necessary methodological details such as the full questionnaire and descriptive statistics. We look forward to the findings from the new ARF study conducted by the FOQ2 which are expected to be made available to the entire research community (Therhanian, 2013).

Our taxonomy can help researchers to understand what conclusions can be drawn depending on the research design. The multiple focus on point estimates, relationships across variables, and replicability is the key to scientific advancement in this area. Together with weighting, data modeling, and learning from the successful case of pre-election polls, these aspects of the debate on online panels data quality should be on the agenda of research on online panel samples.

References

Anich, B. (2002). Trends in marketing research and their impact on survey research sampling. *Imprints*, May.

Baim, J., Galin, M., Frankel, M. R., Becker, R., & Agresti, J. (2009). Sample surveys based on Internet panels: 8 years of learning. *Worldwide Readership Symposium*. Valencia, Spain. Retrieved January 1, 2013, from: http://www.gfkmri.com/PDF/WWRS-MRI_SampleSurveysBasedOnInternetPanels.pdf.

Baker, R., Blumberg, S. J., Brick, J. M., Couper, M. P., Courtright, M., Dennis, J. M., Dillman, D. A., et al. (2010). Research synthesis. AAPOR report on online panels. *Public Opinion Quarterly*, 74, 711–781.

Baker, R., Brick, M. J., Bates, N., Battaglia, M. P., Couper, M. P., Dever, J. A., Gile, K. J., et al. (2013, May). Report of the AAPOR task-force on non-probability sampling.

Berrens, R. P., Bohara, A. K., Jenkins-Smith, H., Silva, C., & Weimer, D. L. (2003). The advent of internet surveys for political research: A comparison of telephone and internet samples. *Political Analysis, 11*, 1–22.

Bourque, C., & Lafrance, S. (2007). Web survey and representativeness: Close to three in ten Canadians do not have access to the Internet. *Should we care? Canadian Journal of Marketing Research, 24*, 16–21.

Bradley, N. (1999). Sampling for Internet surveys. An examination of respondent selection for Internet research. *Journal of the Market Research Society, 41*, 387–395.

Bryan, C. J., Walton, G. M., Rogers, T., & Dweck, C. S. (2011). Motivating voter turnout by invoking the self. *Proceedings of the National Academy of Sciences, 108*, 12653–12656.

Callegaro, M., & DiSogra, C. (2008). Computing response metrics for online panels. *Public Opinion Quarterly, 72*, 1008–1032.

Casdas, D., Fine, B., & Menictas, C. (2006). Attitudinal differences. Comparing people who belong to multiple versus single panels. *Panel research 2006*. Amsterdam: ESOMAR. Retrieved from: internal-pdf://Casdas-Fine-Menictas_2006_ESOMAR-0928660738/Casdas-Fine-Menictas_2006 _ESOMAR.pdf.

Cavallaro, K. (2013). Theory of adaptation or survival of the fittest? *Quirk's Marketing Research Review, 27*, 24–27.

Chan, P., & Ambrose, D. (2011). Canadian online panels: Similar or different? *Vue, (January/February)*, 16–20. Retrieved from: http://www.mktginc.com/pdf/VUE%20JanFeb%202011021.pdf.

Coen, T., Lorch, J., & Piekarski, L. (2005). The effects of survey frequency on panelist responses. *ESOMAR World Research Conference 2005*. Budapest: ESOMAR.

Comley, P. (2007). Online market research. In M. van Hamersveld, & C. de Bont (Eds.), *Market research handbook* (5th ed., pp. 401–419). Chichester: John Wiley & Sons, Ltd.

Crassweller, A., Rogers, J., Graves, F., Gauthier, E., & Charlebois, O. (2011). In search of a new approach to measure newspaper audiences in Canada: The journey continues. Paper presented at the Print and Digital Research Forum, San Francisco, CA. Retrieved from: http://nadbank .com/en/system/files/PDRFNADbankEKOS.pdf.

Crassweller, A., Rogers, J., & Williams, D. (2008). Between random samples and online panels: Where is the next .lily pad? Paper presented at the ESOMAR panel research 2008. Retrieved from: http://www.nadbank.com/en/system/files/Final%20paper.pdf.

Crassweller, A., Williams, D., & Thompson, I. (2006). Online data collection: Solution or band-aid? Paper presented at the Worldwide Readership Research Symposia, Vienna. Retrieved March 15, 2012, from: http://www.nadbank.com/en/system/files/Presentation%20WRRS%20Final.pdf.

Das, M. (2012). Innovation in online data collection for scientific research: The Dutch MESS project. *Methodological Innovations Online*, *7*, 7–24.

Dillman, D. A., & Messer, B. L. (2010). Mixed-mode surveys. In P. V. Marsden, & J. D. Wright (Eds.), *Handbook of survey research* (2nd ed., pp. 551–574). Howard House: Emerald Group.

Duffy, B., & Smith, K. (2005). Comparing data from online and face-to-face surveys. *International Journal of Market Research*, *47*, 615–639.

EKOS. (2010, January). Ekos' observation of MRIA study: Canadian online panels: similar or different? Retrieved August 28, 2011, from: http://www.ekos.com/admin/articles/MRIA-Comparison -Panel-Study-2010-01-27.pdf.

Farrell, D., & Petersen, J. C. (2010). The growth of Internet research methods and the reluctant sociologist. *Sociological Inquiry*, *80*, 114–125.

Felix, J., & Sikkel, D. (1991). Attrition bias in telepanel research. *Kwantitatiewe Methoden*, *61*.

Fulgoni, G. (2005). The professional respondent problem in online panel surveys today. Paper presented at the Market Research Association Annual Conference, Chicago, IL. Retrieved from: http: //www.docstoc.com/docs/30817010/Partnership-Opportunity-Use-of-comScores-Survey-Panels.

Gailey, R., Teal, D., & Haechrel, E. (2008). Sample factors that influence data quality. Paper presented at the Advertising Research Foundation Online Research Quality Council conference (ORQC). Retrieved January 1, 2013, from: http://s3.amazonaws.com/thearf-org-aux-assets/downloads/cnc /orqc/2008-09-16_ARF_ORQC_WaMu.pdf.

Gittelman, S., & Trimarchi, E. (2010). Online research … and all that jazz!: The practical adaption of old tunes to make new music. *ESOMAR Online Research 2010*. Amsterdam: ESOMAR.

Grey Matter Research. (2009). Dirty little secrets of online panels: And how the one you select can make or break your study. Retrieved from: http://greymatterresearch.com/index_files/Online_Panels.htm.

Grey Matter Research. (2012). More dirty little secrets of online panel research. Retrieved from: http://www.greymatterresearch.com/index_files/Online_Panels_2012.htm.

Henrich, J., Heine, S. J., & Norenzayan, A. (2010). The weirdest people in the world? *Behavioral and Brain Sciences*, *33(2–3)*, 61–83.

Henry, P. J. (2008). College sophomores in the laboratory redux: Influences of a narrow data base on social psychology's view of the nature of prejudice. *Psychological Inquiry*, *19*, 49–71.

Jones, E. E. (1986). Interpreting interpersonal behavior: The effects of expectancies. *Science*, *234(4772)*, 41–46.

Malhotra, N., & Krosnick, J. A. (2007). The effect of survey mode and sampling on inferences about political attitudes and behavior: Comparing the 2000 and 2004 ANES to internet surveys with nonprobability samples. *Political Analysis*, *15*, 286–323.

Markus, H. R., & Kitayama, S. (1991). Culture and the self: Implications for cognition, emotion, and motivation. *Psychological Review, 98,* 224–253.

Martinsson, J., Dahlberg, S., & Lundmark, O. S. (2013). Is accuracy only for probability samples? Comparing probability and non-probability samples in a country with almost full internet coverage. Paper presented at the 68th Annual Conference of the American Association for Public Opinion Research, Boston, MA. Retrieved from: http://www.lore.gu.se/digitalAssets/1455/1455221 _martinsson--dahlberg-and-lundmark--2013--aapor-is-accuracy-only-for-probability-samples.pdf.

Miller, J. (2007). Burke panel quality R&D summary. Retrieved from: http://s3.amazonaws.com /thearf-org-aux-assets/downloads/cnc/orqc/09-10-07_ORQC_Miller.pdf.

Miller, J. (2008). Burke panel quality R&D.

Morgan, S. L., & Winship, C. (2007). *Counterfactuals and causal inference: Methods and principles for social research* (1st ed.). Cambridge: Cambridge University Press.

Pasek, J., & Krosnick, J. A. (2010, December 28). Measuring intent to participate and participation in the 2010 census and their correlates and trends: Comparisons of RDD telephone and non-probability sample internet survey data. U.S. Census Bureau. Retrieved from: http://www.census.gov/srd/papers/pdf/ssm2010-15.pdf.

Peterson, R. A. (2001). On the use of college students in social science research: Insights from a second-order meta-analysis. *Journal of Consumer Research, 28,* 450–461.

Petty, R. E., & Cacioppo, J. T. (1996). Addressing disturbing and disturbed consumer behavior: Is it necessary to change the way we conduct behavioral science? *Journal of Marketing Research, 33,* 1–8.

Postoaca, A. (2006). *The anonymous elect. Market research through online access panels.* Berlin: Springer.

Sanders, D., Clarke, H. D., Stewart, M. C., & Whiteley, P. (2007). Does mode matter for modeling political choice? Evidence from the 2005 British Election Study. *Political Analysis, 15,* 257–285.

Scherpenzeel, A., & Bethlehem, J. (2011). How representative are online-panels? Problems of coverage and selection and possible solutions. In M. Das, P. Ester, & L. Kaczmirek (Eds.), *Social and behavioral research and the internet: Advances in applied methods and research strategies* (pp. 105–132). New York: Routledge.

Sears, D. O. (1986). College sophomores in the laboratory: Influences of a narrow data base on social psychology's view of human nature. *Journal of Personality and Social Psychology, 51,* 515–530.

Sikkel, D., & Hoogendoorn, A. (2008). Panel surveys. In E. De Leeuw, J. Hox, & D. A. Dillman (Eds.), *International handbook of survey methodology* (pp. 479–499). New York: Lawrence Erlbaum Associates.

Silver, N. (2012, November 10). Which polls fared best (and worst) in the 2012 Presidential Race. Retrieved from: http://fivethirtyeight.blogs.nytimes.com/2012/11/10/which-polls-fared-best-and -worst-in-the-2012-presidential-race/.

Smith, T. W. (1995). Little things matter: A sampler of how differences in questionnaire format can affect survey responses. In American Statistical Association (Ed.), *Proceedings of the Joint Statistical Meeting, Survey Research Methods Section* (pp. 1046–1051). Washington, DC: AMSTAT.

Smyth, J. D., & Pearson, J. E. (2011). Internet survey methods: A review of strengths, weaknesses, and innovations. In M. Das, P. Ester, & L. Kaczmirek (Eds.), *Social and behavioral research and the internet. Advances in applied methods and research strategies* (pp. 11–44). New York: Taylor and Francis.

Sudman, S. & Wansink, B. (2002). *Consumer panels.* Chicago: American Marketing Association.

Taylor, H. (2007, January 15). The case for publishing (some) online polls. Retrieved January 1, 2013, from: http://www.pollingreport.com/ht_online.htm.

Taylor, H., Bremer, J., Overmeyer, C., Siegel, J. W., & Terhanian, G. (2001). The record of internet-based opinion polls in predicting the results of 72 races in the November 2000 U.S. Elections. *International Journal of Market Research, 43,* 127–136.

Terhanian, G. (2013). Comment to the summary report of the AAPOR task force on non-probability sampling. *Journal of Survey Statistics and Methodology*, *1*, 124–129.

Tourangeau, R., Conrad, F. C., & Couper, M. P. (2013). *The science of web surveys*. Oxford: Oxford University Press.

Van Ossenbruggen, R., Vonk, T., & Willems, P. (2006). Results of NOPVO. Paper presented at the Online panels, close up, Utrecht, Netherlands. Retrieved from: http://www.nopvo.nl/page0 /files/Results_NOPVO_English.pdf.

Vonk, T., van Ossenbruggen, R., & Willems, P. (2006). The effects of panel recruitment and management on research results: A study across 19 online panels. Paper presented at the Panel research 2006, ESOMAR, Barcelona.

Walker, R., Pettit, R., & Rubinson, J. (2009). A special report from the Advertising Research Foundation: The foundations of quality initiative: A five-part immersion into the quality of online research. *Journal of Advertising Research*, *49*, 464–485.

Wells, W. D. (1993). Discovery-oriented consumer research. *Journal of Consumer Research*, *19*, 489–504.

West, R. F., Toplak, M. E., & Stanovich, K. E. (2008). Heuristics and biases as measures of critical thinking: Associations with cognitive ability and thinking dispositions. *Journal of Educational Psychology*, *100*, 930–941.

Williams, J. (2012). Survey methods in an age of austerity. Driving value in survey design. *International Journal of Market Research*, *54*, 35–47.

Yeager, D. S., & Krosnick, J. A. (2011). Does mentioning "some people" and "other people" in a survey question increase the accuracy of adolescents' self-reports? *Developmental Psychology*, *47*, 1674–1679.

Yeager, D. S., & Krosnick, J. A. (2012). Does mentioning "Some People" and "Other People" in an opinion question improve measurement quality? *Public Opinion Quarterly*, *76*, 131–141.

Yeager, D. S., Krosnick, J. A., Chang, L., Javitz, A. S., Levendusky, M. S., Simpser, A., & Wang, R. (2011). Comparing the accuracy of RDD telephone surveys and internet surveys conducted with probability and non-probability samples. *Public Opinion Quarterly*, *75*, 709–747.

Yeager, D. S., Larson, S. B., Krosnick, J. A., & Thompson, T. (2011). Measuring Americans' issue priorities: A new version of the most important problem question reveals more concern about global warming and the environment. *Public Opinion Quarterly*, *75*, 125–138.

YouGov. (2011). YouGov's record. Public polling results compared to other pollsters and actual outcomes. Retrieved January 1, 2013, from: http://cdn.yougov.com/today_uk_import/yg-archives -pol-trackers-record2011.pdf.

Part I
COVERAGE

Introduction to Part I

Mario Callegaro[a] and Jon A. Krosnick[b]
[a]*Google UK*
[b]*Stanford University, USA*

I.1 Coverage bias in online panels

From a survey methodology point of view, coverage bias for online panels is a result of an inequality in Internet access (*undercoverage*) for some segments of the population. The concept of coverage is related to the target population for whom we want to generalize our results. In many cases, it is the general population; in some cases, it is the online population only. In the two of the three chapters in this Part, the online panels discussed have the goal of generalizing their survey findings to the general population, and therefore, are comprised of online and offline households and individuals.

When the primary mode[1] of completing a survey is via an online panel on the Web, it is important to assess which individuals are excluded or have less of a chance to join the panel. Although Internet access is growingly rapidly in many countries, with very few exceptions (e.g., the Netherlands has nearly 100% access), most do not have all of their general population online (International Telecommunication Union, 2013). This gap between those with and without access and the resulting inequalities is called the *digital divide*. Internet access alone is not the only determining factor in online panel membership. How often one goes online, from where (home versus other venues) the Internet is accessed, and a person's skills in using the Internet are other key factors in terms of likelihood of joining an online panel. These inequalities in Internet skills and usage (Brandtzæg, Heim, & Karahasanović, 2011) are part of what is called the *second-level divide* (Hargittai, 2002).

Findings from the literature on the digital divide tell us that when looking just at access, the offline population has specific characteristics generally correlated with low income, low education, and other socio-demographics that are country-specific (Ragnedda & Muschert, 2013). For example, Callegaro (2013) compared the digital divide in the United States using

---

[1] This is not the case when Internet access is provided to non-Internet households by the panel company, as we explain later.

Online Panel Research: A Data Quality Perspective, First Edition.
Edited by Mario Callegaro, Reg Baker, Jelke Bethlehem, Anja S. Göritz, Jon A. Krosnick and Paul J. Lavrakas.
© 2014 John Wiley & Sons, Ltd. Published 2014 by John Wiley & Sons, Ltd.
Companion website: www.wiley.com/go/online_panel

data from File (2012) to that of the United Kingdom by looking at Internet access from home. Using a logistic regression model that predicts access from home and controls for demographic variables, File concluded that the non-Internet population in the United States tended to be female, older, non-white, of lower income, and living in the South. In contrast, Callegaro showed how gender and race were not a factor in the United Kingdom when controlling for other demographic characteristics. The non-Internet population tended to be older, of lower social class, and living in Northern Ireland.

As to the second level divide, the literature reveals that a substantial percentage of those with Internet access are low-frequency users, i.e., going online only two or three times a week or less. In addition, Zillien and Marr (2013) found that Internet skills play a substantial role in determining Internet usage and frequency.

When we connect the findings from the digital divide literature to coverage bias in survey methodology, we realize how the digital divide and the second-level divide can help us understand who can and cannot participate in online panel research.

In building an online panel, the issue of coverage error is handled differently depending on the type of panel. Nonprobability online panels do not provide Internet access to non-Internet respondents, nor do they survey them in another mode. In other words, almost by definition and due to the recruiting methods (see Chapter 1), nonprobability online panels do not have non-Internet users and, as we see in Chapter 2, have very few low-frequency Internet users.

In probability-based online panels built to produce surveys with results that are representative of the overall population of a specific country, coverage bias is approached in one of four ways (e.g. DiSogra & Callegaro, 2010):

1. Give everybody who is being recruited (Internet or non-Internet households) the same device and an Internet connection. For example, give each household a tablet with Internet connection.

2. Give a device and an Internet connection to non-Internet households. For example, give them a desktop or laptop computer and pay for their Internet connection.

3. Survey the non-Internet population and/or very infrequent Internet users in another mode, such as by mail or by telephone.

4. Do not recruit the non-Internet population by design and attempt to compensate for their absence using weighting techniques.

Each of these design decisions has non-trivial costs associated with them in terms of recruiting, management, and survey implementation, not to mention the issue of representativity of the general population.

I.2　The chapters in Part I

The chapters in Part I present three different probability-based panels (based in Germany, Finland, and the United States) and how the issue of coverage bias was dealt with in each.

In Chapter 3, Bella Struminskaya, Lars Kaczmirek, Ines Schaurer, and Wolfgang Bandilla present the GESIS Online Panel Pilot, a pilot study conducted in Germany to build a probability-based panel. The panel was built using random digit dialing with a dual-frame sampling design of cell and landline numbers. Because of the pilot nature of the study and also budget considerations, persons who did not use the Internet were excluded in the recruitment. The study's approach to bias is therefore an example of design 4 although the chapter focuses on the Internet population only, and does not attempt to compensate for their lack of non-Internet households. In this chapter, the authors assess the quality of the data coming from the online panel in comparison to external benchmarks. Struminskaya et al. present a great discussion of the difficulties in comparing data from an online panel to external benchmarks. Scholars and researchers attempting to do so face many challenges, namely, locating high quality studies and official statistics benchmarks and dealing with issues of weighting, question-wording, and potential mode effects. In this specific case, the authors had to focus on the comparison between their panel and benchmarks to the Internet population only (given that non-Internet households were not recruited by design). Second, they had to locate benchmark studies where question(s) about Internet access and use were asked in order to restrict the findings to the online population. Then, even the official high-quality surveys needed to be weighted to account for nonresponse and sampling design. Lastly, the authors had to deal with a relatively small sample size (747 cases) available for their data analysis.

Data collected in the GESIS Online Panel are compared and contrasted with two high-response-rate gold standard surveys: the German General Social Survey (ALLBUS) and the German dataset of the European Social Survey (ESS). When comparing point estimates, the authors find differences between the GESIS Online Panel and the benchmarks, although of few percentage points for most variables. When comparing regression models coefficients across sources, the direction of the coefficients is almost always the same. The authors conclude that the data collected by the GESIS online panel are fairly comparable with the two benchmarks.

Chapter 4 by Kimmo Grönlund and Kim Strandberg presents a comparison of results from a probability-based panel in Finland with two surveys conducted at the same time: one a telephone survey and the other a face-to-face survey. Finland is a country with a very high Internet penetration, estimated at 89% at a person level at the time of the study (International Telecommunication Union, 2013). The panel was built in conjunction with the parliamentary election of 2011, and thus the analysis and topic of the surveys are political in nature. The panel was recruited using the Finnish population registry, which contains addresses, names, gender, year of birth, and other information. We remind the reader that in the survey world, it is a luxury to have such a sampling frame. Most countries do not have a population-based registry. The panel was recruited via a mail invitation, and no other survey mode was offered for the offline population. This chapter presents another example of design 4. The mail recruitment yielded a cumulative recruitment rate of 5.3% with a total sample size of 692 people who joined the panel at wave 1. The panel lasted for three months of data collection with an overall attrition rate of 25% at the last wave. When comparing the point estimates of the online panel with the two benchmarks, the authors' findings reproduce the results of other research looking at online panels and political surveys (Malhotra & Krosnick, 2007). For example, younger panelists were under-represented. That is definitely a nonresponse issue more than a coverage issue, as young adults tend to be almost entirely online. Online panelists were also more interested in politics, more likely to have voted in the past election, and heavier Internet users in comparison to the two benchmarks. Interestingly, demographic weighting helped in the right direction but did not fully bring the point estimates in line with the benchmarks.

This Finnish experience reminds the reader how, even in an ideal situation with a great sampling frame, actually building a high-response/low-attrition rate panel is definitely challenging and requires many resources and a great deal of panel management.

The German and Finnish studies are two examples of how to compare data collected from a probability-based panel to known benchmarks, attempting to keep everything else constant. On the other hand, in Chapter 5, by Allan McCutcheon, Kumar Rao, and Olena Kaminska a different issue is presented: the effects of varying recruitment methods on attrition rates in the US Gallup Panel. The Gallup Panel's approach to coverage bias is an example of design 3. Members were recruited via RDD, and non-Internet households or low-frequency Internet users were assigned to participate via mail surveys. Of particular interest is the fact that the Gallup Panel attempted to recruit every eligible member of the selected household. This study is a follow-up attrition analysis of an experiment varying recruitment modes (mail versus RDD) advance letters, a prepaid monetary incentive ($2), and a phone follow-up (Rao, Kaminska, & McCutcheon, 2010). An examination of the demographics of recruited members by mode reveals that the mail mode elicited a sample composition more likely to be non-white, lower-educated, and of lower income than their counterpart recruited by telephone.

By looking at a time span of about three years in terms of attrition rates, the authors find that the combination of recruitment by mail and assigning non-Internet/low-frequency-user households to mail surveys produced the highest attrition rate of the entire panel. Overall, the phone-recruited panel members retired at a lower rate than those recruited by mail. The entire picture is, however, not complete because these attrition comparisons do not take into account the initial recruitment mode which, as we just reported, elicited a different sample composition in terms of demographics. However, with the use of survival analysis, the authors are able to control for all these factors. This chapter is therefore a good example of the complexities of studying attrition in an online panel and of how numerous variables can have an effect on the overall attrition rates. When looking at recruitment mode by incentive and controlling for everything else, the authors find that mail-recruited respondents tended to leave the panel at a higher rate than their phone-recruited counterparts. The effect of the incentive was practically non-significant. When looking at the effect of mode of recruitment by assignment (Web versus mail surveys), then the distinction among the four groups is even stronger: mail-recruited/mail-assigned respondents had the highest and sharpest attrition rate. Then attrition is less for phone-recruited/mail-assigned respondents, followed by the mail-recruited/Web-assigned group, and lastly for phone-recruited/Web-assigned group, which had the lowest attrition rate. Even when controlling for demographics, the overall findings show how difficult is to retain non-Internet or low-frequency Internet user households even if they are asked to complete the survey via mail.

Finally, the authors perform a unique analysis of within-household attrition that shows a domino effect, i.e., family members tended to leave the panel together more or less as a unit. This chapter presents a rare investigation of attrition in an online (and offline) panel and makes us (survey methodologists) think about the effect that different strategies for recruiting and surveying panel members have on: (1) the initial composition of the panel in terms of demographics; and (2) the overall attrition rates. We hope more studies like this will follow.

References

Brandtzæg, P. B., Heim, J., & Karahasanović, A. (2011). Understanding the new digital divide: A typology of Internet users in Europe. *International Journal of Human-Computer Studies, 69,* 123–138.

Callegaro, M. (2013). Web coverage in the UK and its potential impact on general population web surveys. Paper presented at the Web Surveys for the General Population: How, Why and When? conference, London. Retrieved from: http://www.natcenweb.co.uk/genpopweb/documents/Mario -Callegaro.pdf.

DiSogra, C., & Callegaro, M. (2010). Computing response rates for probability-based online panels. In AMSTAT (Ed.), *Proceedings of the Joint Statistical Meeting, Survey Research Methods section* (pp. 5309–5320). Alexandria, VA: AMSTAT.

File, T. (2012, September 19). Digital divides: A connectivity continuum for the United States. Data from the 2011 Current Population Survey. Retrieved from: http://paa2013.princeton.edu/papers/130743.

Hargittai, E. (2002). Second-level digital divide: Differences in people's online skills. *First Monday*, 7. Retrieved June 17, 2013, from: http://firstmonday.org/ojs/index.php/fm/article/view/942.

International Telecommunication Union. (2013). The world in 2013: ICT facts and figures. ITU. Retrieved June 17, 2013, from: http://www.itu.int/en/ITU-D/Statistics/Documents/facts/ICTFacts Figures2013-e.pdf.

Malhotra, N., & Krosnick, J. A. (2007). The effect of survey mode and sampling on inferences about political attitudes and behavior: Comparing the 2000 and 2004 ANES to internet surveys with non-probability samples. *Political Analysis*, *15*, 286–323.

Ragnedda, M., & Muschert, G. (Eds.). (2013). *The digital divide: The internet and social inequality in international perspective*. New York: Routledge.

Rao, K., Kaminska, O., & McCutcheon, A. L. (2010). Recruiting probability samples for a multi-mode research panel with Internet and mail components. *Public Opinion Quarterly*, *74*, 68–84.

Zillien, N., & Marr, M. (2013). The digital divide in Europe. In M. Ragnedda & G. Muschert (Eds.), *The digital divide: The internet and social inequality in international perspective* (pp. 55–66). New York: Routledge.

3

Assessing representativeness of a probability-based online panel in Germany

Bella Struminskaya, Lars Kaczmirek, Ines Schaurer, and Wolfgang Bandilla

GESIS–Leibniz Institute for the Social Sciences, Germany[1]

3.1 Probability-based online panels

Advances in online survey research methods have led to a substantial increase in the use of online panels, mainly of nonprobabilistic nature. According to the AAPOR Report on Online Panels, in 2009, about 85% of research, which might otherwise have been performed by traditional methods, was done online (Baker et al., 2010).

In academia, cost-efficiency benefits of web-based interviewing are paired with concerns about the quality of data gained from nonprobability online panels. Increasing nonresponse in traditional methods and research showing that claims of representativeness by nonprobability online surveys are hardly confirmed, have led to an increased demand for probability-based online panels. Among the established online panels based on probability samples worldwide are the LISS Panel (in the Netherlands), the RAND American Life Panel, the Gallup Panel, and the KnowledgePanel of Knowledge Networks, a GfK Company (in the United States). In Europe, new online probability-based panels are emerging, examples include ELIPSS in France and GIP in Germany (Das, 2012).

[1] The authors thank Dr. Siegfried Gabler, Dr. Matthias Ganninger, and Kai Weyandt for their help with preparing the manuscript. This work was supported by the German Federal Ministry of Education and Research.

Online Panel Research: A Data Quality Perspective, First Edition.
Edited by Mario Callegaro, Reg Baker, Jelke Bethlehem, Anja S. Göritz, Jon A. Krosnick and Paul J. Lavrakas.
© 2014 John Wiley & Sons, Ltd. Published 2014 by John Wiley & Sons, Ltd.
Companion website: www.wiley.com/go/online_panel

Studies comparing probability and nonprobability online panels continue to show that probability-based online surveys yield more accurate results (Chang & Krosnick, 2009; Scherpenzeel & Bethlehem, 2010; Yeager et al., 2011). Probability-based panels and their data are nonetheless constantly being evaluated because, like all surveys, they are prone to threats to representativeness.

Of particular importance are errors of nonobservation: first, noncoverage or failure to include all eligible members of the target population in the sampling frame and, second, unit nonresponse or failure to gather data from all members of the sample. Undercoverage becomes less of a concern with the inclusion of cell-phone only households in the Random Digit Dial (RDD) telephone sampling frames for telephone recruitment and increased use of address-based sampling for face-to-face and mail recruitment strategies.

Nonresponse at a unit level in online panels poses a problem since online panels usually employ multi-step recruitment protocols in which the initial contact and panel membership request are made face-to-face or via telephone (Vehovar, Lozar Manfreda, Zaletel, & Batagelj, 2002). The second threat to accurate representation of the target population is wave nonresponse or attrition when persons who took part in an initial online interview fail to do so for one, several or all the following waves. If factors influencing wave nonresponse or attrition are correlated with the outcomes of interest, losing these respondents can lead to biased data.

One way to assess representativeness of the data from online panels or online surveys in general is to compare estimates with external benchmarks. Usually government statistics or data from surveys with high response rates are used as a reference. Both sources have their advantages and flaws. Government statistics do not commonly go beyond demographic information, whereas researchers and clients are more interested in opinions and attitudes. Survey data from face-to-face or telephone surveys contain nondemographic information but are themselves subject to nonresponse.

In this chapter we employ the latter strategy. We compare data from the probability-based online panel of Internet users in Germany to data from two high-quality face-to-face surveys. After the description and assessment of the recruitment process, we study sample composition employing several demographic measures as well as including attitudinal variables in our analysis. Finally, we address the challenges of having two benchmark surveys, which have variation within them, and the consequences of drawing inferences from our data.

3.2   Description of the GESIS Online Panel Pilot

3.2.1   Goals and general information

The GESIS Online Panel Pilot (GOPP) was a methodological project primarily aimed at developing effective recruitment and panel-maintaining strategies of a probability-based online panel in Germany for the academic research purposes of social and political scientists.

Respondents were recruited into the panel via telephone using a probability sample. To avoid the issues of decreasing landline coverage, cell phone numbers were included in the sample. With the project having mainly methodological purposes, persons who did not use the Internet were screened out during the recruitment interview. Internet users were asked at the end of the recruitment interview to provide their email address to participate in an online panel. Those who agreed were sent email invitations to online surveys of durations between 10–15 minutes every month for eight months in total.

GESIS Online Panel Pilot is a longitudinal panel in the sense that it collects data from the same sample units on multiple occasions over time (Lynn, 2009). Every monthly questionnaire had a leading topic and included some demographical questions in order not to burden the respondent with all the factual questions in a single survey. A small proportion of repeated measures were collected over the course of the panel. Those measures were mainly included for methodological reasons rather than to study social change.

3.2.2   Telephone recruitment

The target population for the GESIS Online Panel Pilot were German-speaking persons living in the Federal Republic of Germany aged 18 and older, who used the Internet for non-work-related purposes. With the growing mobile-only population in Germany, it was decided to use the dual frame approach for the telephone recruitment. Samples were drawn independently from landline and cell phone frames. The goal was to have a final sample with 50% of eligible cell phone numbers and 50% of eligible landline numbers. In order to handle the overlap between two frames, we calculated the inclusion probabilities for target persons using the formulas of Siegfried Gabler, Sabine Häder and their colleagues under the assumption that the two samples are independent (Gabler, Häder, Lehnhoff, & Mardian, 2012). The inclusion probabilities account for the sizes of the samples and frames in both cell phone and landline components as well as the number of landline and mobile numbers at which a respondent can be reached, and for the landline part, the size of the household. Within the landline part of the dual frame, the contact person was asked to state the number of adults residing in the household and the target person was the person with the most recent birthday. In the cell phone part of the frame no selection procedures were implemented, since cell phone sharing in Germany is estimated to be about 2% (Gabler et al., 2012). For both landline and cell phone samples, persons contacted on institutional telephone numbers or numbers used for non-private use were excluded from the sample. In addition, respondents were asked to provide the total number of landline and cell phone numbers at which they could be reached. The questionnaire employed a complex routing scheme to take into account the various combinations of landline, cell phone, and special telephone contracts available in Germany (e.g., landline numbers with cell phone contracts). The project was executed in three sequential parts, each of those employed the strategies, which proved to be efficient during the previous part or corrected for deficiencies of the previous part. The first part, preceded by a short pre-test, started the recruitment in February 2011 and continued until April 2011. Recruitment of the second part was done between June and July 2011. The recruitment of the third part lasted from June 2011 to August 2011 (for differences between study parts, see Table 3.1). Up to 15 call attempts were made to reach the number. In the second and third part of the recruitment the number of call attempts was reduced to 10.

After a target respondent was identified, he or she was asked to participate in a short interview (about 10 minutes). At the beginning of the interview respondents were asked about non-work-related Internet use. For non-users, the recruitment interview ended at this point. It is good practice to collect information about non-users as well. However, as studying the differences between Internet users and non-users was not a central part of the project, such information was only collected for a subsample. Internet users were asked several attitudinal questions about living in Germany, leisure activities, basic demographic information, frequency of Internet use and online survey experience, as well as questions on household composition and number of landline and mobile numbers.

Table 3.1 Overview of the features of the GESIS Online Panel Pilot.

Feature	Study 1	Study 2	Study 3
Fieldwork start	February 2011	June 2011	July 2011
Fieldwork agency	In-house, own telephone lab	In-house, own telephone lab	External, specialized on market research surveys
Internet-screener	The very first question with target sample unit	Preceded by a general attitudinal question	Preceded by a general attitudinal question
Incentive experiment	10 EUR plus 20 EUR bonus for completion of all eight questionnaires, 5 EUR plus 20 EUR	5 EUR, 2 EUR, 0 EUR	5 EUR, 2 EUR, 0 EUR
Announced panel duration	8 months	No limitation	No limitation
Length of recruitment interview	Regular version	Regular version (90%), shorter version (10%)	Regular version (90%), shorter version (10%)
Experiment on household income question	Household income question asked in 50% of cases	Household income question asked in 50% of cases	Household income question asked in 50% of cases

At the end of the interview, respondents were requested to provide an email address to participate in the online panel. Interviewers were instructed to describe the concept of the online panel and state the amount of incentive the respondent would receive as a result of participating in each online survey. Incentive conditions varied experimentally from 0–10 Euros per completed online questionnaire.

Apart from providing the incentive information, interviewers also answered any additional questions which respondents asked about the panel. This part of the interview was less standardized, interviewers were encouraged to perform it as a dialogue with the respondent and to apply tailoring strategies to maximize the recruitment rate. With interviewers making an effort to persuade respondents to participate in the online panel, hesitant respondents were offered the opportunity of a lagged decision. This meant that individuals having concerns about the scientific nature of the study were referred to the project homepage and followed up by call-backs a few days later.

To explore opportunities of increasing the recruitment rate, two experiments were integrated into the interview process. The first experiment tested the effect of asking a question about household income on the recruitment success. In a split-half experiment we asked half of the respondents to report their monthly household income, providing bracketed categories for their answers. The second half of respondents were not asked about their income.

The second experiment was implemented in the second and the third recruitment parts. The experiment varied the announced length and content of the telephone interview. To test the effect of the interview length on recruitment success, for the shorter version, the recruitment interview was reduced to an introductory question about living in Germany, basic demographic information, and the panel participation request.

For certain conditions such as when a target person refused to participate in telephone surveys in general, multiple appointments had failed before, and/or a target person seemed to be near an interview break-off, the option of only providing an email address was given to the respondent. Interviewers asked about Internet use, briefly described the project and probed for an email address. No demographic information was collected during the interview. For those respondents who joined the online panel, the required additional demographic information and information needed for calculation of design weights was collected in the second online survey.

3.2.3   Online interviewing

Respondents who agreed to participate in the online panel and provided their email address were sent an email invitation which contained a link to the first online survey. The invitation also included incentive information for experimental groups in non-zero incentive conditions.

After clicking on the link in the invitation email, respondents were redirected to the online survey. Each online questionnaire started with a welcome screen, the content of which stayed unchanged during the entire course of the panel except for announcing the number of the survey that respondents were about to complete. The welcome screen contained general information about GESIS as the sponsor of the survey, a link to the data protection policy, and contact information.

In the first online survey respondents were fully informed about data protection issues. In all the following surveys a shorter version of the data protection clauses was employed with a link to the full text. The first screen in all online surveys was followed by a page with instructions on filling in the questionnaire.

Each of the monthly online questionnaires had a main topic which most of the questions addressed. Additionally, demographic questions were spread over questionnaires 2–4. We used this approach instead of starting with a profile survey to make the experience of filling out the first online survey as pleasant and interesting as possible. The lower respondent burden in the beginning aimed to increase the likelihood of respondents staying in the panel.

The main topics of online questionnaires were:

- Questionnaire 1: Multi-topic introductory wave

- Questionnaire 2: Education and employment

- Questionnaire 3: Family life

- Questionnaire 4: Religion and values

- Questionnaire 5: Environment, ecology

- Questionnaire 6: Social networks

- Questionnaire 7: Politics

- Questionnaire 8: Multi-topic (with a focus on gender roles and personality)

Most of the questions used in the online questionnaires were originally conducted as part of other German and international surveys. The panel had a strong focus on replicating questions for two reasons: first, rather complex constructs were implemented to demonstrate their feasibility in an online setting; second, to assess data quality, comparisons with external benchmarks were planned.

For these reasons, a substantial part of questions originated from the German General Social Survey ("ALLBUS") 2010 and the German version of the European Social Survey 2010. Other questions were replicated from a wave of the German Socio-Economic Panel (GSOEP 2008), the Eurobarometer, and the International Social Survey Programme (ISSP). Generally, these questions were implemented using the original question wording. In some cases the wording was adjusted to match the self-administered mode.

Apart from questions on specific topics, each questionnaire ended with the same set of questions to evaluate the current survey. Respondents were also asked whether they had interrupted the completion of the survey and if so, how long the distraction lasted. There was a text field for comments offered each month at the end of the questionnaires. Providing comments was not obligatory for the respondent. Several items of paradata, i.e., data about the process of survey production (Kreuter, 2013), were collected in the course of the online survey. These included, for example, the duration of questionnaire completion, response latencies, and browser information.

The last page of every questionnaire contained a thank you note and information about the amount of incentive respondents received for completing the survey. The control group, which received no incentives, was presented with a thank you note only. Respondents could redeem incentives in various ways: paid to their bank account in Euros or as an Amazon-voucher, in which case no bank account information was required. In the study parts 2 and 3, participants could also choose to have the money donated to various charity organizations.

The panel, as well as the telephone recruitment interview, included experiments about survey methodology. One of those experiments was aimed at studying panel conditioning, the phenomenon when the act of participating in surveys changes attitudes and behavior or the way respondents report attitudes and behavior (Kalton, Kasprzyk, & McMillen, 1989). It employed rotating questionnaires: while some respondents received a specific questionnaire in the beginning of their panel participation, the same questionnaire was completed by others after having completed a few surveys.

Another experiment aimed to capture changes in social desirable responding. It included a repeated measurement design for a subsample of the online respondents recruited during the third recruitment round.

In case of no answer by the respondent for a week, a reminder was sent to the respondent with a text and a link to the online survey. Overall, between the waves of the online surveys a respondent could receive a maximum of three reminders, one each week, if he or she failed to start the online survey. After a month, the respondent received an invitation to the next online survey; however, a chance to complete non-answered questionnaires was given in the course of the entire eight months. For this purpose, respondents could log in to the panel with the authentication information they had received after completion of the first online survey.

Panel members who decided to stop participation could opt out by sending a request to be removed from the panel. In this case they received no further invitations or reminders.

3.3   Assessing recruitment of the Online Panel Pilot

In this section we describe the response metrics for the telephone recruitment and online participation. The presentation of the response rates is twofold. We first present the metrics from the telephone recruitment and then proceed to the online part of the panel. Although all metrics were also calculated separately for each of the three parts, we only report the average values over the three main study parts. These values are also the basis for our analysis later in the chapter. The following response metrics refer to the formulas by Callegaro and DiSogra (2008) and to the AAPOR final disposition codes for RDD telephone surveys (American Association for Public Opinion Research, 2011).

For completed telephone interviews, irrespective of whether a respondent joined the panel or not, the overall response rate AAPOR RR3 was 17.8%, with the landline sample resulting in 18.5% and the mobile sample reaching 17.1%. The response rate is comparably high to other telephone surveys in Germany and also to surveys in the US. The Pew Research Center estimated an average AAPOR RR3 of 9% for a typical telephone survey (Kohut, Keeter, Doherty, Dimock, & Christian, 2012). Our sample consisted of more than 95000 telephone numbers. The contacts were nearly evenly distributed among landline and cell phone respondents.

Respondents to the telephone interview were defined as eligible if they used the Internet for non-work-related purposes and were, among other criteria, of age (18+) and able to participate in a German-language telephone interview. Respondents were eligible if they used the Internet at home or outside of the home, but were ineligible if they only used the Internet at work. Overall, 4840 telephone interviews were completed, of which 3514 respondents confirmed Internet-usage (72.6%). Of those, 1665 respondents were willing to participate in the panel and provided valid email addresses which could successfully be used to send out the link to the first online survey. From those, 1010 persons started the first survey and 934 persons completed it.

From the available figures, we can calculate the response metrics. The recruitment rate is the proportion of respondents who provided initial consent over all eligible respondents (i.e., among other criteria, using the Internet for private purposes). Here, we follow a conservative approach and only count those respondents as having provided their consent who not only said that they would like to join the panel but also provided a valid email address.

$$\text{Recruitment rate (RECR)} = 9\%.$$

As part of quality assurance, email addresses were screened while it was still possible to re-contact respondents via telephone so that they could correct any misspellings. If the welcome email which was sent to respondents shortly after the interview did not go through, interviewers called the respondents again to correct the email address. Roughly 10% of all emails were misspelled, either accidently or on purpose in the first attempt and thus proved non-working. All respondents with a misspelled email address were called back in order to correct the email address. Possible outcomes of these calls were correction of the email address, refusal to provide the interviewer with an email address thus refusing to participate in the panel or non-contact. If corrected emails still did not work, CATI fieldwork managers could decide if further follow-up was needed for further clarification. Those cases for which still no clarification was possible were treated as refusals in response rates calculations. The respondents, who provided a working email address, could still reconsider their choice and not complete the first online questionnaire. Therefore, a respondent was only defined as an active panel member after having completed the first online questionnaire. The corresponding metric was termed the profile rate (Callegaro & DiSogra, 2008), which is the proportion of all participants who completed or partially completed the first online questionnaire over all invited persons. Since the panel did not use a profile survey as a first survey but distributed the collection of demographic information across the first few online questionnaires, the profile rate is identical to the completion rate of the first survey.

Completion rate (COMR, started the first online survey) = 60.7%.

Completion rate (COMR, completed the first online survey) = 56.1%.

The cumulative response rate follows directly from multiplying the recruitment rate (RECR) and the completion rate of the first survey.

Cumulative response rate 1 (CUMRR1) = 9% × 56.1% = 5%.

When considering this last figure, it is necessary to keep in mind that the panel study was comprised of three separate studies, all of which were geared towards assessing the effect of different design features. Naturally, this included many suboptimal recruitment conditions (e.g., a control group without any incentive), which lowered the overall response metrics. In the next section we assess the data quality of the panel by comparing estimates derived from respondents' answers during the recruitment process and the first online questionnaire to estimates from other surveys.

3.4   Assessing data quality: Comparison with external data

In the remainder of this chapter we assess the quality of data, which was collected in the GESIS Online Panel Pilot. Data quality is an aspect of representativeness, which is determined by the quality of recruitment procedures as well as online nonresponse and online response quality. In the previous section we described the recruitment success, reporting the recruitment rates and online response rates. Response rates are not necessarily indicative of nonresponse bias (Groves & Peytcheva, 2008) and nonresponse analyses are not the focus of this chapter. Instead we define data quality as the goodness of our estimates. Estimates are the end product of a survey, which researchers use to perform their analyses, and the results of which are used for decision- or policy-making. Possible biases resulting from the telephone recruitment, unwillingness to join the panel, and online nonresponse are all included in the online estimates.

Assessing representativeness by comparing estimates from an online survey to official records or other external sources of information where one particular statistic is present is not uncommon (Bandilla, Kaczmirek, Blohm, & Neubarth, 2009; Chang & Krosnick, 2009; Scherpenzeel & Bethlehem, 2010; Smith, 2003; Yeager et al., 2011). The gold standard is to compare online survey results to administrative records and/or population and government registers. These sources are not free of problems: they are difficult and costly to obtain in practice, can be outdated, can relate to different reference populations, and contain different operational definitions (Lynn, 2008). An alternative course of action is to use estimates from offline general population surveys with high response rates.

In our case, population registers do not contain the information about the Internet usage: the proportion and the characteristics of the Internet-using population in Germany are themselves subject to estimation (Destatis [the German Federal Statistical Office], 2011). This does not allow us to use official sources since our target population is restricted to the Internet-using population. Moreover, attitudinal measures are usually not included in official statistics reports, what makes them less valuable for assessing the quality of the online panel if one wishes to look beyond demographic measures.

We use two reference surveys to assess data quality: the German General Social Survey (ALLBUS) and the German part of the European Social Survey (ESS). They both include information on private Internet usage and had fieldwork performed within a similar timeframe as the fieldwork of the online panel. The reasons for treating these surveys as benchmarks in terms of data quality are: (1) relatively high response rates and, more importantly, (2) selective nonresponse in the multistep recruitment of the online panel.

3.4.1 Description of the benchmark surveys

3.4.1.1 The German General Social Survey (ALLBUS)

The German General Social Survey (ALLBUS) is a cross-sectional, interviewer-administered, face-to-face survey on attitudes, behavior, and social structure in Germany. It was first conducted in 1980 and has been conducted by GESIS every two years. For our analysis, we use the 2010 data. In 2010, ALLBUS employed a two-stage disproportionate random sample of all persons who resided in private households in the Federal Republic of Germany on the day of the interview and were born before 1 January 1992 with oversampling households in East Germany. In the first sample stage, municipalities in western Germany and municipalities in eastern Germany were selected with a probability proportional to their number of adult residents; in the second sample stage, individual persons were selected at random from the municipal registers of residents. The reported response rate in 2010 was 34.4%,[2] which was not unusual for Germany following the European trend of decreasing response rates (de Leeuw & de Heer, 2002). The initial number of interviews in ALLBUS 2010 was 2827. We excluded from our main analysis all persons, who did not use the Internet, which decreased the sample size to 1869.

3.4.1.2 German subsample of the European Social Survey (ESS)

Like our first reference survey, the European Social Survey is a cross-sectional, face-to-face, biennial survey. It was first conducted in 2002. ESS is a multi-country study, which includes over 30 countries. It is funded by the European Commission, the European Science Foundation, and national funding councils in participating countries.[3] For our analyses we use the German subsample of the 5th ESS round conducted in 2010. The sampling procedure for Germany is the same as it is for the ALLBUS, however, the target population includes individuals residing in private households aged 15 and older. For our analysis we further excluded all individuals aged 15–17 and the non-Internet-using population. The resulting sample size of the ESS was 2038.

Further information about both reference surveys and the GESIS Online Panel Pilot is presented in Table 3.2.

3.4.2 Measures and method of analyses

We examined demographic and attitudinal measures. Among demographic variables were gender, age, educational attainment, legal marital status, employment status (whether working for pay or not), and immigration background. Operationalization of the employment status is not exactly the same in ESS and ALLBUS with ESS asking about being in paid work

---

[2] http://www.gesis.org/en/allbus/study-profiles/2010/.
[3] http://www.europeansocialsurvey.org/ and http://ess.nsd.uib.no/ess/round5/.

Table 3.2 Sample description for GESIS Online Panel Pilot and two reference surveys.

Survey feature	GOPP	ALLBUS	ESS
Sample source	Adjusted Random Digit Dial sample (landline and cell phones)	Two-stage probability: municipalities with the probability proportional to the population size of municipality, individuals from municipal registers (random)	Two-stage probability: municipalities with the probability proportional to the population size of municipality, individuals from municipal registers (random)
Number of sampling points (East and West Germany)	N/A	162	168
Target population	Individuals residing in households in the Federal Republic of Germany aged 18 and older who use the Internet	Individuals residing in households in the Federal Republic of Germany aged 18 and older	Individuals residing in households in the Federal Republic of Germany aged 15 and older
Pre-notifier	N/A	Advance letter, advance brochure	Advance letter, advance brochure
Recruitment mode	CATI	CAPI	CAPI
Mode of data collection	Online, CATI	CAPI	CAPI
Response rate	CATI AAPOR3: 17.1%, Online Completion rate of the first online survey: 56.1%	34.4%	30.5%
Field dates	February 2011–May 2012	May–November 2010	September 2010–February 2011
Incentives	Experimental groups of 0, 2, 5 EUR, 5+bonus 20 EUR, 10+bonus 20 EUR	0, 10 EUR (Interviewers' tailoring)	20 EUR (cash), conditional upon completion
N	794[a]	2827	3031
N Internet users (18+)	794[a]	1869	2038
Percentage of Internet users, weighted	71.5[b]	67.1	72.5

[a] Only regular (nonexperimental) telephone interviews, overall 934 respondents completed the first online survey.
[b] For the Online Panel, only the unweighted proportion of Internet users is available, since non-users were not interviewed and thus answered no questions on the basis of which weights were constructed.

during the last seven days and ALLBUS differentiating between full-time, part-time, less than part-time, and not working. Not stating a reference period, the ALLBUS operationalization may slightly overestimate those in paid employment compared to the ESS. The operationalization in GOPP is comparable to the ALLBUS. The binary variable "working for pay" contrasts those in paid work regardless of the type of employment.

Attitudinal variables used for comparison were political interest, satisfaction with the government, generalized trust, self-rated health status,[4] assessment of the state of health services in Germany, rating of current state of German economy, rating of German economy in one year, rating of respondent's own financial situation, rating of respondent's own financial situation in one year, and general life satisfaction.

In order to eliminate the effect of attrition in the panel data, we only consider measures collected during the recruitment interview and during the first online questionnaire. Data collection for the first online survey was mainly finished in August 2011. Information on the specific mode in which the variables were collected can be found in Table 3.A.1 in Appendix 3.A.

Data quality is assessed by comparing the estimates from GOPP to the two benchmark surveys (Table 3.3). Depending on the scaling of the variable, we compared proportions or means. Statistical significance of difference between pairs of estimates from each benchmark survey was tested using t-tests with the hypothesis that a difference between two proportions/two means equals zero.

Data in both benchmark surveys were weighted using the design weights which accounted for unequal probabilities of selection. For GOPP, the weights were constructed by GESIS. The reference surveys provided the respective weights. The ALLBUS documentation states that since it is based on a sample of individuals drawn from the municipal registers of residence, it needs no weighting when conducting individual-level analysis. However, individuals residing in the former East Germany are oversampled. The ALLBUS design weights correct for that.[5] Design weights for the European Social Survey are computed for each country, depending on the number of stages in the sampling design. For Germany, this number equals 2. The weights are calculated as the inverse of the combined probability of being selected during each stage.[6] For both surveys the weights are rescaled so that the sum of final weights equals the sample size. The GOPP weights correct for the unequal inclusion probabilities as described in Section 3.2.2. The design weight was calculated as the inverse of the inclusion probability for each respondent. For the inclusion probability we used the following formula:

$$\pi_i \approx k_i^L \frac{m^L}{M^L} \times \frac{1}{z_i i} + k_i^C \frac{m^C}{M^C},$$

where: k is the number of telephone numbers where a respondent can be reached, m is the sample size of numbers, M equals frame size, and z is the size of the household the target person resides in. L and C refer to the landline and cell phone components (Gabler et al., 2012).

---

[4] Scale labels for the variable self-assessed health differ somewhat between ALLBUS and ESS. Labels used in ALLBUS were: very good, good, satisfactory, poor, bad; labels used in ESS were: very good, good, fair, bad, very bad. For comparison, a dummy variable was constructed which contrasts reports of very good and good health taking the value of 1 and other values when it takes the value of 0. GOPP used the same operationalization as ALLBUS.

[5] Sample design and weighting used in ALLBUS: http://www.gesis.org/fileadmin/upload/dienstleistung/daten/umfragedaten/allbus/dokumente/Weighting.

[6] Sampling design and weighting procedures are described in the ESS Data Documentation Report for Round 5: http://ess.nsd.uib.no/ess/round5/.

Table 3.3 Comparison of GESIS Online Panel Pilot to two benchmark surveys: demographic and attitudinal variables.

Variable for comparison	GOPP			ALLBUS Internet users			ESS Internet users		
	Estimate	SE	99% Conf. Int.	Estimate	SE	99% Conf. Int.	Estimate	SE	99% Conf. Int.
Male	51.0%	0.020	[45.8%,56.1%]	51.3%	0.012	[48.2%,54.4%]	53.7%	0.012	[50.7%,56.7%]
Age groups									
18–24	16.6%	0.016	[12.4%,20.8%]	13.9%	0.008	[11.8%,16.0%]	14.8%	0.008	[12.6%,16.9%]
25–34	20.6%	0.017	[16.3%,24.8%]	18.7%	0.009	[16.3%,21.1%]	17.4%	0.009	[15.1%,19.7%]
35–49	34.0%	0.018	[29.2%,38.8%]	35.5%	0.011	[32.5%,38.4%]	33.3%	0.011	[30.5%,36.1%]
50–64	21.2%	0.016	[17.2%,25.3%]	23.0%	0.010	[20.4%,25.6%]	25.7%*	0.010	[23.1%,28.3%]
65+	7.6%	0.010	[5.0%,10.1%]	8.9%	0.007	[7.2%,10.7]	8.8%	0.007	[7.1%,10.5%]
Education									
low	12.8%	0.014	[9.1%,16.5%]	23.3%***	0.010	[20.7%,26.0%]	20.1%***	0.010	[17.7%,22.6%]
medium	30.0%	0.019	[25.1%,34.8%]	35.7%**	0.011	[32.8%,38.7%]	37.9%***	0.011	[35.0%,40.8%]
high	57.2%	0.020	[52.0%,62.4%]	41.0%***	0.012	[37.9%,44.0%]	42.0%***	0.012	[39.0%,44.9%]
Married	48.5%	0.020	[43.4%,53.7%]	56.9%***	0.012	[53.8%,59.9%]	57.1%***	0.012	[54.1%,60.1%]
Immigration background	9.6%	0.012	[6.6%,12.6%]	13.4%**	0.008	[11.3%,15.6%]	11.6%	0.008	[9.7%,13.6%]
Working	68.0%	0.019	[63.1%,72.9%]	72.4%*	0.011	[69.7%,75.2%]	65.8%	0.011	[63.0%,68.7%]
Good/very good health	60.5%	0.020	[55.4%, 65.5%]	68.6%***	0.011	[65.7%, 71.4%]	64.4%	0.011	[61.5%, 67.3%]
Life satisfaction (11-pt)	8.56	0.066	[8.39, 8.73]	8.48	0.042	[8.37, 8.59]	8.33**	0.046	[8.21, 8.45]
Trust others									
Yes	19.4%	0.016	[15.3%, 23.4%]	23.7%*	0.010	[21.0%, 26.3%]	–	–	–
No	19.3%	0.016	[15.2%, 23.4%]	35.1%***	0.011	[32.2%, 38.1%]	–	–	–

Depends									
German economy (5-pt)	61.3%	0.019	[56.3%, 66.3%]	41.2%***	0.012	[38.2%,44.2%]	—	—	—
	3.39	0.028	[3.32, 3.46]	3.09***	0.019	[3.04, 3.14]	—	—	—
Self-rated financial situation (5-pt)	3.43	0.030	[3.36, 3.51]	3.47	0.020	[3.42, 3.52]	—	—	—
German economy in 1 year (5-pt)	2.90	0.032	[2.82, 2.98]	3.08***	0.021	[3.02, 3.13]	—	—	—
Self-rated financial situation in 1 year (5-pt)	3.20	0.031	[3.12, 3.28]	3.18	0.017	[3.14, 3.23]	—	—	—
Interest in politics (4-pt)	2.66	0.029	[2.58, 2.74]	—	—	—	2.77**	0.019	[2.72, 2.82]
Satisfaction with government (11-pt)	4.78	0.087	[4.56,5.01]	—	—	—	4.77	0.051	[4.64, 4.90]
Satisfaction with health services (11-pt)	5.38	0.087	[5.16,5.61]	—	—	—	5.81***	0.055	[5.67, 5.96]

*$p < 0.05$,
**$p < 0.01$,
***$p < 0.001$. Differences between ALLBUS and ESS at the significance level of at least 0.05 in bold italic font; SE short for standard error.

As the GOPP estimates are obtained after several stages of nonresponse, which can affect the resulting differences between the three surveys, we use post-stratification weighting to correct for this. Ideally, we would have taken Internet population benchmarks to calculate post-stratification weights. However, neither the German Census which was carried out in 2011, nor the Microcensus (the official representative 1% sample survey of the German population) contains indicators of Internet use. The study by the German Federal Statistical Office "Information and communication technologies (ICT) in households,"[7] which provides Internet usage benchmarks for Germany, uses a quota sample, which makes it unsuitable for calculating weights as the quality of the sample cannot be assessed. None of the other studies with a specific focus on Internet which we reviewed (a listing of which can be found in Kaczmirek and Raabe, 2010) meets the criteria for one or the other reason (e.g., usage of telephone surveys to collect information about Internet users but exclusion of cell phone users). For these reasons, we use ALLBUS to calculate post-stratification weights. Specifically, we used the distributions of age, gender, and education of the subgroup of Internet users to construct the weights. Ideally we would have reported the results for ALLBUS and ESS using post-stratification weights as well. However, ALLBUS does not provide post-stratification weights and ESS post-stratification weights were not available at the time of writing. Therefore, unweighted data from ALLBUS was used to construct the weights. As ALLBUS is itself a sample survey prone to nonresponse, one needs to be careful interpreting the results. We address this issue in the results section.

On the level of individual estimates, if respondents refused to answer the question or provided a "don't know" response, those cases were not included in the calculation of means or proportions. For design weights in GOPP we imputed missing values for variables used to calculate weights (number of landline and cell phone numbers where a respondent can be reached, number of household members over 18 years of age, and other variables) with means of respective variables, which were calculated based on the entire sample. For the two reference surveys, weighting variables did not contain missing data.

3.5 Results

3.5.1 Demographic variables

A straightforward procedure of comparing estimates from the panel with the reference surveys and applying t-tests for statistical significance of the pairwise comparisons for demographic variables showed that the panel did not differ from the two benchmark surveys on gender composition. It further did not differ substantially with respect to age groups. Although respondents in the panel were somewhat younger than in the reference surveys (mean age in GOPP$=41.1$, in ALLBUS$_{(A)} = 42.6$, ESS$_{(E)} = 43.0$; $t_A = -2.20$, $p_A = 0.028$, $t_E = -2.76$, $p_E = 0.006$) and overall age distribution of ESS differed from GOPP ($\chi^2_A = 8.41$, $p_A = 0.078$; $\chi^2_E = 12.95$, $p_E = 0.012$),[8] when looking at the groups, the only significantly different age group in GOPP was the one of 50–64 years. It differed from ESS by 4.5 percentage points.

Educational level was the one variable among demographics, which demonstrated the greatest differences from the reference surveys. Particularly overrepresented were higher-educated Internet users: the difference from ALLBUS was 16.2 percentage points,

---

[7] https://www.destatis.de/EN/Meta/abisz/IKTPrivateHaushalte_e.html.

[8] For χ^2-tests, ALLBUS and ESS distributions are treated as the expected ones, GOPP distribution is treated as observed.

the difference from ESS was 15.2 percentage points. Consequently, the group comprising low-educational level was highly underrepresented with a 10.5 percentage point difference from ALLBUS and a 7.3 percentage point difference from ESS ($\chi^2_A = 88.13$, $p_A < 0.001$; $\chi^2_E = 71.70$, $p_E < 0.001$). The proportion of respondents with low educational level also differed between the two reference surveys: the estimate from ESS differed from ALLBUS by 3.2 percentage points ($p = 0.024$). That is the first of several differences between the reference surveys, which will be discussed later in the section.

Three other demographic variables demonstrated dissimilar patterns across reference surveys but all somehow differed in GOPP from either or both of the comparing surveys. The proportion of married respondents was significantly higher in both reference surveys compared to GOPP ($p < 0.001$). Fewer individuals with an immigration background were found in the panel compared to ALLBUS and compared to ESS. Fewer respondents working for pay were found in the panel than in ALLBUS but no such difference was present when compared to ESS. This is particularly interesting since question wordings varied in ALLBUS and ESS and the ALLBUS wording was used in the panel. Therefore it was reasonable to expect differences both between ALLBUS and ESS (which is the case) and between GOPP and ESS.

Overall, for demographic variables the average absolute percentage point difference between GOPP and ALLBUS was 4.88 percentage points, between GOPP and ESS: 4.78 percentage points, and 2.05 percentage points between ALLBUS and ESS (not presented in Table 3.3). The difference between the online survey and either of the face-to-face surveys was roughly twice as high as between the two reference surveys.

3.5.2 Attitudinal variables

Among the attitudinal variables, two were present in both reference surveys: self-rated health status and life satisfaction. There was a significant difference in self-rated health between responses from the panel and ALLBUS (8.1 percentage points, $p < 0.001$). Furthermore, the estimate from ALLBUS is significantly different from the ESS estimate (4.2 percentage points difference, $p = 0.008$).

Life satisfaction was measured with an 11-point scale from 0 "completely dissatisfied" to 10 "completely satisfied." There was no significant difference in means for ALLBUS and GOPP, however, ESS differed from both GOPP ($t = 2.83$, SE $= 0.081$, $p = 0.005$) and ALLBUS ($t = 2.42$, SE $= 0.062$, $p = 0.016$). Respondents of the panel and ALLBUS demonstrated significantly higher level of life satisfaction than the ESS respondents.

Some concepts were operationalized differently in ESS and ALLBUS: two variables measuring generalized trust and interest in politics, though present in both reference surveys, were not directly comparable. We had to decide which survey had to be used as a reference. The question to measure trust asked whether most people could be trusted, one couldn't be careful enough, or it depended [on the situation]. Respondents in ALLBUS were less trusting than respondents in the panel: the difference on the first value (people can be trusted) of 4.3 percentage points ($p = 0.022$) was substantially lower than differences on categories "one can't be careful enough" (15.8 percentage points, $p < 0.001$) and "it depends" (20.1 percentage points, $p < 0.001$) with respondents in the panel having a larger share of those saying trust is situational. The second variable present in both surveys but measured differently was interest in politics. It was replicated from ESS. The analysis showed that respondents in ESS were significantly more interested in politics than respondents in GOPP.

For other attitudinal measures included in the present analysis, only pairwise comparisons with either of the reference surveys could be performed. The next block of measures concerned the economic situation in Germany and the respondent's personal financial situation. Online respondents rated the current German economy somewhat better than ALLBUS respondents, and next year's economy worse. No significant differences were found in ratings of the respondents' own economic situation for the current and next year.

With respect to ESS, no significant differences were found in satisfaction with government. However, online respondents were less satisfied with the health system ($t = -4.19$, $p < 0.001$).

3.5.3   Comparison of the GESIS Online Panel Pilot to ALLBUS with post-stratification

From the previous analysis it remains unclear what caused the differences between the surveys. The bias may stem from nonresponse or coverage issues. It is therefore needed to rebalance the GOPP to the Internet-using population. In the absence of official benchmarks or other suitable data, the ALLBUS subsample of Internet users was taken to construct the post-stratification weights. Variables used to construct the weights included age, gender, and educational level. Table 3.4 presents the comparison of weighted estimates (design and post-stratification weights) from the panel to the design-weighted ALLBUS data.

Post-stratification did not substantially change the estimates for the demographic variables of marital and employment status, and immigration background: differences

Table 3.4   Comparison of the GESIS Online Panel Pilot to the subsample of ALLBUS Internet users, with post-stratification.

Variable for comparison	GOPP (weighted on ALLBUS Internet users)			ALLBUS Internet users		
	Estimate	SE	99% Conf. Int.	Estimate	SE	99% Conf. Int.
Married	47.0%	0.025	[40.6%, 53.4%]	56.9%***	0.012	[53.8%, 59.9%]
Immigration background	8.4%	0.013	[5.1%, 11.7%]	13.4%**	0.008	[11.3%, 15.6%]
Working	67.8%	0.025	[61.3%, 74.2%]	72.4%	0.011	[69.7%, 75.2%]
Good/very good health	54.9%	0.025	[48.4%, 61.4%]	68.6%***	0.011	[65.7%, 71.4%]
Life satisfaction	8.38	0.087	[8.16, 8.61]	8.48	0.042	[8.37, 8.59]
Trust others						
Yes	17.8%	0.018	[13.1%, 22.4%]	23.7%**	0.010	[21.0%, 26.3%]
No	23.1%	0.022	[17.3%, 28.8%]	35.1%***	0.011	[32.2%,3 8.1%]
Depends	59.1%	0.025	[52.8%, 65.5%]	41.2%***	0.012	[38.1%, 44.2%]
German economy	3.28	0.038	[3.18, 3.38]	3.09***	0.019	[3.04, 3.14]
Self-rated financial situation	3.34	0.037	[3.25, 3.44]	3.47**	0.020	[3.42, 3.52]
German economy in 1 year	2.85	0.043	[2.74, 2.96]	3.08	0.021	[3.02, 3.13]
Self-rated financial situation in 1 year	3.20	0.042	[3.09, 3.31]	3.18	0.017	[3.14, 3.23]

*$p < 0.05$,
**$p < 0.01$,
***$p < 0.001$. SE short for standard error. GOPP uses overall weight (design*post-stratification constructed on the basis of ALLBUS Internet users), ALLBUS is design weighted.

Table 3.5 Comparison of the GESIS Online Panel Pilot to ALLBUS, with post-stratification.

Variable for comparison	GOPP (weighted on ALLBUS complete)			ALLBUS Internet and non-Internet users		
	Estimate	SE	99% Conf. Int.	Estimate	SE	99% Conf. Int.
Married	48.1%	0.036	[38.9%, 57.4%]	57.5%*	0.010	[55.1% , 60.0%]
Immigration background	7.8%	0.014	[4.1%, 11.5%]	16.7%***	0.007	[14.8%, 18.6%]
Working	62.0%	0.038	[52.3%, 71.8%]	57.8%	0.010	[55.3%, 60.3%]
Good/very good health	48.7%	0.028	[41.5%, 55.8%]	56.9%**	0.010	[54.4%, 59.4%]
Life satisfaction	8.29	0.101	[8.03, 8.55]	8.34	0.037	[8.24, 8.43]
Trust						
Yes	15.9%	0.019	[11.0%, 20.8%]	22.7%***	0.008	[20.6%, 24.8%]
No	26.2%	0.037	[16.7%, 35.8%]	42.3%***	0.010	[39.8%, 44.8%]
Depends	57.9%	0.036	[48.5%, 67.3%]	35.0%***	0.009	[32.6%,37.4%]
German economy	3.27	0.038	[3.18, 3.37]	3.04***	0.016	[3.00, 3.08]
Self-rated financial situation	3.31	0.044	[3.20, 3.43]	3.41*	0.017	[3.37, 3.45]
German economy in 1 year	2.82	0.052	[2.68, 2.95]	3.01***	0.017	[2.97, 3.06]
Self-rated financial situation in 1 year	3.18	0.052	[3.04, 3.31]	3.09	0.014	[3.06, 3.13]

*$p < 0.05$,
**$p < 0.01$,
***$p < 0.001$. SE short for standard error. GOPP uses overall weight (design*post-stratification constructed on the basis of ALLBUS complete sample), ALLBUS is design weighted.

between proportions in Table 3.3 and Table 3.4 range between 0.2–1.5 percentage points for these variables. The most pronounced effect was on the variable self-assessed health with 5.6 percentage point difference. Means varied in the range of 0–0.18 scale points. Taken together, post-stratification caused no improvement. On the contrary, with the exception of two categories of trust, status of German economy and self-rated financial situation in one year, for all other variables not used for weighting the differences to the ALLBUS estimates increase. Changes in significance can be documented for employment status and self-rated financial situation.

In Table 3.5 we present the comparison between the weighted Online Panel data and the ALLBUS complete sample. Here, the complete ALLBUS sample was used to construct the weights. The results of this exercise are primarily to illustrate the effect of the design decision to exclude non-Internet users. With the exception of life satisfaction and self-rated financial situation, all other variables differ significantly between the two surveys. These differences are largely attributed to the differences between Internet users and nonusers, which are present in ALLBUS for all variables except for marital status (calculations not presented in Table 3.5).

3.5.4 Additional analysis: Regression

It may be hypothesized that discrepancies of attitudinal variables between the surveys may be attributed to differences in sample composition described earlier in this chapter. All demographic variables of the panel with the exception of gender varied from the estimates of the reference surveys by lesser or greater degree. Post-stratification with "rebalancing" the online sample to the Internet using population from ALLBUS did not solve the problem. Another method to assess differences in attitudinal variables is to compare how well the surveys perform in the model estimation. To illustrate this point, life satisfaction, a variable measured equally in all three surveys was chosen for the following analysis (Table 3.6). Satisfaction with life is an important indicator for welfare and happiness research and is used widely in international comparative studies.

Headey and Wearing (1992) originally identified seven domains which matter most to well-being: marriage and sex, friendship and leisure, material living standards, work, and health. The demographic variables used for comparison to the reference surveys cover at least to some extent five of the domains with the exception of material living standard and friendship and leisure. Further, Headey and Wearing describe the importance of major life events, which we also are unable to include in the analysis. The main purpose, however, being the comparison of three surveys for the present chapter, would allow for bias due to omitted variables as long as the list of control variables stays unchanged for each of the surveys in question.

As the goal of the regression analysis was not to interpret substantial outcomes, we concentrate on the part of the model, which accounts for survey control variables. One thing worth noting, however, is the level of R^2, which equals 11.1%. Such a figure is not uncommon

Table 3.6 Linear regression on life satisfaction by reference surveys.

	Standardized β	Standard error
Male	-0.039**	(0.055)
Age group		
18–24	Ref.	
25–34	-0.092***	(0.101)
35–49	-0.168***	(0.103)
50–64	-0.098***	(0.110)
65+	0.036+	(0.129)
Education		
low	Ref.	
middle	0.057*	(0.086)
high	0.142***	(0.082)
Married	0.166***	(0.068)
Working	0.077***	(0.074)
Immigration background	-0.047**	(0.096)
Health (very good/good)	0.237***	(0.062)
ALLBUS	-0.032	(0.079)
ESS	-0.063***	(0.079)
N	4596	
R^2	0.110	

$^+p < 0.10$,
$^*p < 0.05$,
$^{**}p < 0.01$,
$^{***}p < 0.001$.
All surveys combined, weighted, GOPP is reference category.

for models with socio-demographic variables including age, gender, education, and occupation, which often explain no more than 10% of the variance in individual life-satisfaction (Veenhoven, 1996) in nations like Germany.

Controlling for the sample composition, differences between surveys in life satisfaction do not disappear. To further investigate whether the panel is comparable to the reference surveys on the life satisfaction variable, we fitted three regression models, separately for each survey (Table 3.7).

Here, we employ three criteria for comparison of the models' estimates: direction, magnitude, and statistical significance. The regression model, whose results are presented in Table 3.6, suggests that the panel differs significantly from ESS on the life satisfaction variable. The comparison of ALLBUS and ESS models reveals that all estimates have the same direction, however, the magnitude differs. Additionally, statistical significance ($p < 0.05$) fails to match on three variables: two age group variables and immigration background.

The direction of correlation in GOPP matches these of the reference surveys on all estimates except for the immigration background variable (which most likely is caused by the small number of respondents with immigration background in GOPP) and the oldest age group, probably for the same reason of this group being relatively small. With the exception of the employment status variable (working), variables reaching statistical significance match with the variables reaching statistical significance in ESS. This indicates that as far as modeling of life satisfaction goes, GOPP performs not worse in comparison to ESS than ESS

Table 3.7 Regressions on life satisfaction by survey.

	GOPP		ALLBUS		ESS	
	β	SE	β	SE	β	SE
Male	-0.039	(0.132)	-0.028	(0.083)	-0.043+	(0.088)
Age group						
18–24	Ref.		Ref.		Ref.	
25–34	-0.125*	(0.221)	-0.081*	(0.155)	-0.085**	(0.161)
35–49	-0.272***	(0.230)	-0.101*	(0.155)	-0.188***	(0.166)
50–64	-0.246***	(0.268)	-0.023	(0.159)	-0.107**	(0.177)
65+	-0.004	(0.297)	0.089**	(0.204)	0.012	(0.201)
Education						
low	Ref.		Ref.		Ref.	
middle	0.046	(0.251)	0.064+	(0.122)	0.053	(0.137)
high	0.187**	(0.223)	0.113***	(0.119)	0.152***	(0.131)
Married	0.220***	(0.156)	0.134***	(0.101)	0.171***	(0.110)
Working	0.054	(0.175)	0.093**	(0.121)	0.076**	(0.111)
Immigration background	0.016	(0.229)	-0.066**	(0.131)	-0.049+	(0.160)
Health (very good/good)	0.172***	(0.140)	0.245***	(0.096)	0.249***	(0.095)
N	746		1831		2019	
R^2	0.128		0.104		0.120	

+$p < 0.10$,
*$p < 0.05$,
**$p < 0.01$,
***$p < 0.001$; data are weighted, beta coefficients are standardized, SE is short for standard error.

does to ALLBUS or ALLBUS does to ESS. The R^2 of the GOPP model also is not far off from the ESS model's one. This suggests that the data, which come from GOPP, could be used to fit models (on life satisfaction at the least) and would provide adequate results. The shortcoming of GOPP with the immigration background or oldest age group estimates having an opposite direction would not be relevant if one wished to interpret only significant coefficients, as in ESS both coefficients of immigration background and age group older than 65 do not reach statistical significance of 0.05 or lower.

The results presented in Tables 3.6 and 3.7 may be dependent on the choice of one particular variable: life satisfaction. Therefore, we fitted another model using another variable present in all of the reference surveys: self-assessed health status. As mentioned in Section 3.4.2, self-assessed health was recoded into a binary variable "very good/good health." The results of the logistic regression are presented in Table 3.A.2 in Appendix 3.A. Controlling for demographics, significant differences between GOPP and both reference surveys are found.

It cannot be inferred from the above whether GOPP will prove to be successful in comparing other variables or not. This would have to be decided for other variables or models case by case, which is beyond the scope of this chapter.

3.5.5   Replication with all observations with missing values dropped

Pairwise comparisons between surveys, results of which are presented in Table 3.3, were replicated when all observations were dropped, which contained a missing value on at least one of the variables used in the analysis. The resulting estimates did not differ substantially from the results in Table 3.3. The highest difference was 1.3 percentage points for proportions of demographic variables (mean difference 0.42; 0.54 and 0.48 percentage points for GOPP, ALLBUS and ESS respectively). The highest difference for variable means was 0.06 scale points. No changes of significance, except for ALLBUS and ESS immigration background ($p < 0.05$), took place. Although levels of significance changed for life satisfaction in GOPP and interest in politics in ESS.

3.6   Discussion and conclusion

In some countries and research settings probability-based online panels have established themselves as a cost-effective alternative to face-to-face and telephone general population surveys. However, in other countries such interviewing methods are just beginning to appear. In this chapter we described an online panel pilot study whose goal was to identify and test the conditions under which a probability-based panel can be recruited and maintained in Germany. The target population was restricted to the Internet-using population and recruitment was performed by telephone using an RDD dual-frame approach.

Response metrics in the recruitment process reached satisfactory levels, beginning with moderate response rates for the telephone interviews. Due to the experimental nature of various recruitment approaches, the average recruitment rate was lower than one would expect with the optimal design approach.

The major focus of this chapter has been on assessing the quality of the data generated by the panel. Quality was defined in terms of deviation from estimates produced by two major face-to-face surveys, the German General Social Survey (ALLBUS) and the German subsample of the European Social Survey (ESS).

The results indicate that significant differences exist between the online panel and either of the face-to-face reference surveys on most of the demographic and attitudinal variables. Out of five demographic measures used for comparison, only one was not significantly different from either of the face-to-face surveys. Most attitudinal variables differed when compared to the reference surveys, though for a number of attitudinal measures comparisons had to be restricted to one reference survey due to a lack of data. This was the case when concepts were measured differently between the two reference surveys or the concept was only present in one of them.

When estimating data quality we also need to take into account that the face-to-face reference surveys also showed significant differences between each other. Such differences were found for two demographic and on both attitudinal measures. Not only did these two surveys differ on means or proportions, but also on magnitude and significance of predictors when performing a regression analysis. Taken together, the results suggest that the panel is comparable to traditional surveys in many aspects of sample composition and when modeling social phenomena. Taking into account the estimation of average absolute error calculated for demographics of about 2 percentage points between the two surveys and about 5 percentage points between the GOPP and either face-to-face survey, we can conclude that the GOPP estimates still are not as close to the face-to-face survey as one would wish. However, at the present stage of research, when probability-based online panels are still in their beginning stages, the results presented are rather encouraging.

This research has a number of limitations. First, a design decision to exclude non-Internet users made comparisons with official statistics impossible as the German Census does not include a question on Internet use. This forced us to use other general population surveys as benchmarks restricting these surveys to the subsample of Internet-using population and use one of the surveys for the calculation of post-stratification weights. Using survey data as benchmarks was problematic because both reference surveys had considerable nonresponse. The response rates were not atypical for the current survey climate in Germany, this does not, however, exclude the possibility of bias.

Absence of official statistics is not a problem unique to the group of Internet users. It may apply, for example, to special (hard-to-reach) population groups. One important direction for future research is to employ strategies, which would evaluate bias in reference surveys.

Second, estimates from the online panel are the product of multiple selection processes, each of which is prone to nonresponse: the telephone interview, providing the email address, and response to the first online questionnaire. The mode switch between contacting the respondent and data collection online was an additional source of nonresponse, which neither of our reference surveys had. Respondents, who had agreed to the initial recruitment interview, may have used the time lag between the telephone interview and the invitation to the online survey as an opt-out option. More detailed studies on the causes of these processes and how they influence the differences in estimates would be beneficial.

Third, there is a risk of potential mode effects on measurement. The telephone interview was not only aimed at recruiting participants into the panel, but also at collecting data about the respondents to study nonresponse effects. In such scenarios mode effects are to be expected (de Leeuw, 2005). It is also worth mentioning that the reference surveys were interviewer-administered whereas the GOPP was self-administered in the online surveys. For our comparison we did not use questions that were considered to produce socially desirable answers and for which interviewer-administered and Internet modes have demonstrated bias (Chang & Krosnick, 2009; Kreuter, Presser, & Tourangeau, 2008), but there is still a possibility of mode effects on other items due to visual and aural administration of the questions.

Finally, the results of the overall data quality assessment are dependent on the choice of the set of variables to be examined. The approach of assessing the data quality, which we undertook, of comparing means and reporting absolute errors with respective significance tests, has been performed in a number of studies which examine the quality of online panels. However, apart from demographic variables, covariates considered by researchers range broadly from voting behavior and attitudes to immigration or smoking behavior. The inherent problem of performing such multiple comparisons with various measures is the concern about finding more differences when more variables are added to the set. It might be useful to find an appropriate number of measures to be examined.

The limitations of the approach we undertook for the quality assessment of data produced by the online panel, nevertheless, cannot undermine our overall conclusion that the data and performing model-based analysis with this data are comparable to those of the traditional surveys. In our comparisons we concentrate mostly on biases, the magnitude and the direction thereof. In doing this, one needs to take into account that estimates are a product of multiple processes. As Biemer (1988) points out, "a sample survey is an integrated system of activities and operations for data collection" (p. 273), so the comparison takes place between several systems of data collection. We try to take this into consideration by comparing not only final estimates but also by looking at how surveys perform when modeling social phenomena. However, this strategy also deals with final "survey products." As we show, consequences of certain design decisions cannot be solved with post-stratification either. In future research, this should deserve special attention. As the GESIS Online Panel Pilot, data from which we compared to the general population surveys, is a methodological project, in our opinion, it served the purpose well. When conducting general population surveys in order to assess their representativeness, one would want the design to mirror the procedures implemented in the reference surveys: comparable recruitment mode, comparable experimental conditions and, in the best case scenario, inclusion of the offline population. For Germany, all these strategies will be implemented in the GESIS Panel (http://www.gesis-panel.org), a mixed-mode omnibus access panel survey of the general population, which includes both the Internet and the non-Internet population.

References

American Association for Public Opinion Research. (2011). *Standard definitions: Final dispositions of case codes and outcome rates for surveys*. Deerfield. IL: AAPOR.

Bandilla, W., Kaczmirek, L., Blohm, M., & Neubarth, W. (2009). Coverage und Nonresponse-Effekte bei Online-Bevölkerungsumfragen [Coverage and nonresponse effects in online population-surveys]. In N. Jackob, H. Schoen, & T. Zerback (Eds.), *Sozialforschung im Internet: Methodologie und Praxis der Online-Befragung* (pp. 129–143). Wiesbaden: VS Verlag.

Baker, R., Blumberg, S. J., Brick, J. M., Couper, M. P., Courtright, M., Dennis, J. M., ... Zahs, D. (2010). Research synthesis. AAPOR report on online panels. *Public Opinion Quarterly*, *74*, 711–781.

Biemer, P. P. (1988). Measuring data quality. In R. M. Groves, P. P. Biemer, L. E. Lyberg, J. T. Massey, W. L. Nicholls, & J. Waksberg (Eds.), *Telephone survey methodology* (pp. 273–283). New York: John Wiley & Sons, Inc.

Callegaro, M., & DiSogra, C. (2008). Computing response metrics for online panels. *Public Opinion Quarterly*, *72*, 1008–1031.

Chang, L., & Krosnick, J. A. (2009). National surveys via RDD telephone interviewing versus the Internet: Comparing sample representativeness and response quality. *Public Opinion Quarterly*, *73*, 641–678.

Das, M. (2012). Innovation in online data collection for scientific research: The Dutch MESS project. *Methodological Innovations Online, 7*, 7–24.

De Leeuw, E. D. (2005). To mix or not to mix data collection modes in surveys. *Journal of Official Statistics, 21*, 233–255.

De Leeuw, E. D., & de Heer, W. (2002). Trends in household survey nonresponse: A longitudinal and international comparison. In R. M. Groves, D. A. Dillman, J. L. Eltinge, & R. J. A. Little (Eds.), *Survey nonresponse* (pp. 41–54). New York: John Wiley & Sons, Inc.

Destatis [German Federal Statistical Office]. (2011). *Wirtschaftsrechnungen: Private Haushalte in der Informationsgesellschaft. Nutzung von Informations- und Kommunikationstechnologien* [Budget surveys: Private households in the information society: use of information and communication technology.]. Wiesbaden: Statistisches Bundesamt.

Gabler, S., Häder, S., Lehnhoff, I., & Mardian, E. (2012). Weighting for the unequal inclusion probabilities and nonresponse in dual frame telephone surveys. In S. Häder, M. Häder, & M. Kühne (Eds.), *Telephone surveys in Europe: Research and practice* (pp. 147–167). Heidelberg: Springer.

Groves, R. M., & Peytcheva, E. (2008). The impact of nonresponse rates on nonresponse bias. *Public Opinion Quarterly, 72*, 167–179.

Headey, B., & Wearing, A. (1992). *Understanding happiness: A theory of subjective well-being*. Melbourne: Longman Cheshire.

Kaczmirek, L., & Raabe, J. (2010). Datenquellen und Standarduntersuchungen zur Online -Kommunikation [Data sources and standardized stuies on online communication]. In W. Schwieger, & K. Beck (Eds.), *Handbuch Online-Kommunikation* [Handbook of Online Communication] (pp. 518–540). Wiesbaden: VS Verlag.

Kalton, G., Kasprzyk, D., & McMillen, D. B. (1989). Nonsampling errors in panel surveys. In D. Kasprzyk, G. Duncan, G. Kalton, & M. P. Singh (Eds.), *Panel surveys* (pp. 249–270). New York: Wiley.

Kohut, A., Keeter, S., Doherty, C., Dimock, M., & Christian, L. (2012). *Assessing the representativeness of public opinion surveys*. Washington, DC: The Pew Research Center for the People and the Press.

Kreuter, F. (2013). Improving surveys with paradata: Introduction. In F. Kreuter (Ed.), *Improving surveys with paradata: Analytic uses of process information* (pp. 1–12). Hoboken, NJ: John Wiley & Sons, Inc.

Kreuter, F., Presser, S., & Tourangeau, R. (2008). Social desirability bias in CATI, IVR, and Web surveys: The effects of mode and question sensitivity. *Public Opinion Quarterly, 72*, 847–865.

Lynn, P. (2008). The problem of nonresponse. In E. D. de Leeuw, J. J. Hox, & D. A. Dillman (Eds.), *International handbook of survey methodology* (pp. 35–55). London: Taylor & Francis.

Lynn, P. (2009). Methods for longitudinal surveys. In P. Lynn (Ed.), *Methodology of longitudinal surveys* (pp. 1–19). Chichester: John Wiley & Sons, Ltd.

Scherpenzeel, A., & Bethlehem, J. D. (2010). How representative are online panels? Problems of coverage and selection and possible solutions. In M. Das, P. Ester, & L. Kaczmirek (Eds.), *Social and behavioral research and the Internet* (pp. 105–130). New York: Routledge.

Smith, T. (2003). An experimental comparison of Knowledge Networks and the GSS. *International Journal of Public Opinion Research, 15*, 167–179.

Veenhoven, R. (1996). Developments in satisfaction research. *Social Indicators Research, 37*, 1–46.

Vehovar, V., Lozar-Manfreda, K., Zaletel, M., & Batagelj, Z. . (2002) . Nonresponse in Web surveys. In R. M. Groves, D. A. Dillman, J. L. Eltinge, & R. J. A. Little (Eds.), *Survey nonresponse* (pp. 229–242). New York: John Wiley & Sons, Inc.

Yeager, D., Krosnick, J. A., Chat, L., Javitz, H. S., Levendusky, M. S., Simpser, A., & Wang, R. (2011). Comparing the accuracy of RDD telephone surveys and internet surveys conducted with probability and nonprobability samples. *Public Opinion Quarterly, 75*, 709–747.

Appendix 3.A

Table 3.A.1 Variables for comparison from the GOPP and mode they have been surveyed in.

Variable	Mode of data collection	Measured online in the first questionnaire
Gender	CATI recruitment	Yes
Year of birth/age	CATI recruitment	Yes
Educational attainment	CATI recruitment	No
Marital status	CATI recruitment	No
Employment status	CATI recruitment	No
Immigration background	CATI recruitment	No
Trust	CATI recruitment	Yes (different operationalization)
Political interest	Online first questionnaire	–
Health status	Online first questionnaire	–
Satisfaction with government	Online first questionnaire	–
Satisfaction with health services in Germany		
German economy	Online first questionnaire	–
German economy in a year	Online first questionnaire	–
Financial situation in household	Online first questionnaire	–
Financial situation in household in a year	Online first questionnaire	–

Table 3.A.2   Logistic regression on very good/good health by reference surveys.

	Odds ratio	Standard error
Male	1.031	(0.072)
Age group		
18–24	Ref.	
25–34	0.681**	(0.094)
35–49	0.468***	(0.064)
50–64	0.267***	(0.037)
65+	0.358***	(0.058)
Education		
low	Ref.	
middle	1.331**	(0.124)
high	2.138***	(0.200)
Married	1.175*	(0.093)
Working	1.384***	(0.118)
Immigration background	1.097	(0.120)
ALLBUS	1.681***	(0.172)
ESS	1.418**	(0.143)
Constant	1.320	(0.213)
N	4600	
R^2	0.051	

$^+p < 0.10$,
$^*p < 0.05$,
$^{**}p < 0.01$,
$^{***}p < 0.001$.
All surveys combined, design weighted, GOPP is reference category.

4

Online panels and validity

Representativeness and attrition in the Finnish eOpinion panel

Kimmo Grönlund and Kim Strandberg
Åbo Akademi University, Turku, Finland

4.1 Introduction

The latter part of the 20th Century was characterized by a rapid scientific, technological and economic modernization. The first decade of the 21st Century shows no signs of this process slowing down. People are more dependent on complex technology and a global market than ever before. At the same time, the average level of education has increased and, especially in post-industrialized societies, most people handle a variety of information and are able to make autonomous decisions in their work. Inglehart and Welzel (2005, pp. 2–3, 134) depict the development as a sequence of human development; socio-economic development has brought about changes in individuals' values and attitudes and this has, indirectly, led to new forms of political activity and demands for changes in democratic institutions (e.g., Cain Dalton & Scarrow, 2003; Swanson & Mancini, 1996). Increased economic welfare and higher levels of education have made people more independent and also more focused on individual self-expression instead of basic human needs.

Societal modernization and changes in values have been echoed in the political realm. There is now an increasing pressure for participatory institutional reforms in order to accommodate these new civic values and an increased need for self-expression. The demands have been most evident among the younger generations (Dalton, 2007; Inglehart & Welzel, 2005). Voter turnout and other forms of electoral participation as well as party membership have

Online Panel Research: A Data Quality Perspective, First Edition.
Edited by Mario Callegaro, Reg Baker, Jelke Bethlehem, Anja S. Göritz, Jon A. Krosnick and Paul J. Lavrakas.
© 2014 John Wiley & Sons, Ltd. Published 2014 by John Wiley & Sons, Ltd.
Companion website: www.wiley.com/go/online_panel

been in decline (Blais, 2000; Wattenberg, 2002). Especially younger citizens are increasingly involved in civic action networks whose goal is to challenge elites, not to support them (Dalton, 2007; Inglehart & Welzel, 2005, p. 118; Kippen & Jenkins, 2004; Norris, 2001, p. 217). There has been a steep rise in civic movements, non-governmental organizations and public interest groups (Cain et al., 2003, pp. 2–3).

To simplify, societal modernization seems to have transformed politics from focusing on collective interests though traditional political agents toward meeting the demands of individuals engaged in new types of political movements and activities (see Swanson & Mancini, 1996, for a summary). Kies (2009, p. 70), for instance, demonstrates that young Europeans are clearly engaging to a larger extent in new types of political activities instead of conventional participation.

The aforementioned development provides a methodological challenge to political science scholars seeking to study citizen engagement; on the one hand, citizens' activity is constantly evolving and diversifying and, on the other hand, the channels through which this is taking place are changing even more rapidly. Already in the Web 1.0 phase, studying political behavior online was a case of chasing a "moving target" (McMillan, 2000) and the rapid developments in the Web 2.0 phase have only served to underline this fact (Strandberg, 2012).

Simultaneously with the changing patterns of political participation, nonresponse to household surveys has grown (Groves, 2006). This may be an effect of a decreased feeling of civic duty (Dalton, 2007), which is mainly reflected in a decline in electoral turnout especially in mature democracies (López Pintor & Gratschew, 2002, 2004).[1] Hence, there might be a need to apply new methods when studying citizen participation in the 21st Century. Van Selm and Jankowski (2006) remark that online surveys might be particularly useful for reaching young age groups who are accustomed to computers (cf. Sills & Song, 2002). Pertaining to the developments concerning political participation, the shifts have been most apparent among young citizens (Dalton, 2007; Inglehart & Welzel, 2005; Kies, 2009) which is also true regarding the main user base of online channels for political purposes (e.g., Smith, 2011; Strandberg, 2012).

The aim of this chapter is to examine how suitable an online panel is for tracing people's political opinions and behavior. For this purpose, we present details pertaining to validity from a probability-based online panel survey conducted in conjunction with the 2011 Finnish parliamentary election. The findings in the online panel are compared with two simultaneously gathered probability-based surveys: a face-to-face and a telephone survey.

4.2   Online panels: Overview of methodological considerations

The main benefits of online panels pertain to their accessibility for both researchers and respondents. For researchers, online panels can be comparably inexpensive, easy to conduct, fast to disseminate, and save time on data entry since survey replies are entered electronically (Kellner, 2004; O'Muircheartaigh, 2008; Vehovar, Manfreda, & Koren, 2008), and they can be used for targeting respondents – i.e., certain age groups or specific subgroups of citizens – who are hard to reach via other survey methods (Van Selm & Jankowski, 2006). Moreover, they contain fewer of the problems found in interview-based surveys such as the feeling of social desirability on the part of the participants prohibiting honest answers

---

[1] It has also been pointed out that the evidence of a declining turnout is mixed and might partly be caused by the widely used measure of "voting age population" instead of registered voters (Aarts & Wessels, 2005, pp. 65–66).

to certain delicate questions (Baker et al., 2010, p. 740; Kellner, 2004; O'Muircheartaigh, 2008). However, online panels are subject to various methodological and practical challenges, especially because they have mostly relied on nonprobability based recruitment (Baker et al., 2010; Van Selm & Jankowski, 2006; Vehovar et al., 2008, p. 275).

Since nonprobability recruitment methods are more vulnerable to the element of volunteering than probability samples (Sparrow & Curtice, 2004; Van Selm & Jankowski, 2006), the representativeness of an online panel sample might be weak due to biases toward certain types of citizens, such as heavy Internet users, young people and people with high education (Baker et al., 2010; O'Muircheartaigh, 2008; Sparrow & Curtice, 2004; Vehovar et al., 2008). Baker and colleagues (2010) also remark that nonresponse in the recruitment stage is "considerable" in online panels; according to some estimates, volunteering rates among those who have been invited to participate in online surveys are about 6–15% lower than in offline surveys. Others also note that online panels using probability sampling often have even lower rates of volunteering than those using nonprobability-based sampling (see Göritz, 2004; Vehovar et al., 2008).

Nonresponse in the course of the panel – i.e., panel attrition – should not be overlooked either since it is relatively high in online panels (Baker et al., 2010), potentially due to perceived anonymity and ease of quitting (Vehovar et al., 2008). Baker and colleagues (2010) stress the need to carefully study panel drop-outs in order to learn more about how to address online panel attrition in future panels (cf. Sparrow & Curtice, 2004). Besides the issues of nonresponse at various stages of the panel, the base population from which online samples are to be drawn is also often hard to estimate (cf. Sparrow & Curtice, 2004; Van Selm & Jankowski, 2006). This also poses challenges for probability-based online panels since it is hard for researchers to build the sampling frame of the entire online population due to a lack of reliable registers of the online population (Baker et al., 2010; O'Muircheartaigh, 2008; Sparrow & Curtice, 2004; Van Selm & Jankowski, 2006; Vehovar et al., 2008).

Vehovar and colleagues furthermore remark (2008, p. 276) that online surveys may suffer from limited coverage due to the Internet penetration among the population being limited (cf. Baker et al., 2010). However, in many advanced Western democracies the Internet penetration rate is approaching 80-85% (Internet World Stats, 2011). In fact, Vehovar and colleagues (2008) suggest that a benchmark of 80% Internet penetration – the same benchmark used for telephone surveys upon their arrival – is sufficient for online surveys. However, even though coverage is no longer a major issue in certain national contexts, there are still between-country variations. In Western Europe, for instance, the average Internet penetration rate ranges from 42% in Greece to 93% in the Netherlands (European Commission, 2012). The lack of an interviewer in online panels may also cause an increase in so-called satisficing, i.e., respondents not sufficiently contemplating their answers, since responding to online surveys is quick, easy and not supervised by an interviewer (Baker et al., 2010). The lack of an interviewer also places high importance on the design of the survey questionnaire since it replaces many functions usually handled by the interviewer (Van Selm & Jankowski, 2006; Vehovar et al., 2008). Finally, even though online surveys and panels reduce some practical issues, they might bring forth new mode-specific problems, such as high panel maintenance and recruitment costs, instead.

4.3 Design and research questions

Online panels that recruit using traditional probability-based methods have been slow to appear and are fewer in numbers than volunteer panels (Baker et al., 2010). To a large degree, this has been a consequence of biases in Internet penetration among the general public. Thus, this chapter raises the general question: Has the second decade of the 21st Century brought the Internet to a large enough share of the population in order to use the online sphere as a reliable source for surveying? More specifically, we are interested in knowing how valid the inferences drawn from a probability-based online panel are in a contemporary democratic system with high ICT-usage. In order to answer this overarching question, we compare an online panel with two other simultaneously conducted surveys: a face-to-face survey (the Finnish National Election Study (FNES 2011), see Borg & Grönlund, 2013), and a telephone survey (Grönlund, 2012a, 2012b). All three surveys were based on probability sampling, and since the research team behind the online panel was also involved in the design of the face-to-face and the telephone surveys, the questionnaires include identical questions, making comparisons possible. The comparisons are done using both raw sample data and data which have been post-hoc weighted according to population-based gender and age distributions. The two-stage process deepens our understanding of the possible biases in the three modes of surveying.

We are particularly interested in (1) panel representativeness and (2) attrition. In terms of representativeness, we cover (a) socio-demographic characteristics; (b) attitudinal variables; and (c) political behavior (cf. Groves, 2006, p. 655). Despite the fact that we were involved in planning all three surveys, each of them was designed as an independent survey whose primary aim was to cast light on different aspects of the Finnish national election of 2011. Therefore, the present chapter does not attempt to be an *experimental* study of comparing modes (see Baker et al., 2010).

When planning the online panel, we were curious about the possibility of using a simple random sample (SRS) as a base for an online panel. SRS is the most unrestricted form of sampling and even though it has its limitations (O'Muircheartaigh, 2008, pp. 296–297), we wanted to test its appropriateness in an online panel. The recent AAPOR report on online panels recommends that "researchers should avoid nonprobability online panels when one of the research objectives is to accurately estimate population values" (Baker et al., 2010, p. 48). Naturally, we were aware of the possible bias problems caused by SRS in recruiting to an online panel. The advantage of being able to generate a high quality sample through SRS from the population registry and the fear that no other (probability) method would have worked out better given the online set-up, convinced us. We did, however, have a need for stratification according to language. Therefore, two separate simple random samples according to the mother tongue of the citizens were generated. The process is described in detail below.

In the discussion thus far in this chapter, we can make several central observations: first, political participation is transforming, especially among younger generations (cf. Kies, 2009, p. 70). Second, the Internet has provided a new, and rather distinct, arena in which these participatory activities can be carried out. Third, online panels can provide solutions to the increasing survey-reluctance among citizens in general and younger generations in particular. The main research question can be formulated as follows:

How valid are the population inferences drawn from a probability-based online panel?

This question can be divided into two more specific sub-questions:

RQ$_1$ How representative are the participants in an online panel?

RQ$_2$ Are there any systematic patterns to be found in panel attrition?

4.4   Data and methods

The chapter covers data that have been gathered through three distinct channels. The main focus is on our online panel survey. This survey is compared with a face-to-face (FNES 2011, see Borg & Grönlund, 2013) and a telephone survey. All three surveys were carried out in conjunction with the Finnish parliamentary election of 2011.

Mainly researchers conduct online surveys in order to (1) reach a population with Internet experience; (2) find potential respondents with special (Internet-related) interests; (3) get in touch with persons with "deviant" or covert behaviors who might be difficult to survey otherwise (e.g., sexual minorities), or with the purpose of (4) attracting particular age groups, especially the young (van Selm & Jankowski, 2006). We had several of the aforementioned goals. Based on the theoretical literature and earlier empirical evidence on societal modernization and new forms of participation, we were primarily interested in reaching a population possibly not reachable via the standard method of face-to-face interviewing. We wanted to see whether the "issue-oriented" generations were better reachable in the online mode.

4.4.1   Sampling

As discussed earlier, sampling biases and limited coverage are common problems in online panels (Baker et al., 2010). Both of these mostly arise from limited Internet penetration among the general population, which in turn produces skewed samples (Baker et al., 2010; O'Muircheartaigh, 2008; Sparrow & Curtice, 2004; Vehovar et al., 2008). The focus of this chapter is on surveys in Finland, a country with an overall online penetration rate as high as 89% (Internet World Stats, 2011), even though it should be noted that access from home in Finland is lower at 76% (European Commission, 2012), the national context appears *a priori* suitable for comparing online and offline surveys. In fact, Finland is ranked among the top ten countries in Internet penetration in the world, 14th in terms of fixed broadband penetration, and second in wireless broadband penetration (Internet World Stats, 2011; OECD, 2010).

In creating our raw sample, a stratified random sample of 13000 Finnish citizens aged 18–75 was invited to participate. Stratification was enforced according to mother tongue. Finland is a bilingual country and 5.7% of the population speaks Swedish as their maternal language. For reasons beyond the scope of this chapter, we needed a large enough sample of Swedish speakers in the panel. Therefore, stratification was applied. In fact, two probability samples were generated randomly by the Finnish population registry which provided us with the citizens' postal addresses, names, gender, year of birth, and area of residence. The Finnish-speaking sample consisted of 10000 people, and the Swedish-speaking sample of 3000 people.

Naturally, all citizens within the samples did not necessarily have Internet access, but given the rather high Internet penetration in Finland, that share was expected to be small. Therefore, we mailed out invitation cards to everyone included in the samples. We opted for traditional mail since the population registry does not contain email addresses of the Finnish citizenry. Moreover, emailed invitations have been found to suffer from higher shares of non-delivery than mailed invitations (see Vehovar et al., 2002, 2008). It would have been possible to purchase telephone numbers of the people in the sample, but experimental evidence shows that the mail-out mode for recruitment works better than telephone recruitment (Rao, Kaminska, & McCutcheon, 2010). Personalized invitations have also been found to increase response rates in offline surveys (see Van Selm & Jankowski, 2006). In summary, thus, our sampling and solicitation method was probability-based with personal invitations (Vehovar et al., 2008).

The invitation card explained the basic set-up of the panel and the fact that it was a scientifically designed survey, organized by Åbo Akademi University. Each participant was guaranteed anonymity in all research reports. Moreover, no compensation was offered at this stage. As the recruitment process proved to be slow, one reminder card to participate was sent out. At this stage, we decided it was necessary to apply an incentives program (cf. Göritz, 2004; Van Selm & Jankowski, 2006; Vehovar et al., 2008) and potential participants were now informed that they would be part of a lottery of a year-long subscription to a Finnish newspaper of their choice. The same information was sent out to those who had already volunteered. Everyone who filled in all our surveys took part in the lottery. Eventually, the final number of panel members was a mere 692 people (5.3% of the raw sample). Even though we were from the very beginning fully aware of the problems related to probability sampling in online surveys (cf. van Selm & Jankowski, 2006; Baker et al., 2010), we wanted to test how well random sampling and mail invitation work together with online surveying. Concerning this, 5.3% of the raw sample actually registering as panelists is actually a rather satisfactory rate (see Göritz, 2004). Naturally, even more aggressive recruitment methods, such as prepaid incentives rather than a lottery, could have been applied to further improve the response rate (p. 422). Citizens who accepted the invitation did so by logging on to the project's website and responding to the recruitment survey in which additional background information was collected.

Even though we are able to make comparisons between the three survey modes, there are four aspects in particular to consider. First, the online survey was conducted as a panel study whereas the face-to-face and the telephone surveys were cross-sectional. Second, the sampling method varied between the surveys. The online panel was a stratified random sample (or two simple random samples). Also the telephone sample was initially a stratified random sample – drawn from both landline and mobile phones (i.e. dual-frame) – but the contractor applied geographical, age, and gender quotas in order to guarantee socio-demographic representativeness. The data collection of the phone survey started immediately after the election on April 17th 2011 and continued throughout May.

The face-to-face FNES was based on clustered random sampling, where a starting address was first randomly selected. From this starting address, five interviews were conducted. This method also applied age- and gender-based quotas but no stratification according to mother tongue. The survey also included a drop-off questionnaire which was answered by 71% (N = 806) of those respondents who had been interviewed. The FNES was conducted between April 18th 2011 and May 28th. Third, the surveys were conducted by different organizations. The online panel was fully maintained by the research team at Åbo Akademi University. The other two surveys were contracted to commercial companies that specialized in population-based surveys. The dual-frame phone survey was ordered from TNS Gallup and the face-to-face from Taloustutkimus. Both firms are leading and certificated survey institutes in Finland. Fourth, the sample size varies between the surveys. The effective sample sizes in the three surveys were as follows: Online Finnish 341–464, Swedish 163–198; Face-to-face Finnish 1178, Swedish 114, Phone Finnish 1000, Swedish 1000. Below, we present in detail the processes of sampling and data collection in the online panel.

4.4.2   E-Panel data collection

The main data for the present analysis derive from an online panel survey – called eOpinion 2011 (Grönlund & Strandberg, 2013) – conducted during a period of two months in

Table 4.1   Overview of the eOpinion 2011 online panel survey.

Survey	Published	No. Questions	No. items	Respondents N
Recruitment survey	February	4	4	692
Survey 1	February 23rd	9	16	661
Survey 2	March 9th	8	44	650
Survey 3	March 23rd	8	39	622
Survey 4	March 30th	8	46	630
Survey 5	April 6th	6	35	494
Survey 6	April 13th	10	32	513
Survey 7	April 20th	11	47	507

conjunction with the Finnish parliamentary election held on April 17th 2011. The online panel consisted of a recruitment survey followed by seven actual surveys. Each survey was intentionally restricted in terms of the number of questions and items, bearing in mind that the length of online surveys has been considered inversely influential to the response rate (Van Selm & Jankowski, 2006, p.441; Sheehan & McMillan, 1999). An overview of the e-panel questionnaires is presented in Table 4.1.

As evident in Table 4.1, there was an attrition rate of roughly 25% after seven surveys in the three months of data collection. We will further investigate the characteristics of the participants dropping off during the course of the panel in the findings section (Table 4.2).

Except for the mail-out invitation cards, all data were processed electronically. At the registration stage, each respondent provided us with an email address. We also asked for their age and gender, even though they were already known by us. That way, we could find fraudulents (Baker et al., 2010), i.e., one person receiving the invitation, another taking on the task of responding – in our case, often the respondent's spouse. Less than five such "gender changes" were recorded. Since the panel did not receive a great number of participants in the recruitment stage, we let these occur and changed the background data for these persons accordingly. In any case, education and occupation were only provided by the respondents themselves, not by the Population Registry.

Participants received email notifications asking them to respond to each questionnaire. We were careful not to disclose the email addresses of all the recipients in order to maintain anonymity (cf. Van Selm & Jankowski, 2006). The surveys themselves were answered online. We used a dedicated online survey tool, DimensionNet version 3.5, by SPSS which has all of the features deemed important in the literature on online panels (see Van Selm & Jankowski, 2006). We found that especially the amount of manual labor needed by the research team, as well as the occurrence of missing values was minimized through the use of this dedicated software. Each survey was accessible to the participants for one week, but they could only answer it once. Altogether, seven surveys were conducted after the registration. The surveys were designed to be answered in 10–15 minutes.

4.5   Findings

We shed light on the following aspects where the participants of the online panel are compared to those of FNES and the phone survey: socio-demographics, political interest, satisfaction

Table 4.2   Socio-demographics of the three surveys and the electorate of Finland.

| | Raw data[1] | | | | | Weighted data[2] | | | | |
| | | | E-panel Participants | | | | E-panel Participants | | | |
	Popu-lation	Raw sample	T1	T7	FNES	Phone-survey	T1	T7	FNES	Phone-survey
Gender										
Man	49	51	58	57	50	48	49	48	48	49
Woman	51	49	42	43	50	52	51	52	52	51
Age										
18–24	11	9	5	5	8	8	11	12	11	11
25–34	16	16	10	9	16	16	14	13	15	15
35–44	15	17	14	14	14	14	15	15	15	15
45–54	17	20	23	24	15	18	17	17	17	17
55–64	18	21	27	26	18	19	19	18	18	18
over 64 years	23	18	21	23	29	24	24	25	23	23
Education										
Only compulsory education	28	-	7	7	22	19	7	8	29	19
Vocational or upper secondary	42	-	27	25	29	36	28	27	22	36
Lowest tertiary (institute)	11	-	26	24	24	16	22	21	24	15
Polytechnic or bachelor	10	-	9	9	9	17	10	10	9	17
Master's degree or higher	9	-	31	34	16	13	30	32	16	12
N	4092061	13000	649	473	1286	2000	649	473	1292	2000

Notes:
[1] Weighted according to mother tongue.
[2] Weighted according to mother tongue, age and gender.

with democracy, social and institutional trust, and voting behavior. We also examine two aspects for which the online panel participants could only be compared to the FNES:[2] polit-ical participation and activity; use of the Internet and media. All comparisons are done both using raw data and weighted data. It should be noted, that because the samples in the online panel and the telephone survey were stratified in order to over-sample Swedish-speaking Finns, also the comparisons labeled "raw data" are weighted according to the respondents' mother tongue. All comparisons which are labeled "weighted data" are post-hoc weighted not only according to language, but also according to gender and age. The corresponding shares

---

[2] This is due to the fact that the corresponding question was not asked in the phone survey.

per cell according to mother tongue, gender and age are calculated with data from Statistics Finland.

4.5.1   Socio-demographics

We begin by examining, in Table 4.2, the socio-demographic representativeness of the online panels' participants. This is carried out by putting the raw- and weighted data of the online panel side by side to the general population statistics, the raw- and weighted data concerning the FNES as well as the phone survey. We also address panel attrition in Table 4.2 through comparing the online panel at T1 to those answering the final survey at T7 (see Table 4.1 for a description of the surveys).

Judging from Table 4.2, though the raw sample for the online panel is generally representative of the population, the representativeness of the online panel's participants, in the raw data, is not quite equivalent to that of the FNES or to that of the phone survey. Most likely, this could be due to the element of volunteering (cf. Sparrow & Curtice, 2004; Van Selm & Jankowski, 2006). Specifically, the distribution of gender, age and degree of education was skewed among the participants of the online panel at both the T1 and T7 surveys. Looking at age, the youngest groups were specifically rather surprisingly under-represented in the online panel (cf. Baker et al., 2010; O'Muircheartaigh, 2008; Sparrow & Curtice, 2004). Perhaps this reflects the fact that electoral participation is lower among the young and justifies Milner's (2010) analysis of the Internet generation as "political dropouts." On the other hand, it is possible that the combination of "snail mail" recruitment and online registration was more prone to attract older generations. The age group of 55–64-year-olds was most overrepresented among the respondents. The most striking bias is found in the comparisons of raw data to the general population, however, was regarding education. Even though it is not surprising (see Baker et al., 2010; O'Muircheartaigh, 2008; Sparrow & Curtice, 2004), people with a university degree are highly over-represented at the e-panel. Almost a third of the e-panel participants had at least a Master's degree, compared to 9% among the general public. Still, it is noteworthy that the FNES and the phone survey also struggled, albeit not to the same extent as the online panel, with representativeness concerning education. Turning to the weighted data (the right-hand part of Table 4.2), then, the representativeness of the weighted variables (gender and age) is naturally very good in all three survey types. The skewness regarding education, even with post-hoc weights applied, is more-or-less unaltered for all surveys.[3]

Turning to panel attrition, we note that the slate of participants finishing the online panel was practically identical to those starting it. Hence, there are no systematic patterns regarding socio-demographic characteristics to be found among participants dropping out of the panel (cf. Baker et al., 2010; Sparrow & Curtice, 2004). In fact, a further examination of the other aspects covered in our study (not shown in detail here) reveals that this was also the fact concerning all comparisons. Hence, e-panelists dropping out did not differ systematically concerning any aspect – i.e., socio-demographics; political interest, satisfaction with democracy, trust; political participation and activity; media and Internet use[4] – from those staying on. Consequently, panel attrition will not be further addressed in the remaining tables and corresponding analyses.

---

[3] Ideally, weights according to education should also be applied. However, the total N of the e-panel was not sufficient for weighting according to all of the variables gender, language, age and education as the n-count in certain cells were so low it made the weighting unreliable.

[4] This further analysis could not be conducted for voting behavior since that question had been asked only in the T7 survey of the online panel.

4.5.2   Attitudes and behavior

Moving beyond socio-demographics, then, we focus on validity in terms of attitudinal and behavioral aspects. We start by looking at interest in politics, satisfaction with democracy, as well as social and institutional trust. Table 4.3 shows a comparison of raw and weighted data respectively for the participants in the eOpinion panel, the FNES and the phone survey.

The general picture emerging from Table 4.3 is that the online panel participants are quite similar to those of the FNES and the phone survey. Some minor differences do, however, exist. The online panelists – concerning both raw and weighted data – are somewhat more interested in politics than the respondents of both the FNES and the phone survey. Furthermore, they are

Table 4.3   Comparisons of political interest, satisfaction with democracy, social and political trust.

	Raw data[1]			Weighted data[2]		
	E-Panel	FNES	Phone-survey	E-Panel	FNES	Phone-survey
Interest in politics						
Very interested	29.0	23.2	23.1	25.7	22.9	22.8
Quite interested	58.8	51.4	50.6	61.0	51.5	51.0
Not very interested	10.9	20.3	20.4	11.5	20.5	20.4
Not at all interested	1.3	5.1	5.5	1.8	5.2	5.6
Satisfaction with democracy						
Very satisfied	6.1	5.6	-	5.8	5.6	-
Quite satisfied	56.5	69.4	-	58.0	69.1	-
Quite unsatisfied	29.4	20.3	-	28.6	20.7	-
Very unsatisfied	7.0	3.3	-	6.3	3.1	-
Generalized social trust (average on scale 0–10)	6.6	7.2	-	6.6	7.2	-
Trust in institutions (average on scale 0–10)*						
The president	7.2	8.3	7.0	7.2	8.4	7.0
The parliament	5.8	7.6	5.7	5.7	7.5	5.8
The government	5.5	7.9	5.8	5.4	7.9	5.8
Politicians	4.3	6.4	4.3	4.3	6.4	4.3
The police	7.6	8.3	8.2	7.5	8.3	8.2
The legal system	7.0	8.1	6.9	7.0	8.1	6.9
Political parties	4.8	7.1	4.7	4.8	7.1	4.7

Notes:
*The question in the phone-survey did not ask respondents to assess their trust on a scale, rather it provided fixed answers. For instance: "How much do you trust the following institutions?" The president could be replied to with "Very much," "Quite much," "Not so much," and "Not at all." In order to compare the answers to these questions to those in the FNES and online panel (using the 0–10 scale), we converted these into a value on a scale between 0–10.
Notes:
[1]Weighted according to mother tongue.
[2]Weighted according to mother tongue, age and gender.

also slightly less trusting, concerning both generalized social trust and trust in institutions, than the respondents of the FNES. This is reflected when satisfaction with democracy is compared between the two surveys. Online panelists have a higher share of unsatisfied respondents than the FNES. However, the e-panel respondents are equivalent concerning trust in institutions to the phone survey respondents. So, our overall impression is that the online panel is not less representative than the two offline surveys in terms of interest, democratic satisfaction and trust. Additionally, applying post-hoc demographic weights does not alter the general picture when it comes to differences between the three survey modes. Institutional trust deviates upwards in the FNES compared to both the online panel and the phone survey. A check with a fourth comparable survey, the fifth round of the high quality European Social Survey (ESS) indicates that the online panel in fact provides a *better* estimate of the Finnish population than the FNES. The fifth round of the ESS, conducted as a probability-based face-to-face survey by Statistics Finland (N = 1878) in September–December 2010 reveals that institutional trust means come closer to those of the online panel and the phone survey than the FNES. The ESS uses the same 0–10 scale as we did and the means for trust in Finland were: the parliament 5.4, politicians 4.4, the police 8.0, the legal system 6.9, and political parties 4.5.

Moving on with the analysis of panel representativeness, we turn to actual behavior. Two elements are analyzed here: voting behavior and political participation and activity. Table 4.4 compares both turnout and party choice among the online panelists as well as respondents in the FNES and the phone survey. Again, this is done for both the raw data and weighted data. Also, we assess the validity of the surveys' raw and weighted data through direct comparisons with the actual election results (cf. Kellner, 2004). This is indicated in the "diff." columns in Table 4.4.

Turnout shares in Table 4.4 show that according to self-reported data, practically all of the online panelists voted in the election. This applies for both the raw- and weighted samples of data. Whether this is a matter of sample bias or merely a case of over-reporting is impossible to judge. Most likely this is a combination of both, especially bearing in mind that educated people have been found to have a stronger need to be "good" citizens and exaggerate their voting activity (Bernstein, Chadha & Montjoy, 2001). Also, it should be noted that the respondents in the FNES and the phone survey show higher turnout rates, albeit not as high as for the e-panel, than the actual turnout rate of 70.5% (Statistics Finland). There is also a possibility that some of the persons who did not want to reveal their vote choice were in fact over-reporters and did not vote, finding it more difficult to name the party of the candidate they had voted for when asked.

Looking at the distribution of votes between parties, in the raw data, we find that a higher share of online panelists voted for the conservative National Coalition Party and the Green League than the respondents in the other two surveys. The average difference from the actual election result was also the largest, 2.8 percentage points, for the online panel. In the weighted data we find that the online panelists differ most from the election results – and the other two surveys – concerning the very high support for the Green League and, to some extent, the low support for the Centre Party. The pattern for weighted online panel data becoming more deviant from the electoral result is rather easy to understand. Especially the youngest age group (18–24), which is clearly under-represented in the e-panel, vote to a large extent Green (16%) whereas they do not support the Centre Party (10%) to the same extent (Grönlund & Westinen, 2012, p. 159). Therefore, when these cells are weighted upwards, the deviance increases in the vote choice column of the e-panel.

Table 4.4 Comparisons of voting behavior.

	Election result	Raw data[1]						Weighted data[2]					
		E-panel	diff.*	FNES	diff.	Phone-survey	diff.	E-Panel	diff.	FNES	diff.	Phone-survey	diff.
Voted in the election (turnout)	70.5	97.7	27.3	87.0	16.6	88.4	18.0	98.0	27.6	86.6	16.2	88.1	17.7
Does not want to reveal party choice		5.2		9.6		17.9		5.8		7.3		16.6	
Party choice (among those who revealed it)													
Centre Party	15.8	11.7	4.1	13.5	2.3	16.1	0.3	9.7	6.1	13.3	2.5	15.9	0.1
National Coalition Party	20.4	25.2	4.8	17.0	3.4	22.0	1.6	24.2	3.8	16.3	4.1	22.2	1.8
Social Democratic Party	19.1	14.8	4.3	21.7	2.6	18.7	0.4	15.7	3.4	21.3	2.2	18.1	1.0
Left-wing Alliance	8.1	9.0	0.9	9.3	1.2	9.3	1.2	10.0	1.9	9.6	1.5	9.0	0.9
Green League	7.3	13.6	6.3	9.5	2.2	10.0	2.7	17.4	10.1	10.0	2.7	10.5	3.2
Swedish Peoples' Party	4.3	5.0	0.7	4.9	0.6	4.9	0.6	5.0	0.7	5.2	0.9	5.2	0.9
Christian Democrats	4.0	2.2	1.8	4.0	0.0	4.4	0.4	1.9	2.1	4.0	0.0	4.4	0.4
other	2.0	1.2	0.8	1.6	0.4	1.3	0.7	1.1	0.9	1.6	0.4	1.4	0.6
Mean difference:			2.8		1.5		1.5		3.7		1.7		1.6

Notes:
* Difference to election result (absolute values).
[1] Weighted according to mother tongue.
[2] Weighted according to mother tongue, age and gender.

However, the fact that the share of people not willing to disclose their party choice is clearly lowest in the online panel, can be considered as a positive finding from the point of view of the online method. Only roughly 5% among the panelists have chosen not to report their party choice in the online panel. The highest reluctance is found in the phone survey. Perhaps having the feeling of not being directly observed is the explanation to the higher share of cooperation in the online panel. Generally, concerning the weighted data in Table 4.4, we see that the e-panel here struggles marginally more with validity in light of average difference to the actual election results than in the raw data. Nonetheless, it is quite equivalent to the two other surveys in this regard.

Turning to political action, we present the frequencies of various forms of political activity among the online panel's participants as well as among the FNES participants in the raw and weighted data (Table 4.5).

As is evident in Table 4.5, participants in the raw data of the online panel are some-what more active and potentially active than the FNES respondents. The general impression concerning this is, with some exceptions, that the online panelists are more prone towards engaging (or potentially engaging) in activities which can be considered rather "untraditional" or new (Dalton, 2007; Inglehart & Welzel, 2005, p. 118; Norris, 2003). The most striking difference is perhaps the potential for taking part in demonstrations with a history of violence. Bearing in mind the overrepresentation of older citizens among the participants, this is some-what surprising since such activities are often found among younger citizens (e.g., Inglehart & Welzel, 2005, p. 118; Wass, 2008). Still, as Table 4.3 showed earlier, the online panelists were slightly less trusting of political institutions than the FNES respondents. Tentatively, this could be what is reflected in the proneness on the part of the online panelists to engage in untraditional activities (this, of course, would need to be confirmed with further analyses not necessary for the purposes of this chapter). In the analysis using post-hoc weights, also in Table 4.5, there are few differences compared to the raw data and the same patterns in differences between the e-panel participants and those of the FNES remain. Thus, even when applying post-hoc weights, the e-panelists appear more prone to unconventional forms of participation than those of the FNES.

4.5.3   Use of the Internet and media

The final part of our findings section compares the use of the Internet and the use of other media for political purposes between the raw and weighted samples of online panelists and the respondents of the FNES (Table 4.6). Clearly, in light of other findings concerning online panelists' representativeness, it would be reasonable to expect a skewed representativeness favoring the use of Internet-based channels among the online panel's participants, at least in the raw sample not applying demographic weights (e.g., Baker et al., 2010).

Looking at the figures presented in Table 4.6, we see that the online panelists are indeed more frequent general Internet users than their counterparts in the FNES. This observation can be made for both raw and weighted data. Quite oddly, though, this more frequent Internet use is not reflected as a more frequent use of online channels for following the election. In fact, concerning both offline and online channels, the participants of the e-panel mostly use all types of media channels less frequently for following the elections than the FNES respondents. Only concerning the use of online candidate selectors (voting advice applications) are the online panelists more frequent users than their FNES counterparts. Applying weighted data to the same distributions, additionally, does not alter much; the same patterns evident in the raw data remain here concerning the use of media channels in conjunction with elections.

Table 4.5 Comparisons of political action: participants of the eOpinion 2011 panel and the Finnish national election study.

	Raw data[1]				Weighted data[2]			
	E-panel activity	Potential	FNES activity	Potential	E-panel activity	Potential	FNES activity	Potential
Write letters to editor in newspapers	18.9	66.1	15.9	55.6	16.9	67.6	16.1	56.7
Contact politicians	21.2	65.8	21.9	59.0	18.8	67.1	21.1	60.3
Sign a petition	50.0	39.8	47.5	40.7	56.3	34.7	48.9	40.3
Take part in the activity of political party	13.7	41.7	11.0	41.9	14.1	41.9	10.9	43.0
Take part in the activity of other organizations	42.2	39.5	40.7	41.6	41.4	40.0	40.3	42.6
Environmentally conscious consumption	72.4	22.5	62.9	29.4	75.0	20.7	63.9	28.9
Politically conscious consumption	56.2	34.8	38.9	43.5	56.0	34.8	39.5	43.7
Take part in a boycott	28.1	47.4	19.2	48.7	30.2	47.1	19.7	49.2
Take part in a peaceful demonstration	9.5	52.0	8.8	44.6	11.2	54.7	9.3	45.9
Civil disobedience through taking part in illegal action	3.1	30.3	2.7	18.3	3.4	32.2	3.0	19.0
Take part in demonstrations with a history of violence	0.2	11.5	0.3	6.5	0.2	12.6	0.4	7.0
Use violence to achieve one's goals	0.0	4.6	0.0	1.8	0.0	4.8	0.0	2.0

Notes:

N = 1298 for the FNES and N=695 for e-panel

"Activity" means that the respondent has engaged in the activity during the last four years. "Potential" means the respondents report that they have not yet taken part in respective activity but would consider taking part in the future.

[1] Weighted according to mother tongue.

[2] Weighted according to mother tongue, age and gender.

Table 4.6 Comparisons of media usage, participants of the eOpinion 2011 panel and the Finnish national election study.

	Raw data[1]		Weighted data[2]	
	E-panel	FNES	E-panel	FNES
Use of the Internet				
Daily, more than 2 hours	33.0	33.5	35.3	35.3
Daily, less than two hours	52.3	32.9	50.3	33.5
Couple of times per week	12.4	10.4	11.9	10.5
Once a week or less	2.3	23.2	2.5	20.7
Followed the election through media*				
TV-debates or interviews	42.8	53.8	39.9	52.8
News or current affairs on TV	62.5	68.3	59.7	67.7
TV entertainment feat. politicians	17.3	22.3	16.1	22.1
Radio programs	14.9	18.3	14.5	18.1
Newspaper columns or articles	55.3	54.3	53.3	53.7
TV advertisements	7.3	20.0	7.1	20.1
Newspaper advertisements	17.2	32.0	17.1	32.0
Election news online	22.6	27.2	25.3	29.1
Parties or candidates web pages	6.4	10.1	7.4	10.6
Blogs or online diaries	5.0	5.4	5.9	5.8
Online candidate selectors	27.0	20.5	31.0	21.8
Social media	4.5	9.0	5.2	9.9
Online videos posted by party/candidate	0.9	3.1	0.9	3.3

Notes:

*Figures represent the share of respondents who followed the election through each media channel either very much or quite much.

[1] Weighted according to mother tongue.

[2] Weighted according to mother tongue, age and gender.

4.6 Conclusion

Online panels are here to stay. They provide a fast channel for gathering data on attitudes and behavior, are fairly easy to maintain and minimize the need for coding by the researchers. The obvious fear of attrition – it is easy to shut down the computer whenever bored compared to getting rid of an interviewer at your home – was somewhat surprisingly not a big issue in our case. Moreover, we found no systematic patterns in attrition. Therefore, based on the comments received by the online panelists it appears that organizers of online panels can avoid attrition by keeping each survey short and coherent. Clearly formulated claims and questions, together with easily understood answering options and scales are the keys to success.[5]

Attrition is far more likely to occur when respondents find inconsistencies in online surveys than in face-to-face interviewing. There is no one to ask for precision. Even in mail-in

---

[5] This recommendation is backed up by the comments written to us in the online panel. Respondents want a simple layout where scrolling is not needed per page, they comment negatively on lengthy question formulations and surveys. Also, even though we only had six substantial survey waves in the panel, some respondents started writing comments like: "I wish these surveys would stop," indicating survey fatigue.

surveys unclear wordings are less of a problem, you can just skip them. Online surveys, on the other hand, try to minimize missing data by enforcing response. Online survey software allows for mandatory responding before you can continue, making sure that the respondent chooses at least "don't know" instead of leaving a field unanswered. This can cause irritation and lead to attrition very fast. Online surveys cannot be too long either. That is why online surveys tend to be designed as panels instead of cross-sections. The nature of the online world is in many cases "quick and dirty." Hence, adapting the design of the survey according to this reality is probably an important precaution in order to avoid attrition.

In terms of representativeness, our findings are somewhat mixed. It should be noted, though, that our findings are limited by our rather small final sample size and also due to the fact that we did not include non-Internet households – i.e., through providing them with necessary equipment (cf. Luskin, Fishkin & Iyengar, 2004) – in the online panel. Thus, looking at raw data, the online panelists were not entirely representative regarding socio-demographic aspects, voting, participation and activity, and, naturally, use of the Internet. On the other hand, though, regarding interest in politics, generalized social trust and trust in institutions, the representativeness of the panel's participants was in order. In fact, the Finnish National Election Study seems to have overestimated institutional trust, whereas both the online panel and the telephone survey show trust values that are identical to those measured in the European Social Survey in Finland. This indicates that the level of validity in relation to the Finnish electorate was satisfactory. The overarching observations regarding the impact of the post-hoc demographic weights are that they mainly improved demographic representativeness (which was to be expected). The effects of the weights on the other areas of comparison were modest. However, it was unfortunate that our data did not allow for weighting according to education which could have augmented the effect of weighting. It is, however, difficult to make any conclusions on external validity based on these findings.

Our general impression is that online panelists, at least when recruited in conjunction with an election, are citizens who are interested in politics and to turn out to vote. Yet, at the same time, they are somewhat more critical of political institutions and political parties than e.g., the participants of the National election study. Altogether, concerning recommendations on how to achieve satisfactory representativeness in an online panel, our findings do not readily dismiss the use of probability-based sampling with personal postal invitations as a viable option (cf. Vehovar et al., 2008). Conversely, though, neither do our findings provide decisive proof of the suitability of probability-based sampling as a means for battling non-representativeness in online panels. The element of volunteering is always present in recruiting online panelists (cf. Sparrow & Curtice, 2004; Van Selm & Jankowski, 2006), and hence, it is our belief that some degree of sample skewness – in either demographic or other variables – is to be expected regardless of the sampling method.

Based on our findings and experience, what did we learn? First, even though this study tested probability-based sampling and achieved rather encouraging results, there is nonetheless a further need to compare probability-based and other recruitment methods. Perhaps there is valuable knowledge to gain through applying systematic variations in how participants are solicited from the sampling frame in order to pinpoint the optimal sampling procedure. Second, it is our belief that the field needs to explore ways to enhance representativeness even further than we did here; for instance by applying weights on more variables or through providing computer equipment for those citizens lacking it. Third, the fact that panel attrition was not a problem in this study is comforting to researchers who design online panels in the future. Online panels provide a cost-efficient and quick way to learn how people feel about political issues and act politically.

References

Aarts, K., & Wessels, B. (2005). Electoral turnout. In J. Thomassen. (Ed.), *The European voter: A comparative study of modern democracies* (pp. 64–83). Oxford: Oxford University Press.

Baker, R. P., Blumberg, S. J., Brick, J. M., Couper, M. P., Courtright, M., Dennis, J. M., … Zahs, D. (2010). Research synthesis. Aapor report on online panels. *Public Opinion Quarterly, 74*, 711–781.

Bernstein, R., Chadha, A., & Montjoy, R. (2001). Overreporting voting: Why it happens and why it matters. *Public Opinion Quarterly, 65*, 22–44.

Blais, A. (2000). *To vote or not to vote: The merits and limits of rational choice theory.* Pittsburgh, PA: University of Pittsburgh Press.

Borg, S., & Grönlund, K. (2013). *Finnish national election study 2011* [computer file]. FSD2653, version 2.1 (2013-01-22). Helsinki: Taloustutkimus [data collection], 2011. Election Study Consortium [producer], 2011. Tampere: Finnish Social Science Data Archive [distributor], 2013.

Cain, B., Dalton, R., & Scarrow, S. (2003). *Democracy transformed? Expanding political opportunities in advanced industrial democracies.* New York: Oxford University Press.

Dalton, R. (2007). *The good citizen: How a younger generation is reshaping American politics.* Washington, DC: CQ Press.

ESS Round 5: European Social Survey Round 5 Data (2010). Data file edition 3.0. Norwegian Social Science Data Services [distributor of ESS data].

European Commission. (2012). *E-communications household survey.* Special report, *Eurobarometer,* 381.

Göritz, A. S. (2004). Recruitment for online panels. *International Journal of Market Research, 46,* 411–425.

Grönlund, K. (2012a). *Finnish national election study 2011: Telephone interviews among Finnish-speaking voters* [computer file]. FSD2673, version 1.0 (2012-02-01). Turku: Åbo Akademi University [producer], 2011. Tampere: Finnish Social Science Data Archive [distributor], 2012.

Grönlund, K. (2012b). *Finnish national election study 2011: Telephone interviews among Swedish-speaking voters* [computer file]. FSD2726, version 1.0 (2012-04-18). Turku: Åbo Akademi University [producer], 2011. Tampere: Finnish Social Science Data Archive [distributor], 2012.

Grönlund, K., & Strandberg, K. (2013). Finnish eOpinion Panel 2011 [computer file]. FSD2826, version 1.0 (2013-07-23). Tampere: Finnish Social Science Data Archive [distributor], 2013.

Grönlund, K., & Westinen, J. (2012). Puoluevalinta. In S. Borg (Ed.), *Muutosvaalit 2011* (pp. 156–188). Helsinki: Ministry of Justice 16/2012.

Groves, R. M. (2006). Nonresponse rates and nonresponse bias in household surveys. *Public Opinion Quarterly, 70,* 646–675.

Inglehart, R., & Welzel, C. (2005). *Modernization, cultural change and democracy: The human development sequence.* New York: Cambridge University Press.

Internet World Stats (2011). Top 50 countries with the highest Internet penetration rate. Retrieved June 10, 2013, from: http://www.internetworldstats.com/top25.htm.

Kellner, P. (2004). Can online polls produce accurate findings? *International Journal of Market Research, 46,* 3–21.

Kies, R. (2010). *Promises and limits of web-deliberation.* New York: Palgrave Macmillan.

Kippen, G., & Jenkins, G. (2004). The challenge of e-democracy for political parties. In P. Shane (Ed.), *Democracy online: The prospects for political renewal through the Internet* (pp. 253–265). New York: Routledge.

López Pintor, R., & Gratschew, M. (2002). *Voter turnout since 1945: A global report.* Stockholm: International Institute for Democracy and Electoral Assistance (International IDEA).

López Pintor, R., & Gratschew, M. (2004). *Voter turnout in Western Europe since 1945: A regional report*. Stockholm: International Institute for Democracy and Electoral Assistance (International IDEA).

Luskin, R. C., Fishkin, J. S., & Iyengar, S. (2004). Considered opinions on U.S. foreign policy: Face-to-face versus online deliberative polling. Unpublished manuscript, retrieved July 9, 2013, from: http://cdd.stanford.edu/research/papers/2006/foreign-policy.pdf.

McMillan, S. (2000). The microscope and the moving target: The challenge of applying content analysis to the World Wide Web. *Journalism and Mass Communication Quarterly, 77*, 80–98.

Milner, H. (2010). *The Internet generation: Engaged citizens or political dropouts*. Medford, MA: Tufts University Press.

Norris, P. (2001). *A virtuous circle: Political communication in post-industrial societies*. New York: Cambridge University Press

Norris, P. (2003). *Democratic phoenix: Reinventing political activism*. Cambridge: Cambridge University Press.

OECD (Organisation for Economic Co-operation and Development) (2010). *Total fixed and wireless broadband subscriptions by country*. Retrieved June 10, 2013, from: http://www.oecd.org/sti/broadband/oecdbroadbandportal.htm.

O'Muircheartaigh, C. (2008). Sampling. In W. Donsbach, & M. Traugott (Eds.), *The Sage handbook of public opinion research*. (pp. 294–308). London: Sage.

Rao, K., Kaminska, O., & McCutcheon, A. L. (2010). Recruiting probability samples for a multi-mode research panel with Internet and mail components. *Public Opinion Quarterly, 74*, 68–84.

Sheehan, K. B., & McMillan, S. J. (1999). Response variation in email surveys: An exploration. *Journal of Advertising Research, 39*, 45–54.

Sills, S. & Song, C. (2002). Innovations in survey research: An application of web-based surveys. *Social Science Computer Review, 20*, 22–30.

Smith, A. (2011). *The Internet and campaign 2010*. Report, Pew Internet and American Life Project.

Sparrow, N., & Curtice, J. (2004). Measuring the attitudes of the general public via Internet polls: an evaluation. *International Journal of Market Research, 46*, 23–44.

Statistics Finland (2013). The official statistics of Finland. Retrieved June 10, 2013, from: www.stat.fi.

Strandberg, K. (2012). Sosiaalisen median vallankumous? Ehdokkaiden ja äänestäjien sosiaalisen median käyttö vuoden 2011 eduskuntavaalien kampanjassa. In S. Borg (Ed.), *Muutosvaalit 2011*. Helsinki: Ministry of Justice.

Swanson, D., & Mancini, P. (1996) *Politics, media, and modern democracy: An international study of innovations in electoral campaigning and their consequences*. Westport, CT: Praeger Publishers.

van Selm, M., & Jankowski, N. W. (2006). Conducting online surveys. *Quality & Quantity, 40*, 435–456.

Vehovar, V., Manfreda, K. L., & Koren, G. (2008). Internet surveys. In W. Donsbach, & M. Traugott (Eds.), *The Sage handbook of public opinion research*. (pp. 271–283). London: Sage.

Wass, H. (2008). *Generations and turnout: The generational effect in electoral participation in Finland*. Helsinki: Department of Political Science, University of Helsinki.

Wattenberg, M. (2002). *Where have all the voters gone?* Cambridge, MA: Harvard University Press.

5

The untold story of multi-mode (online and mail) consumer panels[1]

From optimal recruitment to retention and attrition

Allan L. McCutcheon[a], Kumar Rao[b], and Olena Kaminska[c]

[a]*University of Nebraska-Lincoln, USA*

[b]*Nielsen, Sunnyvale, CA, USA*

[c]*University of Essex, UK*

5.1 Introduction

Online and mixed-mode panels have become a popular data source for information on various consumer characteristics and behaviors over the past two decades (see e.g., Baker et al., 2010, Callegaro & DiSogra, 2008). Data are obtained from samples of individuals who are recruited from an underlying population and are surveyed on an ongoing basis. The resulting panel enables researchers to examine consumer behavior patterns and trends over time. These also allow us to estimate the effect of events over the life course. However, these advantages

---

[1] This material is based, in part, upon work supported by the National Science Foundation under Grant No. SES - 1132015. Any opinions, findings, and conclusions or recommendations expressed in this material are those of the author(s) and do not necessarily reflect the views of the National Science Foundation.

---

of consumer panels come with a number of challenges, such as *recruiting* a representative sample of individuals from an underlying population of interest. An approach that is gaining popularity among panel researchers is the use of multiple modes of recruitment, sometimes accompanied by response inducements, to maximize recruitment rate and minimize recruitment cost (see Rao, Kaminska, & McCutcheon, 2010).

A significant challenge that confronts all longitudinal panels, regardless of the recruitment strategy, is *retaining* the recruited sample of individuals in the panel. The most extreme form of retention failure is sample attrition,[2] in which panel respondents are permanently lost to the panel through loss of contact or because of their withdrawal of cooperation. In this chapter, we examine panel attrition in a mixed-mode panel, and we suggest an alternative way that researchers can think about attrition in consumer panels. Specifically, we apply the distinction between respondent attrition and data attrition (Olsen, 2012) to consumer panels. That is, we compare and contrast the amount of calendar time that respondents remain as part of the panel – the usual measure of respondent attrition – with the number of surveys actually completed by the respondents. We argue that the number of surveys actually completed by panel respondents is a better measure of consumer panel efficiency than is the actual number of days or months that respondents remain as panel members.

The attrition data for our study comes from the Gallup Panel whose members were recruited by a multi-mode recruitment experiment (Rao et al., 2010, see also Tortora, 2009) that contrasted phone and mail recruitment. The experimental design permitted us to measure the effect on the recruitment rate of various response-inducement techniques, such as advance letter, prepaid monetary incentive, and phone follow-up. The observational period of this study spans the period of time from when the member was recruited into the panel (i.e., membership start dates in March and April, 2007) to the end of 2009. Figure 5.1 shows the timeframe of the recruitment experiment, along with a few noteworthy milestones such as experiment start and end dates, date of first and last case of recruitment, first case of attrition, and the end date of the observational period of this study.

In the recruitment experiment, apart from the control group that received no response inducements, there were eight treatment groups that received different combinations of the three response inducements (advance letter, prepaid monetary incentive, and phone follow-up). We focus the first part of our analysis on those treatment groups that received prepaid monetary incentives, which in the recruitment experiment were meant to be a one-time payment for agreeing to become a panel member. While the acceptance of this monetary incentive payment may have persuaded recipients to join the panel at the time of solicitation, it is also likely that the effect of the incentive would, over time, wear off for some of these initially incentivized panel members. This is especially likely for those initially incentivized panel members who do not have an intrinsic interest in surveys, once the responsibilities of panel participation (i.e., the commitment to participate in multiple surveys without receiving any further monetary benefits) start to sink in. At that point, the initial one-time incentive payment may no longer be viewed as a counterweight for participating in multiple panel surveys (i.e., the economic exchange argument, see Groves et al., 2000); instead it may lead to expectations for further such incentive payments in the future (as noted by Singer et al., 1998). Given that the Gallup Panel has a no-monetary

---

[2] Here, the term "attrition," in the context of consumer panels, refers to the situation in which the panel member refuses or requests not to participate in future surveys. This is different from the situation in some longitudinal studies in which a sample member may skip one or several waves, but then resumes participation in later waves of the study (see e.g., Zagorsky & Rhoton, 2008).

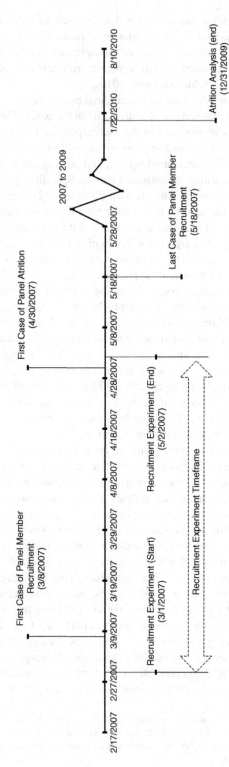

Figure 5.1 Timeframe of the recruitment experiment and the observational end date of the study.

compensation policy for survey participation; such expectations are not fulfilled, possibly evoking a negative response from incentivized members that leads them to drop out of the panel. With this perspective in mind, we examine whether incentivized panel members are at a higher (or lower) risk of attriting from the panel compared to their non-incentivized counterparts within the two recruitment modes (mail and phone), as well as in both modes combined. We also examine the impact of member-related (i.e., demographic characteristics) and panel-related (i.e., number of surveys assigned, number of surveys completed, token gifts sent, and number of non-response reminder postcards sent) predictors of attrition, comparing how these influences differ between the mail and phone recruitment modes.

In the final set of analyses presented in this chapter, we bring together a wide range of demographic, recruitment, and panel participation factors in a multivariate analysis to examine their role in the attrition of panel respondents, specifically drawing the distinction between respondent attrition and data attrition. We argue that consumer panel respondents who complete more surveys are more valuable panel members, even if they spend less calendar time in the consumer panel. This final set of analysis also includes an examination of "shared frailty," that is, the degree to which attrition by one (or more) panel members in multi-member panel households is "contagious." Specifically, we examine whether the attrition of one (or more) panel members in multi-member households lead to the increased likelihood of attrition by others in the household.

5.2 Literature review

Attrition in household panels has been historically developed in the context of the face-to-face mode. Studies on attrition have been conducted on such major panels as the Panel Study of Income Dynamics in the United States (which started in 1968), the German Socio-Economic Panel (from 1984), the British Household Panel Survey (from 1991), and the Household, Income and Labour Dynamics in Australia (from 2001). While the attrition rate largely varies between countries and time period of the panel beginning, there are some general trends of attrition that have been detected across panels.

Specifically, a number of respondent characteristics are found to be related to attrition in household panels, with the chance of attriting found to be higher among males, young, ethnic-minority group members (mainly due to noncontact), single individuals, single-member households (due to noncontact), no presence of children in the household (due to noncontact mainly), lower education, non-home-owners, economically inactive, living in urban location (for review, see Watson & Wooden, 2009). In addition to demographic characteristics, some personality characteristics have also been found to predict continuous participation. Attrition has been found to be less likely among agreeable and conscientious people and people with a high "need for cognition," as well as a "need to evaluate." In addition, attrition has been found to be more frequent among extraverts, and those respondents who are feeling survey burden (see Lugtig, 2012, for a review).

One important difference between face-to-face panels and panels using web mode is the frequency of contact and the number of waves of a panel. Typically, face-to-face panels are conducted yearly or biannually, while web panels tend to take place monthly, e.g., the Longitudinal Internet Studies for the Social sciences (LISS) panel in the Netherlands. This high frequency allows for an analysis of a large number of waves of data. The limited number of attrition studies for general population online panels, however, may contribute to inconsistent results. An early analysis of the LISS panel found higher likelihood of attrition among older panelists, and those who lived in households with disabled people. No differences, however, were found in attrition related to employment, and presence of pensioners in a household (de Vos, 2009). Interestingly, Kruse et al. (2010), analyzing an online panel conducted by Knowledge Networks, finds that attrition within the first 4 waves of participation was not related to age, race, education, employment, gender, marital status, or income. She concludes that this is likely due to differentially higher incentives for these groups at recruitment and throughout the first four waves of the panel, which most likely contributed to retaining these groups in the study. Attrition in a mixed-mode panel, such as the Gallup Poll Panel, where web mode was mixed with mail or telephone modes, has been found to be higher among males, younger adults, non-whites, and those with lower education (Tortora, 2009). Additional variables, including political preference (Lugtig, 2012; Kruse et al., 2010), have been found to increase attrition, as has the number of requests for marketing studies in comparison to political polls (Tortora, 2009).

Importantly, the behavior of panel members in previous waves has been found to be predictive of subsequent attrition. According to Kruse et al. (2010), late responding in a previous wave is related to future attrition, as is not providing complete income information, not responding to a health questionnaire, and dropping out (i.e., breakoff) in a previous wave (de Vos, 2009).

Finally, an interesting concept that appeared with online panels is the idea of "sleepers" (Scherpenzeel, 2013) and "lurkers" (Lugtig, 2012). Lurkers are panel members who participate in a panel irregularly, while sleepers are those who have not participated in a panel for a few waves, but are still sent invitations and have the potential to come back. Exploring participation patterns in the LISS panel using latent class analysis Lugtig (2012) differentiated four classes of panel members: "loyal stayers" (around 12% of original panel members), gradual attriters (65%), lurkers (7%) and fast attriters (17%). While attrition in the LISS panel was found to include approximately 10% of households (12% of individuals) per year, a somewhat larger issue is nonparticipation in any particular wave. Skipping a few subsequent waves has been found to lead to attrition in LISS panel, and a number of experimental studies have been conducted on how to prevent "sleeping," and how best to bring "sleepers" back into the panel. Interestingly, the panel found that neither gifts, nor hard copy or electronic newsletters had any impact on the chance of either sleeping or attrition (Scherpenzeel, 2013). But the researchers report finding a way to bring the sleepers back to the panel, by calling them after two months of nonparticipation and offering an incentive to restart completing interviews.

5.3   Methods

5.3.1   Gallup Panel recruitment experiment

The Gallup Panel conducted its recruitment experiment in spring 2007 using a list-assisted random-digit-dialing (RDD) sampling technique involving both phone and mail recruitment modes. As noted earlier, the recruitment experiment, in addition to the control group, had eight treatment groups each of which received different combinations of the three (advance letter, prepaid monetary incentive, and phone follow-up) response inducements. Table 5.1 shows the various treatment groups with their description and respective response rates[3]. Two of the four mail recruitment treatment groups (i.e., Groups A and B), and two of the five

---

[3] The information in Table 5.1 was originally presented as Table 2 in Rao, Kaminska, and McCutcheon (2010).

Table 5.1 Response rates (household level) by treatments by recruitment stages.

Mode	Treatment groups	Treatment description	Total eligible* sample size (N)	Stage 1: Agreement to join the panel (phone only)		Stage 2: Returning the welcome packet questionnaire (phone and mail)		
				(A) RDD Response rate (%)	(B) RDD Agreement rate (% of A)	(C) Before 21 days (%)	(D) After 21 days (%)	(C+D) Final Panel recruitment rate (%) (n)
Mail	A	(WPQ+ $2) + TF	1423			11.9 a	2.6	14.5 (206)
	B	(WPQ+ $2)	1435			11.4 b	2.0	13.4 (192)
	C	WPQ + TF	1439			10.1 c	2.2	12.3 (177)
	D	WPQ	1428			7.0 a,b,c	1.6	8.6 (123)
	E	(AL + $2) + Phone Survey + WPQ + TF	1172	37.7 a	68.3	4.9	8.7 a	13.6 (159)
	F	(AL + $2) + Phone Survey + WPQ	1187	35.6 b	62.6	4.6	6.4 b	11.0 (131)
Phone	G	AL + Phone Survey + WPQ + TF	1178	30.1 a,b	67.5	3.1	5.6 a	8.7 (102)
	H	AL + Phone Survey + WPQ	1154	28.2 a,b	69.6	3.5	6.3	9.8 (113)
	I	Phone Survey + WPQ + TF	1183	25.6 a,b	59.4	3.0	3.6 a,b	6.7 (79)

*For mail treatments the entire address-matched sample selected was deemed eligible whereas for phone treatments ineligible numbers (i.e., non-working, non-residence, and fax/data line numbers) were removed from the analysis; Treatment descriptions within brackets indicate they were mailed together; AL – Advance Letter; TF – Telephone Follow-up; WPQ – Welcome Packet Questionnaire; Refer to Figure 5.1 for an illustration and Section 5.3.3 for description of recruitment stages; Stage 1 RDD Response Rate is the proportion of completed interviews in the total number of eligible respondents; Stage1 RDD Agreement rate is the proportion of respondents who agreed to join the panel in the total number of completed interviews; Rates marked with the same individual subscript significantly ($p < .05$) differ from one another within each mode. For example, the RDD response rate of Group F is significantly different from Groups G, H, and I, but not with Group E. In the same way, the RDD response rate of Group H is significantly different from Groups E and F.

phone treatment groups (i.e., Groups E and F), received a $2 prepaid monetary incentive. As Table 5.1 shows, the experiment netted 1282 households (right-most column in the table, i.e., C+D) that included 2042 panel members. The incentivized treatments (i.e., Groups A, B, E, and F) recruited 1089 (53.3%) panel members within 688 households.

5.3.2   Panel survey mode assignment

Upon joining the panel, recruited members were assigned to one of the two main survey modes of participation in the Gallup Panel: *mail* or *web*. It is important to note that the assignment of members to a particular mode is not random and is based on a set of questions regarding household Internet access and use. Members who reported using the Internet twice or more per week, and who provided an email address, were assigned to receive panel surveys via web survey mode. Respondents who reported using the Internet less than twice a week, or who did not provide an email address, were assigned to receive paper questionnaires by mail.

Table 5.2 shows the distribution of recruited panel members across the two assigned survey modes. For analysis purposes, the members who were recruited from mail and phone modes were combined into two distinct groups: *incentivized* and *non-incentivized*. The incentivized respondents in both the mail and phone groups include those whose recruitment included a prepaid $2 monetary incentive. The non-incentivized respondents, on the other hand, consist of those who did not receive a prepaid monetary incentive. The mail incentivized group includes recruitment treatments A and B, whereas the non-incentivized group includes treatments C and D. Similarly, the phone incentivized group includes recruitment treatments E and F, whereas the non-incentivized group includes treatments G, H, and I.

The reason we create these eight groups, and subsequently analyze them in their entirety, is because panel attrition among members is likely to be a function not only of how they were *recruited* in the experiment (i.e., the combination of recruitment mode with/without incentive), but also a function of how they were *surveyed* in the panel (i.e., the mode assigned to take the surveys). The task of taking multiple surveys (i.e., waves) can be difficult for both

Table 5.2   Distribution of recruited panel members across assigned survey modes.

Groups	Description	Recruitment experiment treatments*	A: Number of recruited members (n)	Assigned survey modes	
				Web (% of A)	Mail (% of A)
1	Mail Incentivized Group	A and B	634	33.4	66.6
2	Mail Non-Incentivized Group	C and D	499	41.9	58.1
3	Phone Incentivized Group	E and F	455	43.5	56.5
4	Phone Non-Incentivized Group	G, H and I	454	44.9	55.1
		Total:	2,042	40.3	59.7

*Refer to Figure 5.1 for treatment descriptions.

mail- and web-assigned panel members, though this differential difficulty is likely to be due to differing reasons, including the cognitive ability of the members. For instance, while most mail-assigned members have some familiarity with paper questionnaires, web-assigned members on the contrary require some degree of familiarity with computers and their peripherals (such as keyboard and mouse) and manual dexterity in scrolling through the web survey questionnaire. Therefore, taking panel surveys via web or mail is likely to be a different experience for members who are assigned to these two modes. Furthermore, incentives can affect this experience in a more positive way by helping to invoke the norm of reciprocity among those who receive them (Dillman, 2007). Hence, the combination of the use of incentives and mode assignments are likely to affect member attrition rates differently, and this aspect is considered in creating the eight groups analyzed in this study.

In the mail-incentivized and non-incentivized groups, the number of members assigned to mail survey mode was significantly higher than those assigned to web mode (x^2 (1) = 8.52, $p < 0.01$), suggesting that the mail recruitment mode netted members who are less likely to have and use the Internet in their daily lives. The higher likelihood of non-availability and/or non-usage of the Internet for these members made them eligible for mail survey mode assignment in higher numbers. Even though more members in the phone-recruited group, both among the incentivized and the non-incentivized, were assigned to mail survey mode than to web mode, the difference in mode assignments in this case was not significant ($p = 0.67$). In sum, the phone recruitment mode, compared with the mail recruitment mode, appears to have netted a more diverse sample of panel members in terms of Internet availability and usage. Thus, the four groups in Table 5.2, in conjunction with two assigned survey modes, lead to the following eight recruitment-assignment groups we examine in this study:

1. Mail-I-web (Mail Incentivized, web-assigned group)

2. Mail-I-Mail (Mail Incentivized, mail-assigned group)

3. Mail-N-web (Mail Non-incentivized, web-assigned group)

4. Mail-N-Mail (Mail Non-incentivized, mail-assigned group)

5. Phone-I-web (Phone Incentivized, web-assigned group)

6. Phone-I-Mail (Phone Incentivized, mail-assigned group)

7. Phone-N-web (Phone Non-incentivized, web-assigned group)

8. Phone-N-Mail (Phone Non-incentivized, mail-assigned group)

5.3.3   Covariate measures used in this study

As noted earlier, we incorporate four kinds of member-level predictors of attrition into our multivariate analysis: demographic characteristics and measures related to panel burden, compensation, and nonresponse effort. While it is common for consumer panel organizations to offer rewards (such as incentives, vouchers, or token gifts) to their members for participating in their surveys and to have some sort of controls/procedures to monitor, measure, and manage unit non-response, little is known or has been documented about use of such rewards or nonresponse procedures. Before we include covariates in our multivariate analysis, we provide a brief overview of how our three panel-related measures were computed, and also give readers some context and meaning regarding their use in the Gallup Panel.

- *Number of surveys assigned*: This individual-level measure is simply a count of the total number of surveys the member has been assigned since joining the panel. Consequently, survey assignment is a function of a member's active status; that is, surveys are assigned only to active panel members. Also, while most of the panel surveys were assigned to all active panel members, some were assigned solely to certain targeted demographic groups (e.g., young adults – aged 18–24 years) or psychographic groups (e.g., had shopped online within the past three months). Therefore, this measure can vary from one panel member to another, depending on these two factors (i.e., active status and targeted assignment).

- *Number of surveys completed*: This is also an individual-level measure and is simply a count of the total number of surveys the member has completed since the time he or she joined the Gallup Panel. Note that survey completion is a function of survey assignment and *unit* nonresponse. That is, members chose to complete or not complete the surveys that were assigned to them. Internal studies conducted by the Gallup Panel and even a few published ones (e.g., Lee, Hu, & Toh, 2004) have reported that certain demographic groups such as those with high educational attainment and income are more likely to participate in panel surveys. Thus, substantial variability exists in this measure also.

- *Number of token gifts sent*: Although Gallup Panel members receive no monetary incentives for participating in the panel, several token thank you gifts (such as magnetic calendars, notepads, pens, and so forth) were sent throughout the year. These token gifts are given either at the member-level (i.e., every active panel member receives the gift) or at the household-level (i.e., one token gift is given per active member household). For this study, we created an individual-level measure by counting the token gifts sent to the member and his/her household. This measure has less variability across members since most token gifts are sent to all active members or households. However, there are instances when token gifts were sent to targeted individuals and households; that is, respondents in demographic groups who are thought to have a high risk of attrition (e.g., minorities, young adults, those with low education) were likely to receive more gifts.

5.3.4   Sample composition

First, we explore the demographic characteristics of recruited members in the eight recruitment-assignment groups. The results of the analysis are given in Table 5.3, which shows the sample composition of the eight recruitment-assignment groups and totals for mail- and phone-incentivized and non-incentivized groups. We conducted three separate chi-square tests to detect significant group differences on measures of demographic characteristics. In the first test, demographic comparisons were made between the panel members in the eight recruitment-assignment groups (i.e. columns A, B, C, D, E, F, G and H). In the second test, comparisons were made between total mail-incentivized panel members and total mail non-incentivized members. Finally, in the third chi-square test, we compare respondents in the total phone incentivized and the total phone non-incentivized groups. Since the process of panel recruitment by mail and phone differ somewhat (see Figure 1 in Rao et al., 2010), by conducting these separate tests, we explore the possibility of observing a differential incidence of high-risk demographic groups in incentivized and non-incentivized groups, not only within each recruitment mode (i.e., mail and phone), but also across both modes (i.e., all eight recruitment-assignment groups).

Table 5.3 Demographics of recruited members, by recruitment assignment groups.

Demographic Characteristics	Mail Recruitment x Assigned Mode Groups						Phone Recruitment x Assigned Mode Groups						Significance		
	A: Mail-I-web (%)	**B:** Mail-I-Mail (%)	Total (Mail-Incentivized) (%)	**C:** Mail-N-web (%)	**D:** Mail-N-Mail (%)	Total (Mail-Non-incentivized) (%)	**E:** Phone-I-web (%)	**F:** Phone-I-Mail (%)	Total (Phone-Incentivized) (%)	**G:** Phone-N-web (%)	**H:** Phone-N-Mail (%)	Total (Phone-Non-incentivized) (%)	[1]	[2]	[3]
Gender															
Male	48.1	41.0	43.4	45.0	45.9	45.5	50.5	44.7	47.3	47.1	45.2	46.0			
Female	51.9	59.0	56.6	55.0	54.1	54.5	49.5	55.3	52.7	52.9	54.8	54.0	n.s.	n.s.	n.s.
(n)	(212)	(422)	(634)	(209)	(290)	(499)	(198)	(257)	(455)	(204)	(250)	(454)			
Age groups															
<18	2.4	2.9	2.7	2.9	4.2	3.6	2.1	1.2	1.6	2.0	2.4	2.2			
18-29	4.7	6.9	6.2	8.7	6.6	7.5	4.6	7.2	6.1	5.9	9.3	7.8			
30-41	16.6	12.2	13.7	14.0	6.6	9.7	19.6	7.2	12.6	14.7	2.8	8.2	***	n.s.	n.s.
42-60	46.0	34.4	38.3	44.4	32.8	37.7	41.2	28.7	34.2	41.2	34.0	37.3			
61+	30.3	43.5	39.1	30.0	49.8	41.5	32.5	55.8	45.6	36.3	51.4	44.6			
(n)	(211)	(418)	(629)	(207)	(287)	(494)	(194)	(251)	(445)	(204)	(247)	(451)			
Race															
White	82.3	65.8	71.4	78.8	66.9	72.0	86.2	80.3	82.9	84.2	81.9	83.0			
Other	17.7	34.2	28.6	21.2	33.1	28.0	13.8	19.7	17.1	15.8	18.1	17.0	***	n.s.	n.s.
(n)	(209)	(407)	(616)	(208)	(284)	(492)	(196)	(249)	(445)	(203)	(249)	(452)			

Table 5.3 *(continued)*

Demographic Characteristics		Mail Recruitment x Assigned Mode Groups						Phone Recruitment x Assigned Mode Groups						Significance
		A: Mail-Mail-I-web (Incentivized) (%)	B: Mail-I-Mail (%)	Total (Mail-Incentivized) (%)	C: Mail-N-web (%)	D: Mail-N-Mail Non-incentivized (%)	Total (Mail-Non-incentivized) (%)	E: Phone-Phone-I-web (%)	F: Phone-I-Mail (%)	Total (Phone-Incentivized) (%)	G: Phone-N-web (%)	H: Phone-N-Mail Non-incentivized (%)	Total (Phone-Non-incentivized) (%)	[1] [2] [3]
Education														
High School or less		21.3	39.0	33.1	22.6	42.2	33.9	24.7	33.5	29.6	21.1	35.7	29.1	
Some college or more		78.7	61.0	66.9	77.4	57.8	66.1	75.3	66.5	70.4	78.9	64.3	70.9	*** n.s. n.s.
(n)		(211)	(415)	(626)	(208)	(287)	(495)	(198)	(254)	(452)	(204)	(249)	(453)	
Income														
< $50,000		27.3	54.2	45.2	23.7	49.1	38.4	26.0	51.3	40.5	23.9	48.9	37.9	
$50,000 +		72.7	45.8	54.8	76.3	50.9	61.6	74.0	48.7	59.5	76.1	51.1	62.1	*** * n.s.
(n)		(209)	(415)	(624)	(207)	(285)	(492)	(169)	(224)	(393)	(184)	(233)	(417)	
Work status														
In the labor force		68.4	55.8	60.1	71.3	51.2	59.7	67.2	44.9	54.8	61.4	52.2	56.4	
Not in the labor force		31.6	44.2	39.9	28.7	48.8	40.3	32.8	55.1	45.2	38.6	47.8	43.6	*** n.s. n.s.
(n)		(212)	(414)	(626)	(209)	(287)	(496)	(198)	(247)	(445)	(202)	(245)	(447)	

*p <.05
**p <.01
***p <.001; (two-tailed); Statistical significance based on chi-square test with p-values for the following comparisons: (1) All 8 recruitment-assignment groups (i.e., groups A, B, C, D, E, F, G, and H), (2) Total Mail-incentivized vs. Total Mail Non-incentivized groups, and (3) Total Phone-Total incentivized vs. Phone Non-incentivized groups.

In the first chi-square test with comparisons made among all recruitment-assignment groups, we found a significant difference among the groups for all demographic characteristics, except for gender. Although we must be mindful that chi-square tests are non-directional (see e.g., Agresti, 2012), we see from Table 5.3 that the mail-assigned groups within phone recruitment (i.e., Groups F and H) had a higher percentage of elderly panel members (ages 61+; 55.8% and 51.4%), compared with their counterpart groups in the mail recruitment (i.e., Groups B and D). Internal Gallup Panel research has shown that older panel members are more likely than their younger counterparts to remain in the panel. With respect to race, all mail recruitment groups (i.e., Groups A, B, C, and D) had a higher percentage of non-whites compared with the respective counterpart groups in the phone recruitment (i.e., Groups E, F, G, and H). For education and income, an interesting pattern emerges. All mail-assigned groups, irrespective of the recruitment mode (i.e., Groups B, D, F, and H), had a higher percentage of lesser-educated (high school or less) and lower income (less than $50K) members, compared with their web-assigned counterparts. Lastly, for work status, a sharp contrast from the previous finding emerges. All web-assigned groups, irrespective of the recruitment mode (i.e., Groups A, C, E, and G) had a higher percentage of working members, compared to their mail-assigned counterparts (i.e., Groups B, D, F, and H).

In the second chi-square test with comparisons made between mail total incentivized and mail total non-incentivized groups, we found no significant difference between these two groups on any demographic characteristics except for income. The mail-incentivized group had a higher percentage of low income members compared with the non-incentivized group (45.2% vs. 38.4%). In the third and last chi-square test with comparisons made between phone total incentivized and phone total non-incentivized groups, none of the demographic characteristics examined showed significance.

All in all, these results suggest that the phone recruitment-assignment groups (i.e., Groups E, F, G, and H), compared with their mail counterparts, are demographically predisposed to exhibiting lower levels of attrition by having a higher percentage of elderly and a lower percentage of non-white, less educated, and lower-income panel members. This insight will help in better understanding the dropout, attrition, and survival results, to which we turn our attention next.

5.4   Results

5.4.1   Incidence of panel dropouts

In this study, we ascribe panel dropout status to anyone who either requested to be removed from the panel, or who failed to continue his or her participation in the panel by refusing to complete the assigned surveys. There are several ways for panel members to drop out of the Gallup Panel. Members can voluntarily drop out by calling the toll-free support phone number and requesting removal or requesting removal on any questionnaire they receive (i.e., in the comment box). Members can also be dropped from the panel involuntarily, if they fail to respond to five consecutive surveys. In this study, we do not distinguish between voluntary or involuntary dropouts, but instead analyze all forms of dropouts together.

As noted earlier, the recruitment experiment in total netted 2042 panel members; by the end of the observation period of this study, 967 (47.4%) members dropped out of the panel. Table 5.4 shows the demographic characteristics of members who dropped out of the panel across the eight recruitment-assignment groups.

Table 5.4 Demographics of attrited panel members, by recruitment-assignment groups.

Demographic Characteristics	A: Mail-I-web (%)	B: Mail-I-Mail (%)	Total (Mail-Incentivized) (%)	C: Mail-N-web (%)	D: Mail-N-Mail (%)	Total (Mail-Non-incentivized) (%)	E: Phone-I-web (%)	F: Phone-I-Mail (%)	Total (Phone-Incentivized) (%)	G: Phone-N-web (%)	H: Phone-N-Mail (%)	Total (Phone-Non-incentivized) (%)	Significance [1] [2] [3]
Gender													
Male	43.2	44.4	44.1	47.3	47.0	47.1	53.8	42.7	46.9	46.9	45.6	46.1	
Female	56.8	55.6	55.9	52.7	53.0	52.9	46.2	57.3	53.1	53.1	54.4	53.9	n.s. n.s. n.s.
(n)	(81)	(259)	(340)	(93)	(181)	(274)	(65)	(110)	(175)	(64)	(114)	(178)	
Age groups													
<18	1.2	1.6	1.5	2.2	2.2	2.2	1.6	1.8	1.7	1.6	1.8	1.7	
18–29	8.6	10.5	10.1	12.9	9.4	10.6	8.1	12.7	11.0	9.4	13.2	11.8	*** n.s. n.s.
30–41	19.8	16.0	16.9	19.4	8.3	12.0	25.8	10.0	15.7	23.4	5.3	11.8	
42–60	53.1	35.8	39.9	46.2	34.8	38.7	30.6	26.4	27.9	35.9	33.3	34.3	
61+	17.3	36.2	31.7	19.4	45.3	36.5	33.9	49.1	43.6	29.7	46.5	40.4	
(n)	(81)	(257)	(338)	(93)	(181)	(274)	(62)	(110)	(172)	(64)	(114)	(178)	
Race													
White	75.3	62.2	65.5	78.3	65.1	69.7	76.6	79.0	78.1	79.7	73.7	75.8	** n.s. n.s.
Other	24.7	37.8	34.5	21.7	34.9	30.3	23.4	21.0	21.9	20.3	26.3	24.2	
(n)	(81)	(249)	(330)	(92)	(175)	(267)	(64)	(105)	(169)	(64)	(114)	(178)	

Education															
High School or less	31.3	43.5	40.6	26.1	46.4	39.5	35.4	38.0	37.0	15.6	40.7	31.6			
Some college or more	68.8	56.5	59.4	73.9	53.6	60.5	64.6	62.0	63.0	84.4	59.3	68.4	***	n.s.	n.s.
(n)	(80)	(255)	(335)	(92)	(179)	(271)	(65)	(108)	(173)	(64)	(113)	(177)			
Income															
<$50,000	26.3	57.1	49.7	26.1	50.0	41.9	36.2	56.8	49.6	27.8	49.0	41.7			
$50,000 +	73.8	42.9	50.3	73.9	50.0	58.1	63.8	43.2	50.4	72.2	51.0	58.3	***	n.s.	n.s.
(n)	(80)	(254)	(334)	(92)	(178)	(270)	(47)	(88)	(135)	(54)	(102)	(156)			
Work status															
In the labor force	74.1	57.7	61.7	77.4	53.4	61.6	67.7	45.6	54.2	64.1	48.6	54.3			
Not in the labor force	25.9	42.3	38.3	22.6	46.6	38.4	32.3	54.4	45.8	35.9	51.4	45.7	***	n.s.	n.s.
(n)	(81)	(253)	(334)	(93)	(178)	(271)	(65)	(103)	(168)	(64)	(111)	(175)			

*p <.05;

**p <.01;

***p <.001; (two-tailed); Statistical significance based on chi-square test with p-values for the following comparisons: [1] All 8 recruitment-assignment groups (i.e., groups A, B ,C ,D ,E ,F ,G, and H), [2] Total Mail-incentivized vs. Total Mail Non-incentivized groups, and [3] Total Phone-incentivized vs. Total Phone Non-incentivized groups.

The chi-square test groups in this analysis were similar to ones used in the sample composition analysis. While the incentivized and non-incentivized groups within mail and phone recruitment are similar on all demographics, the eight recruitment-assignment groups, on the other hand, are different on all demographics, except for gender. The dropout rates for various age groups had one common trend across all recruitment-assignment groups: a sizable proportion came from members who are aged 42 years and older. The age of attrited members differentiate the recruitment-assignment groups ($x^2(28) = 69.69$, $p < 0.001$), with mail-assigned groups (Groups D, F, and H) having a higher proportion of elderly (ages 61 and above) members dropping out of the panel. The racial composition among the recruitment-assignment groups differs significantly ($x^2(7) = 22.20$, $p < 0.01$) as well. Two groups in particular stand out in terms of high proportion of minority dropouts: Groups B and D. On education, the recruitment-assignment groups differ significantly ($x^2(7) = 29.86$, $p < 0.001$), with the web-assigned groups (Groups A, C, E, and G), irrespective of the incentive, within mail and phone recruitment had a high proportion of dropouts among highly educated panel members. The results also indicate a possible three-way interaction for this demographic. Within mail and phone recruitment, non-incentivized web-assigned groups have a higher proportion of dropouts compared with their incentivized counterparts (i.e., Groups A and C; Groups E and G). Household income also differentiates the recruitment-assignment groups significantly ($x^2(7) = 54.54$, $p < 0.001$); web-assigned groups, in general, had more members with higher income than the mail-assigned groups. Lastly, the work status of attrited members vary across the recruitment-assignment groups ($x^2(7) = 38.38$, $p < 0.001$), with a pattern that is similar to the one observed for income. The attrited members in the web-assigned groups (Groups A, C, E, and G) were more likely to be employed than their respective mail-assigned counterparts.

5.4.2   Attrition rates

Table 5.5 shows the attrition analysis results for the entire sample and across the eight recruitment-assignment groups. For the entire sample, the overall attrition rate is 21.4%, 20.5%, and 15.7% in 2007, 2008, and 2009, respectively. The survival rate dropped from 78.6% in 2007 to 52.6% by the end of 2009. In general, the attrition rate's decline over time is indicative of the sample increasingly consisting of fatigue-and-attrition-resistant panel members who are more interested in the subject matter of the panel than those not interested enough to remain in the panel.

By examining the attrition rates in 2007 closely, we find that the rates for mail recruitment-assignment groups are higher than their respective counterparts in phone recruitment. Two groups, in particular stand out: incentivized and non-incentivized mail-assigned Groups B and D, with attrition rates of 35.3% and 34.5% respectively. Not only for 2007, but for the next two years as well, these two groups had the highest attrition rates. It is interesting to note that these two groups had the highest percentage of non-white (see Table 5.3, Group B with 34.2% and Group D with 33.1%) and less educated (Group B with 39.0% and Group D with 42.2%) panel members at recruitment stage. These factors may have played a role in high attrition rates for these two groups, since historically these two demographic groups have been difficult to retain in the Gallup Panel. Turning to the remaining two groups in the mail-recruitment cadre (i.e., the incentivized and non-incentivized web-assigned groups A and C), we find that the attrition rates of Group A are consistently lower than in Group C across all three years. Finally, among the phone recruitment-assignment groups, an

Table 5.5 Attrition rates of the entire sample and across recruitment-assignment groups.

Year	Entire Sample (N = 2042)			Recruitment-Assignment Groups											
				Mail Recruitment-Assignment Groups						Phone Recruitment-Assignment Groups					
	Survival rate (%)	Drop-outs	Attrition rate (%)	A: Mail-I-Wed (n=212) Attrition rate (%)	B: Mail-I-Mail (n=422) Attrition rate (%)	Total (Mail-Incentivized) (n=634) Attrition rate (%)	C: Mail-N-web (n=209) Attrition rate (%)	D: Mail-N-Mail (n=290) Attrition rate (%)	Total (Mail-Non-incentivized) (n=499) Attrition rate (%)	E: Phone-I-web (n=198) Attrition rate (%)	F: Phone-I-Mail (n=257) Attrition rate (%)	Total (Phone-Incentivized) (n=455) Attrition rate(%)	G: Phone-N-web (n=204) Attrition rate (%)	H: Phone-N-Mail (n=250) Attrition rate (%)	Total (Phone-Non-incentivized) (n=454) Attrition rate (%)
No. of panel members															
2042															
Start															
2007 1604	78.6	438	21.4	14.2	35.3	28.2	15.3	34.5	26.5	12.6	16.3	14.7	8.8	16.8	13.2
2008 1275	62.4	329	20.5	16.5	26.0	22.2	22.0	30.0	26.2	12.7	17.7	15.5	14.5	21.6	18.3
2009 1075	52.6	200	15.7	13.8	19.3	16.9	15.9	18.0	17.0	11.9	16.9	14.6	11.9	16.6	14.3
		967													

identical pattern in attrition rates was observed: the rates of mail-assigned groups (Groups F and H) are consistently higher than web-assigned groups (Groups E and G) across all three years.

Taken as a whole, these results suggest the possibility that the attrition rate for a particular recruitment-assignment group might be more of a function of the recruitment mode (mail or phone recruitment) than incentive (i.e., whether the group received an incentive or not). Since multiple factors are likely to simultaneously influence attrition rates, we analyze attrition rates among recruitment-assignment groups more systematically from a multivariate perspective in the analyses that follow.

5.4.3   Survival analysis: Kaplan–Meier survival curves and Cox regression models for attrition

In this section, we examine attrition among recruitment groups and recruitment-assignment groups to investigate the survival patterns of attrited members from the time they were recruited into the panel until they dropped out of the panel. The objective in this analysis is to examine the differential impact of incentives, recruitment mode, assigned survey mode, and demographic characteristics on survival probabilities of panel members. Note that the panel members who drop out early may differ from those who drop out later; that is, those who drop out early may be those who value extrinsic benefits of panel participation (such as incentives) more than the intrinsic benefits (such as having their voice heard), compared with their longer-tenured counterparts. If that is the case, then the survival curves would indicate a sharper drop in survival rates for the incentivized groups during the first year, and it would be interesting to see if this effect reverberates all the way from the recruitment mode to the assigned mode. The survivor functions were estimated by the Kaplan–Meier (KM) method (the product-limit estimation) (Cleves, Gould, Gutierrez, & Marchenko, 2010). Survival functions among the groups were compared by the logrank test. In the analysis, panel members who were active during the entire observational period (i.e., from their corresponding membership start date to the end of 2009) are considered as right-censored observations,[4] and those who dropped out are considered as event observations. Figure 5.2 shows the KM survival curves for panel members by mode of recruitment (mail and phone) and whether they were incentivized.

As these four survival curves indicate, there is a substantial difference between the survival times of those panel members who were recruited by phone and those who were recruited by mail; though the four groups are statistically distinguishable from each other ($x^2(3) = 53.86$, $p < 0.001$), this distinction appears mostly to be a function of recruitment mode rather than incentive. Overall, dropout seems to occur more quickly in mail-incentivized and non-incentivized groups. Notice that the graph appears as a distinct "island" of incentivized and non-incentivized curves lying between the mail and phone recruitment modes. Also, while the phone recruitment curves taper off slowly after 200 days, the mail recruitment curves take a steep descent. Later on, the widening gap between the two islands of curves suggests that the dropouts in the mail recruitment groups increased at a higher rate than the phone recruitment groups.

---

[4] These right-censored cases are a case of type III censoring in which study participants enter the study once at different times during the observation period and remain event-free at the end of the observation period (Lee & Wang, 2003). Since the entry times are not simultaneous, the censored times for these participants are different. Lee and Wang (2003) also refer to this type of censoring as progressive censoring or random censoring.

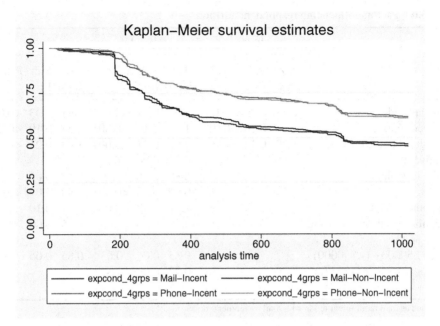

Figure 5.2   Recruitment mode by incentivized and non-incentivized factor levels.

In Table 5.6 we report three models for response time of drop outs from the panel during the period of the investigation – the first model (Model 1) compares the panel duration of web-assigned panelists who were recruited by phone, against the other three combinations (i.e. mail-recruited/mail-assigned, mail-recruited/web-assigned, and phone-recruited/mail-assigned). As the results of this model clearly illustrate, all three combinations of recruitment and panel participation result in more rapid attrition than among those panel members who were recruited by phone and assigned to participate in Gallup Panel surveys administered via the web.

In Model 2, we investigate the impact of the $2 recruitment incentive and the demographic factors of gender, age, race, education, income, and work status. Given the nonrandom assignment of panelists to the web and mail modes of the panel data collection – recalling that only those with web access could be assigned to the web mode – it is important to include such controls for a more accurate assessment of the impact of mode. Still, even after including these controls, the influence of recruitment mode (phone vs. mail) and assigned mode (mail vs. web) of panel participation remains highly significant; even once these additional factors are included in the model – each of these three combinations result in significantly higher attrition rates than does the phone recruitment with web participation. Of interest, the results for Model 2 clearly indicate that, once these demographics are taken into account, neither the $2 incentive at the recruitment, nor the respondent's labor force status, appear to play a significant role in panel retention. It appears, however, that age, education, and income are inversely correlated with attrition, though the latter two are measured as dichotomies. Also, the results of this model indicate that non-white respondents are significantly more likely to drop out of the panel more quickly. Finally, when the incentives and labor force participation variables are eliminated from the estimation (see Model 3), the model suggests that men are more likely to attrite at a faster rate than are women panel members.

Table 5.6 Factors influencing respondent attrition.

	Model 1			Model 2			Model 3		
	Haz. Ratio	Std. Err.	P>z	Haz. Ratio	Std. Err.	P>z	Haz. Ratio	Std. Err.	P>z
Mail-Mail* (Phone-web)	2.63	0.26	<0.001	2.92	0.34	<0.001	2.91	0.33	<0.001
Mail-web (Phone-web)	1.36	0.16	0.01	1.54	0.19	<0.001	1.54	0.19	<0.001
Tel-Mail (Phone-web)	1.50	0.17	<0.001	1.65	0.21	<0.001	1.68	0.21	<0.001
Incentive (none)				0.91	0.06	n.s.			
Gender (female)				0.88	0.06	0.06	0.87	0.06	0.04
Age				0.99	0.00	<0.001	0.99	0.00	<0.001
Race, other (White)				1.31	0.10	<0.001	1.31	0.10	<0.001
Education less than HS (some college +)				0.84	0.06	0.03	0.85	0.06	0.03
Income $50000+ (<$50000)				0.85	0.06	0.03	0.85	0.06	0.03
Work, not in Labor Force (in Labor Force)				0.97	0.08	n.s.			

*Interpreted as mail recruitment mode with mail assignment mode.

Figure 5.3 shows the KM survival curves for panel members by mode of recruitment (mail and phone) and mode of assignment (mail and web). Interestingly, the mode by which panel members were recruited appears to have an enduring influence; those panel members who were recruited by phone tended to have higher "survival rates" in the panel, than did those who were recruited by mail, as compared to those respondents within the same mode of data collection. Importantly, however, these survival curves clearly indicate that those respondents who participated in the web portion of the Gallup Panel enjoyed a far higher panel survivor rate than those who participated in the mail portion of the Gallup Panel.

In the next section, we again turn to an analysis of attrition among these panel respondents, though with a somewhat different view. In these analyses, we ask two rather different types of questions than in the analyses thus far reported. First, we examine whether respondents' continuing panel participation is likely to be impacted by the continued participation and/or attrition of other household members who are in their household. Second, we examine an alternative survival model that focuses on the number of completed surveys, rather than on the traditional model of measuring survival by the calendar time respondents participate in panels.

5.4.4 Respondent attrition vs. data attrition: Cox regression model with shared frailty

In this section, we turn our focus to an analysis of 514 multi-member panel households in which at least one panel member completed at least one survey. This nets a total of 971 panelists in the 514 multi-member households. Our focus in this analysis is on the impact of "shared frailty" on panel attrition – that is, the degree to which attrition behavior on the part of one (or more) panel members is "contagious" to other members in multi-member panel households (see e.g., Cleves, Gutierrez, & Marchenko, 2010). Further, we explore the difference

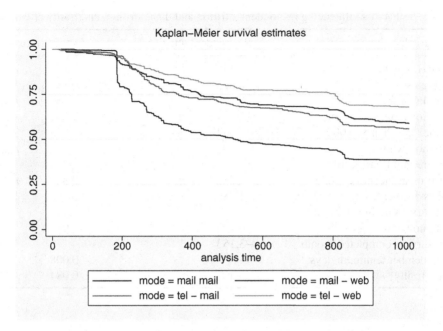

Figure 5.3   Recruitment mode by assigned mode factor levels.

between *respondent* attrition and *data* attrition (Olsen, 2012), investigating whether measuring panel survival in terms of surveys completed (data attrition) yields different conclusions than the usual focus of panel retention in terms of calendar time in the survey (respondent attrition). Our interest, here, is in determining whether surveys completed or calendar time in the panel is of greater consequence for consumer panels.

The data reported in Table 5.7 indicates that, with the exceptions of age and labor force participation, demographics do not play a significant role in respondent attrition. Respondents who are not in the labor force appear to remain members of the panel for a shorter period of time (i.e., attrite sooner), but they show no significant difference in their survey completions (i.e. data attrition) than do those who are members of the labor force. Also, the age of the panel members shows no difference in the overall length of time that they remain in the panel, though younger ones appear to complete fewer surveys before attriting from the panel than do older panel members. Net of the other variables in the model, none of the remaining demographic variables appears to have much influence on either respondent attrition or data attrition.

The recruitment and assigned survey mode variables, on the other hand, show a somewhat more consistent pattern with respect to both respondent and data attrition. Recruitment incentives show no significant influence on either form of attrition. On the other hand, panel members who responded by mail (rather than web), showed a far higher likelihood of attriting in both of the two attrition measures. A comparison of the average completion rates for the two groups of respondents in the entire sample shows that web respondents are likely to complete nearly three times as many surveys as are the mail respondents; the 823 web respondents average 20.96 (SD = 15.50) completes, while the 1219 mail respondents average 7.13 (SD = 6.51) completes. Finally, those who were recruited by phone show a significantly higher number of completions than do those who were recruited by mail.

Table 5.7   Factors influencing respondent attrition and data attrition, and frailty effect.

Demographic characteristics	Respondent attrition		Data attrition	
Gender (Female)	−.144	n.s.	.155	n.s.
Age (years)	0.003		−0.116	***
Not in labor force	0.325	***	0.05	n.s.
Education (some college)	−0.144	n.s.	−0.121	n.s.
Income (less than $50,000)	−0.044	n.s.	−0.055	n.s.
Race (non-white)	−0.29	n.s.	0.112	n.s.
Phone recruit (mail recruit)				
Recruitment Incentive (none)	−0.042	n.s.	0.027	n.s.
Mail assignment (web assignment)	1.989	***	0.845	***
Total surveys assigned/month				
Total number of gifts	−0.199	***	−0.047	***
Total surveys completed/month	−3.153	***	−	
Panel member tenure, in days	−		−0.008	***
Theta (frailty)	0.49	***	0.084	n.s.

$^*p < .05$;
$^{**}p < .01$;
$^{***}p < .001$.

The ongoing panel activities present somewhat of a mixed story. An increase in the number of symbolic gifts that panel respondents are given increases their panel longevity, whether measured by calendar time or number of survey completes. The higher the number of partici-pation requests sent per month, however, significantly decreases the length of time that panel respondents remain in the panel, though this variables significantly increases the number of survey completes – that is, the rate of participation requests increases respondent attrition, but decreases data attrition. The final estimates in this block of variables indicate that a higher number of completes per month leads to a longer duration in the panel, suggesting that the habit of survey completion contributes to calendar retention. Similarly, the longer a respondent is part of the panel, the greater the likelihood that he or she will complete a greater number of surveys.

Finally, and importantly, the statistically significant shared frailty effect (theta) for respondent attrition suggests that family members tend to attrite more-or-less as a unit – that respondent attrition is contagious within the family. Interestingly, the lack of a significant shared frailty effect for data attrition suggests that the attrition of high- and low-performing panel members within the family are not related. This suggests an important conclusion for panel researchers – dropping low-performing family panel members may contribute to the attrition of high-performing members. Thus, it might be advisable for panel researchers to tolerate those relatively low-performing family members in families with a high panel performer, in order to retain the relatively higher-performing panel members.

5.5   Discussion and conclusion

The retention of panel participants in survey panels – whether consumer or non-consumer pan-els – presents one of the major obstacles encountered by panel researchers. Without continued efforts at panel recruitment, differential panel attrition contributes to the eventual nonrepresentativeness of panels, and calls into question the inferential capabilities of panel survey data. In addition, for consumer panels, attrition leads to a decline in the economic viability of the panel – the loss of panel membership through attrition may contribute to a loss of revenue.

In the current study, we find that the recruitment mode – phone versus mail – appears to have an enduring importance for panel retention. Even in the multivariate analyses, while holding several other factors constant, the data attrition was significantly lower among those recruited by phone. Interestingly, respondents who were recruited by phone appear to remain in the panel longer than those respondents who were recruited by mail; this influence on retention appears to remain true independent of the *panel* mode of data collection (web versus mail) used for the respondents. Thus, the increased cost of panel recruitment characteristic of phone versus mail (phone tends to be more expensive) may be offset by the longer period of panel retention by those respondents who are recruited by phone.

Our findings also clearly underscore the conclusion that while gifts to panel members appear to have positive consequences for panel retention, recruitment incentives do not have enduring consequences for panel retention. Although recruitment incentives may have an important, positive effect on recruitment (see e.g., Rao et al., 2010), these positive effects appear to wear off once the respondent begins his or her participation in such mixed-mode panels.

Clearly, the data from this experiment also support a conclusion that retention is superior for web participants as compared to those panel participants whose survey data were collected by mail. Although the assignment of panel respondents' mode was not randomized – those who reported having access to the Internet were assigned to the web mode of data collection, while the others were assigned to the mail mode of data collection – the effects of panel mode remained quite large, even when several key demographic variables were controlled for. Thus, this finding suggests that web survey data collection is likely to be preferable to mail survey data collection for panel retention.

Our panel "frailty" analysis results indicate that multi-member panel households are likely to attrite *in time* as a unit. Thus, while it may prove more cost-effective to recruit multiple members of a household during the recruitment phase, this may result in a "contagion" effect in which one household member's attrition may prove contagious, leading others – including high-performing panel members – in the household to attrite as well. An important consequence of this pattern suggests that it may be worthwhile for panel researchers to endure the poor performance of one (or more) member in households in which there is one (or more) high panel performer.

Finally, the analyses presented in this study indicate that the logic of *data* attrition and *respondent* attrition may lead to important differences in the conclusions we draw regarding the characteristics of ease of panel respondent retention. For example, while our analyses suggest that age has little impact on the length of time (respondent attrition) that respondents remain in the panel, age appears to be clearly related to the number of survey completes (data attrition) by panel respondents. Also, not being in the labor force decreases the length of time that a respondent is in the panel, though it does not decrease the number of surveys that he or she completes. Importantly, the distinction between respondent and data attrition leads to our important conclusion that there is a frailty – that is, household contagion – effect for the length of time that a panel members in multi-member households participate, but this frailty effect is not significant in the data attrition model – leading us to conclude that panel researchers may wish to allow lower performers to remain in the sample, when a high performer is in the same household.

References

Agresti, A. (2012). *Categorical data analysis*, 3rd ed. New York: John Wiley & Sons, Inc.

Baker, R., Blumberg, S. J., Brick, J. M., Couper, M. P., Courtright, M., Dennis, J. M., Dillman, D. A., et al. (2010). Research synthesis: AAPOR report on online panels. *Public Opinion Quarterly*, *74*, 711–781.

Callegaro, M., & DiSogra, C. (2008). Computing response metrics for online panels. *Public Opinion Quarterly*, *72*, 1008–1032.

Cleves, M., Gould, W., Gutierrez, R. G., & Marchenko, Y. V. (2010). *An introduction to survival analysis using Stata*, 3rd ed. New York: Stata Press.

de Vos, K. (2009). *Panel attrition in LISS*. Tilburg, The Netherlands: CentERdata.

Dillman, D. A. (2007) *Mail and internet surveys: The tailored design method 2007 update with new internet, visual, and mixed-mode guide*, 2nd ed. New York: John Wiley & Sons, Inc.

Groves, R. M., Singer, E., & Corning, A. (2000). Leverage-saliency theory of survey participation: Description and an illustration. *Public Opinion Quarterly*, *64*, 299–308.

Kruse, Y, Callegaro, M., Dennis, J. M., DiSogra, C., Subias, S., Lawrence, M., & Thompson, T. (2010). Panel conditioning and attrition in the AP-Yahoo! new election panel study. In JSM (Ed.) *Proceedings of the Joint Statistical Meeting, American Association for Public Opinion Research Conference* (pp. 5742–5756). Washington, DC: AMSTAT.

Lee, E., Hu, M.Y., & Toh, R. S. (2004) Respondent non-cooperation in surveys and diaries: An analysis of item non-response and panel attrition. *International Journal of Market Research*, *46*, 311–326.

Lee, E. T. & Wang, J. W. (2003) *Statistical methods for survival data analysis*, 3rd ed. New York: John Wiley & Sons, Inc.

Lugtig, P. (2012). I think I know what you did last summer: Improving data quality in panel survey. PhD dissertation, University of Utrecht, The Netherlands.

Olsen, R. J. (2012). Respondent attrition vs data attrition and their reduction. Paper presented at the Future of Survey Research conference, National Science Foundation.October 3–4. Washington, DC.

Rao, K., Kaminska, O., & McCutcheon, A. L. (2010). Recruiting probability samples for a multi-mode research panel with internet and mail components. *Public Opinion Quarterly*, *74*, 68–84.

Scherpenzeel, A. (2013). Experiences from a probability-based Internet panel: Sample, recruitment and participation. Paper presented at the web Surveys for the General Population: How, Why, and When? Conference, organized by the NCRM Network of Methodological Innovation, London, February 25–26.

Singer, E., Van Howeyk, J., & Maher, M. P. (1998). Does the payment of incentives create expectation effects? *The Public Opinion Quarterly*, *62*, 152–164.

Tortora, R. (2009). Attrition in consumer panels. In P. Lynn (Ed.), *Methodology of longitudinal surveys* (pp. 235–249). New York: John Wiley & Sons, Inc.

Watson, N., & Wooden, M. (2009). Identifying factors affecting longitudinal survey response. In P. Lynn (Ed.), *Methodology of longitudinal surveys* (pp. 157–182). New York: John Wiley & Sons, Inc.

Zagorsky, J., & Rhoton, P. (2008). The effects of promised monetary incentives on attrition in a long term panel survey. *Public Opinions Quarterly*, *72*, 502–513.

Part II
NONRESPONSE

Introduction to Part II

Jelke Bethlehem[a] and Paul J. Lavrakas[b]
[a]*Statistics Netherlands, The Netherlands*
[b]*Independent Research Psychologist/Research Methodologist, USA*

II.1   The nonresponse problem

Like any other survey, web panels suffer from nonresponse. It is the phenomenon that persons selected for the panel/survey do not provide the required information; their data is either entirely or partial missing. The focus in this chapter is on unit nonresponse. This occurs if a respondent does not provide any information at all, i.e., the questionnaire remains completely empty.

Nonresponse is a problem because it affects the representativity of the panel. Nonresponse may cause specific groups to be overrepresented or underrepresented. Consequently, wrong conclusions may be drawn from the collected information.

Nonresponse may occur in various ways in a web panel. In the first place, there typically is a great deal of nonresponse during the recruitment phase. Recruitment nonresponse may be high because participating in a panel requires substantial commitment and effort from respondents.

In the second place, if the web panel is used for cross-sectional studies, there may be nonresponse in the specific surveys based on samples from the panel. This survey nonresponse is often low as the panel members have already agreed to participate in these surveys. That is why they became a member of the panel. Nevertheless, there can still be nonresponse. Examples of potential causes are: not available during the field period (e.g., never seeing the survey invitation), not interested in the specific topic, and otherwise not being able (e.g., due to illness). Nonresponse for individual panelists need not be permanent. After skipping one of the specific surveys, a panel member may decide to participate again in a subsequent survey.

In the third place, if the panel is used for longitudinal studies, there may be attrition. People get tired of completing the specific survey questionnaires and decide to stop their cooperation in the panel. Once they stop, they will never start again. Or, because of low participation of an individual panelist or other reasons, the panel company may drop (terminate) the panelist.

Online Panel Research: A Data Quality Perspective, First Edition.
Edited by Mario Callegaro, Reg Baker, Jelke Bethlehem, Anja S. Göritz, Jon A. Krosnick and Paul J. Lavrakas.
© 2014 John Wiley & Sons, Ltd. Published 2014 by John Wiley & Sons, Ltd.
Companion website: www.wiley.com/go/online_panel

II.2   The nonresponse bias

Nonresponse may cause estimates to be biased. To obtain more insight in the nature and magnitude of the nonresponse, often a theoretical framework is used based on the concept of response probability (aka response propensity).

Let the *target population U* of the panel consist of a set of N individuals, which are labeled $1, 2, \ldots, N$. Associated with each individual k is an unknown value Y_k of the *target variable*. The set of all values of the target variable is denoted by $Y_1, Y_2, \ldots, Y_N$. The objective of a specific panel is assumed to be an estimation of the population mean

$$\overline{Y} = \frac{1}{N} \sum_{k=1}^{N} Y_k. \tag{II.1}$$

To construct a panel, a simple random sample of size n is selected without replacement. The sample is represented by the series of indicators $a_1, a_2, \ldots, a_N$, where the k-th indicator a_k assumes the value 1 if individual k is selected in the sample, and otherwise it assumes the value 0. In the case of simple random sampling without replacement, the sample mean

$$\overline{y} = \frac{1}{n} \sum_{k=1}^{N} a_k Y_k \tag{II.2}$$

is an unbiased estimator of the population mean. Now suppose there is nonresponse in the recruitment phase. It is assumed that each individual k in the population has a certain, unknown probability ρ_k of response. If individual k is selected in the sample, a random mechanism is activated that results in probability ρ_k in response and in probability $1-\rho_k$ in nonresponse. In this model, a set of response indicators $R_1, R_2, \ldots, R_N$ can be introduced, where $R_k = 1$ if the corresponding element k responds, and where $R_k = 0$ otherwise. So, $P(R_k = 1) = \rho_k$, and $P(R_k = 0) = 1-\rho_k$.

The recruitment phase response only consists of those elements k for which $a_k = 1$ and $R_k = 1$. Hence, the number of available individuals is equal to

$$n_R = \sum_{k=1}^{N} a_k R_k, \tag{II.3}$$

The number of nonrespondents is denoted by n_{NR}, where $n = n_R + n_{NR}$. The values of the target variable only become available for the n_R responding individuals. The mean of these values is denoted by

$$\overline{y}_R = \frac{1}{n_R} \sum_{k=1}^{N} a_k R_k Y_k. \tag{II.4}$$

Bethlehem (2009) shows that the expected value of the response mean is approximately equal to

$$E(\overline{y}_R) \approx \frac{1}{N} \sum_{k=1}^{N} \frac{\rho_k}{\overline{\rho}} Y_k \tag{II.5}$$

where

$$\overline{\rho} = \frac{1}{N} \sum_{k=1}^{N} \rho_k \tag{II.6}$$

is the mean of all response probabilities in the population. Expression (II.5) shows that, generally, the expected value of the response mean is unequal to the population mean to be estimated. Therefore, this estimator is biased. This bias is approximately equal to

$$B(\bar{y}_R) = \frac{R_{\rho Y} S_\rho S_Y}{\bar{\rho}},$$ (II.7)

where $R_{\rho Y}$ is the correlation coefficient between the values of the target variable and the response probabilities, S_ρ is the standard deviation of the response probabilities, and S_Y is the standard deviation of the variable Y. From this expression of the bias, the following conclusions can be drawn:

- The bias vanishes if there is no relationship between the target variable of the survey and the response behavior ($R_{\rho Y} = 0$). The stronger the correlation between target variable and response behavior, the larger the bias will be.

- The bias vanishes if all response probabilities are equal ($S_\rho = 0$). Indeed, in this situation the nonresponse is not selective. It just reduces the final sample size. The more the values of the response probabilities vary, the larger the bias will be.

- The magnitude of the bias increases as the mean of the response probabilities decreases. Translated into practical terms, this means that lower response rates, in theory (if not always in practice), will lead to larger biases.

The formulae described above apply not only to the recruitment nonresponse. They can also be used for nonresponse in specific surveys and for attrition in the waves of a longitudinal panel.

II.3   Exploring nonresponse

To obtain insight in the possible negative consequences of nonresponse, one must attempt to estimate the magnitude of the various components in bias expression (II.7). The first component is the correlation between the target variable and the response behavior. This correlation cannot be computed in practical situations, because the values of the target variable are missing for the nonrespondents.[1] The way out is to use auxiliary variables. These are variables that are measured in the panel and for which also a complete population distribution (or complete sample distribution) is available. It is possible to establish for these auxiliary variables if the nonresponse is selective. Usually, three nonresponse mechanisms are distinguished:

- *Missing Completely at Random (MCAR)*. The response behavior is completely unrelated to the target variables and auxiliary variables. As a consequence, estimators are unbiased. This is the situation the researcher would like to be in. Unfortunately, s/he almost never is.

- *Missing at Random (MAR)*. There is no direct relation between the target variables and the response behavior, but there are relations between auxiliary variables and response behavior. This means estimates will be biased, but this bias can be removed or reduced by applying some kind of correction technique like adjustment weighting.

[1] However, if a nonresponse follow-up survey is conducted, it is possible to gather such information.

- *Not Missing at Random (NMAR)*. There is a direct relationship between target variables and response behavior. Estimates will therefore be biased and it will be impossible to do anything about the bias. This is the situation the researcher does not want to be in, but s/he often is.

Thus, it is important to have a set of relevant auxiliary variables. The variables should be relevant in two respects: (1) they must have power to explain response behavior; and (2) they must have power to explain the target variables of the survey. For example, if an auxiliary variable can explain response behavior, but is completely unrelated to a target variable, corrections using this auxiliary variable will not reduce the bias of estimates for the target variable.

The auxiliary variables can always be used to explore relationships with response behavior. If such relationships exist, and these variables are known to have relationships with target variables, one must fear for biased estimates, and a correction procedure is called for. Such a procedure will only be successful in case of Missing at Random (MAR).

The auxiliary variables can also be used to estimate response probabilities. Usually logit models are use for this, see e.g., Bethlehem, Cobben, and Schouten (2011). Note that these models can only be applied if the individual values of the auxiliary variables are available for all nonrespondents. If it becomes clear that there is a lot of variation in the (estimated) response probabilities, the researcher must fear there are biased estimates.

The response rate of the survey is often used as an indicator of the quality of the survey response. To some extent this is meaningful, as the average response probability is one of the components in bias expression (II.7). But this indicator does not tell the complete story (cf. Groves, 2006). A low response rate in combination with uncorrelated response behavior ($R_{\rho Y} = 0$) will still produce unbiased estimates; as in the case of many pre-election surveys' ability to accurately forecast election outcomes despite having very low response rates. Therefore, there is a need for additional response quality indicators. One such indicator is the R-indicator, proposed by Schouten, Cobben, and Bethlehem (2009). It is defined by $R = 1 - 2 S_\rho$, where S_ρ is the standard deviation of the response probabilities. $R = 1$ if all response probabilities are equal (complete representativity, no nonresponse bias). The closer R is to 0, the bigger the lack of representativity, and the larger the risk of a serious bias.

A final consideration about nonresponse in online panels relates to the fact that panels generate a great deal of data about each of their panelists from all the questionnaires the panelists complete while members of the panel. These data provide a wealth of valuable behavioral, attitudinal, demographic, and psychographic information that can (and we believe should) be used to study nonresponse bias in the specific surveys the panelists complete for clients of the panel, i.e., comparisons of the possible differential nature of the responders and nonresponders in each specific survey. Our book does not include a chapter about this approach nor are we aware that any online panel does this routinely or, if it does, that it reports it routinely even to the client of a given survey. This remains an important area for future research to investigate the nature and size of nonresponse bias within a panel as its members are sampled to complete questionnaires for specific surveys.

II.4   The chapters in Part II

In Chapter 6, Peter Lugtig, Marcel Das, and Annette Scherpenzeel studied recruitment nonresponse and attrition in a web panel that was constructed by means of probability sampling (the LISS Panel). This is a Dutch panel for which the probability sample was selected from the population register. Every effort was made to achieve high response rates and low attrition rates, and to obtain and maintain a representative panel. People were recruited for the panel using a mixed-mode approach including mail, telephone, and face-to-face contact. There were monetary incentives for the respondents. And those without Internet were offered free Internet access. Nevertheless, notwithstanding all these efforts, there was nonresponse in the recruitment phase and there was attrition in the specific surveys taken from the panel.

Lugtig et al. were able to link respondents and nonrespondents to register data. That gave them a set of possible correlates of response behavior. They show that specific groups are underrepresented among panel members: elderly (over 65 years of age), singles, those living in urbanized areas, and home-owners.

It turns out that the elderly and singles are less likely to attrite. Hence, the underrepresentation of these groups becomes gradually less severe over time. However, problems increase for variables such as household size, degree of urbanization, and singles with children.

To obtain more insight in attrition patterns, they apply latent class analysis. Groups of respondents are formed based on their attrition patterns. Lugtig et al. find nine groups with different nonresponse behavior. Examples are a group that drops out very fast, a group that drops out very slowly, and a group that is very loyal for two years and then vanishes.

In Chapter 7, Anja Göritz studies the nonresponse in a German self-selection panel that had already existed for 12 years. More specifically, she looks at the starting rate (the percentage of panel members starting the questionnaire of a specific survey) and the completion rate (the percentage of panel members completing the specific survey questionnaire). She then explores a possible relationship between these rates and 26 different study characteristics (such as the use of incentives, length of the questionnaire, and the length of the fieldwork period of the specific study).

Göritz shows that some characteristics may have a positive effect on starting rates but not necessarily on completion rates. For example, use of incentives increases the starting rate, but some types of incentives may decrease the completion rate.

On the one hand, Göritz suggest that the starting rate can be increased by conducting the survey in winter, by sending reminders, by announcing the length of the questionnaire, by reducing the length of the questionnaire, by inviting panel members who have already participated in a previous study, and by using incentives. On the other hand, completion rates can be improved by increasing the time intervals between studies, by inviting panel members who have already participated in a previous study, and by using incentives other than payment via PayPal and loyalty points.

It is difficult, and sometimes almost impossible, to set up a web panel based on probability sampling. Therefore, many web panels are based on self-selection of its members. The main methodological question then is whether such a panel is representative of the population. To answer this question, more insight must be obtained into the reasons why people join a web panel. This is what Florian Keusch, Bernard Batinic, and Wolfgang Mayerhofer attempt to do in Chapter 8. They do this for an Austrian self-selection web panel.

Their results show that helping to develop better products and services, and entertainment were mentioned more often than monetary incentives. It turns out, however, that monetary incentives do increase the starting rate and completion rate more than other reasons to join. The decision to become a panel member is also partly determined by the way the people are recruited. People who came into the panel through banner advertisement, Google AdWords campaigns or member-get-member systems show high starting rates.

How to keep the starting rates at a high level? This is the question that Annette Scherpenzeel and Vera Toepoel try to answer in Chapter 9 for the LISS Panel. Panel members were constructed using probability sampling. An incentive valued at 15 Euros per hour is used to encourage the panel members to participate in specific surveys taken from the panel.

Scherpenzeel and Toepoel examined the effect of giving feedback to panel members on response rates. They distinguished two types of feedback: traditional feedback like newsletters, free postcards and booklets, and newer innovative feedback like YouTube-like videos about the surveys, or graphs summarizing the survey results.

In general, feedback does not seem to have meaningful and significant effects on long-term participation in the LISS Panel, neither traditional feedback nor innovative feedback. Moreover, there were also no effects within specific demographic groups. Although there were no effects, the innovative feedback was appreciated more than traditional feedback. The conclusion of this chapter is that one should reconsider giving feedback to respondents, taking into account that some forms of feedback may be expensive.

References

Bethlehem, J. G (2009). *Applied survey methods: A statistical perspective.* Hoboken, NJ: John Wiley & Sons, Inc.

Bethlehem, J. G., Cobben, F., & Schouten, B. (2011). *Handbook of nonresponse in household surveys.* Hoboken, NJ: John Wiley & Sons Inc.

Groves, R. M. (2006). Nonresponse rates and nonresponse bias in household surveys. *Public Opinion Quarterly, 70,* 646–675.

Schouten, B., Cobben, F., & Bethlehem, J.G. (2009). Measures for the representativeness of survey response. *Survey Methodology 36,* 101–113.

6

Nonresponse and attrition in a probability-based online panel for the general population

Peter Lugtig[a,c], Marcel Das[b], and Annette Scherpenzeel[a]

[a]*Utrecht University, The Netherlands*

[b]*CentERdata, Tilburg University, The Netherlands*

[c]*University of Essex, United Kingdom*

6.1 Introduction

Online panel surveys have changed social and market research. Especially in applied market research, online panels are a very important tool for conducting surveys. In the early 2000s, nearly all online panels were based on self-selected samples of respondents who have access to the Internet. These self-selected panels offer quick and cheap data collection. This comes at the price of low external validity.[1] Thus, self-selected panel respondents are likely to differ from the population to which the results from these respondents are meant to generalize. More importantly, the nature and size of these potential biases can never be properly assessed, because sampling theories do not apply to studies that do not rely on a random sampling scheme. This lack of external validity has continuously concerned academic researchers and people working in official statistics.

---

[1] External validity refers to how well the findings of a research study generalize to the population the study purports to represent (cf. Campbell & Stanley, 1966; Borsboom, Mellenbergh, & van Heerden, 2004).

Online Panel Research: A Data Quality Perspective, First Edition.
Edited by Mario Callegaro, Reg Baker, Jelke Bethlehem, Anja S. Göritz, Jon A. Krosnick and Paul J. Lavrakas.
© 2014 John Wiley & Sons, Ltd. Published 2014 by John Wiley & Sons, Ltd.
Companion website: www.wiley.com/go/online_panel

Typically self-selected panels draw quota samples from a large database of self-selected respondents, and they can exclude any panel members who did not respond to earlier survey requests. Biases caused by undercoverage, nonresponse and attrition are ignored to a great extent.

To overcome the problem of self-selection, one may draw randomly from the target population and recruit a panel from this sample. When the target population also includes households without Internet access, providing selected households with access solves the coverage problem. Alternatively, those households without Internet may be interviewed using another mode (e.g., mail). This chapter discusses how problematic nonresponse and attrition are in panels where respondents are recruited offline and those without Internet access are given Internet access.

The most notable examples of probability-based online panels are those run by GfK Knowledge Networks (http://www.knowledgenetworks.com/knpanel/) and RAND (https://mmicdata.rand.org/alp/) in the United States, and by CentERdata in the Netherlands (http://www.centerdata.nl/en). In order to solve the undercoverage problem that not all households have access to the Internet, households in these panels are offered Internet access if they do not yet have access (Scherpenzeel & Das, 2011).[2] New initiatives modeled along the lines of CentERdata's LISS panel (see Section 6.3) have recently been started in France (ELIPSS, Étude Longitudinal par Internet Pour les Sciences Sociales; Centre for Socio-political data at Sciences Po) and Germany (GIP, German Internet Panel). The German GESIS Omnibus Panel also includes non-Internet households, but instead of providing them with equipment, these households are approached by mail.

Nonresponse and attrition are problematic in all panel surveys, especially when nonresponse rates are different for specific sub-groups of the population. From earlier studies into attrition in offline, interviewer-administered panel surveys, we know about some of the correlates of attrition. However, nonresponse and attrition may have a different nature in online self-administered panel studies than in offline panels. Because email and the Internet are the primary survey modes in online surveys, they might attract different types of respondents while repelling other groups of respondents as compared to traditional offline panel surveys. Also, over the course of the panel, because of differences in the survey process, and the absence of an interviewer, attrition may be different between online and offline panel surveys. This may lead to a decrease in statistical power, or different biases in online panel surveys than in offline panel surveys.

This chapter aims to answer the following question: What differences can we expect between the correlates of attrition in online panels and offline panels? Using an ongoing online panel dataset from the Netherlands spanning about 50 waves, we illustrate a method that can be used to study the correlates of attrition. For this dataset, we illustrate the correlates of attrition for various groups of respondents and demonstrate that each group follows a distinct attrition pattern over time. Finally, we compare the correlates of attrition with the correlates of initial nonresponse during the recruitment phase of the online panel, in order to see whether the correlates for initial nonresponse and attrition over time differ or not.

---

[2] Leenheer and Scherpenzeel (2013) find that 50% of LISS-households who have Internet acccess, do generally not use their computer for other things than completing the questionnaires, while the other 50% become very light computer users. Further, people do not change their leisure time after being given Internet access. This implies being given Internet access does not radically change these households.

6.2   Attrition in online panels versus offline panels

The contact mode (email) and data collection mode (Internet) make an online panel different from offline panels, which are mostly recruited and conducted through face-to-face interviewing, often complemented with mail, and telephone as secondary modes (cf. Lynn, 2009).

During the recruitment phase of an online panel study, specific groups of respondents might not be willing to complete surveys on the Internet, because they feel unable to do so. People likely to be negative towards the Internet as a survey mode are those who dislike using computers, and specifically the elderly (De Leeuw & Hox, 2011). An earlier experiment by Olson, Smyth, and Wood (2012) has found that some people do not prefer the Internet as a survey-mode, but that almost everyone was open to a mail survey, independent of preferred survey mode. Those people who not prefer online surveys are likely to come from a specific stratum of the target population (Olson, Smyth, & Wood, 2012). So, online panels face larger challenges in the recruitment phase than offline panel surveys. First, they have to make respondents switch from the recruitment mode (offline) to the interview mode (online). Second, some people dislike the Internet as a survey mode. For these reasons, age might be a more important correlate of initial nonresponse in online panels. Some important correlates of nonresponse in cross-sectional surveys are also likely to affect initial nonresponse in panel surveys: people in single households are less likely to be contacted, as are people in highly urbanized areas (Groves & Couper, 1998).

Even when respondents are capable of participating in an online panel, nonresponse in any particular wave of data collection can be very high. Email invitations are used to invite respondents for new waves, and these can easily be forgotten or ignored, possibly leading to higher levels of wave nonresponse than in offline panel surveys. In offline panels, requests for survey participation are usually communicated by an interviewer, making it far less likely that requests are forgotten or ignored. Furthermore, online panels differ from offline panel surveys in the frequency of data collection. In offline panel surveys, interviews are typically conducted annually, whereas in online panels, data collection occurs far more frequently: monthly, or even weekly. Because of the high frequency of data collection in online panel surveys, there are more occasions for respondents not to respond.

These differences imply that methods to prevent nonresponse and attrition may differ between online and offline panels. In offline panels, it is important to keep contact with respondents in between waves of data collection. This is done using change-of-address cards and recording as many contact details as possible (Laurie & Lynn, 2009). More recently, contingent monetary incentives have been offered upon participating in a wave to make sure respondents keep up at least their extrinsic motivation (Laurie & Lynn, 2009; Lipps, 2009; see also Chapter 1 in this volume). Several offline and online panels also try to use communication strategies to keep respondents intrinsically motivated in panel participation, so as to prevent attrition. For this, they use newsletters, season/birthday greetings, and appreciation of participation cards, all of which strive to communicate the importance of participation (Laurie & Lynn, 2009). Chapter 9 in this volume notes that these methods are not always effective, however.

Interviewers play a key role in contacting respondents in offline panels, and trying to ensure the interview is a pleasant experience for respondents. Using high caliber interviewers for those more likely to drop out (Voorpostel & Lipps, 2011), monitoring the quality of interviewers (Schaan, 2011), and refusal avoidance training for interviewers (Kroh, 2011;

Uhrig, 2008; Voorpostel & Lipps, 2011), have been found to be successful in lowering attrition.

Online panel surveys do not use interviewers, but rely on email as the main mode of contact and the Internet for administration of the survey. Non-contacts may occur when respondents do not read emails, but as long as multiple email addresses and phone numbers are collected for every respondent, survey managers can usually trace and contact respondents.

Earlier studies into the correlates of attrition have all used offline panel surveys. Some of the correlates of attrition found in those studies may for that reason not apply to online panel surveys. In this chapter, we limit our analyses to covariates for which we have validation data at the population level. Of those, it has been shown that women attrite less often than men (Behr, Bellgardt, & Rendtel, 2005; Lepkowski & Couper, 2002). Women are thought to be more conscientious and more committed and thus miss fewer waves, although evidence for this is mixed (Uhrig, 2008). Other correlates of attrition are marital status (never being married), whether someone has moved or is planning to move (Lillard & Panis, 1998), and the size of the household (Lipps, 2009). The fact that household composition is important may be due to persuasion by other household members to stay involved in the panel survey or also drop out. Age has been found not to be related to attrition, although the oldest old and children around the age of 18 are more at risk (Lipps, 2009).

As emails are easily ignored or forgotten, the only method that can effectively be used to prevent attrition is trying to keep respondents' motivation high. As noted above, contingent incentives are routinely offered for every wave that is completed. Because both the contact mode and the interview itself are far less personal in online than in offline panel surveys, it is very likely that attrition is a far more common phenomenon in online panel surveys, than it is in offline surveys. Moreover, because of the high frequency of data collection and greater tendency for nonresponse associated with that, it also is likely that the nature of attrition is far more varied, with respondents dropping out, and coming back to the panel survey regularly. This also has implications for the way attrition should be studied.

When investigating attrition bias in offline panels, some researchers pool all wave-on-wave attrition patterns (Nicoletti & Peracchi, 2005; Sikkel & Hoogendoorn, 2008; Watson & Wooden, 2009), and simply discern two groups: the attriters and stayers. This approach ignores the possibility that attrition is different, for example, between Waves 2 and 3 and Waves 7 and 8. Moreover, it does not acknowledge or allow for individual response propensities to change with time. Another approach is to study nonresponse separately for every wave-to-wave transition (Uhrig, 2008). Apart from the fact that this requires many analyses, it is hard to deal with respondents returning to the survey, which implies that respondents can attrite multiple times. Other authors have focused only on the final state of attrition, and have limited themselves to predicting whether attrition occurs or not (Tortora, 2009), or they use duration models controlling for wave effects (e.g., Lipps, 2009). Sikkel and Hoogendoorn (2008) look at the correlates of psychological background variables with the duration of panel membership (in months) and find no relationship.

Durrant and Goldstein (2010) took a more integrative approach and looked at all possible monotonic attrition patterns in a four-wave panel study. With non-monotonic attrition, and longer panel spans, this approach is also challenging. Finally, Voorpostel (2009) and Behr et al. (2005) followed the example of Fitzgerald et al. (1998) and separate a group of attriting ("lost") respondents from returning ("ever out") respondents, thereby also allowing for non-monotonic attrition. In online panel surveys, this last approach of separating "staying," "lost," and "ever out" respondents seems promising. As online panel studies mature, different groups of "ever-out," and "lost" respondents are likely to appear. Thus, from a theoretical and managerial perspective, it makes sense to separate those respondents who only miss a few waves, from those who miss many. Similarly, it is also useful to separate the very fast attriters from those respondents who attrite after many waves of data collection.

Before we turn to a more technical description of how such groups can easily be separated in any set of panel response data, we will briefly describe the dataset that we are using as an example in this chapter.

6.3   The LISS panel

The data for our study stem from the Longitudinal Internet Studies for the Social sciences (LISS) panel, administered by CentERdata in the Netherlands. More information about the recruitment of the panel, response percentages for all waves, as well as the full questionnaires, can also be found on www.lissdata.nl or in Scherpenzeel and Das (2011).

The LISS panel was started in 2007, and interviews respondents monthly about a wide range of topics. The original sample for the panel was based on a simple random sample of Dutch households, who were contacted and recruited using a mixed-mode design. This design included mail, telephone, and face-to-face as the contact modes. Upon establishing contact with one of the household members, all household members were asked to participate in the panel survey. One member of the household served as a reference person and, as the household informant, provided information about the household. After sample members become panel members, they receive an email invitation on the first Monday of every month with an individualized URL that leads them to the questionnaire of that particular month.

The time required to complete questionnaires in the LISS panel is about half an hour per month, and respondents receive an incentive of €15 for every hour of completing questionnaires.[3] They are reminded in case of initial nonresponse in a specific wave, and occasionally receive information about research findings.

New panel members were added to the panel at the end of 2009 and 2011. Most of the new sample members came from top-up samples drawn at those times. In a top-up sample, new respondents are recruited into the panel using a new random sample. In 2009, the top-up sample was stratified based on the inverse response propensities of the original sample, whereas the 2011 top-up sample was again a simple random sample. Other new sample members who entered in 2009 and 2011 consist of household members who have reached the age of 16 and people who have entered the household as a new partner of a panel member.

As long as respondents complete at least one questionnaire item in a particular month, respondents in our study are considered to have participated in that wave. The binary response/nonresponse data from the first 48 waves of the LISS panel serve as our primary dependent variable for studying attrition patterns. As covariates, we will use population data from Statistics Netherlands (2012). For some of these statistics, we have validation information about every sample member. In particular, the variables that we use from the population register are urbanicity of the geographical area and household size. This information was supplemented with data from the recruitment interview. During the recruitment interview, respondents were asked about the household composition and whether they were a home-owner. They were also asked about the age and gender of all other members of

---

[3] Time to complete the survey is estimated beforehand, and every panel member is paid the same corresponding amount.

Table 6.1 Panel recruitment and initial nonresponse rates at various stages.

Sampled	Households N = 9844*	% of sampled 100
Contact successful	8911	91
Completed recruitment interview or central questions	7335	75
Willing to become panel member	5217	53
Household registered as panel member	4722	48

*After panel recruitment was completed, 454 households that were recruited into the panel during a pilot study in 2007 were added to the panel. These households are excluded from Table 6.1, because the pilot study was mainly used for experimentation into successful recruitment strategies for the panel. Including the pilot households, the total number of registered households in the panel is 5005. Within those households, there are 9831 eligible individuals. Of those eligible individuals, 8026 members finally became panel respondents. Some months later, however, a further 122 eligible respondents who first did not want to participate, changed their minds, and did become panel members. These 8148 respondents will be used in our further analyses.

their household. Upon becoming a member of the study, respondents completed a profile questionnaire asking them about their background characteristics. From this dataset, we use a variable that asked respondents about their voting behavior in the Dutch parliamentary elections in 2006. The covariates that we use will not fully explain either nonresponse during the recruitment of the panel, or panel attrition. We use these covariates here for the following reason: these are the best statistics available at the population level to allow for the study of nonresponse and attrition.

6.3.1 Initial nonresponse

Because of the availability of the register data from Statistics Netherlands, we are able to study the correlates of initial nonresponse in the recruitment phase of the LISS panel. Initial nonresponse can occur for various reasons (noncontact, refusal to participate at all in the recruitment interview, or refusal to become a panel member, inability to physically or intellectually participate, etc.). For details on the correlates of nonresponse in every stage of the recruitment phase, we refer to Leenheer and Scherpenzeel (2013). Table 6.1 shows that non-contacts amounted to 9% of all sampled households. At the next stages, more households either refused to participate at all, reported no willingness to become a panel member, or did not complete any online questionnaire. All in all, 48% of the sampled households registered as panel members. A particular effort was made to offer Internet access to households without Internet access.[4] This effort was successful, and was especially effective in targeting older respondents (Leenheer & Scherpenzeel, 2013).

Because panel attrition is an individual attribute, we have chosen to report all our analyses at the individual level.[5] All the statistics from the recruitment interview as reported in Table 6.2 are therefore weighted for household size, but as a simple random sample was used, no further design weights were used.

---

[4] 350 people (4.5% of panel members) received a computer with Internet connection. Home-Internet coverage in the Netherlands was between 80–90% in 2008 (Statistics Netherlands, 2008), implying that nonrespondents were more likely than respondents not to have Internet at their home.

[5] In the attrition analyses also the households that were recruited in the pilot experiment are taken into account.

Table 6.2 Sample percentages for panel members, initial nonrespondents, and surviving panel members in the LISS panel.

Sample percentages	Initial nonrespondents	Initial panel members	Active in April 2008	Active in April 2010	Active in Dec 2011	Population
Over 65%	**20.3***	10.3*	11.5*	11.2*	**11.8***	15.3
Household composition						
single HH	**16.1***	12.7*	13.4*	13.4*	13.8*	14.9
couple without children	**22.3***	31.7*	**34.8***	33.9*	**35.3***	27.0
couple with children	**51.4**	49.4	**46.6***	47.1*	**45.3***	50.4
single with children	**7.8***	5.6*	**4.7***	4.9*	4.8*	6.7
other	**1.4***	0.6*	0.6*	0.7*	0.7*	1.0
Household size (mean)	**1.91***	2.58	**2.83***	**2.88***	**2.83***	2.24
Home-owner	**61.2***	75.6*	76.3*	76.1*	76.2*	68.4
Urbanicity						
extremely urban	**25.8***	12.9*	**13.4***	13.2*	12.5*	19.4
very urban	**20.6***	25.7*	**26.4***	25.9*	26.4*	23.1
moderately urban	**14.5***	22.2*	22.1*	**22.7***	**23.0***	18.3
slightly urban	**15.9***	23.0*	22.5*	**22.6***	**22.3***	19.4
not urban	**23.4***	16.2*	**15.7***	15.7*	15.9*	19.8
Voting behavior 2006						
CDA (Christian democrats)	–	–	25.2	25.3	25.5	26.5
PvdA (labor)	–	–	19.4*	19.4*	19.9	21.2
VVD (liberals)	–	–	15.7	15.4	15.3	14.7
SP (socialists)	–	–	17.3	17.9	17.7	16.6
other parties	–	–	22.4*	21	21.6	21
N	9878[a]	8148	7627	6892	6248	

Notes: 1 Dutch population statistics correspond to individually based statistics, with January 1st 2009 as the base date. Statistics can be found on http://statline.cbs.nl (Statistics Netherlands, 2008).

2 April 2008 was chosen as the start date for showing panel composition, because of the fact that prior to this date, some respondents were still being recruited into the panel.

3 Panel members are respondents who participated in at least one interview until April 2008. We considered a respondent to be an "active" member at any point in time when he/she completed a survey in that particular month, or any of the two preceding months.

4 Household statistics (in the recruitment phase) represent statistics based on individual household members.

Source: Eurostat http://epp.eurostat.ec.europa.eu/statistics_explained/index.php/Housing_statistics

5 *: significant difference with the population using a Chi-square test with $\alpha = .01$. For the population, a sample size equal to the number of panel members was used in conducting the Chi-square test.

6 Statistics in bold: significant difference with initial panel members with $\alpha = .01$. Independent sample Chi-square tests were run for the difference between panel members and nonrespondents, and paired sample t-tests for proportions for the difference between panel members and active panel members in a particular wave.

[a]This number is computed as the nonrespondents from nonresponding households plus nonrespondents from responding households. The number of nonresponding households (5122 – see Table 6.1) is multiplied by the average household size for nonresponding households minus the number of children living in under 16 (who are ineligible) to estimate the number of eligible individuals in nonresponding households. This amounts to 5122 * (1.91 – 0.31) = 8195 respondents.

The number of nonrespondents in responding households is computed by taking the difference between the panel members and eligible respondents from all responding households (9831 – 8148 = 1683 – see Table 6.1). In total, the number of nonrespondents thus amounts to about 8195 + 1683 = 9878. Note that this excludes nonresponse at the household level during the pilot phase of the LISS panel.

Columns 2 and 3 of Table 6.2 show the percentage of sample members who remained nonrespondents in the panel recruitment phase (column 2) and those who were successfully recruited into the panel (column 3). Because of the fact that a simple random sample of households was drawn, any differences that we find between recruited panel respondents and the general Dutch population, are caused by nonresponse (including noncontact and noncoverage). We find that panel members: (1) are less likely to be older (over 65 years old); (2) are less likely to come from single households; (3) are less likely to come from extremely urban areas; and (4) are more likely to be home-owners than are the nonrespondents. When compared to the population statistics, shown in the last column, it is clear that especially older respondents, respondents from extremely urban areas and single households are missing disproportionally from the panel members at the start of the LISS panel.

In columns 4–6, the same statistics are shown at various stages of the LISS panel for all active panel members at that moment. We see that, although slowly, the initial bias that we find for older respondents is decreasing, meaning that older respondents, when they do become a panel member, are less likely to attrite. A similar trend is found for single households, but biases get larger over time for household size and urbanicity and the proportion of singles living with children.

Apart from focusing on socio-demographic characteristics, we also show statistics for one of the substantive variables in the panel for which we have validation data: voting behavior in the Dutch parliamentary elections of 2006. Online panels often are used for opinion polls, and voting behavior is a common topic in such polls.[6] We see that for voting behavior, minor biases in the LISS panel do exist. The proportion of voters for the Christian Democrats and Labor Party is underestimated, while it is overestimated for the Liberals and Socialists. Differences with the population, however, are statistically insignificant.

From Table 6.2, we see that nonresponse and attrition biases in the LISS panel vary over time. This implies that attrition is differential among respondents, or stated differently, that specific groups of respondents are less likely to drop out than others. Furthermore, we find that developments in biases are not consistent. For example, the bias in the proportion of panel members who are over 65 years of age, does not become consistently better or worse over time. Obviously, an approach that studies attrition patterns in more detail will help to illuminate this process further.

6.4 Attrition modeling and results

To model the binary response data for all 48 waves in our study, we have used a Latent Class model. The advantage of using Latent Classes over other methods is that respondents are categorized on the similarity of their response patterns. This does not mean that all respondents within one latent class follow the exact same response pattern. If our data show classes that either always or never participate, this will mean that all respondents who have a high propensity to be a member of that class will closely resemble these extreme types. However, there will also be classes that vary more in their response data. In fact, because every respondent will be assigned to one or more classes, it is likely that some classes still resemble heterogeneous response patterns.

---

[6] Because voting behavior is an individual attribute, we cannot show statistics for the recruitment phase of the LISS panel, as these statistics are all based on household reports.

We treat a response (i.e., completing at least one item in the questionnaire) to a particular wave as 1, and non-response as 0. We did not have a clear hypothesis regarding the number of classes we expected to find, but it is likely that at least 1 class of attriters, 1 class of staying (i.e., non-attriting), and 2 classes that reflect the refreshment samples in 2009 and 2011 will result from our analyses. In addition to this, we believed it was likely that we would find some additional classes describing different patterns of attrition. The attrition pattern within the Latent Classes can take any form; the model is estimated without any constraints on the attrition patterns, and because of that, the solution is entirely based on grouping respondents with similar attrition patterns.

Principally, the fit of our Latent Class model will improve with any additional class that we add to our model, possibly leading to a situation where we have as many latent classes as respondents. To overcome this problem, we used the Bayesian Information Criterion (BIC) as our heuristic for comparing the quality of models with different numbers of classes. Lower values for BIC indicate a better relative fit of the model to the data. As absolute differences for BIC between competing models can be small, it is desirable to test whether any difference between two models is also significant. A Bootstrapped Likelihood Ratio Test is in this case advisable, but we opted for the Lo-Mendell Rubin test, because of the fact that we simultaneously corrected our standard errors for the fact that individual respondents were clustered in households (Nylund, Asparouhov, & Muthén, 2007). Apart from focusing on the fit values as indicated by the BIC, we also inspected values of the entropy (Celeux & Soromenho, 1996) as a criterion for the classification quality. The entropy lies between 0 and 1, where 1 means that every individual can be assigned to a particular class without any measurement error, and 0 means that respondents are randomly assigned to a latent class. Entropy values above .80 are considered to be good (Celeux & Soromenho, 1996).

Finally, we wanted to avoid substantive solutions that included classes with fewer than 5% of all respondents, reflecting a minimum of about 500 respondents per class. This lower limit is somewhat arbitrary, but it is likely that very small classes present only a variant on a pattern that is found in one of the larger classes. All models are estimated using MPLUS 6.12 (Muthén & Muthén, 2011).

Table 6.3 shows the fit statistics of all models with 1–12 classes we tested. From Table 6.3, we prefer the model with nine Latent Classes. The solutions with 10, 11 and 12 classes, respectively, do fit our data better when looking at the BIC. The smallest class of respondents in those solutions, however, includes fewer than 500 respondents. Closer inspection of the attrition pattern for the smallest classes in the solutions with 10–12 classes showed that these classes followed an attrition pattern that mimicked the attrition pattern of the group of slow attriters, though at a somewhat faster or slower rate. For this reason, we chose the model with nine classes as our final model.

Figure 6.1 shows the results of the model with nine latent classes. For each class the response propensities shown indicate the proportion of respondents within each class that participated in each wave of the panel study.[7] The most loyal respondents together (Class 1 –stayers) comprise about 20% of the total sample. These respondents participated in almost all waves of the panel, and showed response propensities close to 1.

---

[7] From Figure 6.1, one can distill that drop-out occurred disproportionally in waves 8 and, to a lesser extent wave 22 of the study. We believe this temporary drop-out to be caused by the topic of those months' questionnaires. The questionnaire of wave 8 was introduced in the invitation as a complex questionnaire on each household's income and assets. In most of attriting classes, respondents returned after this wave, but for respondents who were already in the process of dropping out, receiving this questionnaire only quickened the attrition process.

Table 6.3   Model fit, entropy and smallest class sizes of the Latent Class models for attrition.

	BIC	Lo-Mendell Rubin test p-value	Entropy	Min. class size
1-class	836566	–	1	12476
2-class	527401	$p<0.01$	.991	6057
3-class	449638	$p<0.01$	.992	1580
4-class	393778	$p<0.01$	.987	1565
5-class	372274	$p<0.01$	.978	1484
6-class	359605	$p<0.01$	.974	1126
7-class	349077	$p<0.01$	.974	874
8-class	343771	$p=0.11$	.972	621
9-class	**338674**	$p<0.01$	**.967**	**573**
10-class	335601	$p<0.01$	.966	438
11-class	332204	$p=0.01$	.965	318
12-class	329224	$p<0.01$	.964	314

Note: The Lo-Mendell Rubin tests whether the class solution with 1 class less, fits the data significantly worse than the model with the current class. The class shown in bold is the preferred model.

A second group (Class 2 – Slow Attriters), comprising 13% of the panel, has a stable response propensity at a slightly lower rate, between 0.8 and 0.9. Towards the end of the period, their response propensities do decline to about 0.6.

Three groups follow a traditional pattern of attrition. The fastest group of this subset (Class 6 – Fast Attriters), about 12% of the sample, starts to drop out from the first wave of the panel study, and essentially have dropped out altogether after one year. A second group of this subset (Class 5; 8%) starts to attrite after 1 year. The final group of this subset attrites after about two years of panel membership (Class 3; 4% of the sample).

In addition, we see a group that participates infrequently throughout the panel, showing consistent response propensities between 0.4 and 0.6, indicating that they participate in about every second wave of the panel survey. We call this group "lurkers" (Class 4).

The three remaining classes consist of respondents who were added to the sample over the course of the panel. Two classes are formed by the group of respondents entering the panel in 2009 (the largest one staying [Class 8], but a smaller one dropping out [Class 9]). The final class (Class 7) consists of respondents who, though consenting to become a panel member during the recruitment interview, never start participating. The increase in response propensities in the last two waves for this class is caused by the fact that this also includes a group of top-up respondents and children reaching the age of 16 at the end of 2011, who both enter the panel in the final two waves of our study.

The value for the entropy in the model with 9 classes is very high (.967), implying that there is very little classification error. This means that almost every respondent can accurately be assigned to be a member of one specific class. From here on, we use the Most Likely Class membership to categorize every respondent and look at the composition of each attrition class for the solution with nine Latent Classes.

Table 6.4 shows the composition of each of the classes on socio-demographic background variables. We focus our analyses here on the original sample members who were recruited into the panel in 2007 and leave the top-up sample members in 2009 and 2011 out of the analyses.

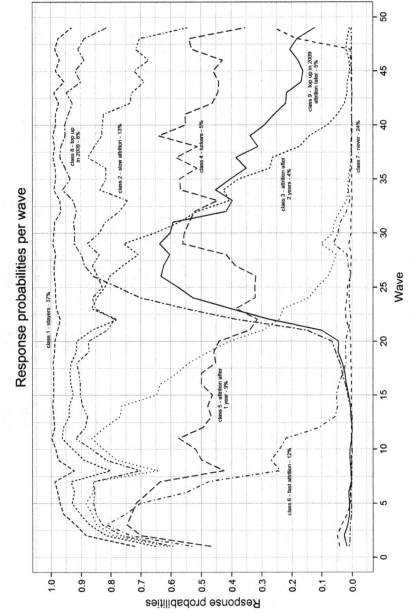

Figure 6.1 Attrition patterns for the nine Latent Classes and their sample fractions.

Table 6.4 Latent Class respondent profiles and population characteristics.

Sample percentages	Class 1 Stayers	Class 2 Slow attriters	Class 3 Attrition after 2 years	Class 4 Lurkers	Class 5 Attrition after 1 year	Class 6 Fast attrition	Class 7 Never participate	Population
N	2465	1572	559	671	1127	1509	245	
sample percent	19.8%	12.6%	4.5%	5.3%	9.0%	12.1%	2.0%	
Male %	47	43	47	48	44	48	50	50
Median age (of over 15)	52	41	51	38	43	43	40	44
Over 65 %	16.2	9.1	13.8	2.4	10.4	8.5	4.7	15.3
Household composition								
single HH	15.7	12.6	14.6	10.8	12.3	11.0	5.2	14.9
couple without children	42.3	30.8	31.6	24.6	33.2	30.8	20.3	27.0
couple with children	37.4	51.1	46.2	57.5	48.3	53.9	67.7	50,4
single with children	3.9	4.9	6.9	6.4	6.0	3.9	5.7	6.7
other	0.7	0.7	0.6	0.7	0.2	0.4	1.0	1.0
Household size	2.60	3.01	2.84	3.05	2.80	2.95	3.52	2.24
Home owner	74.7	77.2	73.1	76.9	79.2	77.1	73.4	68.4

Urbanicity							
extremely urban	10.8	**12.6**	**15.0**	17.9	**12.8**	**14.3**	16.7	19.4
very urban	26.3	**26.6**	**23.2**	27.0	**26.9**	**28.3**	18.1	23.1
moderately urban	22.5	**23.4**	**21.0**	23.2	**21.8**	**19.4**	21.9	18.3
slightly urban	23.8	**22.3**	**23.2**	20.2	**22.6**	**22.5**	25.7	19.4
not urban	16.5	**15.1**	**17.4**	11.8	**15.9**	**15.6**	17.6	19.8
Voting 2006							
CDA (Christian democrats)	26.9	**22.9**	26.0	24.3	28.2	21.6		26.5
PvdA (labour)	20.6	**19.5**	15.5	19.0	20.6	17.5		21.2
VVD (liberals)	14.5	**17.5**	15.8	14.0	15.0	17.4		14.7
SP (socialists)	17.8	**17.1**	18.5	16.5	14.7	**18.4**		16.6
other parties	20.2	**23**	24.2	26.2	21.5	**25.1**		21

Notes: 1 Class 8 – 2009 top-up who stay (N = 949, 8%); class 9 – 2009 top-up who attrite (N = 647 5%) are not shown in the table.

2 The sample size of class 7 is 2975. In computing the statistics for this class, 360 children aged 15 or younger at the start of the panel were not included, as are 2389 respondents who were part of the top-up sample.

3 Voting behavior is not reported for the class of respondents who have never participated, due to too many missing data on this question (> 50%).

Statistics in bold: significant difference between class of stayers (class 1) and other class using a Chi-square test with $\alpha = .01$.

Also, we use the statistics as recorded at the start of the panel, and do not account for any changes in, for example, household composition that occurred over the course of the panel. Household composition and household size do of course change for respondents, but accounting for such changes would make it necessary to work with time-variant covariates at each of our 48 measurement occasions, making model estimation unacceptably time-consuming.

As shown in Table 6.4, large differences exist between the classes. Among our initial sample members, the stayers (Class 1) form the first class in our data (20%). These respondents are more likely to be older, come from single households or live with someone else as a couple without children, and are less likely to live in extremely urban areas than respondents in all other classes.

The class that is most different from the stayers is not surprisingly the group who never participates in the LISS panel. Whereas the median age is highest in the group of stayers, the median age is one of the lowest in the class who never participates (Class 7). People who never participate are least likely to come from single households or be from a couple without children. It is also remarkable that females are overrepresented in all classes, except the class that never participates.

Respondents in all classes differ from each other, but not as much as the stayers differ from respondents who never participate. Those who attrite after two years of panel membership (Class 3) look most like the stayers in terms of age, gender, and household composition. The group of Lurkers (Class 4) resembles the class who never participates quite closely, especially on age and household composition. The other attriting classes sometimes resemble the class of stayers and sometimes the class who never participates; their profile is less clear.

When it comes to differences between the social-demographic composition of every attrition class and the general population, there is no single class that stands out as performing best. All classes seem to differ from the population, but in various ways. Because of attrition, biases in some variables (urbanicity, gender) become worse over time, because the profiles of the attriting classes here come closest to the population estimate. For some other variables (age, household size), biases decrease over time, because for those variables, it is the class of stayers that resembles the population estimate best. The consequences of attrition for estimates of voting behavior are relatively small. Voting percentages for the Christian Democrats were somewhat underestimated at the start of the panel (see Table 6.2), but in the class of stayers, the Christian Democrat vote is well estimated. The only class of respondents that is very different from the population on voting behavior is the class of Fast Attriters. In this class, the percentage of voters for the Christian Democrats and Labor is strongly underestimated.

In conclusion, we find that there are large differences between the different attriting classes. The class of stayers includes respondents who always participate in every wave. We find that every attrition class differs on one or more variables from this class, meaning that attrition itself is selective. When we compare every attriting class to the general population, we see that no class of respondents resembles the population well. In order to answer our last research question, whether initial nonresponse and attrition can be explained by the same process, we take a closer look at the profiles of different groups of nonrespondents in the next section.

6.5   Comparison of attrition and nonresponse bias

Table 6.5 combines information on various types of nonrespondents from Tables 6.2 and 6.4. The second column shows the sample statistics for initial nonrespondents in the recruitment

Table 6.5 Sample percentages of various types of nonrespondents in the LISS panel.

Sample percentages	Initial nonrespondents	Never participate Only recruitment interview	Fast attrition About 6 months of data	Attrition after 1 year About 18 months of data	Attrition after 2 years About 42 months of data	Population
Over 65	20.3	5.7	8.5	10.4	13.8	15.3
Household composition						
single HH	16.1	7.5	11.0	12.3	14.6	14.9
couple without children	22.3	22.3	30.8	33.2	31.6	27.0
couple with children	51.4	61.1	53.9	48.3	46.2	50.4
single with children	7.8	8.4	3.9	6.0	6.9	6.7
other	1.4	0.8	0.4	0.2	0.6	1.0
Household size (mean)	1.91	3.33	2.95	2.80	2.84	2.24
Home owner	61.2	70.1	77.1	79.2	73.1	68.4
Urbanicity						
extremely urban	25.8	16.7	14.3	12.8	15.0	19.4
very urban	20.6	18.2	28.3	26.9	23.2	23.1
moderately urban	14.5	21.9	19.4	21.8	21.0	18.3
slightly urban	15.9	25.5	22.5	22.6	23.2	19.4
not urban	23.4	17.7	15.6	15.9	17.4	19.8

Note: Sample sizes are the same as reported in Table 6.2 (Initial nonrespondents) and Table 6.4. The statistics for the respondents who "Never participate, only recruitment interview" so not correspond exactly to Table 6.4. Here, only respondents are included who participated in zero waves (n = 196).

phase of the LISS panel, while the third to the sixth columns show statistics for four Latent Classes of attriters. There are strong differences between nonrespondents in the panel recruitment phase and the class of respondents who never participate. Although both groups of respondents effectively do not participate in the LISS panel, they differ strongly on age, household composition, household size, and urbanicity. People over 65 years old are commonly found among nonrespondents in the recruitment phase of the panel, but hardly among respondents who did consent to participate, but then never do in practice. This might be related to the fact that about 14% of respondents over 65 years of age received Internet upon consenting to become a panel member; hence quick drop-out among them is unlikely.

All other attrition classes differ from initial nonrespondents as well. One would expect differences between Late Attriters and initial nonrespondents to be larger than for the class that never participates, but we did not find this. The later attrition occurs, the more the class resembles the initial nonrespondents. But even for the most recent attrition class, the differences with initial nonrespondents remain large. In short, *nonrespondents in the recruitment phase are very different from all people who attrite from the panel.*

6.6   Discussion and conclusion

Because of the lack of a sampling frame for sampling the general public via the Internet, several research organizations have established online panels that rely on traditional probability samples. They recruit people using offline survey methods, hoping to overcome the problem of the low external validity of self-selected panels.

This chapter showed how nonresponse and attrition in online panels differ from nonresponse and attrition in offline panels. In general, respondents who cannot use the Internet, or do not like it, are likely to refuse during the recruitment phase to participate in online panels, making initial nonresponse highly selective (nonrandom). The LISS panel that we used in this study addressed this issue by providing households that could not otherwise participate with a computer and an Internet connection. *However, we found that initial nonresponse bias was not entirely eliminated, especially for age.*

Differences in attrition between online and offline panels are related to the mode of contact. Email invitations are easily ignored or forgotten, leading to higher levels of nonresponse. On top of this, the high frequency of data collection requests that online panel surveys use is also likely to lead to higher wave-nonresponse at any given wave. We used a Latent Class model to study the diffuse patterns of attrition that are the results of this. *We found that attrition does not occur gradually or in a linear way.* For example, we find a group of "lurkers", i.e., respondents who participate only in about every second wave of the survey, and do so for the entire period of the panel. We also find differences among attrition patterns. There is a group of respondents that drops out very fast, but also a group that does so very slowly, and a group that stays very loyal for the first two years of data collection, then dropping out after this. We further find that substantively all these classes differ from each other. *Finally, the correlates of nonresponse in the panel recruitment phase and attrition are very different.* This implies that nonresponse during recruitment and panel attrition are different processes. We would welcome studies that try to replicate our findings in other online panel surveys, though such specific characteristics as the frequency of data collection will of course influence the number and size of Latent Classes in every panel survey. This includes replication of our study in a self-selected panel survey. We would suspect that fast attrition does not occur as often in self-selected panels as in our study because of the internal motivation that is necessary to join a self-selected panel, but it would, for example, be interesting to see whether lurkers are present to the same extent in self-selected panels as in our analyses.

The LISS panel does not include sample weights that correct for nonresponse and attrition, and our analyses show why providing them is complicated. The number of separate weights that need to be calculated with 48 waves of data is extraordinarily high. Cross-sectional weights could practically still be computed, but computing and using longitudinal weights becomes complicated when users do not use all waves of the LISS in substantive analyses. One solution to this problem would be to let users calculate nonresponse weights using an interactive (online) module. Multiple imputation of missing data, however, seems a more attractive solution to correct for nonresponse and attrition, because it is more readily available in the major software packages, and can deal with item- and unit nonresponse simultaneously. Missing values because of either item- or wave nonresponse can be imputed using all data that were collected in the same and earlier waves of data collection. Another approach that the LISS panel is undertaking is to try to re-activate specific classes of attriters using a tailored design approach. For example, the lurkers may be activated by contacting them more frequently to keep them involved in the panel. Attriting respondents can perhaps be targeted more effectively by offering an extra incentive, or addressing very specific concerns that led the respondents to drop out.

Although this chapter has only discussed attrition and nonresponse in a probability-based online panel, nonresponse and attrition of course exist in self-selected panels as well. Drawing quota samples, as is often done in self-selected panels (see Chapter 8) does not solve the problem of nonresponse, but rather transforms it into a coverage problem. Self-selected panel managers may stratify their quota samples on past response behavior, effectively only sampling from loyal panel members to overcome the problem of attrition. Response rates calculated for self-selected panels therefore are not informative, and hide the problem of nonresponse that also exists in self-selected panels. Our analyses showed, however, that the profile of loyal panel members is very different from that of the general population, and that those respondents who drop out form important strata of the general population. In self-selected panels, such biases will therefore not appear as much as nonresponse bias, but rather as coverage bias during the self-selection process.

Within the framework of probability-based surveys, coverage and nonresponse biases can be assessed, and corrected by inviting (stratified) top-up samples into the study, or using weighting or imputations. Although this may at times be complicated given the complex data structure of probability-based panel surveys, correcting for undercoverage and nonresponse in self-selected panels is impossible. Due to the absence of a sampling frame and frame variables one cannot accurately weight data from the self-selected sample back to the target population. Even in the case when external validation data at the level of the target population are available for a self-selected panel survey, weighting will only reduce biases when the external validation variables adequately explain both the self-selection process and the variable of interest. In self-selected panels, model-based weighting procedures seldom achieve these dual goals, and for that reason one does not know whether any weighting procedure effectively corrects for coverage and nonresponse bias, or only makes it worse.

References

Behr, A., Bellgardt, E., & Rendtel, U. (2005). Extent and determinants of panel attrition in the European Community Household Panel. *European Sociological Review*, *21*, 489–512.

Borsboom, D., Mellenbergh, G.J., & van Heerden, J. (2004). The concept of validity. *Psychological Review*, *111*, 1061–1071

Campbell, D. T., & Stanley, J.C. (1966). *Experimental and quasi-experimental designs for research.* Skokie, IL: Rand McNally.

Celeux, G., & Soromenho, G. (1996). An entropy criterion for assessing the number of clusters in a mixture model. *Journal of Classification*, *13*, 195–212.

De Leeuw, E. D., & Hox, J. (2011). Internet surveys as part of a mixed mode design. In M. Das, P. Ester, & L. Kaczmirek (Eds.), *Social and behavorial research and the Internet: advances in applied methods and new research strategies.* (pp. 45–76). New York: Routledge.

Durrant, G. B., & Goldstein, H. (2010). Analysing the probability of attrition in a longitudinal survey. University of Southampton Working Paper, (M10/08).

Fitzgerald, J., Gottschalk, P., & Moffitt, R. (1998). An analysis of sample attrition in panel data: The Michigan panel study of income dynamics. *Journal of Human Resources*, *33*, 251–299.

Groves, R. M., & Couper, M. P. (1998). *Nonresponse in household survey interviews.* New York: John Wiley & Sons, Inc.

Kroh, M. (2011) Documentation of sample sizes and panel attrition in the German Socio Economic Panel (1984 until 2010). DIW paper 2011–59. Retrieved February 5, 2013, from: http://www.diw.de/documents/publikationen/73/diw_01.c.385005.de/diw_datadoc_2011-059.pdf.

Laurie, H., & Lynn, P. (2009). The use of respondent incentives on longitudinal surveys. In: P. Lynn (Ed.), *Methodology of longitudinal surveys* (pp. 205–234). Chichester: John Wiley & Sons, Ltd.

Leenheer, J., & Scherpenzeel, A. (2013). Does it pay off to include non-Internet household in an Internet panel? *International Journal of Internet Science*, 8. Retrieved July 2, 2013 from: http://www.ijis.net/ijis8_1/ijis8_1_leenheer_and_scherpenzeel_pre.html.

Lepkowski, J. M., & Couper, M. P. (2002). Nonresponse in the second wave of Longitudinal Household Surveys. In R. M. Groves, et al. (Eds.), *Survey nonresponse.* New York: John Wiley & Sons, Inc.

Lillard, L. A., & Panis, W.A. (1998). Panel attrition from the Panel Study of Income Dynamics: Household income, marital status and mortality. *The Journal of Human Resources*, *33*, 437–457.

Lipps, O. (2009). Attrition of households and individuals in panel surveys. *SOEP Papers*, (164). Retrieved June 28, 2013, from: www.diw.de/documents/publikationen/73/diw_01.c.96125.de/diw_sp0164.pdf.

Lynn, P. (2009). Methods for longitudinal surveys. In P. Lynn (Ed.), *Methodology of longitudinal surveys* (pp. 1–20). Chichester: John Wiley & Sons, Ltd.

Muthén, L. K., & Muthén, B. (2011). *MPLUS.* Los Angeles: Statmodel.

Nicoletti, C., & Peracchi, F. (2005). Survey response and survey characteristics: Microlevel evidence from the European Community Household Panel. *Journal of the Royal Statistical Society Series A*, *168*, 763–781.

Nylund, K. L., Asparouhov, T., & Muthén, B. (2007). Deciding on the number of classes in latent class analysis and growth mixture modeling: A Monte Carlo simulation study. *Structural Equation Modeling*, *14*, 535–569.

Olson, K., Smyth, J. D., & Wood, H. M. (2012). Does giving people their preferred survey mode actually increase participation rates? An experimental examination. *Public Opinion Quarterly*, *76*, 611–635.

Schaan, B. (2011). Fieldwork monitoring in SHARELIFE. In M. Schröder (Ed.), *Retrospective data collection in the survey of health ageing and retirement in Europe* (pp. 29–36). Mannheim: SHARELIFE Methodology, MEA.

Scherpenzeel, A., & Das, M. (2011). True longitudinal and probability-based internet panels: Evidence from the Netherlands. In M. Das, P. Ester, & L. Kaczmirek (Eds.), *Social and behavorial research and the internet: Advances in applied methods and research strategies* (pp. 77–104). New York: Routledge.

Sikkel, D., & Hoogendoorn, A. (2008). Panel surveys. In E. D. de Leeuw, J. J. Hox, & D. A. Dillman (Eds.), *International handbook of survey methodology* (pp. 479–499). New York: Lawrence Erlbaum.

Statistics Netherlands (2008). No Internet access for 1.2 million Dutch people. *web magazine* 18 December 2008. Retrieved June 14, 2013, from: www.cbs.nl/en-GB/menu/themas/vrije-tijd -cultuur/publicaties/artikelen/archief/2008/2008-2641-wm.htm.

Statistics Netherlands (2012). Statline database. Retrieved February 4, 2013, from: http://statline.cbs.nl /statweb/?LA=en.

Tortora, R. D. (2009). Attrition in consumer panels. In P. Lynn (Ed.), *Methodology of longitudinal surveys* (pp. 235–248). Chichester: John Wiley & Sons, Ltd.

Uhrig, S. N. (2008). The nature and causes of attrition in the British household panel survey. ISER Working Paper (5). Retrieved June 28, 2013, from: https://www.iser.essex.ac.uk/publications /working-papers/iser/2008-05.

Voorpostel, M. (2009). Attrition in the Swiss household panel by demographic characteristics and levels of social involvement. *FORS Working Paper, 1_09*. Retrieved June 28, 2013, from: http://aresoas.unil.ch/workingpapers/WP1_09.pdf.

Voorpostel, M., & Lipps, O. (2011). Attrition in the Swiss household panel: Is change associated with drop-out? *Journal of Official Statistics*, 27, 301–318.

Watson, N., & Wooden, M. (2009). Identifying factors affecting longitudinal survey response. In P. Lynn (Ed.), *Methodology of longitudinal surveys* (pp. 157–182). Chichester: John Wiley & Sons. Ltd.

7

Determinants of the starting rate and the completion rate in online panel studies[1]

Anja S. Göritz
University of Freiburg, Germany

7.1   Introduction

Motivating people first to start and then to complete a web-based questionnaire is a challenge. About 15 years ago, marketing research as well as academia began building online panels to tackle this challenge. With a pre-recruited sample from an online panel, field time can shrink to a few days or even to a few hours. Moreover, survey-collected data can be matched to data that have been collected during participants' registration with the panel or to data collected in earlier surveys, thus allowing for richer insights as well as for nonresponse, reliability, and consistency checks. There are also ethical advantages. People in the panel agreed at their sign-up that they could be invited to studies whose purpose could not be disclosed before the research is concluded. Moreover, members of online panels who abandoned a study prematurely could be debriefed because they were accessible via email.

   To gather as much data in panel studies as possible at the lowest possible cost, researchers need to carefully design the panel as well as individual studies that are conducted in the panel. Researchers have tried different measures to achieve this aim, for example, sending reminder emails, keeping questionnaires short, offering a summary of study result, providing rewards. In this review, across the entire life of a university-based online panel (www.wisopanel.net)

---

[1] I thank France-Veronique Scholz for help in coding 138 studies.

Online Panel Research: A Data Quality Perspective, First Edition.
Edited by Mario Callegaro, Reg Baker, Jelke Bethlehem, Anja S. Göritz, Jon A. Krosnick and Paul J. Lavrakas.
© 2014 John Wiley & Sons, Ltd. Published 2014 by John Wiley & Sons, Ltd.
Companion website: www.wiley.com/go/online_panel

Table 7.1   Hypotheses pertaining to 30 predictors of the starting rate and the completion rate.

Hypothesis	Predictor	Starting rate	Completion rate
1	Age of study	+	−
2	No. of panelists in year of study	−	0
3	Growth of panel in year of study	+	−
4	No. of studies in the panel in year of study	+	0
5	Season of study [spring/summer/fall/winter]	+	0
6	Time lag to preceding study [days]	−	+
7	No. of invited panelists to the study at hand	0	0
8	No. of invited panelists to preceding study	0	0
9	Type of study [in-house/commissioned]	0	0
10	Field time [days]	+	−
11	Summary [0,1]	0	0
12	Reminder [no/yes]	+	−
13	Invited because of prior participation [no/yes]	+	+
14	Invited because of demographics [no/yes]	0	0
15	Announcement of study length [no/yes]	+	0
16	Study length [min]	−	−
17	Incentive experiment [no/yes]	0	0
18	Lottery [0,1]	0	0
18_1	Payout [€]		
18_2	No. of lottery prizes		
18_3	Lottery for cash [0,1]		
18_4	Lottery for vouchers [0,1]		
18_5	Lottery for gifts [0,1]		
19	Per-capita reward [0,1]	+	+
19_1	Value [€]		
19_2	Payment via PayPal [0,1]		
19_3	Donation to charity [0,1]		
19_4	Loyalty points [0,1]		
19_5	Bank transfer [0,1]		
19_6	Gift [0,1]		

Notes: + denotes a hypothesized positive effect, − denotes a hypothesized negative effect, 0 denotes a hypothesized null effect

spanning 204 individual studies and 13 years, 30 potential influences (i.e., predictors) on the starting rate and the completion rate were examined (see Table 7.1).

7.2   Dependent variables

The starting rate is the number of invited panelists who called up the first page of a study divided by the number of all invitees. The completion rate is the number of invited panelists who stayed until the final page of a study divided by the number of invitees who called up the first study page. We did not choose the number of invited panelists as denominator with the completion rate to clearly differentiate between the two participation behaviors "starting a study" and "sticking with a study," as study characteristics might influence them differently.

7.3   Independent variables

We coded 30 study characteristics. Of the 30 characteristics, 19 are primary characteristics (i.e., they apply to and were coded in all 204 studies) and 11 are secondary characteristics (i.e., they are hierarchically dependent upon one of two primary characteristics and thus apply to and were coded in only the relevant subset of studies). All primary and secondary characteristics are listed in Table 7.1. The secondary characteristics are indented and listed underneath their hierarchically superior primary characteristic. Five of the 11 secondary characteristics pertain to the reward type *lottery*, and 6 of the 11 secondary characteristics pertain to the reward type *per-capita reward*.

Each study that was conducted in the panel featured either one or more rewards or no reward at all. The rewards in any given study are expressed as percentages, and the total of all rewards per study is 100%. For example, if a particular study was an incentive experiment, 50% of the participants might have been offered a cash lottery, while the other 50% might have been offered no reward. There are two types of reward: First, if every respondent gets a reward we are dealing with a per-capita reward (i.e., loyalty points, payment via PayPal, payment via bank transfer, charitable donation or gift). In each study that relied on a per-capita reward, the percentage of people who received the per-capita reward is coded (e.g., 30%). Moreover, in each study that relied on a per-capita reward the mean of the value of the per-capita reward across the various per-capita rewards offered in this study is coded (e.g., 1.5 €). Second, if only some winners get a reward, then the type of reward is a lottery (in North America often called *sweepstakes*). In each study that relied on a lottery, the percentage of people who were entered into a lottery is coded (e.g., 100%). Forms of a lottery in this panel are a cash lottery, a voucher lottery or a gift lottery. In each study that relied on a lottery, we coded the total payout of the lottery (e.g., 100 €).

7.4   Hypotheses

In the following, we state the hypotheses, which are informed by a review of the literature (see Table 7.1). Each hypothesis has two parts. The part that pertains to the starting rate is denoted by subscript $_a$, and the part that pertains to the completion rate is denoted by subscript $_b$.

> *Hypothesis 1_a: The starting rate is higher, the older a study (i.e., the earlier a study is in the chronological order of all studies conducted in the panel).*
>
> *Hypothesis 1_b: Completion is lower the older a study.*

It is common knowledge that throughout the past decade the public's willingness to respond to survey requests has decreased. One reason might be participation fatigue (Evans & Mathur, 2005). The waning interest in taking part in surveys might be due to the satiation of web users' curiosity as web surveys have now been around for a while, compared to when they were relatively new at the time when this panel was inaugurated. Furthermore, there now is a barrage of survey requests prattling down on web users. In this panel in particular, inactive panelists are not removed from the panel. Thus, the share of inactive panelists in the panel is likely to have risen throughout the life of this panel with the result that the starting rate might have dwindled. In contrast to the starting rate, panelists who already started to respond to a survey are more likely to finish the survey the younger a study as people have improved their computer skills over the years.

> *Hypothesis 2$_a$: The starting rate is lower, the more panelists are in the panel in the year of study.*
>
> *Hypothesis 2$_b$: Completion is unaffected by the number of panelists who are in the panel in the year of study.*

The fewer people are in the panel, the more often an individual panelist is invited into a study. Thus, a high number of panelists who are in the panel in the year of study reflects a low study load. On the one hand, a high study load might tire panelists. On the other hand, if panelists are invited to studies frequently, they might stay familiar with the study and committed to the panel because with each invitation their memory of the panel is refreshed. As the total number of studies conducted in this panel per year is moderate (i.e., panelists get invited to a study every few weeks), we expect the latter effect to prevail: That is, a yearly study load higher than the moderate base rate study load should increase the likelihood that panelists respond to survey requests. Once panelists have responded, there is no obvious reason why their completion should be a function of the number of studies conducted in the given year.

> *Hypothesis 3$_a$: The starting rate is higher, the larger the panel's growth in the year of the study.*
>
> *Hypothesis 3$_b$: Completion is lower, the larger the panel's growth in the year of the study.*

The larger the panel's growth in the year of the study, the higher the number of fresh panelists. Longitudinal research shows that respondents' willingness to respond wanes, the more waves a study lasts, probably mostly due to dwindling interest. Fresh panelists are curious of what awaits them in the panel with which they have signed-up. Thus, fresh panelists should have a higher response probability than seasoned panelists.

Completion in a survey should be influenced by the number of fresh panelists in the panel as well, but in the opposite direction. First, it is more likely that fresh panelists' expectations will differ from what they find once they start a questionnaire, whereas seasoned panelists know what to expect and therefore are less likely to be disappointed, so they are more likely to complete a study. Second, completion depends on panelists' computer skills. On average, fresh panelists should be less Internet savvy than long-term panelists. For example, fresh panelists might be more likely to get stuck in the survey or close the survey inadvertently.

> *Hypothesis 4$_a$: The starting rate is higher, the more studies are conducted in a given year.*
>
> *Hypothesis 4$_b$: Completion is unaffected by the number of studies in a given year.*

The number of studies in a given year reflects the study load on panelists. Given the moderate base rate study load in this panel, a higher study load is expected to increase the starting rate, because the panel stays fresh in panelists' memory. Moreover, Vonk, van Ossenbruggen, and Willems (2006) report that the likelihood of responding to the first survey request after joining the panel decreases with tenure in the panel. The likelihood of panelists responding was 67% when being invited to the first study within the first month of their panel membership. This dropped to 47% for people who had been panel members for longer than 12 months when they were invited to their first study. For the case at hand, the more studies are conducted, the less time elapses before new panelists are invited to their first study, and the less time that elapses until the first invitation, the higher the likelihood of responding.

> *Hypothesis 5_a: The starting rate is higher in winter.*

> *Hypothesis 5_b: Completion is unaffected by the season in which the study is conducted.*

We assume that there are seasonal differences in using the Internet. Despite the rise in mobile Internet usage, using the Internet and consequently participating in online surveys primarily is an indoor activity. Indoor activities are more likely to take place in winter and less likely to take place in the outdoor seasons of spring, summer and fall, with summer having the most outdoor activities. Moreover, in summer a higher share of panelists are on vacation and therefore less inclined or available to respond to a study request. However, once a panelist has sat down for a study, the season of the study should not matter with regard to sticking with the study until the end; in other words, completion should not be affected by season.

> *Hypothesis 6_a: The starting rate is lower, the longer the time lag from the preceding study.*

> *Hypothesis 6_b: Completion is higher, the longer the time lag from the preceding study.*

Another attribute of study load is time lag from the preceding study. Given the moderate base rate study load in this panel, a higher study load is expected to increase the starting rate, because the panel stays fresh in panelists' memory. Moreover, Vonk, van Ossenbruggen, and Willems (2006) report that the likelihood of responding to the first survey request after joining the panel decreases with tenure in the panel. For the case at hand, the smaller the time lag to the preceding study, the less time elapses before a new panelist is invited to his or her first study. By contrast, for people who have started participating in a study, the more time has elapsed since the last study the more eager they are to finish the study at hand, perhaps either to gain the reward on offer or because they feel obliged to take part.

> *Hypothesis 7_{a_b}: The starting rate and the completion are unaffected by how many panelists are invited to the current study.*

There is no obvious reason why the starting rate and the completion rate should depend on the number of panelists invited to the study at hand.

> *Hypothesis 8_{a_b}: The starting rate and the completion rate are unaffected by how many panelists were invited to the preceding study.*

There is no obvious reason why the starting rate and the completion rate should depend on the number of panelists invited to the preceding study.

> *Hypothesis 9_{a_b}: The starting rate and the completion rate are independent of whether a study is conducted in-house or commissioned by an outside sponsor.*

Two types of studies are distinguished: (1) in-house studies initiated by the home unit; and (2) studies commissioned by other universities and noncommercial units. There is no obvious reason why the starting rate and completion rate should depend on the type of study, especially since the type of study is not communicated to invitees.

Hypothesis 10_a: The starting rate is higher, the longer the field time of a study.

Hypothesis 10_b: Completion is lower, the longer the field time of a study.

The longer the field time, the more likely an invitee finds the time to respond to the survey request. This was corroborated in four experiments: Göritz and Stieger (2009) showed that the more days a study is fielded, the higher the starting rate. However, the longer the field time, the more likely people who are not very motivated to take part find nothing better to do than to start the study; but because of their limited motivation, they are less likely to complete the study. Moreover, a long field time renders it more likely that panelists who read their email infrequently read the survey request in time. Panelists who read their email infrequently are less Internet savvy than people who read their email frequently, and hence the former are more likely to drop out of a study for skill reasons (e.g., they close the study inadvertently). While the mechanism remains unclear – be it motivation or skill or both – the effect *per se* has been empirically established: In three of four experiments the longer the study was in the field, the lower the completion rate (Göritz & Stieger, 2009).

Hypothesis 11_{a_b}: The starting and completion rates are unaffected by promising to provide a summary of the results.

Under most circumstances, study results are the least expensive type of incentive. Batinic and Moser (2005) conducted a review of several uncontrolled studies that had been carried out in online panels: On average, studies in which results were offered elicited a lower starting rate. In the same vein, in a commercial online panel study a summary of results decreased the starting rate (Göritz & Luthe, 2013a). Several reasons might account for the deleterious effect of offering to provide study results: First, the invitation email does not specify the content of the result report, so no specific interest is triggered. Second, panelists do not want to keep themselves cognitively occupied beyond the actual study. If a result report is offered, the process of study participation remains unfinished until respondents are sent the report. As receiving results can take quite a while, invited panelists are deterred (Batinic & Moser, 2005). By contrast, in Göritz and Luthe's (2013b) Study 2, which was conducted in a nonprofit panel, giving the panelist study results had no impact on the starting rate and the completion rate. Moreover, in Scherpenzeel and Toepoel (Chapter 9 in this volume) various implementations of giving feedback on survey results had no or inconsistent effects on participation behavior in a probabilistic online panel. However, if topic salience is high (Tuten, Galešic, & Bošnjak, 2004) or if study results are tailored to each participant (Marcus, Bošnjak, Lindner, Pilischenko, & Schütz, 2007), study results do enhance the starting rate as well as the completion rate. In most studies that entered this review, study results were not tailored to each participant, but a sizable share of studies had a salient topic. Overall, offering study results is not expected to have an effect on the starting rate and the completion rate.

Hypothesis 12_a: A reminder increases the starting rate.

Hypothesis 12_b: A reminder reduces the completion rate.

In online panels, reminders are implemented by sending nonrespondents an email to remind them to participate. Except for the time spent sending, reminders are free. Göritz and Crutzen (2012) meta-analyzed 38 studies in three online panels: Reminders increased the starting rate, on average, from 49.5 to 65.6%, but mildly reduced completion rate, on average, from 98.0% among pre-reminder respondents to 97.2% among reminded respondents. In a similar vein, in Kongsved et al. (2007), the starting rate was 17.9% before a reminder and went up to 64.2% after a reminder. In that study, however, reminders were sent by postal mail. In Leopold (2004), the pre-reminder starting rate in an online panel was 39.9% and went up to 53.6% after a reminder was sent via email. With regard to completion, 2.9% of pre-reminder respondents abandoned the survey prematurely compared to 8.0% of post-reminder respondents. Finally, in 68 online panel studies (Batinic & Moser, 2005), the more reminders were used in a study, the higher the starting rate: In the 16 surveys without reminders, the average starting rate was 72%, in the 41 surveys with one reminder, it was 75%, in the five surveys with two reminders, it was 79%, and in the six surveys with three reminders, it was also 79%. Thus, reminders seem to augment the starting rate, which is not surprising as reminders cannot reduce the starting rate. However, reminders seem to reduce completion. Göritz and Crutzen (2012) proposed two mechanisms: First, respondents who participate after receiving a reminder may have lower motivation compared to pre-reminder respondents. This lower motivation is probably the reason why they did not take part prior to receiving a reminder. Second, people who read their email less frequently are less computer-literate. Panelists who participate after a reminder might have lower computer literacy than those who participate before the reminder. Panelists who use the Internet less frequently are less likely to read the initial invitation on time, with the consequence that they are more likely to participate only after the reminder. Having lower computer literacy is associated with dropping out of a study for skill reasons. The two mechanisms – one motivational, the other one skill-based – are not mutually exclusive.

> *Hypothesis 13$_{a\_b}$: The starting rate and the completion rate are higher if only those panelists are invited to a study who have taken part in a previous study versus if panelists are invited independently of having taken part in a previous study.*

Göritz and Wolff (2007) as well as Göritz, Wolff, and Goldstein (2008) showed that panelists who respond or who complete a given wave of a longitudinal study are more likely to respond in the next wave. In addition, Göritz, Wolff, and Goldstein (2008) showed that panelists who complete a given wave of a longitudinal study are more likely to complete a later wave. Furthermore, Göritz (2008) showed that panelists' participation in a given survey wave is more likely if they have taken part in any of three preceding waves.

> *Hypothesis 14$_{a\_b}$: The starting rate and the completion rate are unaffected if people are invited on the basis of their demographics vs. on the basis of chance (i.e., a random sample).*

One would expect particular demographic segments of the panel population (e.g., women vs. men, young panelists vs. old panelists, unemployed panelists vs. employed panelists, fresh panelists vs. seasoned panelists) to differ in their likelihood to heed a survey request and to complete a survey. However, with the dichotomous variable that is at our disposal (i.e., study invitation based on demographics: yes/no), many different demographic selection criteria flow into the yes-category. These different criteria and even complementary values of a single demographic criterion might cancel each other out. For example, in some studies that were coded "yes," people were selected for being fresh in the panel, whereas in other studies that were coded "yes," people were selected for being a seasoned panelist. With this dichotomous predictor it is impossible to single out any criterion and study its isolated effect. On the other hand, as many demographic criteria were used only once or twice across all 196 studies, putting them in a dedicated predictor in an effort to study their effect would open the floodgates to chance outcomes.

> *Hypothesis 15$_a$: The announcement of the length of a study in the survey request increases the starting rate.*

> *Hypothesis 15$_b$: The announcement of the length of a study in the survey request does not affect the completion rate.*

Panelists might appreciate transparency as to what length of study to expect, especially in this panel where announcement of the length of a study is the norm. Moreover, if the survey request does not announce the length of the study, panelists might fear a long study. Once panelists have started to participate in the study, it should not matter whether the study's length was announced because by then they have accepted in an implied-in-fact manner the lack of transparency as well as the possibility that this might be a long study.

> *Hypothesis 16$_{a\_b}$: The starting rate and the completion rate decline with the length of the study.*

Respondent burden rises with the length of a study. The longer the study, the higher the burden, which in turn should lower the starting rate and the completion rate. Accordingly, Galešic and Bošnjak (2009) found that more respondents started a questionnaire if they were told that the study lasts 10 minutes than if they were told that the study lasts 30 minutes. Similarly, according to Deutskens and colleagues (2004), short questionnaires elicit a higher starting rate.

> *Hypothesis 17$_{a\_b}$: The starting rate and the completion rate are unaffected by whether a study incorporates an incentive experiment.*

There is no obvious reason why a study that happens to incorporate an incentive experiment (i.e., a study in which not every participant is offered the same incentive) should elicit a different starting rate and completion rate than a study in which every participant is offered the same incentive. In an incentive experiment each panelist gets his or her own invitation email and hence is offered a particular incentive but does not know that other panelists are offered different incentives.

> *Hypothesis 18$_{a\_b}$: A lottery does not affect the starting rate and the completion rate.*

Lotteries are widespread in web-based research. Compared to per-capita rewards, lotteries are easy to implement and usually cap the costs because most lotteries cost the same regardless of the number of participants. The savings compared to a per-capita reward rise with the number of completers.

With regard to the effect of a lottery of any kind, six cash lotteries did not impact the starting rate and completion rate in a nonprofit panel (Göritz, 2006). Likewise, three cash lotteries staged in a nonprofit panel did not affect the starting and completion rate (Göritz & Luthe, 2013c; Experiments 1, 2 and 4). Furthermore, in Experiment 1 by Göritz and Luthe (2013b) a cash lottery did not affect the starting and completion rate in a nonprofit panel. In Experiment 2 by Göritz and Luthe (2013b), a cash lottery mildly enhanced the starting and completion rate. Moreover, in a commercial panel, different versions of a cash lottery in their entirety increased the starting rate but failed to affect completion (Göritz & Luthe, 2013a). In a commercial panel, Tuten, Galešic, and Bošnjak (2008) tested nine cash lotteries against a no-incentive group and in total observed no impact on the starting and completion rate. Furthermore, in a nonprofit panel, a lottery for vouchers did not affect the starting rate and completion (Göritz & Luthe, 2013c; Experiment 4). Moreover, in a four-wave experiment, a lottery for vouchers increased the starting rate compared with a no-incentive group only in Wave 1 (Göritz & Wolff, 2007), while completion was unaffected. Finally, two gift lotteries conducted in a nonprofit panel by Göritz and Luthe (2013c; Experiments 2 and 3) failed to affect the starting and completion rate.

With regard to lottery payout, in a commercial panel, Göritz and Luthe (2013c) implemented three cash lotteries that differed in payout but did not differ in their impact on the starting and completion rate. Furthermore, in a commercial panel, Tuten, Galešic, and Bošnjak (2008) tested three cash lotteries, but did not find a linear relationship between lottery payout and the starting rate as well as between lottery payout and completion. Finally, Experiment 1 in Göritz (2004), which was conducted in a commercial panel, compared four lotteries that differed in payout. Lottery payout had no impact on the starting rate and completion. Experiment 2 in Göritz (2004), which was conducted in a nonprofit panel, compared two lotteries that differed in payout. Again, lottery payout had no impact on the starting rate and on completion.

With regard to the number of lottery prizes, the starting rate and completion did not differ as a function of the number of lottery prizes in six cross-sectional experiments in a nonprofit online panel (Göritz, 2006). Göritz and Luthe (2013b) conducted two experiments in a nonprofit panel. In both experiments the number of lottery prizes did not significantly affect the starting rate and completion rate. Furthermore, in a commercial panel, Göritz and Luthe (2013c) implemented three cash lotteries that differed in the number of raffled prizes but did not differ in their impact on the starting and completion rate.

To sum up, in most incentive experiments in online panels, no variant of a lottery reliably increased the starting or the completion rate.

Hypothesis 19$_{a\_b}$: The starting rate and the completion rate are higher with a per-capita reward.

Because it is difficult to pay online panelists cash, a monetary per-capita reward is usually paid using online intermediaries such as PayPal or in a proprietary currency such as loyalty points. In PayPal, an option called mass payment makes it possible to pay many recipients simultaneously. However, each transaction as well as the mass payment itself costs a fee. Loyalty points can be redeemed by the panelist against money, upon reaching a threshold.

Experimental findings on per-capita rewards in online panels are sparser than findings on lotteries. In a five-wave experiment, initially there was no difference in the starting rate between a group that was offered loyalty points and a group that was offered a cash lottery. Over time, however, loyalty points outperformed the lottery (Göritz, 2008). Moreover, Göritz (2004) compared loyalty points, a cash lottery and a gift lottery: The starting rate with loyalty points was significantly larger than with the cash lottery. In a five-wave experiment,

Göritz (2008) sent a mousepad to one half of new members of a commercial online panel; the other half did not get any incentive. The prepaid gift significantly increased participation in the first wave of the study. Furthermore, in a three-wave experiment in a nonprofit panel, Göritz, Wolff, and Goldstein (2008) promised one group of participants 1.50 Euros to be paid via PayPal for their participation at each wave. This was the first time that these panelists had been offered a payment via PayPal in this panel. The control group was not offered any reward. The promise of the payment reduced the starting rate in Wave 1 (36% vs. 45%), but increased it in Wave 2 (80% vs. 60%). Completion was not affected.

With regard to the value of a per-capita reward, Göritz (2004) tested four amounts of loyalty points. While there was no significant association between the starting rate and the number of loyalty points, completion was higher, the higher the number of loyalty points.

To sum up, in some experiments a per-capita reward increased the starting rate and the completion rate.

7.5   Method

The WiSo-Panel is an opt-in online panel at the University of Freiburg in Germany. This panel has been in operation since 2000. The panel members are volunteers from all walks of life. During the 13 years of the panel's lifetime, the panel has continuously admitted new panelists and the panel has remained in the same hands. Studies conducted in this panel vary in topic, but are all noncommercial. A total of 204 web-based studies have been conducted in the panel from its inception through the end of 2013. Only one of the 204 studies had missing data on the starting rate and the completion rate. We imputed these two data points by inserting the mean across the ten surrounding studies. With seven studies we knew the number of completers but lacked the somewhat higher number of respondents. We decided to keep those studies in the sample by calculating a conservative starting rate derivative: number of completers divided by number of invitees. These studies' completion rate was considered missing.

Every study begins by emailing a survey request to all or to a selection of panelists. In the invitation, panelists are told the study topic, study length (with the exception of a few studies), the closing date as well as the type and amount of incentive. For their participation in a study, panelists usually are offered one or more incentives from a choice of different incentives such as a donation to a charity, a gift, money paid via PayPal, money transferred into their bank account, a certain number of loyalty points, inclusion in a lottery, study results or no incentive.

The 30 predictors are listed in Table 7.2. The two dependent variables *starting rate* (started/invited) and *completion rate* (finished/started) are expressed as percentages. In Step 1, for each of the two dependent variables a multiple linear regression analysis weighted by sample size of the study (i.e., number of invitees) was conducted using the 18 primary predictors. In this way, larger studies have a higher weight towards the overall result. We sought to identify a parsimonious model using the backward method. To assess multicollinearity, we examine the variance inflation factor (VIF). For each independent variable in our final models, VIF does not exceed 5, which corresponds to the rule of thumb proposed by Urban and Mayerl (2006, p. 232). As "season of study" is a nominal predictor, we entered the three seasons spring, summer and fall, as dummy-coded variables and left "winter" as the reference category.

Step 2 follows only if the primary predictors *lottery* and/or *per-capita reward* were significant in Step 1. Here, we explored which variant of the lottery and/or the per-capita reward worked best. To this end, we regressed each dependent variable on the secondary predictors using the backward method weighted by sample size of the study

Table 7.2 Description of predictors.

Predictor	N	Mean	SD	Min.	Max.
Age of study [days]	204	1,884	–	22	4,759
No. of panelists in year of study	204	7,682	4,111	317	11,928
Growth of panel in year of study	204	928	1,040	−404	4,001
No. of studies in year of study	204	18.5	6.1	6	25
Season of study	204	spring: 59, summer: 55, fall: 48, winter: 42			
Time lag to preceding study [days]	204	31	28	0	197
Invited panelists to current study	204	2,782	2,750	47	12,031
Invited panelists to preceding study	204	3,218	3,119	65	12,031
Type of study	204	in-house: 164, commissioned: 40			
Field time [days]	204	11	8	1	58
Summary [0,1]	204	nobody: 133, some: 47, everybody: 24			
Reminder	204	yes: 16, no: 188			
Invited because of prior participation	204	yes: 51, no: 153			
Invited because of demographics	204	yes: 69, no: 135			
Announcement of study length	204	yes: 198, no: 6			
Study length [min]	204	11.8	5.4	2	45
Incentive experiment	204	yes: 75, no: 129			
Lottery [0,1]	204	nobody: 126, some: 22, everybody: 56			
Payout [€]	78	117	126	5	800
No. of lottery prizes	78	5.1	6.3	1	39
Lottery for cash [0,1]	78	nobody: 19, some: 16, everybody: 43			
Lottery for vouchers [0,1]	78	nobody: 64, some: 5, everybody: 9			
Lottery for gifts [0,1]	78	nobody: 72, some: 2, everybody: 4			
Per-capita reward [0,1]	204	nobody: 112, some: 46, everybody: 46			
Value [€]	92	1.45	0.78	0.50	5.00
Payment via PayPal [0,1]	92	nobody: 56, some: 33, everybody: 3			
Donation to charity [0,1]	92	nobody: 58, some: 34, everybody: 0			
Loyalty points [0,1]	92	nobody: 25, some: 30, everybody: 37			
Bank transfer [0,1]	92	nobody: 61, some: 29, everybody: 2			
Gift [0,1]	92	nobody: 87, some: 2, everybody: 3			

7.6 Results

7.6.1 Descriptives

Across all 204 studies, 567591 survey requests were sent out. Upon these survey requests, 182424 panelists called up the first study page, which is a mean of 894 panelists per study (SD = 709; minimum: 27, maximum: 3144). This amounts to a mean starting rate of 43.2%. Across the 197 studies with available completion rate, 154511 panelists stayed until the final study page, which corresponds to a mean of 784 panelists per study (SD = 617; minimum: 25, maximum: 2836). This amounts to a mean completion rate of 85.9%. A total of 75 studies incorporated an incentive experiment that compared at least two incentive conditions that differed either in kind such as summary vs. no summary or in amount as in 2 × 20 € vs. 4 × 20 €. Some of the 129 studies that did not incorporate an incentive experiment featured a combination of incentives (e.g., a summary alongside a reward), but in contrast to the incentive experiments all invitees received the same incentive(s). A total of 78 studies featured a lottery – sometimes alongside other incentives – whereas 92 studies relied on a per-capita reward – sometimes alongside other incentives. All study characteristics are depicted in Table 7.2.

7.6.2   Starting rate

Table 7.3 depicts the final model pertaining to the starting rate: $F(12, 191) = 28.49, p < .001$, $R^2_{adj} = .62$. The starting rate is higher, if the following conditions apply:

- the fewer panelists that are in the panel in the year of the study;
- the larger the panel's growth in the year of the study;
- it takes place in winter;
- the fewer panelists that are invited to the given study;
- if the study is in-house;
- if a reminder is sent;
- if panelists are invited who took part in a previous study;
- if the study's length is announced in the survey request;
- the shorter the study;
- if a per-capita reward is offered

Table 7.3   Significant predictors of the starting rate and the completion rate.

Predictor	Starting rate		Completion rate	
	β	p	β	p
Age of study			−.17	.02
No. of panelists in the year of study	−.26	<.001		
Growth of panel in the year of study	.10	.05		
No. of studies in the year of study			−.15	.03
Spring study	−.19	.001	.23	<.01
Summer study	−.29	<.001	.20	.01
Fall study	−.14	.02	.17	.03
Time lag to preceding study			.12	.05
Invited panelists to study at hand	−.32	<.001		
Type of study	−.17	<.001		
Reminder	.32	<.001		
Invited because of prior participation	.35	<.001	.14	.02
Announcement of study length	.09	.05		
Study length	−.10	.03	−.42	<.001
Per-capita reward	.16	.002		

As the primary predictor *per-capita reward* was significant, next we regressed the starting rate on the secondary predictors pertaining to a per-capita reward: $F(1, 90) = 8.38, p < .01$, $R^2_{adj} = .08$. Of these secondary predictors (i.e., value of per-capita reward, payment via PayPal, donation to charity, loyalty points, bank transfer and gift), only donation to charity reaches significance: $\beta = .29, p < .01$.

7.6.3   Completion rate

Table 7.3 depicts the final model: $F(8, 188) = 11.17, p < .001, R^2_{adj} = .29$. Completion is higher under the following conditions:

- the younger a study;
- the fewer the number of studies in the year of study;
- in seasons other than winter;
- the longer the time lag to the preceding study;
- if panelists are invited; who took part in a previous study;
- the shorter the study.

7.7   Discussion and conclusion

Twenty-six of the 38 hypotheses were supported (see Table 7.4). Next, we discuss possible reasons for the 12 refuted hypotheses.

$H1_a$: Against expectation, younger studies do not elicit a lower starting rate. Perhaps this is due to two facts: (1) The panel continuously admits new panelists whose initial interest in responding to surveys is comparable across the years. (2) While inactive panelists are not removed by the panel operator, sooner or later most inactive panelists unsubscribe of their own accord or their email address becomes invalid and they are therefore removed from the panel; thus they do not noticeably depress the starting rate.

$H3_b$: Completion in a study is not lower, the larger the growth of the panel in the year of study. The larger the growth, the higher the number of fresh panelists in the panel. Perhaps it is true that fresh panelists' expectations more likely to differ from what they find once they start a questionnaire, whereas seasoned panelists know what to expect and therefore are in this respect more likely to complete a study. Moreover, it might be true that fresh panelists – because they are less Internet-savvy than long-term panelists – might inadvertently drop out of a study more likely. However, these mechanisms – if they apply at all – might be offset by a higher level of curiosity and drive among fresh panelists that makes them more likely to carry through until the end of a study than seasoned panelists.

$H4_{a\_b}$: Conducting more studies in the panel does not elicit a higher starting rate. Perhaps the base rate study load is sufficient to keep the panel in the panelists' memory, or the number of studies in the panel in the year of the study is too indirect an indicator of survey load. Moreover, against expectations, conducting more studies in the panel in the year of the study reduces the completion rate. Perhaps, panelists become more choosy the more often studies are offered.

Table 7.4   Outcome of hypothesis tests.

Hypothesis	Predictor	Starting rate	Completion rate
1	Age of study [days]	✗	✓
2	No. of panelists in the year of the study	✓	✓
3	Growth of panel in the year of the study	✓	✗
4	No. of studies in the panel in year of the study	✗	✗
5	Season of study [spring/summer/fall/winter]	✓	✗
6	Time lag to the preceding study [days]	✗	✓
7	No. of invited panelists to the study at hand	✗	✓
8	No. of invited panelists to the preceding study	✓	✓
9	Type of study [in-house/commissioned]	✗	✓
10	Field time [days]	✗	✗
11	Summary [0,1]	✓	✓
12	Reminder [no/yes]	✓	✗
13	Invited because of prior participation [no/yes]	✓	✓
14	Invited because of demographics [no/yes]	✓	✓
15	Announcement of study length [no/yes]	✓	✓
16	Study length [min.]	✓	✓
17	Incentive experiment [no/yes]	✓	✓
18	Lottery [0,1]	✓	✓
19	Per-capita reward [0,1]	✓	✗

Notes: ✓ denotes a hypothesis that was upheld, ✗ denotes a refuted hypothesis.

H5$_b$: Completion is not unaffected by the season in which the study is conducted. Instead, completion was significantly higher in seasons other than winter. Perhaps this is due to mood effects. Alternatively this might reflect a self-selection effect that carries over from the lower response rate in these seasons. As in spring, summer and fall, people have better things to do than to participate in a web-based study, those who do participate after all must have high motivation. These highly motivated respondents are more likely to complete a study than the on-average less-motivated respondents in winter.

H6$_a$: A larger time lag to the preceding study does not decrease the starting rate. Perhaps the base rate study load is sufficient to keep the panel in the panelists' memory. Moreover, a time lag may reflect study load too indirectly because a time lag codes the minimum interval between studies, but individual respondents' actual interval is often longer as each panelist is not invited to each study. At any rate, further studies could examine if there is an ideal interval between any two studies.

H7$_a$: The starting rate is not unaffected by how many panelists are invited to the current study. Instead, the fewer panelists are invited to a study, the higher the starting rate. A post-hoc explanation is that to more exotic and therefore potentially more interesting studies usually a smaller sample was invited than to ordinary studies.

H9$_a$: The starting rate is not unaffected by whether a study is in-house or commissioned by an external sponsor. Namely, in-house studies elicit a higher starting rate than commissioned studies. In-house studies usually have a psychological topic, whereas commissioned studies are more diverse in topic. Perhaps studies with a psychological topic are more interesting to panelists. After all, these panelists have registered with a panel that is maintained at a university Psychology department.

$H10_{a\_b}$: With regard to the field time of the study, there was a tendency in the expected direction in that a longer field time augmented the starting rate; but this tendency failed to reach a conventional level of significance. Moreover, a longer field time did not reduce the completion rate. Perhaps most panelists take part during the first few days after receiving a survey request so that the actual field time makes no noteworthy difference. Thus, Göritz and Stieger's (2009) experimental evidence was not confirmed.

$H12_b$: Contrary to Göritz and Crutzen (2012) but confirming Batinic and Moser (2005), sending a reminder does not reduce completion. However, the completion-reducing effect of a reminder in Göritz and Crutzen (2012) was small.

$H19_b$: The completion rate was not significantly higher if panelists were offered a per-capita reward. However, a tendency in this direction was found that fell short of a conventional level of significance.

As the offer of a per-capita reward enhanced the starting rate we explored that a donation to charity works best. Also worthy of note: There was no significant linear relationship between the value of a per-capita reward and the starting rate.

7.7.1   Recommendations

When it comes to maximizing the starting rate to a study, clients who enlist the services of online panels have some useful tips to follow: While little can be done if the study is commissioned and about the sample size of the study, if there is a choice of different panels, a client may engage the smallest panel, but one that has recently grown in members. Moreover, if possible, the client may commission the study in winter, order that a reminder be sent to nonrespondents, announce the length of the study in the survey request, keep the study short, have panelists invited to the study who took part in a previous study, and offer a per-capita reward. When it comes to maximizing completion to a study, one might conduct the study in panel where fewer studies have recently been run, in seasons other than winter, one might invite panelists who took part in a previous study, conduct the study after a longer time lag to the previous study and keep the study short. To maximize the starting rate and completion rate at the same time, one should keep the study short and have panelists invited who took part in a previous study.

7.7.2   Limitations

As with all research studies, this study has some limitations. First, only data from the German WiSo-Panel were analyzed. This panel, being a typical panel, and this analysis comprising all web-based studies that have been conducted in this panel (hence there is no sample error), there are no *a priori* reasons why results should not generalize to other opt-in, especially non-commercial, panels. Of course, this should be put to the test. Moreover, as the WiSo-Panel is a German panel, we do not know if results would hold in panels in other countries. However, again, there are good reasons to assume that this would be the case. Germany is a typical country in the group of industrialized, Western countries. Naturally, this assumption of comparability should be tested empirically. Moreover, it is conceivable that results will only partly apply or will not apply in panels in non-Western countries.

Second, weighted linear regression was used to analyze the data. As only linear associations can be detected, this review remains silent about nonlinear associations. For example, it is conceivable that there is a curvilinear relationship between time lag to the preceding study and the starting rate. A very short time lag and a very long time lag might give rise to a low starting rate, whereas a medium time lag might elicit a higher starting rate.

Third, the regression analyses proceeded stepwise. There is a tradeoff between a stepwise and a direct-enter approach. With the stepwise method, the program chooses the model that represents the given data best, of which the researcher tries to make sense, often in a post hoc manner (Backhaus et al., 2010). With the direct-enter method, if there are many predictors, as in the case of this study, there often is high multicollinearity. In view of this tradeoff, we opted for the stepwise approach, thereby keeping multicollinearity low and reducing the downside of the stepwise approach by calculating an adjusted R^2-value and by formulating hypotheses *a priori*.

Fourth, there might be interaction effects among predictors that remain undetected in the present analyses. We did not hypothesize and test for any interactions because prior knowledge as well as theorizing about interactions is almost nonexistent, and there would have been a huge number of interactions to possibly test. As this study has paved the way with regard to main effects, it is a springboard for research on moderators.

Fifth, it is unlikely that this review has captured all the predictors that exert an influence on participation behavior in online panels. Future research should look at additional predictors, such as topic salience. In a similar fashion, the precision with which we could examine whether panelists' demographics determine participation in panel studies was insufficient. Future research should look at demographic predictors, such as proportion of women in the sample and mean age of the sample. The best way of examining demographic predictors is by way of experiment.

Sixth, in this review, the starting rate and the completion rate were the dependent variables. They are quantitative facets of participation behavior. However, there might be tradeoffs with other facets of participation behavior, especially qualitative facets, such as length of open-ended answers, item omissions, straight-lining and no-comment responses. This knowledge gap points to the need for even more comprehensive studies.

Seventh, the availability of funding, length of a study, and clients' requests determined the type and amount of incentives offered for participation in particular studies. Some of these determinants were included in the analysis (i.e., type of study, study length), so they are controlled for. We were unable to test all possible determinants of offered incentives as to whether they had an influence on the outcome of this review, especially on incentive effects.

Finally, the review that was conducted is non-experimental. It summarizes uncontrolled studies that vary in more than one characteristic at the same time. Thus, internal validity is limited. Experimental evidence – for example, with regard to field time (Göritz & Stieger, 2009) – should be given a higher weight. Associations found in this study should be examined in more detail and with more methodological rigor in dedicated experiments. As such, this review is a starting point for further research.

References

Backhaus, K., Erichson, B., Plinke, W., & Weiber, R. (2010). *Multivariate Analysemethoden* (13th ed.). Berlin: Springer.

Batinic, B., & Moser, K. (2005). Determinanten der Rücklaufquote in Online-Panels. *Zeitschrift für Medienpsychologie, 17*, 64–74.

Deutskens, E., De Ruyter, K., Wetzels, M., & Oosterveld, P. (2004). Response rate and response quality of internet-based surveys: An experimental study. *Marketing Letters, 15*, 21–36.

Evans, J. R., & Mathur, A. (2005). The value of online surveys. *Internet Research, 15*, 195–219.

Galešić, M., & Bošnjak, M. (2009). Effects of questionnaire length on participation and indicators of response quality in a web survey. *Public Opinion Quarterly, 73*, 349–360.

Göritz, A. S. (2004). The impact of material incentives on response quantity, response quality, sample composition, survey outcome, and cost in online access panels. *International Journal of Market Research, 46*, 327–345.

Göritz, A. S. (2006). Cash lotteries as incentives in online panels. *Social Science Computer Review, 24*, 445–459.

Göritz, A. S. (2008). The long-term effect of material incentives on participation in online panels. *Field Methods, 20*, 211–225.

Göritz, A. S., & Crutzen, R. (2012). Reminders in web-based data collection: Increasing response rates at the price of retention? *American Journal of Evaluation, 33*, 240–250.

Göritz, A. S., & Luthe, S. C. (2013a). Lotteries and study results in market research online panels. *International Journal of Market Research, 55*, 27–42.

Göritz, A. S., & Luthe, S. C. (2013b). How lotteries and study results influence response behavior in online panels. *Social Science Computer Review, 31*, 371–385.

Göritz, A. S., & Luthe, S. C. (2013c). Effects of lotteries on response behavior in online panels. *Field Methods, 25*, 219–237.

Göritz, A. S., & Stieger, S. (2009). The impact of the field time on response, retention, and response completeness in list-based Web surveys. *International Journal of Human Computer Studies, 67*, 342–348.

Göritz, A. S., & Wolff, H.-G. (2007). Lotteries as incentives in longitudinal Web studies. *Social Science Computer Review, 25*, 99–110.

Göritz, A. S., Wolff, H.-G., & Goldstein, D. G. (2008). Individual payments as a longer-term incentive in online panels. *Behavior Research Methods, 40*, 1144–1149.

Kongsved, S. M., Basnov, M., Holm-Christensen, K., & Hjollund, N. H. (2007). Response rate and completeness of questionnaires: A randomized study of Internet versus paper-and-pencil versions. *Journal of Medical Internet Research, 9*, e25.

Leopold, H. (2004). *Rücklauf bei Online Befragungen im Online Access Panel.* Hamburg: Kovač.

Marcus, B., Bošnjak, M., Lindner, S., Pilischenko, S., & Schütz, A. (2007). Compensating for low topic interest and long surveys: A field experiment on nonresponse in web surveys. *Social Science Computer Review, 25*, 372–383.

Tuten, T. L., Galešic, M., & Bošnjak, M. (2004). Effects of immediate versus delayed notification of prize draw results on response behavior in web surveys: An experiment. *Social Science Computer Review, 22*, 377–384.

Tuten, T. L, Galešić, M., & Bošnjak, M. (2008). Optimizing response rates and data quality in web surveys: The immediacy effect and prize values. In L. O. Petrieff, & R. V. Miller (Eds.), *Public Opinion Research Focus* (pp. 149–157). Hauppauge, NY: Nova Science Publishers.

Urban, D., & Mayerl, J. (2006). *Regressionsanalyse: Theorie, Technik und Anwendung* (2nd ed.). Wiesbaden: Verlag für Sozialwissenschaften.

Vonk, T., van Ossenbruggen, R., & Willems, P. (2006). The effects of panel recruitment and management on research results. *Panel Research*, ESOMAR, Barcelona.

8

Motives for joining nonprobability online panels and their association with survey participation behavior

Florian Keusch[a], Bernad Batinic[b], and Wolfgang Mayerhofer[c]

[a]*University of Michigan, USA*
[b]*University of Linz, Austria*
[c]*WU (Vienna University of Economics and Business), Austria*

8.1 Introduction

The Internet has not only dramatically changed people's daily lives but also had a tremendous effect on how traditional research methods are implemented and the way in which scientists conduct research today (Denissen, Neumann, & van Zalk, 2010). However, one of the major obstacles to conducting surveys online is still the issue of representation (i.e., coverage) of the general population in a country. Although many industrialized countries have Internet penetration rates of 75% and above, there seems to be mutual consent that the results from online surveys cannot be generalized to populations beyond that of Internet users (Bethlehem, 2008; Comley, 2007; Couper & Miller, 2008). Web surveys are a viable mode for well-defined populations, of which an exhaustive sampling list (i.e., frame) of individual email addresses exists (e.g., students of a university, employees of a company, members of a professional organization) (Couper, 2000). However, surveying general populations even with virtually full Internet coverage is not possible because there is no centralized sampling frame that includes all people with Internet access. Compared with dwellings and telephone numbers,

Online Panel Research: A Data Quality Perspective, First Edition.
Edited by Mario Callegaro, Reg Baker, Jelke Bethlehem, Anja S. Göritz, Jon A. Krosnick and Paul J. Lavrakas.
© 2014 John Wiley & Sons, Ltd. Published 2014 by John Wiley & Sons, Ltd.
Companion website: www.wiley.com/go/online_panel

from which exhaustive sampling frames can be created, there is no method for creating a frame that enumerates all Internet users and subsequently drawing random samples from the list. Recruitment via banner ads, links, and other ad-hoc sources is often expensive and suffers from rather unpredictable outcomes (Göritz, 2004a). Furthermore, using the email addresses of individuals with whom the survey sponsor had no prior established relationship (i.e., cold addresses) is not only seen as a violation of netiquette but is also illegal in most countries and therefore considered an absolutely unacceptable survey practice (Dillman, Smyth, & Christian, 2009). Altogether, strict data protection laws, the increased use of mobile phones, and a general growing unwillingness of consumers to voluntarily provide time and information for survey research purposes in recent years have led to the rise of online panels (Bowman, 2007), which have been shown to be a fast, easy, and cost-effective alternative for collecting large amounts of data.

Numerous commercial as well as non-profit research organizations have turned to pre-recruiting consumers into online panels (Göritz, 2004a). In academia, online panels are now used as a source for recruiting respondents to collect survey data in various fields (Paul & Batinic, 2010; Ensher & Murphy, 2011). By opting in for an online panel, willing persons give their consent to occasionally receive invitations for web surveys and provide information on personal interests, behavior, lifestyle, and ownership, as well as basic demographic characteristics, all stored as profile data in the panel's database. Therefore, the costs of locating potential respondents can be substantially reduced and clients of online panels benefit from the immediate availability of large samples with various key characteristics and from shorter field times (Evans & Mathur, 2005; Göritz, 2004b).

Samples drawn from online panels have proven to produce fewer break-offs (O'Neil, Penrod, & Bornstein, 2003) and higher participation rates (Schillewaert & Meulemeester, 2005) than have other online recruiting techniques. However, these relatively high participation rates are not surprising because nonrespondents would have already dropped out during the recruitment stage and only motivated respondents would thus be left in the panel. Although response rates used as measures of the combined success of panel recruitment and invitation efforts cannot be quantified because of the self-selection approach of nonprobability online panels, research on probability online panels shows that nonresponse during recruitment can be extensive (Lee, 2006). In general, it must be noted that participation rates in nonprobability online panels are not necessarily a good indicator of sample quality. These rates instead reflect the effectiveness of panel management because the panel provider can actively influence participation rates by excluding hard-to-reach groups and selecting only responding panelists (Willems, van Ossenbruggen, & Vonk, 2006). Additionally, many commercial online panel providers report that participation rates have decreased noticeably in recent years.

Critics especially denounce the self-selection approach used by nonprobability online panel providers, which allows users to volunteer themselves for panels (Bethlehem & Stoop, 2007; Couper, 2000; Postoaca, 2006), as well as the heavy emphasis on monetary rewards for participation (Baker et al., 2010; Smith & Brown, 2005; Sparrow, 2006). This nonrandomized sampling technique bears the risk of attracting heavy Internet users who are only monetarily motivated. To reduce potential coverage error, Couper (2000) suggests not using such volunteer panels and rather employing pre-recruited online panels that actively contact potential members using established offline probability-based techniques. Still, this latter method does not overcome the problem of nonresponse bias from systematic drop-outs at the stage of panel recruitment and survey invitation. "The online access panel sample will eventually consist of only those people who could be reached through the Internet (coverage), joined the online access panel (selectivity) and responded to the survey invitation ((non-)response)" (Loosveldt & Sonck, 2008, p. 94). However, Bethlehem and Biffignandi (2012) show that the worst case bias of a large self-selection survey can be "a factor 13 x larger" (p. 313) than the bias of a small probability survey.

Despite the methodological flaws with regard to the generalizability of web survey data to general populations, the use of data gathered via nonprobability online panels for marketing decision and public opinion purposes is widespread and further growth is anticipated (Baker et al., 2010). Therefore, it is essential to learn more about the people who readily give their permission to be contacted regularly for research purposes, to understand their intentions and motives for doing so, and especially to examine their participation behavior in these nonprobability online panels.

This chapter seeks to provide a detailed look into the motives of nonprobability online panel members for joining and how these motives are related to survey participation behavior. First, we start with an overview of the relevant literature on theories that can help explain participation behavior in surveys in general and online panels in particular. Additionally, we present the results from studies conducted by panel providers throughout the world who have directly asked their members about their motives for joining the online panels as well as scientific research on survey participation behavior in online panels. We then present data from a new study conducted in an Austrian nonprobability online panel that establishes a link between motives for online panel enrollment and survey participation behavior in the panel. The chapter concludes with a discussion of the outcome of the study and its theoretical and practical implications.

8.2   Motives for survey participation and panel enrollment

Different reasons have been identified to help explain participation decisions regarding surveys in general and online panels in particular. Some authors have conceptualized the participation decision process in formal theories of survey participation (for extensive overviews, see Albaum, Evangelista, & Medina, 1998, or Bosnjak, Tuten, & Wittmann, 2005). Porst and von Briel (1995) list three global motives for participating in surveys: (1) altruistic (e.g., helping the sponsor of the survey or society in general); (2) survey-related (e.g., interest in the topic or method, liking the interviewer); and (3) personal reasons (e.g., personality traits, professional interest). The scientific literature has primarily focused on survey-related factors, such as number of contacts, personalization, questionnaire design, and incentives, and their influence on survey participation behavior. A wide body of literature exists that guides survey researchers in developing effective surveys (for web surveys, see Best & Krueger, 2004, Couper, 2008, or Dillman et al., 2009). Regarding personal factors, the interest of individual respondents in a given topic has been identified as a major contributor to high response rates in offline (Dommeyer, 1985; Groves, Presser, & Dipko, 2004; Groves, Singer, & Corning, 2000; Heberlein & Baumgartner, 1978; van Kenhove, Wijnen, & de Wulf, 2002; Martin, 1994) and web surveys (Galesic, 2006; Marcus, Bosnjak, Lindner, Pilischenko, & Schütz, 2007).

8.2.1   Previous research on online panel enrollment

One line of research on online panels that is primarily pursued by panel providers, mainly to identify potential fraudulent subjects, has concentrated on describing different groups of panel members, for example, in terms of their participation and response behavior

(Fisher, 2007; Smith & Brown, 2005) or their relationships with the provider (Comley, 2007), and on identifying reasons for registering with an online panel (Cape, 2008; Postoaca, 2006; Sparrow, 2006). Another line of research has attempted to determine the factors that influence the participation and response behavior of panel members. Researchers have used different recruitment strategies (Göritz, 2004a), sociodemographics (Coen, Lorch, & Pierkaski, n.d.; Garland, Santus, & Uppal, n.d.; Knapton & Myers, 2005), panel tenure (Coen et al., n.d.; Dennis, 2001; Toepoel, Das, & van Soest, 2008, 2009), and incentives (Göritz, 2004b) as predictors for survey participation behavior in online panels. Less stress has been placed on general personality traits and their influence in the participation decision process. Rogelberg, Conway, Sederburg, Spitzmüller, Aziz, and Knight (2003) used two of the "Big Five" personality traits (Costa & McCrea, 1992; Goldberg, 1990) to predict participation in a survey. They demonstrated that, in a student survey, nonrespondents were less conscientious and less agreeable than respondents. More recently, Brueggen and Dholakia (2010) used the Big Five personality traits together with measures of need for cognition, mindful attention and awareness, and curiosity to examine the participation decision in a Dutch student online panel with a short time span. They showed that students who joined the online panel scored higher on scales measuring the need for cognition, curiosity, extraversion, and agreeableness than did students who decided not to join the online panel. Survey participation in the online panel was driven by the same traits as those for panel entry, except for openness instead of extraversion.

Some research exists that looks directly at individuals' reasons for panel enrollment. Jacobson and Jordan (1993) found that for a paper-and-pencil longitudinal survey of nurses, the major reasons for volunteering were curiosity, financial remuneration, the desire to participate in something important, and the desire to help the researchers. Participants cited breaking monotony, acquiring prestige, peer pressure from participating friends, and the wish to be promoted as less important reasons for volunteering in the study. A number of online panel providers directly asked their members about the motives for joining their panel. Over 52% of the people seeking to join the British IMC online panel said that it was an enjoyable way to earn money or enter prize draws. Only 20% wanted to become panelists because they thought that they would be interested in the topics covered and another 19% indicated that they enjoyed answering questions (Sparrow, 2006). A membership survey of the Australian Opinion World panel, operated by Survey Sampling International (SSI), showed similar results. Again, money was by far the most important main reason for joining (34%), followed by the opportunity to win prizes (16%). Fewer people enrolled because they enjoyed voicing their opinions (15%) or having the opportunity to influence important decisions (15%) (Cape, 2008). With other online panels, monetary motives seemed to play a minor role in joining. The panelist satisfaction study conducted by Ipsos among 5000 of their over 200000 panel members in the most important European markets revealed that only 13% saw panel enrollment as a good way to make more money. By far the largest group of members said that they joined to contribute their opinions and wanted to improve products and services (43%). Other factors that played a major role in becoming an Ipsos panelist were curiosity (15%) and fun (13%) (Postoaca, 2006). Comparable results were found for nonprobability online panels in Germany (Rodenhausen, Kaufmann, & Hartmann, 2007) and Austria (Marketagent, 2009).

Willems et al. (2006) compared 19 online panels in the Netherlands as part of the Dutch Online Panel Study NOPVO 2006 and found that the majority of panel members (54%) cited intrinsic drivers such as curiosity, the opportunity to express one's opinion, or being part of a community as their main reason for participating in web surveys. 26% of Dutch panelists said that surveys were a "nice thing to do" and 17% cited incentives as their main argument. The study also showed that those motives were correlated with survey participation and response behavior in the online panel. According to Willems et al. (2006), "professional respondents" (members of multiple online panels and therefore people who complete a great number of web surveys) mainly thought that web surveys were "nice to do." "Loyal respondents," i.e., respondents who took more time in answering the questionnaire and put more effort into the open-ended questions, were mainly motivated by intrinsic drivers. "Inattentive respondents," characterized by speeding and providing minimum input to open-ended questions, were rather incentive-driven. Walker, Pettit, and Rubinson (2009) found that members of online panels that offer monetary incentives are more likely to participate in surveys to support their income than are members of online panels that offer no monetary incentives. Brueggen, Wetzels, de Ruyter, and Schillewaert (2011) developed a Survey Participation Inventory (SPI) comprising 32 items that measured eight motives for participating in online panel surveys. They tested the SPI with one of the largest online panels in Belgium and identified three clusters of panel members. The largest group included "Intrinsics," who were driven by a number of motives such as enjoyment, interest, and curiosity. This group showed the highest response rate in the online panel. "Voicing assistants" were mainly motivated by motives such as giving an opinion and helping. The smallest group, with the lowest response rate, contained the "Reward seekers," who participated in web surveys primarily because of the incentives they received. Although Brueggen et al. (2011) conclude that "incentives are apparently not the main motivational driver behind online panel participation" (p. 370), there do seem to exist "different levels of primary motivation being intrinsic or extrinsic depending on the local culture and recruitment methods to the panel" (Cape, 2008, p. 4).

8.2.2   Reasons for not joining online panels

Only a few studies have examined people's motives for not joining panels. Olson and Klein (1980) found from the interviewers' documentation that lack of time (43%), invasion of privacy (28%), and dislike of the government as the sponsor of the survey (15%) were the main reasons to refuse the first face-to-face interview for one longitudinal income survey. SSI asked respondents of cross-sectional web surveys for their motives for not joining online panels as well as former panel members who had quit the panel. The main reasons, apart from never having been asked (36%) and not knowing what an online panel was (18%), were fear of commitment (26%), lack of time to take more surveys (13%), and not wanting to receive more emails (13%). The main reasons for leaving the online panel were that participants were upset about not qualifying for any surveys (34%), receiving not enough (24%) or too many survey invitations (13%), and again lack of time (14%) (Cavallaro, 2012).

8.2.3   The role of monetary motives in online panel enrollment

Overall, the results of previous research on motives for enrollment in online panels reveal that monetary reasons seem to play the dominant role for some panels; however, with other online panels, members were more motivated to join from non-monetary motives, such as helping to produce better products and services, curiosity, or just fun. Extensive research on the influence of unconditional cash incentives on participation in self-administered offline surveys has shown that such incentives are one of the most successful boosters of response rates in both cross-sectional (Armstrong, 1975; Brennan, 1992; Edwards et al., 2002; Mizes,

Fleece, & Roos, 1984; Yu & Cooper, 1983) and longitudinal studies (Berk, Mathiowetz, Ward, & White 1987). Unconditional cash incentives (i.e., noncontingent incentives) can be used during the recruitment stage for online panels if potential panel members are contacted via mail or in person. However, the virtual character of the online realm makes it difficult for survey researchers to provide noncontingent incentives, because cash cannot be delivered directly with an email invitation for web surveys and because many Internet users are still suspicious of online services that allow payments and money transfers via the Internet (Bosnjak & Tuten, 2003; Göritz, Wolff, & Goldstein, 2008), such as PayPal. Although Göritz (2005) could demonstrate in a meta-analysis of web surveys conducted with academic and commercial online panels that unconditional incentives (e.g., prize draws, e-coupons for online shops) increased the likelihood of survey participation compared with conditional incentives offered only to those respondents who completed the entire questionnaire, it is common practice among online panel providers to reimburse participants based on a bonus points system. The more often a panelist completes a questionnaire in the online panel, the higher are the payouts in the form of bank transfers, gift certificates, or donations. This bonus point system is highly acceptable to online panel members (Postoaca, 2006).

8.3   Present study

The aim of the present study was twofold. First, it was our intention to describe online panel members in terms of their attitudes toward online panels. Second, the correlation between the motives a panelist had when joining the online panel and the panelists' participation behavior was analyzed. Linking the results of the survey to historic participation behavior data of online panel members helped examine the association between enrollment motives and survey participation behavior. Hence two research questions were formulated and answered in the course of this study: (1) What are the main reasons for joining a nonprobability online panel? and (2) How are these motives associated with participation behavior?

8.3.1   Sample

To answer the research questions, a study was conducted in an online panel environment. In June 2009, members of one of the largest Austrian nonprobability online panels – which recruits new volunteer members via a variety of online sources – were invited to participate in a web survey. Panel members received an email from the panel provider announcing that a new ten-minute web questionnaire had been made available online. For full participation, respondents received an incentive of 100 panel points, equaling € 1. A total of 1803 panelists filled out the entire web questionnaire.

8.3.2   Questionnaire

By clicking on the embedded link in the email invitation, respondents were automatically transferred to the introductory page of the web questionnaire. There, respondents were thanked for their decision to take part in the survey and were informed about the sponsor (the Institute for Advertising and Marketing Research at WU, Vienna University of Economics and Business[1]) and the topic of the survey ("Buying Behavior and Personal Attitudes") as well as

---

[1] The data collection was part of the first author's dissertation project at the Institute for Advertising and Marketing Research at WU, Vienna University of Economics and Business, Austria.

the estimated time to complete the questionnaire (10 minutes). By clicking the Start button, respondents proceeded to the next page, which displayed the first question of the survey. The web questionnaire used an adaptive design including a number of automated skips and tailored questions based on answers given by respondents to previous questions. The questionnaire contained a maximum of 120 items presented on 25 screens, including the introductory and final pages. The questionnaire included several multi-item scales, and whenever multiple items used the same rating scale, they were presented in a grid format on one screen. Other screens just presented individual items, such as questions about age and gender.

The questionnaire covered various topics that can be divided into four main categories. It collected information about (1) the reasons for becoming a member of the online panel; (2) general attitudes and personality (e.g., the Big Five); (3) media usage, consumer behavior, personal interests, and recreational activities; and (4) socio-demographic characteristics of the respondents. In this chapter, we focus on the first topic.

To characterize the participating online panel members in terms of their attitudes toward the panel, respondents were asked to select all applicable items from a list of seven reasons for joining the online panel (see Appendix 8.A.1 for the list of items in German):

1. I was curious what the online panel was about.

2. I want to give my opinion to help create better products and services.

3. Panel membership is a good way of earning some extra money.

4. I am always the first to try out new things.

5. I like to be the first to learn about new products and services.

6. Answering surveys is fun.

7. Other reason, please specify:_____

The questionnaire also asked respondents to indicate the number of other online panels in which they were enrolled.

In addition to directly asking panelists whether their motives for enrolling in the online panel were of a monetary nature (see Item 3 above), the concept of materialism was employed to measure the level of monetary orientation of online panel members. Different scales that claim to capture the materialism of individuals have been developed (see Richins & Dawson, 1992, for an overview of proposed instruments) and used in research on topics, including happiness in life (Belk, 1984; Richins, 1987), general consumer satisfaction (Richins & Dawson, 1992), and saving and spending behavior (Belk, 1985). It has been widely recognized that materialism is a multifaceted construct comprising factors such as possessiveness, nongenerosity, and envy (Belk, 1984; 1985) and success, centrality, and happiness (Richins, 2004; Richins & Dawson, 1992). Because the current study focuses on the monetary factor of materialism, the scale developed by Richins (1987) was used. It not only captures a strong orientation towards earning money but also comprises just seven items, thus limiting the burden for respondents. Respondents had to rate the items on a five-point scale from 1, "agree strongly," to 5, "disagree strongly" (see Appendix 8.A.2 for list of items in German):[2]

---

[2] In the original study, Richins (1987) found the scale to represent a bi-dimensional construct: (1) personal materialism comprising items 1, 2, 3, and 4; and (2) general materialism, represented by items 5 and 6. The original study excluded item 7 because it was the only item that loaded heavily on a third factor.

1. It is important to me to have really nice things.

2. I would like to be rich enough to buy anything I want.

3. I'd be happier if I could afford to buy more things.

4. It sometimes bothers me quite a bit that I can't afford to buy all the things I would like.

5. People place too much emphasis on material things.

6. It's really true that money can buy happiness.

7. The things I own give me a great deal of pleasure.

Respondents who broke off the survey by closing the web browser could log back on to the questionnaire by clicking again on the link in the email invitation and continue from the point where they had left off. The web questionnaire was available online for 12 days.

On average, it took respondents approximately 12 minutes to complete the entire questionnaire. Two senior academic marketing research experts who were otherwise uninvolved with the study independently concluded that the questionnaire could not be thoroughly read, fully understood, and filled out seriously in less than five minutes. Therefore, the data of 51 respondents who had obviously just rushed through the questionnaire to become eligible for the incentive and did not pass this threshold were eliminated from further analysis,[3] reducing the sample to 1752 respondents.

8.3.3 Data on past panel behavior

From the online panel provider's database we extracted additional information about past participation behavior in the online panel, the type of recruitment that brought members into the panel, and the date of entry. For 149 respondents, no information about entry into the panel was available. Therefore, these respondents were excluded from all analyses reducing the final sample size for this study to 1612.

The members' past survey participation behavior in the online panel was measured as the proportion of accepted invitations in the online panel (starting rate) and the proportion of surveys the panelists broke off (break-off rate) during the period of 12 months prior to the current survey.[4] The advantage of those two measures over the actual proportion of entirely filled out questionnaires per respondent (completion rate) is that it is not affected by incomplete survey participation caused by screen-outs and filled quotas, two factors that cannot be influenced by the individual panelist. On average, the responding members had received 20 email invitations from the panel provider during the previous year (min.: 1, max.: 42) and had accepted approximately three out of four (min.: 0%, max.: 100%) by clicking on the link embedded in the invitation and starting to fill out the web questionnaire. Panelists broke off fewer than 13% of all web surveys started during the last year.

---

[3] Additional analysis of the answers from the 51 respondents who were excluded revealed that these respondents were also more likely to show other satisficing behavior than respondents not speeding through the questionnaire, such as nondifferentiation in grid questions (41.2% of excluded sample straight-lined in at least one of six grid questions vs. 2.2% of remaining sample; $\chi^2 = 233.709$, $df = 1$, $p < .001$), selecting neutral responses on rating scales (five t-tests comparing measures for midpoint responding (MRS) by Weijters, Schillewaert, and Geuens (2008); $p < .05$), and item omission (1.2 missing items for excluded sample vs. .3 for remaining sample; $t = 1.652$, $df = 50.3$, $p = .105$).

[4] There is a significantly negative correlation between the starting rate and the break-off rate ($r = -.234$, $p < .001$).

The records in the panel provider's database showed that the majority of the 1612 respondents had joined the online panel through invitations that were sent out by commercial address providers (permission marketing; 27.5%) and through the takeover of another online panel when the panel provider acquired the respondent pool of another provider (21.7%). Some 12.5% of the respondents were recruited at the end of intercept web surveys on various websites (intercept surveys); 11.5% each entered the online panel through cooperation with another online survey platform and a referral system; 8.9% joined through Google AdWords campaigns; and 6.5% joined through banner advertisements. The membership tenure of the panelists spanned from a minimum of 23 days to a maximum of more than three years.

8.3.4 Analysis plan

This current study investigated the relationship between motives for joining the online panel and panelists' subsequent panel behavior in several steps. First, we present a descriptive analysis of the answers to the questions on motives for joining the online panel. Then, bivariate relationships between citing a reason for joining the online panel (1= "citing a motive," 0 = "not citing a motive") and the panelists' gender (0 = "female," 1 = "male"), age (1 = "30 and younger," 2 = "31–50 years," 3 = "51 and older"), and education level (0 = "without high school degree," 1= "with high school degree") were examined using Pearson's Chi-Square test. Next, an exploratory factor analysis helped us analyze the answers to the items on Richins's (1987) materialism scale. Differences in the materialism index between socio-demographic groups were then analyzed with bivariate analysis (t-test and ANOVA). Finally, the correlation of motives for joining the online panel and materialism with participation behavior was examined, first, by employing bivariate analysis (Pearson's Chi-Square test) and then using OLS regression. For both continuous dependent variables (starting rate and break-off rate during the last year), two models each predicted survey participation behavior of the online panelists. The first set of models included predictors for the motives for joining the online panel as six binomial variables and a continuous measure of materialism. Additionally, we controlled for the socio-demographic characteristics of the panel members with one dummy variable for gender (reference group = "female"), two dummy variables for age (reference group = "30 and younger"), and one dummy variable for education (reference group = "without high school degree"). The second set of models additionally included controls for the method of recruitment (six dummy variables with "panel takeover" as the reference group) and continuous variables for panel tenure in days and the number of other online panels in which a member was enrolled.

8.4 Results

8.4.1 Motives for joining the online panel

When asked to name all of their reasons for joining the online panel, 67.2% of the respondents said that they had enrolled in the panel to help develop better products and services. Roughly the same proportion of members cited having fun when participating in surveys as a reason for joining (66.6%). Only 40.1% said that they wanted to earn some extra money, and 38.8% were just curious about what an online panel was all about. Less than one-third of the panel members reported that they had joined because they wanted to be among the first to learn about new products (31.2%) or considered themselves early adopters, who are always among the first to try out new things (24.2%). About 3% of the respondents selected the "Other reason, please specify" option in this question. Because of the substantially low number of useful answers, only six out of 57 respondents who chose this answer option actually typed a reason in the answer box, the responses to this option were not used for further analysis. Almost every second respondent (49.4%) reported that s/he was enrolled in at least one other online panel. Every tenth respondent was a member of three or more other online panels.

Table 8.1 shows that significantly more females said that they had joined the panel for monetary reasons and because they thought that filling out questionnaires was fun. Regarding age, it can be said that older panel members more often reported joining out of curiosity, to be the first to learn about new products, and because surveys are fun, whereas younger panel members cited money significantly more often as a reason for joining the online panel. Panelists without a high school degree stated more often that they wanted to be the first to learn about new products and services than did members with a high school degree.

8.4.2   Materialism

Initial exploratory factor analysis (principal component analysis with varimax rotation) supported Richins's (1987) findings that the materialism scale is not unidimensional. The Kaiser criterion and a scree plot suggested a three-factor solution (eigenvalue Factor 1: 2.60; Factor 2: 1.13; Factor 3: 1.07; Factor 4: 0.69). As in the original study, the third factor had high loadings for one item only. In the present study, however, it was item 5 ("People place too much emphasis on material things") that formed a new factor, whereas in the original study it was item 7 ("The things I own give me a great deal of pleasure"). After excluding item 5, the analysis was performed again producing a two-factor solution with an overall measure of sampling adequacy (Kaiser-Myer-Olkin) of .760 and a significant Bartlett's test of sphericity (Chi square = 2089.76, $df = 15$, $p < .001$). The two extracted factors captured 62.0% of the variance among the items.

Table 8.2 shows the varimax-rotated loading matrix, including the means and standard deviations for the six items. Items 2, 3, 4, and 6, representing "materialism toward money", load highly on the first factor, whereas items 1 and 7 formed the second factor, "materialism toward possessions." The items loading highly on Factor 1 showed high internal consistency with a Cronbach's alpha of .78 (see Table 8.3). This did not apply to the two items forming Factor 2 (Cronbach's alpha = .38). Because it was the intention of this study to capture solely the monetary side of materialism in the first place, Factor 2 was excluded and we used only the summed score of the scale measuring materialism toward money for further analysis.

Differences in materialism toward money between socio-demographic groups were found for age and education but not for gender. Materialism toward money seemed to decrease with age (F = 14.81, $df = 2/1591$, $p < .001$). With regard to education level, panelists without a high school degree scored higher on the materialism toward money scale than did those with a degree (t = −4.74, $df = 1570$, $p < .001$).

8.4.3   Predicting survey participation behavior

Table 8.4 shows the regression models for predicting the starting rate for the online panel during the previous year. The coefficient of determination for Model 1, including motives for joining the online panel and controlling for respondents' socio-demographics, is low but significant ($R^2 = .03$, $p < .001$). Controlling for age, gender, and education, respondents who had joined the online panel because they were curious what it was all about had almost a 4

Table 8.1 Motives for joining the online panel (in %) by gender, age, and education.

	Overall	Gender		Age			Education	
		male (N = 802)	female (N = 810)	30 and younger (N = 585)	31 to 50 years (N = 725)	51 and older (N = 302)	without high school degree (N = 642)	with high school degree (N = 947)
1. I was curious what the online panel was about.	38.8%	40.6%	36.9%	33.7%	40.1%	45.5%	40.2%	37.7%
			n.s.	$\chi^2=12.50$, $df=2$, $p=.002$			n.s.	
2. I want to give my opinion to help create better products and services.	67.2%	65.0%	69.5%	64.4%	68.8%	68.9%	68.4%	66.4%
			n.s.	n.s.			n.s.	
3. Panel membership is a good way of earning some extra money.	40.1%	34.4%	45.8%	52.6%	36.4%	24.8%	40.7%	39.8%
		$\chi^2=21.75$, $df=1$, $p<.001$		$\chi^2=71.74$, $df=2$, $p<.001$			n.s.	
4. I am always the first to try out new things.	24.2%	24.4%	24.0%	25.1%	24.6%	21.5%	26.8%	22.8%
			n.s.	n.s.			n.s.	
5. I like to be the first to learn about new products and services.	31.2%	32.5%	29.9%	27.0%	31.7%	38.1%	36.8%	27.2%
			n.s.	$\chi^2=11.54$, $df=2$, $p=.003$			$\chi^2=14.70$, $df=1$, $p<.001$	
6. Answering surveys is fun.	66.6%	62.2%	71.0%	62.1%	66.8%	75.2%	69.3%	64.4%
		$\chi^2=13.93$, $df=1$, $p<.001$		$\chi^2=15.42$, $df=2$, $p<.001$			n.s.	

Note: N = 1612.

Table 8.2   Varimax-rotated two factor structure of materialism items including means (M), standard deviation (SD), and factor loadings (F1, F2).

	M	SD	F1	F2
3. I'd be happier if I could afford to buy more things.	2.75	1.24	.85	.04
4. It sometimes bothers me quite a bit that I can't afford to buy all the things I would like.	1.78	1.24	.82	.03
2. I would like to be rich enough to buy anything I want.	2.48	1.27	.77	.14
6. It's really true that money can buy happiness.	2.94	1.11	.58	.38
7. The things I own give me a great deal of pleasure.	1.84	.77	−.04	.82
1. It is important to me to have really nice things.	2.64	1.05	.22	.72
5. People place too much emphasis on material things. (r)[a]	1.98	.95	(excluded)	

Note: N = 1612.

Table 8.3   Number of items (N), mean (M), standard deviation (SD), and reliability coefficient (Cronbach's alpha) of the two materialism scales.

	N	M	SD	Cronbach's alpha
Materialism toward money	4	13.06	3.77	.777
Materialism toward possessions	2	7.52	1.45	.382

Note: N = 1612.

percentage point lower starting rate than did members who did not cite this reason. In contrast, monetarily-motivated panel members seemed to be highly willing to regularly participate in web surveys, showing a starting rate that was almost 7 percentage points higher than that for those who did not cite this motive. The same held true for people who thought that surveys were fun, with a starting rate that was 4 percentage points higher than that of respondents who did not cite this motive. Model 2 added controls for panel recruitment, panel tenure, and number of other online panels in which a respondent was enrolled, increasing R-squared to .06 ($p < .001$). Without substantial changes in the main predictors from Model 1, Model 2 shows that the panel members who had been recruited into the online panel through banner advertisements, member referrals, Google AdWord campaigns, and intercept surveys showed starting rates that were 12, 10, 6, and 5 percentage points higher, respectively, than those for respondents who had become members through a panel takeover. Additionally, the higher the number of other online panels in which a member was enrolled, the higher was the proportion of started surveys in the online panel.

Table 8.4 Estimated coefficients from OLS regressions of starting rate in the online panel during the last year (%).

	Model 1			Model 2		
	b(s.e.)	beta	p	b(s.e.)	beta	p
Intercept	73.72 (3.40)		<.001	67.34 (4.42)		<.001
Motive: Curiosity	-3.73 (1.41)	-.07	.008	-2.95 (1.40)	-.05	.036
Motive: Better products/services	1.42 (1.50)	.02	.344	1.59 (1.49)	.03	.287
Motive: Incentives	6.77 (1.46)	.12	<.001	5.30 (1.47)	.10	<.001
Motive: Novelty	-2.03 (1.71)	-.03	.233	-1.52 (1.69)	-.02	.367
Motive: First to know about new products/services	1.74 (1.61)	.03	.279	.57 (1.60)	.01	.722
Motive: Entertainment	4.04 (1.48)	.07	.007	2.96 (1.48)	.05	.045
Materialism toward money	-.21 (.19)	-.03	.263	-.23 (.19)	-.03	.253
Gender (Female)[a]						
Male	-.93 (1.41)	-.02	.508	.56 (1.43)	.01	.695
Age (30 and younger)[a]						
31–50 years	.65 (1.59)	.01	.683	1.53 (1.58)	.03	.332
51 and older	-1.54 (2.09)	-.02	.461	-.55 (2.12)	-.01	.796
Education (Without high school degree)[a]						
With high school degree	.51 (1.44)	.01	.721	1.42 (1.44)	.03	.326
Recruitment (Panel takeover)[a]						
Permission marketing				1.65 (2.43)	.03	.497
Intercept surveys				5.13 (2.61)	.06	.049
Survey platform				2.08 (2.53)	.02	.410
Google AdWords				6.44 (3.24)	.07	.047
Banner advertisement				12.42 (3.40)	.11	<.001
Referral				10.37 (3.26)	.12	.002
Panel tenure (in days)				-.01 (.01)	-.01	.768
Number of other online panels enrolled				2.30 (.58)	.10	<.001
R²		.03			.06	
N		1,612			1,612	

[a]Reference categories in parentheses.

Table 8.5 Estimated coefficients from OLS regression of break-off rate in the online panel during the last year (%).

	Model 3			Model 4		
	b(s.e.)	beta	p	b(s.e.)	beta	p
Intercept	17.85 (2.05)		<.001	11.81 (2.68)		<.001
Motive: Curiosity	.60 (.85)	.02	.481	.77 (.85)	.02	.362
Motive: Better products/services	−1.08 (.91)	−.03	.233	−1.09 (.90)	−.03	.224
Motive: Incentives	−2.52 (.88)	−.07	.004	−2.30 (.88)	−.07	.010
Motive: Novelty	−.28 (1.03)	−.01	.786	−.25 (1.01)	−.01	.806
Motive: First to know about new products/services	−.68 (.97)	−.02	.483	−.43 (.96)	−.01	.657
Motive: Entertainment	−.61 (.90)	−.02	.498	−.20 (.88)	−.01	.824
Materialism toward money	−.30 (.11)	−.01	.793	−.01 (.11)	−.01	.898
Gender (Female)[a]						
Male	−1.24 (.851)	−.04	.146	−2.56 (.86)	−.08	.003
Age (30 and younger)[a]						
31–50 years	−1.61 (.96)	−.05	.094	−1.81 (.95)	−.06	.056
51 and older	−1.99 (1.27)	−.05	.117	−1.17 (1.28)	−.03	.362
Education (Without high school degree)[a]						
With high school degree	−1.10 (.87)	−.03	.204	−1.86 (.87)	−.06	.032
Recruitment (Panel takeover)[a]						
Permission marketing				1.15 (1.47)	.03	.434
Intercept surveys				2.12 (1.57)	.04	.176
Survey platform				2.53 (1.52)	.05	.097
Google AdWords				2.27 (1.96)	.04	.247
Banner advertisement				.258 (2.04)	.01	.899
Referral				1.34 (1.99)	.03	.499
Panel tenure (in days)				.01 (.01)	.22	<.001
Number of other online panels enrolled				.11 (.34)	.01	.737
R^2		.01			.05	
N		1,612			1,612	

[a]Reference categories in parentheses.

Table 8.5 shows the results of the regression models predicting the break-off rate for the online panel during the previous year. The coefficient of determination for Model 3, including motives for joining the online panel and controlling for respondents' socio-demographics, was again low and not significant ($R^2 = .01, p = .075$). Adding controls for panel recruitment, panel tenure, and the number of other online panels in Model 4 increased the R-squared significantly to .05 ($p < .001$). Controlling for age, gender, education, panel recruitment, and panel tenure, the proportion of broken-off surveys was negatively correlated with the intention to earn money. Respondents who cited incentives as a motive for joining the online panel showed a break-off rate that was over 2 percentage points lower than that for respondents who did not cite that motive. Neither the form of recruitment nor the number of other online panels in which a member was enrolled showed an association with break-offs. However, the longer a respondent has been a member of the panel, the lower was the break-off rate. Additionally, gender and education showed a significant correlation with the percentage of break-offs in Model 4. Male members and panelists with a high school degree had break-off rates that were more than 2.5 and 2 percentage points lower, respectively, than the break-off rates of female panel members and respondents without a high school degree.

8.5 Conclusion

The results of the study presented in this chapter show that the motives a person has when joining an online panel are highly significant predictors of survey participation behavior. However, because this correlational study was not based on an experiment, it analyzed associations between motives and behaviors only. Therefore, whether these motives cause those behaviors or are merely correlated with them remains unclear.

8.5.1 Money as a leitmotif

Although other enrollment motives were cited more often, and fewer than 40% of all respondents stated that money had been a reason for joining, monetary motives had the strongest correlation with survey participation behavior in the online panel among the motives for joining the online panel. Respondents who said that they had joined because of the incentives they would earn by filling out web questionnaires had, on average, a starting rate of 5 percentage points higher and a break-off rate of almost 2 percentage points lower than that of panelists who had cited non-monetary reasons. However, does this finding confirm the widely articulated worry of many researchers who fear that online panelists are just money-grubbing mercenaries who are not representative of the general population and are therefore not valuable respondents for survey research?

Not necessarily. First, although money is obviously of appeal when a person joins an online panel, members learn very quickly that they will not get rich just by filling out web questionnaires. On average, a member of the online panel was invited to participate in 20 web surveys during one year and accepted approximately 14 such invitations. If the mean length of a questionnaire is assumed to be 15 minutes, participation therefore yields 150 bonus points or € 1.5. Hence, during the last year, the average panelist earned € 30 from being a member of the online panel (ignoring screen-outs and break-offs). Even with multiple online panel memberships – the respondents in this study were on average enrolled in one more panel – this type of remuneration does not add up to an income that would make other paid work obsolete. Obviously, this does not answer the question of what would happen if either the value of the incentive were dramatically increased or no incentive were provided at all. More research is needed that examines the influence of the monetary value of the incentive.

Second, the results from this study do not lead to the conclusion that panel members, though monetarily motivated, disregard their duties by just rushing through the questionnaires and producing bad-quality data. After excluding less than 3% of the respondents, i.e., those who obviously did not conscientiously read the questions and gave random answers, the results had high face validity. The answers to the Big Five inventory (not reported here) were highly comparable to those of other studies. Additionally, the percentage of respondents who showed nondifferentiation in grid questions (i.e., straight-lining) was very low. Paradata, such as response time and click behavior, which are automatically logged during filling out the questionnaire, make it much easier for the researcher to identify fraudulent behavior compared with postal surveys, for which this type of data is not available. This method also offers an advantage over interviewer-administered methods, which may suffer from harder-to-detect interviewer falsification.

Third, other reasons such as the intention to help develop better products and services or having fun when filling out questionnaires were cited as very important factors in the decision to enroll in the online panel. Respondents who said that they enjoyed taking part in web surveys showed a starting rate during the last year that was more than 3 percentage points higher than that of respondents who did not name this motive. This is a very promising finding for researchers because the attitude towards surveys is increasingly negative in the general population. To ensure that panelists are still enthusiastic after filling out yet another questionnaire on insurances or any other rather boring topic, panel providers have to ensure that questionnaires are easy to fill out, interesting, and attractive. A first step in achieving this goal is to provide a respondent-friendly questionnaire that obeys the basic principles of survey research methodology with regard to question structure and wording, answer category, scale selection, and respondents' guidance, and to offer a visually-attractive questionnaire that motivates the respondent to give thoughtful answers throughout the questionnaire. Although this seems to be more-than-obvious advice that applies to any questionnaire in any mode, real-life experience, especially in the field of commercial marketing research, shows that online panel providers seem to put more weight on pleasing their clients than on providing the respondent with a pleasant survey experience.

Overall, the results of the current study suggest that though money is often a decisive motive for becoming an online panel member, other factors play a similarly important role in predicting survey participation behavior in the online panel. In particular, the form of recruitment by which a member had joined the online panel proved to have a significant relationship with participation decisions. Members who had joined the panel through banner advertisement or Google AdWord campaigns and those who had been referred through a member-find-member system showed particularly high starting rates. More experimental research is needed that not only directly supports the causal reasoning in this current study but also examines the interaction between recruitment for online panels and the motivational factors that drive people to join them.

8.5.2   Limitations and future work

As with any study, limitations exist that are sure to promote more research in the future and encourage new studies to counterbalance the overgeneralization of findings. First, this study is limited in that the population surveyed came from one particular Austrian nonprobability online panel. Whether the results of the current study are applicable to the online panels of other providers who recruit their members with different techniques or in other countries or who offer incentives in different forms and values is questionable. Particularly in regard to the influence of enrollment motives and recruitment techniques on survey participation behavior, the fact that previous research shows that different motives dominate in different online panels must be considered. Because the composition of online panels varies, panel providers should explicitly examine which factors affect survey participation behavior in their particular panel.

Second, the independent psychographic variables used as predictors in the regression models were only in part able to explain the variance in survey participation behavior. Even when including controls for member's socio-demographic characteristics and additional information on panel recruitment, tenure, and multiple panel membership the proportion of variability in the data set that was accounted for by the statistical model did not exceed much more than 6%. The limited explanatory power of the presented models raises the question of what other factors influence the decision to participate in a given survey in an online panel environment. Future research should therefore consider other psychographic variables and survey-related factors, including the personal interest in the topic of a given survey and the time and place when and where a respondent receives an invitation, in an experimental setting. Additionally, to reveal potential bias, further research should examine how the enrollment motives and other psychographic characteristics of panel members correlate with the target variables of a given survey and other auxiliary variables.

Third, as did other researchers before us (e.g., Brueggen et al., 2011; Walker et al., 2009; Willems et al., 2006), we used self-reported measures to capture the motives that had led people to join online panels. Therefore, we cannot rule out that the findings of the current study are affected by social desirability measurement biases. For example, it is noble to say that one is joining the panel to help develop better products and services when it may not at all be true for a given panelist.

Finally, although the relationship between monetary motives and response behavior has been demonstrated, the scale measuring monetary materialism failed to significantly explain variation in survey participation behavior. This finding might be attributed to the scale's heavy focus on the desire for wealth, something that obviously cannot be obtained by becoming an online panel member.

References

Albaum, G. S., Evangelista, F., & Medina, L. (1998). Role of response behavior theory in survey research: A cross-national study. *Journal of Business Research, 42,* 115–125.

Armstrong, J. S. (1975). Monetary incentives in mail surveys. *Public Opinion Quarterly, 39,* 111–116.

Baker, R., Blumberg, S., Brick, J. M., Couper, M. P., Courtright, M., Dennis, M., ... Zahs, D. (2010). *AAPOR report on online panels.* Deerfield, IL: AAPOR.

Belk, R. W. (1984). Three scales to measure constructs related to materialism: Reliability, validity, and relationships to measure happiness. *Advances in Consumer Research, 11,* 291–297.

Belk, R. W. (1985). Materialism: Trait aspects of living in the material world. *Journal of Consumer Research, 12,* 265–280.

Berk, M. L., Mathiowetz, N. A., Ward, E. P., & White, A. A. (1987). The effect of prepaid and promised incentives: Results of a controlled experiment. *Journal of Official Statistics, 3,* 449–457.

Best, S. J., & Krueger, B. (2004). *Internet data collection.* Thousand Oaks, CA: Sage Publications.

Bethlehem, J. (2008). *How accurate are self-selection web surveys?* (Discussion paper 08014). The Hague/Heerlen: Statistics Netherlands.

Bethlehem, J., & Biffignandi, S. (2012). *Handbook of web surveys*. Hoboken, NJ: John Wiley & Sons, Inc.

Bethlehem, J., & Stoop, I. (2007). Online panels: A paradigm theft? In M. Trotman, T. Burrell, L. Gerrard, K. Anderton, G. Basi, M. Couper, … A. Westlake (Eds.), *ASC 2007. The challenges of a changing world. Proceedings of the 5th International Conference of the Association for Survey Computing* (pp. 113–131). Berkeley, UK: ASC.

Bosnjak, M., & Tuten, T. L. (2003). Prepaid and promised incentives in web surveys: An experiment. *Social Science Computer Review, 21*, 208–217.

Bosnjak, M., Tuten, T. L., & Wittmann, W. W. (2005). Unit (non)response in web-based access panel surveys: An extended planned-behavior approach. *Psychology & Marketing, 22*, 489–505.

Bowman, J. (2007). Hands up, who's on an access panel? *Research World, April*, 24.

Brennan, M. (1992). The effect of monetary incentive on mail survey response rates: New data. *Journal of the Market Research Society, 34*, 173–177.

Brueggen, E., & Dholakia, U. M. (2010). Determinants of participation and response effort in web panel surveys. *Journal of Interactive Marketing, 24*, 239–250.

Brueggen, E., Wetzels, M., de Ruyter, K., & Schillewaert, N. (2011). Individual differences in motivation to participate in online panels: The effect on response rate and response quality perceptions. *International Journal of Market Research, 53*, 369–390.

Cape, P. (2008). *Multi-panel membership*. (Survey Sampling International White Paper).

Cavallaro, K. (2012). *Revealing the answer to life's most difficult market research question: Why don't people join panels?* (Survey Sampling International White Paper). Retrieved from: http://www.surveysampling.com/ssi-media/Corporate/white_papers/SSI_Research_Resisters_WP.image.

Coen, T., Lorch, J., & Piekarski, L. (n.d.). *The effects of survey frequency on panelists' responses*. (Survey Sampling International White Paper). Retrieved from: http://www.surveysampling.com/ssi-media/Corporate/white_papers/The-Effects-of-Survey-Frequency-on-Panelists-Responses.image.

Comley, P. (2007). Online Market Research. In M. van Hamersveld, & C. de Bont (Eds.). *Market research handbook* (pp. 401–419; 5th ed.). Chichester: John Wiley & Sons Ltd.

Costa, P. T., & McCrea, R. R. (1992). *Revised NEO personality inventory (NEO PI-R) and NEO five factor inventory (NEO-FFI)* (Professional manual). Odessa: Psychological Assessment Resources.

Couper, M. P. (2000). Web surveys: A review of issues and approaches. *Public Opinion Quarterly, 64*, 464–494.

Couper, M. P. (2008). *Designing effective web surveys*. Cambridge: Cambridge University Press.

Couper, M. P., & Miller, P. V. (2008). Web survey methods: Introduction. *Public Opinion Quarterly, 72*, 831–835.

Denissen, J. J. A., Neumann, L., & van Zalk, M. (2010). How the Internet is changing the implementation of traditional research methods, people's daily lives, and the way in which developmental scientists conduct research. *International Journal of Behavioral Development, 34*, 564–575.

Dennis, J. M. (2001). Are Internet panels creating professional respondents? *Marketing Research, 13*, 34–38.

Dillman, D. A., Smyth, J. D., & Christian, L. M. (2009). *Internet, mail, and mixed-mode surveys. The tailored design method* (3rd ed.). Hoboken, NJ: John Wiley & Son, Inc.

Dommeyer, C. J. (1985). Does response to an offer of mail survey results interact with questionnaire interest? *Journal of the Market Research Society, 27*, 27–38.

Edwards, P., Roberts, I., Clarke, M., DiGuiseppi, C., Partap, S., Wentz, R., & Kwan, I. (2002). Increasing response rates to postal questionnaires: Systematic review. *British Medical Journal, 324(7347)*, 1183–1191.

Ensher, E. A., & Murphy, S. E. (2011). The mentoring relationship challenges scale: The impact of mentoring stage, type, and gender. *Journal of Vocational Behavior, 79*, 253–266.

Evans, J. R., & Mathur, A. (2005). The value of online surveys. *Internet Research, 15*, 195–219.

Fisher, S. (2007). How to spot a fake. *Quirks Marketing Research Review*, January. Retrieved from: http://www.quirks.com.

Galesic, M. (2006): Dropouts on the web: Effects of interest and burden experienced during an online survey. *Journal of Official Statistics*, *22*, 313–328.

Garland, P., Santus, D., & Uppal, R. (n.d.). *Survey lockouts: Are we too cautious?* (Survey Sampling International White Paper). Retrieved from: http://www.surveysampling.com/ssi-media/Corporate/White%20Paper%202012/Survey-Lockouts.image.

Goldberg, L. R. (1990). An alternative "description of personality": The Big-Five factor structure. *Journal of Personality and Social Psychology*, *59*, 1216–1229.

Göritz, A. S. (2004a). Recruitment for online access panels. *International Journal of Market Research*, *46*, 411–425.

Göritz, A. S. (2004b). The impact of material incentives on response quantity, response quality, sample composition, survey outcome, and cost in online access panels. *International Journal of Market Research*, *46*, 327–345.

Göritz, A. S. (2005). Contingent versus unconditional incentives in www-studies. *Metodološki zvezki. Advances in Methodology and Statistics*, *2*, 1–14.

Göritz, A. S., Wolff, H.-G., & Goldstein, D. G. (2008). Individual payments as a longer-term incentive in online panels. *Behavior Research Methods*, *40*, 1144–1149.

Groves, R. M., Presser, S., & Dipko, S. (2004). The role of topic interest in survey participation decisions. *Public Opinion Quarterly*, *68*, 2–31.

Groves, R. M., Singer, E., & Corning, A. (2000). Leverage-salience theory of survey participation: Description and an illustration. *Public Opinion Quarterly*, *64*, 299–308.

Heberlein, T. A., & Baumgartner, R. (1978). Factors affecting response rates to mailed questionnaires: A quantitative analysis of the published literature. *American Sociological Review*, *43*, 447–462.

Jacobson, S. F., & Jordan, K. F. (1993). Nurses' reasons for participating in a longitudinal panel survey. *Western Journal of Nursing Research*, *15*, 509–515.

Knapton, K., & Myers, S. (2005). Demographics and online survey response rates. *Quirk's Marketing Research Review*, January. Retrieved from: http://www.quirks.com.

Lee, S. (2006). An evaluation of nonresponse and coverage errors in a prerecruited probability web panel survey. *Social Science Computer Review*, *24*, 460–475.

Loosveldt, G., & Sonck, N. (2008). An evaluation of the weighting procedures for an online access panel survey. *Survey Research Methods*, *2*, 93–105.

Marcus, B., Bosnjak, M., Lindner, S., Pilischenko, S., & Schütz, A. (2007). Compensating for low topic interest and long surveys: A field experiment on nonresponse in web surveys. *Social Science Computer Review*, *25*, 372–383.

Marketagent (2009). *Panel insights.* (Company Folder). Retrieved from: http://www.marketagent.com/webfiles/pdf/Panel_Insights_Jan_2010.pdf.

Martin, C. L. (1994). The impact of topic interest on mail survey response behaviour. *Journal of the Market Research Society*, *36*, 327–338.

Mizes, J. S., Fleece, E. L., & Roos, C. (1984). Incentives for increasing return rates: Magnitude levels, response bias, and format. *Public Opinion Quarterly*, *48*, 794–800.

Olson, J. A., & Klein, R. E. (1980). Interviewers' perceptions of reasons for participation refusal in a national longitudinal survey, 1979–1980. In American Statistical Association (Ed.), *Proceedings of the Survey Research Methods Section*, 552–557.

O'Neil, K. M., Penrod, S. D., & Bornstein, B. H. (2003). Web-based research: Methodological variables' effects on dropout and sample characteristics. *Behavior Research Methods, Instruments, & Computers*, *35*, 217–226.

Paul, K. I., & Batinic, B. (2010). The need for work: Jahoda's latent functions of employment in a representative sample of the German population. *Journal of Organizational Behavior*, *31*, 45–64.

Porst, R., & von Briel, C. (1995). *Wären Sie vielleicht bereit, sich gegebenenfalls noch einmal befragen zu lassen? Oder: Gründe für die Teilnahme an Panelbefragungen.* [Would you be willing to be surveyed again? Or: Reasons for participation in panel surveys]. (ZUMA-Working Paper Nr. 95/04). Mannheim: ZUMA.

Postoaca, A. (2006). *The anonymous elect: Market research through online access panels.* Berlin: Springer.

Richins, M. L. (1987). Media, materialism, and happiness. *Advances in Consumer Research, 14,* 352–356.

Richins, M. L. (2004). The material values scale: Measurement properties and development of a short form. *Journal of Consumer Research, 31,* 209–218.

Richins, M. L., & Dawson, S. (1992). A consumer values orientation for materialism and its measurement: Scale development and validation. *Journal of Consumer Research, 19,* 303–316.

Rodenhausen, T., Kaufmann, G., & Hartmann, A. (2007). Der Reiz des Neuen. Teilnahmemotivation von Online-Panelisten. [The appeal of novelty: Participation motivation of online panelists]. *Research & Results, 1/2007,* 30–31

Rogelberg, S. G., Conway, J. M., Sederburg, M. E., Spitzmüller, C., Aziz, S., & Knight, W. E. (2003). Profiling active and passive nonrespondents to an organizational survey. *Journal of Applied Psychology, 88,* 1104–1114.

Schillewaert, N., & Meulemeester, P. (2005). Comparing response distributions of offline and online data collection methods. *International Journal of Market Research, 47,* 163–178.

Smith, R. S., & Brown, H. H. (2005). *Assessing the quality of data from online panels: Moving forward with confidence.* (Harris Interactive White Paper). Retrieved from: http://www.websm.org /uploadi/editor/1175495342HI_Quality_of_Data_White_Paper.pdf.

Sparrow, N. (2006). Developing reliable online polls. *International Journal of Market Research, 48,* 659–680.

Toepoel, V., Das, M., & van Soest, A. (2008). Effects of design in web surveys: Comparing trained and fresh respondents. *Public Opinion Quarterly, 72,* 985–1007.

Toepoel, V., Das, M., & van Soest, A. (2009). Relating question type to panel conditioning: Comparing trained and fresh respondents. *Survey Research Methods, 3,* 73–80.

van Kenhove, P., Wijnen, K., & de Wulf, K. (2002). The influence of topic involvement on mail-survey response behavior. *Psychology & Marketing, 19,* 293–301.

Walker, R., Pettit, R., & Rubinson, J. (2009). A special report from the Advertising Research Foundation: The foundations of quality initiative: A five-part immersion into the quality of online research. *Journal of Advertising Research, 49,* 464–485.

Weijters, B., Schillewaert, N., & Geuens, M. (2008). Assessing response styles across modes of data collection. *Journal of the Academy of Marketing Science, 36,* 409–422.

Willems, P., van Ossenbruggen, R., & Vonk, T. (2006). The effects of panel recruitment and management on research results. Dutch Online Panel Study NOPVO 2006. Paper presented at the ESOMAR Panel Research 2006, Barcelona.

Yu, J., & Cooper, H. (1983). A quantitative review of research design effects on response rates to questionnaires. *Journal of Marketing Research, 20,* 36–44.

Appendix 8.A

8.A.1 German translation of motives for joining the online panel as used in the study

Inwiefern stimmen Sie persönlich den folgenden Aussagen zu?

> *Geben Sie bitte anhand der Skala 1 = ,,stimme voll zu" bis 5 = ,,stimme überhaupt nicht zu" an, wie sehr diese Aussagen auf Sie persönlich zutreffen.*

1. Es ist mir wichtig, schöne Dinge zu besitzen.

2. Ich wäre gerne so reich, dass ich mir alles kaufen kann, was ich gerne möchte.

3. Ich wäre glücklicher, wenn ich mir mehr leisten könnte.

4. Manchmal stört es mich ziemlich, dass ich mir nicht all die Dinge leisten kann, die ich gerne hätte.

5. Die Leute legen zu viel Wert auf materielle Dinge.

6. Es stimmt, dass Geld glücklich macht.

7. Die Dinge, die ich besitze, bereiten mir große Freude.

8.A.2 German translation of Materialism Scale (Richins, 1987) as used in the study

Es gibt zahlreiche Gründe, warum man sich für ein Online Panel anmelden kann. Welche der folgenden Gründe treffen auf Sie persönlich zu?

Mehrfachantworten möglich.

1. Ich war neugierig und wollte wissen, worum es in einem Panel geht.

2. Ich will mit meiner Meinung zur Verbesserung von Produkten und Dienstleistungen beitragen.

3. Die Mitgliedschaft im Panel ist ein guter Weg, um Geld zu verdienen.

4. Ich bin gerne der Erste, der neue Dinge ausprobiert.

5. Ich weiß gerne früher über neue Produkte und Dienstleistungen Bescheid als der Rest der Gesellschaft.

6. Die Beantwortung der Befragungen macht mir Spaß.

7. Sonstige Gründe und zwar:_____

9

Informing panel members about study results

Effects of traditional and innovative forms of feedback on participation

Annette Scherpenzeel and Vera Toepoel

Utrecht University, The Netherlands

9.1 Introduction

Maintenance of high participation rates is a crucial issue for panels in which long-term surveys of the general population are carried out (Lynn, 2009). To be representative of the target population, probability-based panels need to have a high initial response rate across all sample units but also to ensure a continuing high participation rate from year to year. Many panels try to keep respondents interested by regularly sending them letters, gifts or information about research findings. However, little is known about the effectiveness of these measures with respect to the participation of panel members. In this chapter, we evaluate the effects of different forms of feedback, both traditional and more innovative, on the monthly response rates in the online LISS panel and on the appreciation of the panel members. We try to answer the question whether it is worthwhile to send these kinds of feedback to panel members. Moreover, we investigate the benefit of using different materials or different information channels for

---

[1] The LISS panel data were collected by CentERdata (Tilburg University, The Netherlands) through its MESS project. This work was supported by the Netherlands Organisation for Scientific Research.

Online Panel Research: A Data Quality Perspective, First Edition.
Edited by Mario Callegaro, Reg Baker, Jelke Bethlehem, Anja S. Göritz, Jon A. Krosnick and Paul J. Lavrakas.
© 2014 John Wiley & Sons, Ltd. Published 2014 by John Wiley & Sons, Ltd.
Companion website: www.wiley.com/go/online_panel

different subgroups of panel members, since this form of tailoring can be used to try to reduce possible nonresponse bias.

9.2   Background

9.2.1   Survey participation

Nonresponse arises in different phases in a panel: at the initial contact, the request for participation, the survey itself, or between waves in the form of panel attrition. When full participation fails to be obtained, the inferential value of the panel is threatened. The accumulation of nonresponse losses from one year to the next can result in a vast reduction of the sample size and the statistical power of studies. The attrition rate of the LISS panel, for example, is about 10% per year for households and about 12% per year for persons (Scherpenzeel, in press). Although these attrition rates might seem low, the accumulation over years leads to a retention of only 61% of the panel households originating from the main 2007 sample that were still participating in the panel in January 2012.

In addition, if nonresponse is not random, it can introduce a bias in the resulting estimates. Therefore, finding tools for dealing with nonresponse effects is an important and on-going challenge for panel managers. Survey researchers often use an array of methods in the hope of minimizing nonresponse. Although attempts are made to reduce nonresponse effects through post-survey adjustments (e.g., weighting), increasing cooperation is the most logical and appealing goal to strive for. On the part of the researcher, persuasion strategies are used to increase response rates. On the part of the respondent, decision-making strategies are used in the response process.

The effectiveness of tools for improving response rates depends on the ability of the researchers to step inside the respondent's mind. Often, potential respondents do not devote a lot of cognitive energy to deciding whether to participate in a survey. Groves, Cialdini, and Couper (1992) argue that the survey request situation most often favors a heurist approach because the potential respondent typically does not have a large personal interest in survey participation. Consequently, they argue that the decision to participate in a survey for any particular respondent is likely to depend on one or two highly prominent and normally diagnostic considerations. Groves et al. (1992) borrow concepts from social psychology and extent these issues to the domain of survey participation.

- *Norm of reciprocity*: One important heuristic with regard to "compliance with requests" is the norm of reciprocation: individuals are motivated to provide to others the general form of behavior that they have received from those others (Gouldner, 1960). In survey terms: potential respondents should be more willing to comply with a survey request to the extent that the compliance constitutes the "repayment" of a perceived favor. For example, the payment of a noncontingent incentive (prior and unconditional of response) should show higher participation rates compared to no provision of an incentive because people are expected to give something in return for the incentive. In addition, Groves et al. (1992) argue that the provision of informational letters, brochures, etc. will gain higher compliance than not providing such materials. This will be the case to the extent the information has value for the respondent.

- *Positive regard*: Providing incentives (material and non-material) also corresponds to the "liking" heuristic: people are favorably inclined toward those individuals they like. Thank you notes and postcards are ways of providing rewards and can be used to show positive regard (Dillman, 2007). "Helping tendencies" are proven to depend on emotions such as anger, happiness, and sadness. People who are happy are more inclined to help others (Clark & Waddell, 1983). In addition, happiness generally enhances compliance (Groves et al., 1992).

- *Public interest*: Budowski and Scherpenzeel (2005) use the results and public interest of the panel study in an attempt to stimulate interest. Budowski and Scherpenzeel argue that emphasizing the public interest of the results of the survey might be seen as a special form of the norm of reciprocity, making people feel obliged to provide information to social institutions (i.e., a coercive factor). Stoop (2012) also argues that the decision to participate in a survey depends on how important the participation is for someone else, e.g., society, science, etc. Participation is sometimes viewed as serving the common good. Following her reasoning, emphasizing the importance of the survey to respondents (e.g., what will be done with the survey results) can be a way of reducing nonresponse.

- *Involvement*: Sue and Ritter (2007) argue that involvement is also an important estimate of one's likelihood of participating in an online survey. The more involved a potential respondent is, the more likely he or she is to participate. Having additional contacts and providing panel members with information about the panel or the survey outcomes (e.g., newsletters) are ways to try to increase panel members' involvement with the panel.

- *Interaction*: Groves et al. (1992) argue that maintaining interaction (contact) is of crucial importance when persuading (potential) respondents to comply with a survey request. Maximizing the number of contacts is an important and widely used rule in the survey world. For example, announcement letters, newsletters, and thank you notes are commonly used, as promoted by the Tailored Design Method (Dillman, 2007). It is relatively unknown whether advance letters are successful in increasing response rates because they are special or simply because they create an additional contact with the potential respondent. The effect of sending newsletters or thank you notes on (future) response rates is relatively unknown.

- *Maximizing rewards*: The Tailored Design Method also refers to maximizing rewards for respondents to increase response rates. Rewards can be tangible (monetary or non-monetary), but also intangible (Albaum & Smith, 2012). Intangible rewards can be the provision of a graph with the cumulative results of all answers, to help position a respondent's answer in reference to the general mean (benchmark), at the end of the survey.

9.2.2   Methods for increasing participation

Researchers often provide respondents with incentives to increase response rates. Although considerable effort and money is spent on these incentives, the effects are often not known. Material incentives are commonly used and well documented. Research shows that *monetary incentives* are more powerful than non-monetary incentives, and that noncontingent (i.e., prepaid and unconditional) incentives work better than contingent (i.e., postpaid and conditional) incentives (Toepoel, 2012). Although *lotteries* are commonly used as a way to distribute incentives, there is little evidence that they work. In addition, *donations to charity* do not work or might even decrease response rates (Göritz & Neumann, 2013).

There has been little research on the effect of *non-material incentives* on response rates (Manzano & Burke, 2012). Larson and Poist (2004) asked respondents how much more or less likely they would be to complete a survey if they were to receive monetary incentives, or communications such as a results summary, pre-notifications and follow-up mailings. Most respondents indicated that the promise of a results summary would most strongly increase their willingness to participate, followed by a monetary incentive, pre-notifications and follow-up mailings. Göritz (2005), however, found that response rates did not improve in the web panel when respondents were told that they could review a summary of the results of the survey they had just taken. This result was confirmed by Edwards et al. (2009) who did not find any effect of offering survey results versus not offering them in a meta-analysis of 12 randomized studies using postal questionnaires.

Similar results were reported by Scherpenzeel and Toepoel (2012), who found that response rates in a telephone and a face-to-face recruitment interview were strongly affected by prepaid incentives enclosed with the advance letter but not affected by the content of the advance letter. Budowski and Scherpenzeel (2005) found a 20% increase in participation due to an announcement letter. Iredell et al. (2004) also found significant higher response rates while sending an introductory postcard. Hembroff et al. (2005) investigated the effect of an advance letter, postcard, or nothing in an independent cross-section telephone survey. Their results clearly indicated that sending advance letters is more effective than sending postcards, which, in turn, is more effective than sending nothing. The effectiveness of advance letters in one-time telephone surveys is also demonstrated by the meta-analysis of De Leeuw et al. (2007). However, Singer, Van Hoewyk, and Maher (2000) did not find any differences between an advance letter and no-letter condition in a telephone survey, whereas in a large national RDD study, Camburn et al. (1996) did. Yet, Singer at al. (2000) did find a significant and large effect for prepayment enclosed with an advance letter. Motte and Brault (2008) sent postcards and stickers to a random group of participants in the period between survey contacts. They found no differences between the response rates of the group that received the *additional mailing* and the group that did not.

In conclusion, advance letters seem to have effect on response rates in many studies, but not in all. In general, the effects of incentives are stronger than the effects of an advance letter alone. We suspect that the mixed results are found because the effect size of an advance letter on response rates depends on the topic of the study, the content of the letter, and the target population. This implies that pre-tests and pilot studies of advance letters are necessary for each specific survey.

Budowski and Scherpenzeel (2005) sent a *brochure* to keep in touch with participants and reminded them of their involvement in a specific study in the Swiss Household Panel. A lot of information and results were given in the brochure and a strong emphasis was put on the public interest of the study. This also is a common strategy for the British Household Survey. Laurie, Smith, and Scott (1999) note that panel members appreciate receiving *feedback* about the survey. In addition, the Panel Study of Income Dynamics (PSID) sends panel members an annual respondent newsletter describing the scientific value of the study (McGonagle & Schoeni, 2013). It is intended to help panel members feel they are contributing to a worthwhile project. Their experience (confirmed by our experience in the panels we work with) is that many respondents request more information about the surveys they take part in. However, neither Budowski and Scherpenzeel nor Laurie, Smith, and Scott implemented experiments to test the effect of these strategies.

In general, the literature on non-material incentives is not well developed, hence the effects of various forms of such incentives are relatively unknown. With the rise of the Internet, many multimedia stimuli can be included in the survey and in the communication with panel members. *Real-time results* can be offered to respondents, as well as the possibility of showing the position of a respondent's answer relative to the total survey outcomes. In addition, YouTube- type of *movies and pictures* can be added to the survey. However, it is unknown whether these communication strategies have any effect on response rates.

Furthermore, the effectiveness of communication tools might be panel- or survey-specific. Effectiveness might be related to the authority the survey comes from, the population targeted, the topic of the survey, etc. In addition there might be a difference between one-time surveys and panel surveys where there is already some relationship between the sponsor and the respondent. It is not guaranteed that something that can work on a one-time survey would work on a panel, and vice versa. Mode differences could also affect results. Thus, a pilot study is always a good idea before implementing a previously untried technique.

9.2.3 Nonresponse bias and tailored design

There is a difference between nonresponse and nonresponse bias. Higher response rates do not necessarily improve sample composition (cf. Groves, 2006). For example, if the techniques applied to reduce nonresponse have a differential impact on subgroups, disproportionately raising response rates among already-represented subgroups, they will exacerbate the selective nonresponse and thereby have a detrimental effect on nonresponse error. Therefore, one should be careful with applying techniques for reducing overall nonresponse. Link and Mokdad (2005) note that an advance letter biased their sample towards older, white respondents, and people with higher socio-economic status. These are groups who typically already show high response rates. The advance letter might have reduced nonresponse, but increased nonresponse *bias*.

Thus, different target groups require different approaches (Trussell & Lavrakas, 2004). Advance letters may be too complicated for people with a lower education level or better appreciated by people with a higher education level. In addition, monetary incentives may not be uniformly effective, for example, recruiting more people with lower income or less education. Singer et al. (2000) investigated whether sending an incentive and/or advance letter results in "biased" recruitment. With the exception of education, they did not find any significant differences in the effect of these tools on different demographic subgroups. Singer (2002) found that incentives increased response rates by persuading those demographic groups that would normally not participate in the survey. She concluded that dependent variables would have been seriously mismeasured if incentives had not been used.

In the study presented here, one of the targets was to develop motivational strategies tailored to different demographic respondent groups. The idea of tailoring was induced by the need to preserve representativeness, in addition to keeping up retention rates: If we know the variation in response probabilities and retention probabilities across groups, we can focus on specifically motivating the groups with the lowest response probabilities and highest risk of attriting (see Trussell & Lavrakas, 2004). Therefore, we have estimated the effects of the feedback materials specifically for people with the lowest response probabilities and for people with the highest response probabilities.

9.3 Method

9.3.1 Sample

The data for this chapter were collected in the LISS panel (Longitudinal Internet Studies for the Social Sciences). This panel is the principal component of the MESS project, operated by the CentERdata research institute in Tilburg, the Netherlands. It consists of almost 8000 individuals who complete online questionnaires every month. All data collected in the LISS panel are available to all researchers at www.lissdata.nl and all analyses reported can be replicated.

The LISS panel is based on a true probability sample of households drawn from the population register by Statistics Netherlands.[2] Households that could not otherwise participate are provided with a computer and an Internet connection. LISS panel members complete online questionnaires every month, for which they get an incentive of €15 per hour of interview time. This per hour incentive is the most important encouragement for long-term participation in the LISS panel. In several experiments, we have shown the effectiveness of monetary incentives in a probability-based online panel (Scherpenzeel, 2011; Scherpenzeel & Toepoel, 2012).

Between the start of the panel in October 2007 and June 2009, 7% of the households that had ever completed the initial household questionnaire left the panel. The overall monthly response varies between 65%–79%. A considerable part of the monthly nonresponse is associated with the same panel members every month. Panel members who participated before but have not completed a questionnaire for three months or longer are defined as "sleepers." During the two experimental periods, in 2009 and 2011, sleepers were never forced out of the panel nor stimulated to restart participating in any other way than by the feedback materials they were assigned to receive.[3]

Two studies (De Vos, 2009; Lugtig et al., 2010) have shown that attrition and inactive periods are much more related to respondents' past response behavior than to demographic characteristics. Skipping a questionnaire or completing questionnaires irregularly turn out to be the best predictors for future drop-out. This suggests a need for a general strategy to keep panel members "awake" and motivated by regularly attracting their attention and raising their interest. For the materials used in this study, differential effects were expected, since not all panel members are equally interested in study results, and the different types of materials probably appeal to different population groups. To estimate these differential effects, we carried out separate analyses within specific demographic groups. These subgroup analyses were intended to give a basis for tailored motivation strategies that will balance response probabilities across groups.

9.3.2   Experimental design

To provide a basis for a general communication strategy, a series of experiments with different forms of communication was conducted. The experiments were done among active and inactive (also called sleeping) panel members at the same time. Each experiment included a randomly assigned control group of panel members who were not sent the specific communication material tested in that experiment, but who did receive all other materials and incentives. A baseline "zero-condition" group of randomly assigned panel members did not receive any of the communication materials or incentives. We conducted two series of experiments. In the first series (Experiment 1), we used more traditional forms of information. In the second series (Experiment 2), we used more innovative forms of feedback information. The experimental design for each series of experiments is discussed below.

---

[2] For a more detailed description of the panel, the sample, recruitment, and response, see the website www.lissdata.nl.

[3] In 2010, in between the two experimental periods, sleepers were contacted to re-activate them and, if that was unsuccessful, sometimes forced out of the panel (see Scherpenzeel, 2011).

9.3.2.1 Experiment 1: Traditional forms of information

In the first experiment we used traditional forms of feedback. The communication materials varied from gifts to information and incentives and from paper versions to electronic versions. They were all designed by a professional communication agency (see online appendix for illustrations of the materials, at: www.wiley.com/go/online_panel_research) and included:

- *Cards*: A set of five free cards, each showing a statistical fact from the panel data and a funny text based on that statistic. Sent in December 2008 to panel members with a letter to thank them for participating in the past year.

- *Ring binder*: A small ring binder with tabs and information pages containing some frequently asked questions, tips for answering questionnaires, and general information. Sent end of April 2009.

- *Advance letter and incentive*: A letter announcing the LISS core questionnaire about income, housing and assets (economic situation), in combination with experimental variation in a promised, unconditional incentive.[4] Sent in May 2009.

- *E-cards*: A set of e-cards on the same basis as the paper free cards mentioned above, each showing a statistical fact from the panel data and a funny text based on that statistic. Email with link to the e-cards sent beginning of July 2009.

- *Newsletter*: A short newsletter describing some research results and including some nice-to-know facts, answers to some frequently asked questions, and a day in the life of a LISS panel helpdesk employee. Sent in September 2009, to half of the panel members in paper form, to the other half in electronic form with the exclusion of the control groups.

9.3.2.2 Experiment 2: innovative forms of information

In 2011, a new series of experiments was conducted in which more innovative and interactive ways of informing respondents were used. The material consisted of short, YouTube-like videos, in which a researcher talked about the results he had obtained with the data collected in the LISS panel. The second new material consisted of graphs that were generated immediately at the end of an online questionnaire, representing the real-time answer distribution on one of the questions the respondent had just answered (see the online appendix at: www.wiley.com/go/online_panel_research for an impression of the materials; the videos can also be watched, in Dutch, at the URL: http://www.lisspanel.nl/website/onderzoeken). Similar to the first experiment, each experiment included a randomly selected group of panel members that was not exposed to any video or graph, and a baseline "zero-condition" group of randomly assigned panel members not presented any of this type of feedback. The materials, timing, and design of the second experiment are as follows:

- *Video (1)* interview with professor of Sociology, included in core questionnaire "Family and Household." Shown in March 2011 (N = 700).

- *Graphs (1)* showing panel results on desired retirement age; hours of care-giving to family members; satisfaction with atmosphere at work. Included in core questionnaire "Work and Schooling" in April 2011 (N = 700).

---

[4] The extra incentive was either €2.75 or €5.50 depending on the experimental condition. This variation was part of another study but had no significant effect on the response rates.

- *Video (2)* interview with professor of Psychology, included in core questionnaire "Personality." Shown in May 2011 (N = 700).

- *Graphs (2)* showing panel results satisfaction with financial situation; opinion on minimally needed household income; difficulty in making ends meet. Included in core questionnaire "Income" in June 2011 (N = 700).

- *Video (3)* interview with researcher in Econometrics, included in questionnaire on risk aversion.[5] Shown in October 2011 (N = 700).

- *Graphs (3)* showing panel results on satisfaction with government; trust in science,[6] and acceptance of female managers. Included in core questionnaire "Politics and values" in December 2011 (N = 700).

- Baseline group panel members presented with each material (N = 700).

- Control group panel members not presented with any of the materials (N = 700).

9.4   Results

9.4.1   Effects of information on response

We use logistic regression for the analyses, in which there are only two values to predict: that probability (p) is 1 rather than 0, i.e., the person participated in following month or the month thereafter (Table 9.1) or not; participated all months (Table 9.2); or was still an active panel member at the end of the period (Table 9.2). Similar to linear regression, the B coefficients in our logistic regression models indicate the rate of change in the dependent variables as the predictor variable changes, but the logistic coefficient is interpreted as the change in the "log odds" of the dependent variable (for more details, see, for example, Burns & Burns, 2008; or Hosmer et al., 2013). In addition to the B coefficients, the tables show the response percentages as predicted by the model. The predicted response percentages differ from the observed response percentages in that the effect of other variables of influence are controlled for,[7] but the interpretation of the percentages is the same. Finally, we include the Nagelkerke R-squared in our tables to indicate how much of the variance in the response rates can be explained by the models used. Nagelkerke R-squared is a pseudo-R-squared used for logistic models, which ranges from 0–1. Higher values indicate a better model fit (Nagelkerke, 1991).

Table 9.1 shows the effects of each communication form on the response in the month immediately following it and on the response in the second month after. Response in a certain month is here defined as: A panel member answered (either partially or completely) one or more questionnaires that month. The total number of questionnaires for which a LISS panel member is selected and invited in a month can vary, but each panel member is always invited, each month, to complete at least one questionnaire. Since each mailing was sent in a different month, the response rate in the following month was different for each mailing. Therefore,

---

[5] The third video, of October 2011, is not included in the analyses because it was not presented to the complete selection of respondents.

[6] At the time of this questionnaire, a scientific scandal had just been discovered: a well-known researcher and professor of psychology at Tilburg University had been publishing fictional research results for years. The discovery of this fraud, called "the Stapel affair," was national news and expected to harm people's trust in science.

[7] In this case: the effect of demographic variables.

Table 9.1 Logistic regression effects of mailings on monthly response and predicted response percentages.

	N^1	Response in month following mailing		Response in second month after mailing	
		Received all versus Condition: model predicted response (%)[2]	Model B coeff.	Received all versus Condition: model predicted response (%)[2]	Model B coeff.
Experiment 1, 2008/2009					
Received all[3]	3842				
All except cards (Dec.)	268	86 vs 83	−.248	84 vs 80	−.263
All except binder (Apr.)	298	80 vs 81	.073	79 vs 77	−.069
All except letter (May)	300	80 vs 84	.268	75 vs 73	−.138
All except e-cards (July)	317	75 vs 76	.102	73 vs 76	.228
All except newsl. (Sep.)	298	76 vs 76	.027	75 vs 73	−.088
Experiment 2, 2011					
Presented all[2]	683				
All except video 1 (Mar.)	668	70 vs 69	−.024	73 vs 70	−.144
All except video 2 (May)	653	68 vs 69	−.025	69 vs 70	.005
All except graph 1 (Apr.)	669	73 vs 75	.061	68 vs 73	.226
All except graph 2 (June)	646	69 vs 69	.008	73 vs 74	.103
All except graph 3 (Dec.)	627	77 vs 78	.066	74 vs 77	.190
Nagelkerke R-squared[4]		Between .047 and .167		Between .051 and .193	

[*]$p < .05$,
[**]$p < .01$
[1]N in both experiments is the number of panel members in the month following mailing, who have valid data on all demographics included in the analyses. In experiment 1, the N includes only panel members who actively participated (non-sleepers) at the onset of the experiments in December 2008. In experiment 2, the N includes all (active and sleeping) panel members assigned to one of the conditions.
[2]Since each mailing was sent in a different month, the response percentage obtained in the "Received All" control condition was different in each independent model for each material.
[3]Effects were estimated in 10 independent models, one for each material, with the "Received all" or "Presented all" category as reference in each model. The demographic variables gender, age, household composition, education, occupation and having a computer and Internet provided by the Institute were included in each model (effects not shown).
[4]The range of Nagelkerke R-squared values given here is for the 10 models, one for each material, all including demographic variables.

the estimates in the Table 9.1 were obtained with 10 independent logistic regression models, one for each material. In each logistic regression model, the group who received everything is taken as the reference (comparison) group. The response rate of this reference group was different in each of the months in which the different mailings were sent. In addition, each model included a series of demographic variables: gender, age, household composition, education, occupation and having a computer and Internet provided by the Institute. The effects of these demographic variables are not shown in Table 9.1, since they are not the focus of this chapter and only included as control variables.

Table 9.2 Logistic regression effects of mailings and demographics on long-term participation.

	N[1]	Participated all months		Still active at end of period	
		Model predicted (%)	Model B coeff.	Model predicted (%)	Model B coeff.
Experiment 1, 2008/2009					
Received all[2]	3945	44		84	
All except cards (Dec.)	273	41	−.119	82	−.173
All except binder (Apr.)	304	38	−.217	84	.081
All except letter (May)	304	44	−.015	83	−.038
All except e-cards (July)	320	46	.128	86	.219
All except newsletter (Sept.)	298	46	.109	82	−.072
Received nothing	576	42	−.056	84	.042
Nagelkerke R-squared[3]			.094		.074
Experiment 2, 2011					
Presented all[2]	681	39		88	
All except video 1 (Mar.)	669	33	−.197	87	−.084
All except video 2 (May)	663	38	−.049	90	.211
All except graph 1 (Apr.)	675	39	−.009	91	.276
All except graph 2 (June)	670	38	−.013	88	.045
All except graph 3 (Dec.)	678	37	−.038	89	.113
Received nothing	678	38	.017	89	.097
Nagelkerke R-squared[3]			.137		.093

*$p <.05$,
**$p <.01$
[1]N is defined in note 1 in Table 9.1.
[2]Effects were estimated in two models, one for each experiment, with the "Received all" or "Presented all" category as reference in each model. The demographic variables gender, age, household composition, education, occupation and having a computer and Internet provided by the institute were included in each model (effects shown in Appendix A).
[3]The Nagelkerke R-squared given here is for the model including demographic variables (see also Appendix A).

We illustrate the explanations of the logistic regression models given above with the following example: The cards in the first experiment were sent in December 2008, hence the dependent variable in the first model is the response rate in January 2009. For a respondent who did not get the cards, the B coefficient was −.248, indicating that the probability that a respondent in this condition would respond in January was slightly lower than the probability that a respondent who received everything would respond. This small difference in response probability is not significant in this model. When controlling for the effects of gender, age, household composition, education, occupation and having a computer and Internet provided by the Institute, the response rate in January is 86% for respondents who received everything and 83% for respondents who did not receive the cards.[8] The Nagelkerke R-squared for this specific model was .071 (not shown), indicating a very weak relationship between receiving

[8] The observed percentages for this model were also 86% and 83% respectively, indicating that no collinearity existed between the demographic variables in the model and the (randomly allocated) experimental conditions. This was true for all models presented in Tables 9.1 and 9.2.

or not receiving the cards, together with the demographic characteristics, and responding the following month.

In Table 9.1, we summarize the results of the 10 models by showing the reference group only once. The results in Table 9.1 show that none of the materials had any significant effect on the response in the following months. As described in the method section, the newsletter was sent in two forms: to half of the sample in paper form by regular mail and to the other half in electronic form by Email (see the online appendix at www.wiley.com/go/online_panel _research for illustrations). The form did not make any difference: independent logistic regressions show that neither of the two newsletters had any effect on the response in the following months, compared to no newsletter at all.

9.4.2   "The perfect panel member" versus "the sleeper"

Table 9.2 presents the effects of the communication materials on long-term participation. Table 9.2 shows the effects of each material on the probability of staying a "perfect" panel member (defined as completing questionnaires each month without exception) and the probability of not "sleeping," in other words, still actively participating, at the end of the period of experimentation (defined as having completed at least one questionnaire in the last three months of that period). In contrast to Table 9.1, the effects in Table 9.2 could be estimated in only two models, since the dependent variables in Table 9.2 were the same for all conditions. The "Received all" or "Presented all" category was used as reference in each model. No significant effects were found: the materials had no effect on constant participation or on the likelihood of remaining active. The models in Table 9.2 included the effects of respondent demographic characteristics: these effects, shown in Appendix 9.A, appear to be more important than the communication efforts. Older people, people who are unemployed, retired people, students, people who have equipment from the organization, and people with higher education have higher probabilities of participating than counteracting groups. Similar effects of demographic characteristics on participation over time in the LISS panel have been reported by de Vos (2009) and Lugtig et al. (2010). However, interactions of different forms of communication and respondent characteristics could be possible, which we tested as shown in Table 9.3.

9.4.3   Information and nonresponse bias

It has been shown that the long-term response probabilities in the LISS panel do not show large variations across demographic groups (De Vos, 2009; Lugtig et al., 2010). However, specific strategies to motivate panel members could induce biases in the panel composition if they have a differential effect on subgroups. Therefore, we have estimated the effects of the feedback materials specifically for people with the lowest response probabilities and for people with the highest response probabilities.

In Table 9.3, we estimated the effects of the interactive feedback materials on the active participation at the end of the period of experimentation for specific subgroups of panel members. These subgroups had either a higher than average probability of becoming a sleeper over time (young people; people in paid work; lower-educated people; people with children) or a low probability (elderly; people with higher education).

In the first experiment, two significant effects were found in different educational groups: the letter and the newsletter both stimulated the long-term active participation for people with lower education (*not* receiving the letter decreased the probability of remaining active). In

Table 9.3 Effect of feedback on long-term participation, for different demographic groups: logistic regression.

Still active at end of period of:	Younger than 35		Older than 64		Single person household	
	N^1	B	N^1	B	N^1	B
First experiment						
Received all	1088		503		565	
All except cards	67	−.232	38	−.776	46	.167
All except binder	90	.222	33	−.906	38	.930
All except letter	92	.154	45	−.981	47	−.018
All except e-cards	89	.471	34	−.964	31	−.353
All except newsletter	85	−.299	40	−.479	41	.039
Received nothing	161	.372	69	−.618	78	.158
Second experiment						
Presented all	183		127		123	
All except video 1	186	−.317	111	.648	126	−.552
All except video 2	158	−.097	118	1.679	123	.119
All except graph 1	184	.064	136	.032	117	.338
All except graph 2	193	−.285	116	.255	138	−.583
All except graph 3	189	−.217	120	.723	124	−.170
Presented nothing	191	.149	120	−.150	138	−.049

Still active at end of period of:	Household with children		Lower education		College or higher education	
	N^1	B	N^1	B	N^1	B
First experiment						
Received all	1978		1498		1090	
All except cards	133	−.287	104	−.228	76	−.279
All except binder	172	−.055	119	−.173	76	.261
All except letter	129	−.252	117	−.532[*]	84	.232
All except e-cards	160	.461[*]	124	−.009	112	.608
All except newsletter	146	−.120	104	−.544[*]	88	.505
Received nothing	289	.168	206	.097	165	.087
Second experiment						
Presented all	301		247		222	
All except video 1	321	.074	225	−.444	212	.209
All except video 2	293	.362	266	−.006	196	1.091[**]
All except graph 1	315	.511[*]	230	.870[**]	220	.564
All except graph 2	298	.270	250	.181	198	.154
All except graph 3	315	.255	242	.014	219	.207
Presented nothing	302	.245	231	−.042	207	.453

(*continued overleaf*)

Table 9.3 (*continued*)

Still active at end of period of:	Younger than 35		Older than 64		Single person household	
	N^1	B	N^1	B	N^1	B
Still active at end of period of:	Paid work		School, study, housekeeping or other			
	N^1	B	N^1	B		
First experiment						
Received all	2210		937			
All except cards	161	−.075	65	−.471		
All except binder	178	.262	75	−.027		
All except letter	165	.102	77	−.190		
All except e-cards	185	.356	76	−.002		
All except newsletter	176	.130	64	−.403		
Received nothing	323	−.065	140	.553		
Second experiment						
Presented all	365		143			
All except video 1	376	−.115	150	−.035		
All except video 2	362	.217	149	−.176		
All except graph 1	363	.244	159	.372		
All except graph 2	373	.046	143	.159		
All except graph 3	384	.114	132	.129		
Presented nothing	394	.110	134	.006		

$^*p <.05$,
$^{**}p <.01$.
^{1}N is defined in note 1 in Table 9.1.

addition, the e-cards had a negative effect on staying an active panel member for households with children (not receiving the e-cards was associated with an increased in the probability of staying active). The insignificant effects in the other demographic groups could in the first experiment be related to the small number of cases for the subgroups in some of the experimental conditions.

Similar to the first experiment, we also found two significant effects in the second experiment: for the low-educated panel members as well as for the panel members in households with children, the first graph, which showed panel answers on questions about work and care, decreased the active participation at the end of the year (not receiving the first graph increased the probability of staying active). A possible explanation might be that lower-educated people do not like graphical presentation of results. In the next section is shown that indeed the middle- and high-educated respondents found the graphs more interesting than the low-educated panel members. However, it seems somewhat unlikely that people would stop participating in all questionnaires because they do not appreciate a specific graph presented in a single questionnaire.

For panel members with a high education (college degree or more), the second video, in which a psychology professor explained the use of the personality data, significantly reduced the active participation over time (not receiving the video increased the change of remaining active for this group). The professor in this video explained something about individual differences, which might, with some effort, be misunderstood as a threat to the respondent's privacy. The highly educated panel members are perhaps more concerned about their privacy.

No significant effects were found within the youngest and oldest age groups or within the group of panel members in paid work or being at home. The two effects of education are the only significant effects found among the large number of tests of materials within subgroups in Table 9.3. Since these effects are difficult to explain, they might be due to chance. In all, we conclude that the materials did not help in reducing nonresponse bias.

9.4.4   Evaluation of the materials

In January 2010, we asked panel members in an online questionnaire how they liked the different materials we had sent in the first experiment. The letter about the income, housing, and assets questionnaire was not included in this evaluation. In general, only about half of the respondents said they liked the materials. The postcards were liked the best (61% liked it or very much liked it), followed by e-cards (51%), e-newsletter (48%), binder (45%), and paper newsletter (42%).

For the second experiment, we asked respondents in an online questionnaire if they thought the material was clear, interesting, and useful. The results are shown in Table 9.4. The evaluation of panel members who completed the questionnaires is predominantly positive. The third graph experiment, showing the answers of the panel members on satisfaction with the government, trust in science, and female managers are evaluated as the most clear, most interesting, and most useful of all feedback materials presented. The percentages for the innovative materials (second experiment) are considerably higher than the percentages for the traditional materials (first experiment).

In Tables 9.5 and 9.6, the effects of respondent demographic characteristics on the evaluation of the materials are presented. Striking in Table 9.5 is the negative relation of age to the liking of the cards. Furthermore, the higher-educated panel members seemed to least like most of the materials. Single households like the paper newsletter and people without work like the

Table 9.4   Evaluation of the materials by respondents (experiment 2).

	Definitely clear/Clear (%)	Definitely interesting/Interesting %	Definitely useful/Useful %
Video 1 (Family and Household)	67	56	57
Video 2 (Personality)	77	66	65
Graph 1 (Work and Schooling)	72	55	57
Graph 2 (Income and Finances)	79	67	65
Graph 3 (Politics and Values)	83	68	67

Table 9.5 Significant logistic regression effects of demographic characteristics on the evaluation of the cards, binder, e-cards, and newsletters (first experiment).[1]

Liked or liked very much:		Cards	Binder	E-cards	Paper newsletter	E-newsletter
		B	B	B	B	B
Gender[2]	Man					
	Woman					
Age/10		−.250**		−.125**		
Household	Single				.393**	
	With kids					
	Other					
Education	Primary					
	High school or vocational		−.359**	−.466**		
	College or higher	−.386**	−.716**		−.378**	
Occupation	Paid work					
	Unemployed or social benefit		.394**			
	School, study, housekeeping or other					
	Retirement					
Internet/PC	Not provided by LISS					
	Provided by LISS		.655**		.678**	
Nagelkerke R-squared		.054	.059	.033	.038	.014

*p <.05,
**p <.01.
[1]Only significant effects are shown.

information binder. Finally, the panel members who use a computer and Internet provided by the Institute are rather positive about the informational material (binder and newsletter).

Table 9.6 shows which panel members thought the innovative materials in the second experiment were interesting. Women evaluated the first video as more interesting than men, and older panel members, panel members without paid work and perfect panel members were also more interested in the videos than others. The graphs, in contrast, are evaluated as more interesting by men than by women. There is no effect of age on the evaluation of the graphs, but there is a consistent effect of education: the middle- and higher-educated panel members found the graphs more interesting than the low-educated panel members. This could be related to cognitive ability to read graphs.

Two other effects were that the perfect panel members evaluated Graph 2 more often as interesting than other panel members, and households with children were less interested in Graph 3 than single person households (Table 9.A.1 in Appendix 9.A). We do not find strong evidence that the liking of the materials is related to response rates for the subgroups under investigation.

Table 9.6 Significant logistic regression effects of demographic characteristics on the evaluation of the videos (second experiment).[1]

Definitely interesting/Interesting:		Video 1	Video 2	Graph 1	Graph 2	Graph 3
		B	B	B	B	B
Gender[2]	Man					
	Woman	.242**		−.217**	−.183*	−.217*
Age/10		.218**	.186**			
Household	Single					
	With kids					−.341**
	Other					
Education	Primary					
	High school or vocational			.368**	.395**	.494**
	College or higher			.301**	.625**	.717**
Occupation	Paid work					
	Unemployed or social benefit	.343*	.472**			
	School, study, housekeeping or other	.193*				
	Retirement		.331**			
Internet/PC	Not provided by LISS					
	Provided by LISS					
Participation	Not participated all months					
	Participated all months	.220**	.260**		.344**	
Nagelkerke R-squared		.059	.061	.015	.027	.037

*$p <.05$,
**$p <.01$.
[1]Only significant effects are shown.

9.5   Discussion and conclusion

High participation rates are crucial in panels in which long-term surveys are carried out. Although high response rates do not necessarily improve sample composition, it is essential for longitudinal surveys to preserve the number of respondents for which data over a number of waves are available.

In this chapter we investigated the value of sending traditional forms of information to members of an online panel and the effectiveness of new, more innovative forms. As the literature review showed, specific strategies to motivate panel members could induce biases in the panel composition if they have a differential effect on subgroups. Since the different forms of feedback in this study are likely to be differentially appreciated by different types of respondents, we estimated the effects of each type of material on participation and appreciation specifically for people with the lowest response probabilities and for people with the highest response probabilities.

Our results show little effects of advance letters, newsletters (paper or e-letter), postcards or e-cards, and information binders. In addition, more innovative ways of communication, such as feedback graphs and videos, did not show higher response rates. The separate subgroup analyses did not show clear, substantive effects for specific combinations of materials and respondent characteristics.

We can conclude that these materials are not likely to be effective in tailored strategies to motivate respondents with the lowest response rates and highest risk of attriting and thus decrease variation in response probabilities and retention probabilities in the panel. Earlier studies have shown that nonresponse and attrition are weakly related to respondent's demographic characteristics in the LISS panel (De Vos, 2009; Lugtig et al., 2010). Hence, the long-term response probabilities in the LISS panel do not show large variations across demographic groups and it might therefore not be surprising that we were not able to increase the response probabilities of specific demographic groups in this study.

The new interactive materials were appreciated rather positively, more so than the old materials (if we assume we can compare the different evaluation questions). The videos could perhaps be made somewhat more animated, and tailored to specific subgroups. The fact that only some of the new materials worked for specific subgroups suggests that further research is needed to understand how materials such as videos and graphs work. This can lead to improvement of the materials and hence their effect on response rates.

In the second experiment, we included all panel members assigned to the experimental and control groups by design as we did for the traditional materials. The difference is, however, that the traditional materials were sent by mail or email and could thus be seen by any panel member, actively participating or not. We had no information about whether they actually read or looked at these materials. The new videos and graphs were built into the questionnaires, to give more immediate and interactive feedback. This meant that panel members who did not participate that month, or did not complete the questionnaire, did not see the materials. In addition, some of the panel members (6%–10%) who did complete the questionnaire were not able to play the video or to generate the graph. They had, for example, no flash player installed, which was necessary to play the video. The graph was, in addition, not generated for the first 100 panel members who answered the questionnaire, since the number of responses on which the graph was based would then be too small. Finally, a substantial number of respondents reported, in the online evaluation questions immediately following the video or graph that they did not watch it, even though they were able to play/generate it. However, when we selected only those panel members who reported they had actually watched the videos and looked at the graphs, and compared their participation with the panel members in the control conditions, we found exactly the same: no significant overall effects on long-term participation, and very few effects within subgroups.

The results of both series of experiments, with the traditional materials in 2009 and with the more innovative materials in 2011, raise the question whether it is worthwhile continuing giving different forms of feedback to panel members, if it does not have a positive effect on participation rates, and for some education levels perhaps even may have a negative effect. We first note that the members of the LISS are contacted at high frequency rates. The effect of sending or presenting feedback might be different in panels for which respondents are contacted less often, for example, once a year. Second, the general response rates in the LISS panel are relatively high for an online panel, and attrition is low, perhaps related to the substantive monetary incentives paid to panel members for their participation. It therefore is possible that the feedback materials in this case did not increase the response rates any further beyond the level reached by the use of the monetary incentives. We had hypothesized that the feedback materials would appeal to feelings of closeness, interest, and helping science, and expected that these persuasive aspects would add up to the feeling of reciprocity evoked by the monetary incentive, to increase participation. As Groves et al. (2000) describe, the influence of each aspect of a request to participate in a survey depends on the importance an individual respondent assigns to that aspect (leverage) and the salience relative to the salience of other aspects of the survey request. This leverage–salience theory of survey participation would predict that the feedback materials increase participation, for those respondents interested in them, if the salience of the materials is high relative to the monetary incentive and other factors (Groves et al., 2004). Hence, our results suggest that the salience of the feedback information was not very high. To increase the salience, the materials, or a link to the materials, should perhaps have been sent closer in time to the moment when panel members make the decision to participate or not. We could, for example, have sent them immediately after the invitation emails, though we cannot control when exactly the respondent opens and reads or looks at the material.

Since many panel members reported that they appreciated the more innovative forms of feedback, we are still providing electronic newsletters and videos to the LISS panel members, even though they had no significant effect on the response rates. These materials are relatively low cost materials that respondents can choose either to read/watch or not.

The results of this study raise some discussion points for the development of future strategies in the LISS panel:

- Is it cost-effective to send (this type of) communication materials to LISS panel members, if they have little or no effect on participation and are not even appreciated very much? We could consider presenting the graphs only to the middle- and higher-educated panel members, since they report appreciating them more.

- Should we try other materials than the ones in this study (gifts and information)? Which?

- What other motivational measures could we take, other than regularly sending gifts and information, to keep up the panel participation?

Until now research has shown little evidence that feedback has a significant positive effect on response rates. On the other hand, respondents keep asking for feedback on the surveys they have taken. Future research should aim at finding materials that are salient enough to respondents to affect their decision to participate. Experimentation with new technology may be fruitful since it can provide a direct link to other information channels for feedback. In addition, experimentation on tailoring (giving individual respondents exactly what they want) instead of targeting everyone is warranted. Using discussion groups of panel members can be a way to start this type of research. The timing of the feedback materials is another point on the research agenda for the future of online panel research. Presenting feedback at the moment that respondents feel the need to receive it seems a logical thing to strive for, but has consequences for panel management. For now, there is little reason to invest time and money in feedback materials to improve response rates in online panels. Therefore, methodological research is of crucial importance to find ways to increase participation in online panel research. Modern techniques provide us with more tools than ever before, and it is up to survey researchers to continually develop new ways to increase cooperation in panels in which long-term surveys are carried out.

References

Albaum, G., & Smith. S. M. (2012). Why people agree to participate in surveys. In L. Gideon (Ed.), *Handbook of survey methodology for the social sciences* (pp. 179–194). New York: Springer.

Budowski, M., & Scherpenzeel, A. (2005). Encouraging and maintaining participation in household surveys: The case of the Swiss Household Panel. *ZUMA Nachrichten, 56*, 10–36.

Burns, R., & Burns, R. (2008). Business research methods and statistics using SPSS. In *Logistic Regression*. London: Sage. Retrieved June 20, 2013, from: http://www.uk.sagepub.com/burns/website%20material/Table%20of%20Contents.pdf.

Camburn, D., Lavrakas, P. J., Battaglia, M. P., Massey, J. T., & Wright, R. A. (1996). Using advance respondent letters in random digit dialing telephone surveys. In *Proceedings of American Statistical Association 1995, Section on Survey Research Methods* (pp. 969–974).

Clark, M. S. & Waddell, B.A. (1983). Effects of moods on thoughts about helping, attraction, and information acquisition. *Social Psychology Quarterly, 46*, 31–35.

De Leeuw, E., Callegaro, M., Hox, J., Korendijk, E., & Lensvelt-Mulders, G. (2007). The influence of advance letters on response in telephone surveys: A meta-analysis. *Public Opinion Quarterly, 71*, 413–443.

De Vos, K. (2009). *Sleepers in LISS* (Working paper). CentERdata, Tilburg, The Netherlands. Retrieved September 3, 2012, from: www.lissdata.nl/assets/uploaded/Sleeping%20panelmembers%20in%20the%20LISS%20panel.pdf.

Dillman, D. A. (2007). *Mail and internet surveys: The tailored design method*. Hoboken, NJ: John Wiley & Sons, Inc.

Edwards, P. J., Roberts, I., Clarke, M. J., DiGuiseppi, C., Wentz, R., Kwan, R., et al. (2009). *Methods to increase response to postal and electronic questionnaires*. Retrieved March 3, 2013, from: Cochrane Database of Systematic Reviews, Wiley Online Library Home: http://onlinelibrary.wiley.com/doi/10.1002/14651858.MR000008.pub4/pdf.

Göritz, A. S., & Neumann, B. P. (2013). *The Longitudinal Effects of Incentives on Response Quantity in Online Panels*. Manuscript submitted for publication. University of Freiburg, Germany.

Gouldner, A. W. (1960). Norm of reciprocity: A preliminary statement. *American Sociological Review, 25*, 161–178.

Groves, R. M., Cialdini, R. B., & Couper, M. P. (1992). Understanding the decision to participate in a survey. *Public Opinion Quarterly, 56*, 475–495.

Groves, R. M., Presser, S., & Dipko, S. (2004). The role of topic of interest in survey participation decisions. *Public Opinion Quarterly, 68*, 2–31.

Groves, R. M., Singer, E., & Corning, A. (2000). Leverage-saliency theory of survey participation: Description and an illustration. *Public Opinion Quarterly, 64*, 299–308.

Hembroff, L. A., Rusz, D., Rafferty, A., McGee, H., & Ehrlich, N. (2005). The cost-effectiveness of alternative advance mailings in a telephone survey. *Public Opinion Quarterly, 69*, 232–245.

Hosmer, D.W. Jr.,, Lemeshow, S., & Sturdivant, R. X. (2013). *Applied logistic regression*. Hoboken, NJ: Wiley-Blackwell.

Iredell, H., Shaw, T., Howat, P., James, R., & Granich, J. (2004). Introductory postcards: Do they increase response rate in a telephone survey of older persons? *Health Education Research, 19*, 159–164.

Larson, P. D., & Poist, R. F. (2004). Improving response rates to mail surveys: A research note. *Transportation Journal, 43*, 67–74.

Laurie, H., Smith, R., & Scott, L. (1999). Strategies for reducing nonresponse in a longitudinal panel survey. *Journal of Official Statistics, 15*, 269–282.

Link, M., & Mokdad, A. (2005). Advance letters as a means of improving respondent cooperation in RDD studies: A multi-state experiment. *Public Opinion Quarterly, 69*, 572–587.

Lugtig, P., Hox, J., De Leeuw, E. D., & Scherpenzeel, A. (2010, July). Attrition in the LISS panel: Separating "stayers", "sleepers" and other types of panel respondents. Paper presented at the Second Panel Survey Methods workshop, Mannheim, Germany.

Lynn, P. (2009). Methods for longitudinal surveys. In P. Lynn (Ed.), *Methodology of longitudinal surveys* (pp. 1–19). Chichester: John Wiley & Sons, Ltd.

Manzano, A. N., & Burke, J. M. (2012). Increasing response rate in WebBased/Internet surveys. In L. Gideon (Ed.), *Handbook of survey methodology for the social sciences* (pp. 327–344). New York: Springer.

McGonagle, K. A., & Schoeni, R. F. (2013). The panel study of income dynamics: Overview & summary of scientific contributions after nearly 40 years. Retrieved March 6, 2013, from: http://www.ciqss.umontreal.ca/longit/Doc/Robert_Schoeni.pdf.

Motte, A., & Brault, M. C. (2008). Keeping in touch with project participants between surveys: A mailing experiment. In *Proceedings of Statistics Canada Symposium 2008 data collection: Challenges, achievements and new directions*. (Component of Statistics Canada Catalogue no. 11-522-X). Ottawa, Canada: Statistics Canada.

Nagelkerke, N. J. D. (1991). A note on a general definition of the coefficient of determination. *Biometrika, 78*, 691–692.

Scherpenzeel, A. (2011). Why do Internet panel members become inactive and how can they be reactivated? Paper presented at the European Survey Research Association Conference, Lausanne, Switzerland.

Scherpenzeel, A. (In press). Survey participation in a probability-based internet panel in the Netherlands. In U. Engel, B. Jann, P. Lynn, A. Scherpenzeel, & P. Sturgis, P. (Eds.), *Improving survey methods*, London: Taylor & Francis.

Scherpenzeel, A., & Toepoel, V. (2012). Recruiting a probability sample for an online panel: Effects of contact mode, incentives and information. *Public Opinion Quarterly, 76*, 470–490.

Singer, E. (2002). The use of incentives to reduce nonrsponse in household surveys. In R. Groves, D. Dillman, J. Eltinge, & R. Little (Eds.), *Survey nonresponse* (pp. 163–177) New York: John Wiley & Sons, Inc.

Singer, E., Van Hoewyk, J., & Maher, M. P. (2000). Experiments with incentives in telephone surveys. *Public Opinion Quarterly, 64*, 171–188.

Stoop, I. (2012). Unit non-response due to refusal. In L. Gideon (Ed.), *Handbook of survey methodology for the social sciences* (pp. 121–148). New York: Springer.

Sue, V. M., & Ritter, L. A. (2007). *Conducting online surveys*. London: Sage.

Toepoel, V. (2012). Effects of incentives in surveys. In L. Gideon (Ed.), *Handbook of survey methodology for the social sciences* (pp. 209–226.). New York: Springer.

Trussell, N., & Lavrakas, P. J. (2004). The influence of incremental increases in token cash incentives on mail survey response: Is there an optimal amount? *Public Opinion Quarterly, 68*, 349–367.

Appendix 9.A

Table 9.A.1   Logistic regression effect of mailings and demographics on long-term participation.

			Participated all months	Still active at end of period
First experiment		N^1	B	B
Received all[2]		3945		
All except cards		273	−.119	−.173
All except binder		304	−.217	.081
All except letter		304	−.015	−.038
All except e-cards		320	.128	.219
All except newsletter		298	.109	−.072
Received nothing		576	−.056	.042
Gender	Man	2801		
	Woman	3219	−.072	.067
Age / 10			.262**	.260**
Household	Single	856		
	Couple without children or other household	2122	−.126	−.069
	Couple or single with children	3042	−.400**	−.295*
Education	(Extended) primary	2299		
	High school or vocational	2008	.111	.150
	College or higher	1713	−.021	.185
Occupation	Paid work	3445		
	Unemployed or social benefit	310	.367**	.285
	School, study, housekeeping or other	1451	.315**	.403**
	Retirement	814	.234**	.562**
Internet/PC	Not provided by LISS	5634		
	Provided by LISS	386	.036	.957**
Nagelkerke *R*-squared[3]			.094	.074
Second experiment				
		N^1	B	B
Presented all		681		
All except video 1 (Family and Household)		669	−.197	−.084

Table 9.A.1 (*continued*)

			Participated all months	Still active at end of period
All except video 2 (Personality)		663	−.049	.211
All except graph 1 (Work and Schooling)		675	−.009	.276
All except graph 2 (Income and Finances)		670	−.013	.045
All except graph 3 (Politics and Values)		678	−.038	.113
Presented nothing		678	.017	.097
Gender	Man	2190		
	Woman	2524	−.126	.058
Age/10		4714	.342**	.329**
Household	Single	863		
	Couple without children or other household	1731	−.004	−.058
	Couple or single with children	2120	−.297**	−.018
Education	(Extended) primary	1658		
	High school or vocational	1605	.108	.363**
	College or higher	1451	.087	.404**
Occupation	Paid work	2580		
	Unemployed or social benefit	347	.317**	.021
	School, study, housekeeping or other	990	.248**	.534**
	Retirement	797	.184	1.145**
Internet/PC	Not provided by LISS	4384		
	Provided by LISS	330	.033	.795**
Nagelkerke R-squared[3]			.137	.093

*$p <.05$,
**$p <.01$.

[1]N includes all panel members assigned to one of the groups, having valid data on all demographics included in the analyses. In March 2012, the number of perfect panel members was 2594 and the number of "sleeping" panel member was 824 N is defined in note 1 in Table 9.1.

[2]Effects were estimated in one model.

[3]Nagelkerke R-squared is a pseudo-R-squared used for logistic models, which ranges from 0 to 1. Higher values indicate better model fit (Nagelkerke, 1991).

Part III
MEASUREMENT ERROR

Introduction to Part III

Reg Baker[a] and Mario Callegaro[b]

[a]*Market Strategies International, USA*

[b]*Google UK*

III.1 Measurement Error

There are many potential sources of error in surveys. Data quality discussions about online panels have tended to focus heavily on coverage error, the failure of the sample to accurately represent the target population. This is understandable given that no country in the world has 100% Internet penetration, although it is sufficiently high in some countries that the potential error may be ignorable. Further, and as we have discussed in other Parts, most online panels are not recruited using probability-based methods. It is generally not known whether everyone with Internet access has the chance to join the panel and the response to recruitment efforts typically is very low, often less than 1% (Baker et al., 2010).

Measurement error occurs when the observed value in a survey differs from the underlying or true value in the target population. Groves (1989) describes four potential sources of measurement error: (1) the questionnaire; (2) the mode of data collection; (3) the respondent; and (4) the interviewer. All surveys, regardless of the sampling method used, are subject to measurement error and this is often cited as a major reason why the estimates from two well-designed and well-conducted surveys disagree.

The two chapters in Part III focus on the respondent as a source of error in online panel surveys. In Chapter 10, Hillygus and her colleagues consider the impact of professional respondents, that is, people who complete many surveys, sometimes by joining multiple panels to increase their survey opportunities. Their findings suggest that the often-cited link between frequent survey participation and undesirable behaviors such as satisficing is a weak one, and that the motivation to participate may be more important.

In Chapter 11, Greszki and his colleagues focus on a very specific respondent behavior: speeding through surveys and (presumably) failing to cognitively engage with the questionnaire in any meaningful way. Their work contrasts the prevalence and impact of speeding in

Online Panel Research: A Data Quality Perspective, First Edition.
Edited by Mario Callegaro, Reg Baker, Jelke Bethlehem, Anja S. Göritz, Jon A. Krosnick and Paul J. Lavrakas.
© 2014 John Wiley & Sons, Ltd. Published 2014 by John Wiley & Sons, Ltd.
Companion website: www.wiley.com/go/online_panel

probability versus nonprobability panels. Their research shows that speeding occurs, though not at the high levels that some have assumed. They find that speeding generally is not so prevalent or systematic that it affects the estimates, nor were there major differences between the probability and nonprobability surveys they compared.

References

Baker, R., Blumberg, S. J., Brick, J. M., Couper, M. P., Courtright, M., Dennis, J. M., Dillman, D. A., et al. (2010). Research synthesis. AAPOR report on online panels. *Public Opinion Quarterly, 74,* 711–781.

Groves, R. M. (1989). *Survey errors and survey costs.* New York: Wiley-Interscience.

10

Professional respondents in nonprobability online panels

D. Sunshine Hillygus[a], Natalie Jackson[b], and McKenzie Young[c]

[a]*Duke University, USA*

[b]*Marist College, USA*

[c]*Global Strategy Group, USA*

10.1 Introduction

Among the concerns raised about nonprobability online panels is the presence of so-called "professional" respondents – well-trained or experienced survey-takers who seek out large numbers of surveys, typically for the cash and incentives offered (Baker et al., 2010, 756–757). There is accumulating evidence that there exists a large number of frequent survey-takers who participate in many different online panels. A 2006 comScore study concluded that fewer than 1% of panel members in the ten largest market research online survey panels in the United States were responsible for 34% of the completed questionnaires (Grover & Vriens, 2006). An analysis using 16 different online panels in the United States and the United Kingdom found that the average panelist belonged to four different survey panels (Gittelman & Trimarchi, 2009). Similarly, a study of 19 different panels in the Netherlands found that 62% of panelists belonged to multiple online panels (Willems, Vonk, & Ossenbruggen, 2006b). As opt-in panel surveys become more prevalent in academic and policy research, there is a need to explore if and how frequent survey-takers affect the reliability and validity of the collected data.

Online Panel Research: A Data Quality Perspective, First Edition.
Edited by Mario Callegaro, Reg Baker, Jelke Bethlehem, Anja S. Göritz, Jon A. Krosnick and Paul J. Lavrakas.
© 2014 John Wiley & Sons, Ltd. Published 2014 by John Wiley & Sons, Ltd.
Companion website: www.wiley.com/go/online_panel

The presence of frequent survey-takers is widely recognized, but researchers are only just beginning to assess their influence on data quality. Thus far, few academic journal articles have been published on the topic, though there are a handful of marketing firm white papers and conference presentations that scrutinize the survey-taking behavior of professional respondents. Much of the existing research has focused on the concern that frequent survey-takers will be more likely to lie or rush through a survey for the incentives, jeopardizing the integrity of their responses. In other words, in an effort to complete many surveys, frequent survey-takers might be inattentive to individual surveys and thus more likely to engage in satisficing behavior, which results in less reliable data. Beyond measurement error, however, we believe there is a broader question about if and how professional respondents might differ systematically from other respondents. In other words, are the attitudes, opinions, and beliefs of frequent survey-takers different from those of less experienced respondents?

In this chapter, we compare frequent and nonfrequent survey-takers in a survey from the 2010 YouGov Cooperative Congressional Election Study. In contrast to the expectations of many, we do not find overwhelming and consistent evidence that frequent survey-takers are significantly more likely to satisfice. We do, however, find that professional respondents are less politically interested, engaged, and knowledgeable than other respondents in the survey. We posit that this difference might reflect the contrasting motivations of those volunteering to respond to a political survey, with professional survey-takers motivated by incentives and nonprofessional survey-takers motivated by interest in the survey topic. Given the well-established finding that nonprobability online surveys have overall samples that are more politically attentive and engaged on average than probability-based samples (Malhotra & Krosnick, 2007), these results suggest that eliminating professional respondents from nonprobability political surveys – as some have recommended – could actually result in a *more* biased sample on these dimensions. Although the biases we identify may well not extend to other survey topics, firms, or samples, these results do suggest that measurement error might not be the only (or most important) concern regarding professional respondents.

10.2   Background

The term "professional respondent" is commonly used in the survey and marketing profession, but the way it is defined and operationalized varies widely. Professional respondents are typically identified as having a high level of survey-taking activity, but the specific measure might include the number of individual surveys completed in a given time frame (Garland et al., 2012), the number of online panel memberships (Comley, 2005; Baker & Downes-LeGuin, 2007), duration in a particular survey panel (Dennis, 2001), or some combination of these metrics. The specific threshold used to classify a respondent as professional based on each measure also differs; for example, some consider completion of four or more surveys per month to be excessive (Garland et al., 2012; Honda & Motokawa, 2004), while others use a threshold of 30 or more surveys a month to define professional respondents (Gittelman & Trimarchi, 2009; Frede et al., 2006). Clearly, there is no consensus about how many surveys is too many. Given these differing definitions, it is perhaps no surprise that estimates about the prevalence of professional respondents in online nonprobability samples also varies across studies and across panels. In an early study of seven prominent survey panels, Krosnick et al. (2005) found that the median number of surveys taken in the previous year ranged from six to 31. Moreover, the presence of professional respondents in any given panel largely depends on the specific recruitment and management practices of the particular survey panel (Willems, Vonk, & van Ossenbruggen, 2006a). Indeed, the prevalence of professional respondents likely stems not only from the combination of self-selection and incentives that characterizes nonprobability online panels, but also the fact that many of the online survey firms recruit panelists from the same websites (Fulgoni, 2005).

There are also two distinct conceptual frameworks for thinking about professional respondents. The most relevant line of research is focused on sample selection, with the concern that frequent survey-takers volunteer or "opt in" to answer a lot of surveys in search of monetary incentives (Gittelman & Trimarchi, 2009). A second line of research, however, is focused on panel conditioning, whereby extended duration in a panel means that respondents learn about the survey process, questionnaire, or topic. Panel conditioning is an issue not only for online, nonprobability surveys, but also longitudinal surveys of all modes, so there exists a more extensive academic literature on the topic.

These are quite different notions of professionalism – for the first, professional respondents are recruited; for the second, professional respondents are created. Panel conditioning research typically considers participation in a single survey panel (and often a single longitudinal study), while those concerned with sample selection recognize that the same respondent might participate in multiple survey panels. For some, the difference between the two types of professional respondents boils down to motivation. Toepoel et al., (2009) found that respondents who were experienced due to repeated surveying in a panel study are distinct from recruited professional respondents in that they do not necessarily participate because of the incentives offered (Mason & Watts, 2010). Despite this potential theoretical differentiation, the operationalization of professional respondents and experienced respondents is often identical. It is difficult to measure respondent motivation, so it is more common for both lines of research simply to rely on levels of survey-taking activity. The number of surveys taken – a common measure of professionalism – is also the preferred metric for assessing panel conditioning (Adams, et al., 2012; Coen, et al., 2005). Moreover, there are typically parallel concerns about the impact of professional or experienced survey-takers on the reliability and validity of the data. So, though this chapter uses the first concept of professional respondents, we reference panel conditioning literature where relevant.

10.3   Professional respondents and data quality

The primary concern about professional respondents has been that they will threaten data quality by providing inaccurate or fraudulent responses. Respondents introduce measurement error if they do not take a survey seriously and simply speed through the questions to get to the end. Respondents who are taking a lot of surveys might be inattentive to any one survey, and thus more likely to engage in satisficing (Krosnick, 1991). That is, they give less cognitive effort to answering the survey, as evidenced through fast response times, use of "don't know" responses, item nonresponse, random response selection, open-ended gibberish, or straight-lining responses (Baker et al., 2010).

Existing research on the extent of satisficing among frequent survey-takers is mixed, however. Many have found that frequent survey-takers complete questionnaires more quickly than those with less experience (Frede et al., 2006; Knapton & Garlick, 2007; Toepoel et al., 2008; Yan & Tourangeau, 2008), though some find no such differences (Coen et al., 2005). Less clear is whether speedier completion time results in more measurement error. More broadly, Toepoel et al., (2008, p. 985) conclude that experienced respondents are more likely to take "shortcuts in the response process" than fresh respondents based on higher interitem correlations for multiple-item-per-screen formats and higher likelihood of selecting first response options among the more experienced survey-takers. Garland et al. (2012) similarly found that respondents who had previously taken surveys were more likely to give a "don't know" response than fresh respondents. Others conclude that experienced respondents are more likely to answer questions strategically to avoid follow-ups (Mathiowetz & Lair, 1994; Meurs et al., 1989; Nancarrow & Cartwright, 2007). According to Miller (2007), opt-in, online panelists are more likely to satisfice than online respondents in general – in other words, the observed satisficing is not simply an issue with mode.

On the other hand, other research suggests that professional respondents are actually *less* likely to satisfice (Chang & Krosnick, 2009; Schlackman, 1984; Waterton & Lievesley, 1989). In a study of 17 online panels for the Advertising Research Foundation, Walker et al. (2009) examined the the presence of professional respondents and concluded there was "little evidence that it impacted data quality to any significant degree" despite their large numbers and some attitudinal differences. In fact, those belonging to multiple panels and taking more surveys per month were *less* likely to exhibit "bad behaviors." Likewise, even though professional respondents made up 17% of the Dutch Online Panel Comparison Study (NOPVO) sample, Matthijsse et al. (2006) concluded that there were only slight implications for data quality; any differences between professional and nonprofessional respondents disappeared after controlling for gender, income, and urbanization. Others have found that experienced panelists are less likely to answer "don't know" (Binswanger et al., 2006; Smith & Brown, 2006; Waterton & Lievesley, 1989) and have higher consistency across related survey questions, evidence of higher convergent validity (Garland et al., 2012). According to Smith and Brown (2006), frequent survey-takers are no more likely to straight-line than inexperienced survey-takers and are more likely to answer sensitive questions about income and race. Finally, De Wulf and Berteloot (2007) showed that professional respondents are more positive towards the survey process and more willing to complete subsequent surveys. Thus, the evidence is decidedly mixed as to whether professional respondents provide more or less reliable data.

Beyond measurement error, however, there is a broader question about if and how professional respondents might differ from other respondents in terms of attitudes, opinions, and beliefs. If they differ on these metrics, the presence of professional respondents could bias estimates of quantities of interest even in the (unlikely) scenario in which there was no measurement error in their responses. Here again the existing research is inconclusive. Researchers have found that frequent survey-takers are demographically different from other panelists and from probability samples; they are less likely to be employed full-time, have lower incomes, lower levels of home ownership, and belong to smaller households, for example (Casdas et al., 2006; Frede et al., 2006; Gittelman & Trimarchi, 2009). Perhaps more importantly, some have observed significant differences in the attitudes and behaviors of professional respondents, even controlling for demographic differences (Knapton & Garlick, 2007; Willems, Ossenbruggen, & Vonk 2006a; Casdas et al., 2006; Gittelman & Trimarchi, 2009; Walker et al., 2009). In a study by Casdas et al. (2006), multiple panel respondents drank less wine, invested less, smoked more, read more magazines, and owned more pets. Gittelman and Trimarchi (2010) found that frequent survey-takers exhibit different buying and media behavior than less frequent survey-takers. Miller (2007) concluded that professional respondents were less likely to be impressed by new products, while Walker et al. (2009) found that more professional respondents were more likely to report "purchase interest" in new product concepts.

In contrast, others conclude the differences are not substantial. In an analysis of 3054 different measures, Smith and Brown (2006) found that just 4.6% of items showed significant differences for respondents who were "hyperactive," defined as individuals active in two or more online panels. Interestingly, two items for which there were significant differences are relevant to the current analysis of political outcomes: signing a political petition and paying attention to the news. In another study, Frede et al. (2006) found significant differences between those answering more than 10 surveys a month and those answering fewer, but they concluded there were simply too few of the heavier responders in the Ipsos, NPD Group and TNS panels to bias the overall results. Going one step further, Coen et al. (2005) argued in an SSI White Paper that "responses from frequent responders are *more* in line with actual consumer behavior than responses from less frequent responders"(emphasis added).

The relevant panel conditioning research is similarly mixed. Chang and Krosnick (2009, p. 14) conclude that "accumulating experience at doing surveys makes panel members less and less like the general public they are intended to represent." Others, however, find little evidence that panel experience biases outcomes of interest (Kruse et al., 2010; Toepoel et al., 2009). For example, Pineau et al. (2005) found in an analysis of 30 different survey outcomes that less than 10% of items showed differences in attitudes and behaviors based on tenure in the panel. Clearly, the existing research does not offer a consistent picture of the impact of professional respondents on data quality.

10.4 Approaches to handling professional respondents

Even if there is no consensus about the impact of professional respondents on the reliability and validity of the data, the concerns remain widespread. Panel companies have adopted a variety of procedures and management practices to try to eliminate professional respondents, reduce over-surveying, and validate respondents are who they say they are (Baker et al., 2010). Panels actively search for false identities and routinely embed trap or "red herring" questions (Conrad et al., 2005; Kapelner & Chandler, 2010; Miller, 2007; Oppenheimer et al., 2009). For example, Downes-Le Guin et al. (2006) found that 14% of panelists in his study reported owning a Segway (which has less than 0.1% incidence in population), whereas telephone surveys had no such overreporting. Others limit the number of surveys a panelist takes within a specified time period (Dennis, 2001).

In addition to such efforts to avoid including professional respondents in the sample in the first place, companies frequently toss out data from "undesirable" respondents after data collection (Knapton & Garlick, 2007; Rogers & Richarme, 2009). Some firms collect more than the target number of respondents with the expectation that some will be eliminated. Peruzzi (2010, July 8) advises that "between 1 and 5% of survey data from panel sample is garbage. Garbage – throw it out; don't bring it into your final dataset to analyze." Others, however, caution that it is inadequate to attempt correction through purging problematic respondents (Harlow, 2010) or demographic weighting (Casdas et al., 2006).

In sum, panel companies and clients are obviously concerned about the quality of the data that professional respondents will provide, but it remains unclear how they might affect the data or what can be done about it. In this chapter, we offer one more analysis of professional respondents to this growing body of research. We examine the data quality implications of frequent survey-taking and multiple panel participation among respondents in the 2010 Cooperative Congressional Election Study. Admittedly, our analysis is limited to a single sample from a single survey firm on the topic of political attitudes and behaviors. But given the prominence of political surveys in the polling field, we believe this analysis is of interest even if any patterns observed here cannot be generalized to other survey topics or panels.

10.5   Research hypotheses

Given the topic of the survey and the timing during the 2010 congressional campaigns, we hypothesize that professional respondents might differ from other respondents in their political attitudes and behaviors, reflecting differences in motivation to participate in the study. It is well established that people with more interest in the survey topic respond at higher levels than those less interested in the topic (Goyder, 1987; Groves et al., 2000, 2004). This effect might be amplified in opt-in online panels where the number of survey invitations is large. As the 2010 American Association of Public Opinion Researchers (AAPOR) report notes,

> People who join panels voluntarily can differ from a target population in a number of ways (e.g., they may have less concern about their privacy, be more interested in expressing their opinions, be more technologically interested or experienced, or be more involved in the community or political issues). For a specific study sample, this may be especially true when the topic of the survey is related to how the sample differs from the population

(Baker et al., 2010, p. 746). Although Yan and Tourangeau (2008) found no evidence that survey topic matters, others have found such biases, especially in online political surveys, where online panelists have been found to be more politically engaged, interested, and knowledgeable than the general population (Malhotra & Krosnick, 2007).

By contrast, we expect professional respondents are more likely to be motivated by the incentives being offered. Research consistently finds that those participating in more surveys are more likely to be doing so for monetary reasons (Frede et al., 2006; Sparrow, 2007). Paolacci et al. (2010) reported that more than 61% of respondents in a Mechanical Turk sample said that earning additional money was the primary reason they participated, and almost 14% said it was their primary source of income. These statistics might seem surprising since the incentives offered for completing online surveys are often very small amounts of money – sometimes less than a dollar per survey – but that is precisely why professional respondents complete large numbers of surveys. Online message boards routinely recommend signing up for dozens of panels at a time to maximize cash and prizes, and a Google search for the terms "online surveys for money" yields more than 300 million hits.

Of course, researchers have long pointed out that there are many different reasons why respondents join panels (see Dillman et al. (2009) for a thorough discussion of motivation theory). Comley (2005), for example, classifies four groups of respondents: the helpers, the opinionated, the incentivized, and the professionals. Less clear, however, is how motivation, frequency of survey-taking, and data quality all interact. This is clearly a topic that deserves further study, not only for panel research, but across all longitudinal studies that use respondent incentives. Unfortunately, the survey used in our analysis does not include a measure of motivation for participation (indeed, motivation is a notoriously tricky notion to measure). In a separate Mechanical Turk survey, however, we did find support for the assumption that frequent survey-takers were significantly more likely to say their primary reason for taking the survey was the monetary incentive, confirming the conclusions of earlier research. It thus seems reasonable to assume that professional respondents may differ from nonprofessional respondents in how the survey topic factors into their motivation and decision to take part in the survey. If professional respondents choose to participate in a survey because of the incentives offered, the survey topic is likely of less consequence. Thus, the pattern we expect to find in a political survey is that more frequent survey-takers will be less politically informed, engaged, and knowledgeable than less frequent survey-takers.

It is worth noting that this expectation contrasts with the findings for respondents "trained" through panel conditioning. Previous research has found that long-term participation in longitudinal political studies results in a sample that is *more* politically interested on average. Political knowledge questions, in particular, are sensitive to panel conditioning effects (Battaglia et al., 1996; Das et al., 2011; Kruse et al., 2010; Nancarrow & Cartwright, 2007; Toepoel et al., 2009). For example, it is thought that participation in repeated political studies might induce respondents to pay attention to the campaign (Bartels, 1999). In contrast, professional respondents in our analysis have participated in a large number of surveys of varying topics and foci. Indeed, participation in a large number of surveys might actually mean that any one survey leaves less of a lasting impression on the respondent.

10.6 Data and methods

We examine the relationship between survey-taking frequency and political attitudes and behaviors using a survey of 1000 American adults conducted as part of the 2010 Coopera-tive Congressional Election Study (CCES). The survey was administered over the Internet by YouGov, with respondents drawn from their opt-in panel using a stratified (by age, race, gen-der, and education) sample that is then matched to a random sample from the 2008 American Community Survey on a set of demographic and (imputed) political variables.

The CCES had two waves, pre and post election, with the pre-election phase conducted in October and the post-election phase conducted in November. The study was a collaboration between 40 research teams; half of the questionnaire consisted of Common Content asked of all respondents, and half of the questionnaire consisted of Team Content designed by each individual participating team and asked of a subset of 1000 people. Overall, the CCES had a final matched sample size of 55400 respondents answering the Common Content questions in the pre- and post-election waves. The sample used 196235 email addresses for the study, of which 9262 were determined to be ineligible, 79723 did not respond, and 27155 had par-tial responses. The study had 75450 completed interviews (Ansolabehere, 2012, August 10). YouGov recruits respondents for their panel by advertising short surveys about entertaining topics on popular websites and then inviting those who respond to join the panel.[1]

The Duke University team questionnaire included questions about the number of surveys completed in the past four weeks and the number of survey panels to which the respondent belongs. Detailed question wording for all questions in the analysis can be found in Appendix 10.A. The mean number of self-reported surveys in the past four weeks is 4.54, with about one-quarter of respondents reporting participation in more than one survey per week. The mean number of self-reported panels is 2.25, with 53% reporting participation in more than one online panel and 36.5% in 3 or more online panels. The full distribution of responses are shown in Figure 10.1. To avoid applying arbitrary thresholds to define "professional" respon-dents, our analysis uses these continuous measures in the analysis that follows.

---

[1] The invitation sent to respondents did not explicitly mention it was a political survey, instead calling it "a survey on national and community affairs conducted by PollingPoint in conjunction with 35 of the nation's leading univer-sities and research institutes." Arguably, however, such national affairs in October of election year might be thought to be about politics. Moreover, we might expect that the large number of partial responses came disproportionately from individuals who were either not politically interested or not as motivated by the compensation.

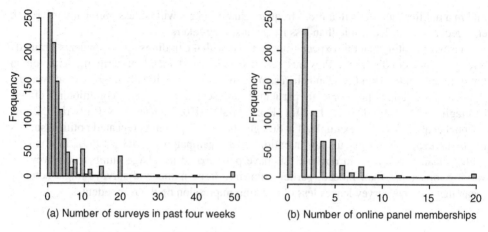

Figure 10.1   Self-reported survey-taking behavior in CCES.

One obvious weakness of these self-report measures is that they they could be subject to overreporting or underreporting. While this would not be a problem if such error were random, we might expect there to be systematic underreporting since some panel companies have procedures to discourage frequent survey-taking. As a robustness check, we replicated our results excluding those individuals who answered that they completed "0" other surveys, on the assumption this response may have been the most clearly dishonest response. In almost every case, the results provide even stronger support for our expectations. Although we did not have access to the number of surveys each respondent has completed for YouGov, the self-report measure has the advantage that it captures survey-taking in all online panels. Not surprisingly, those who belong to multiple panels report a higher number of surveys completed in the previous month (correlation of .615).

We examine the relationship between these survey frequency measures and a variety of different political attitudes and behaviors, including political knowledge, interest, turnout, engagement, and ideological extremism. We estimate a series of multivariable models that include either the number of surveys completed in the last four weeks or the number of panel memberships as the key independent variable. For the ease of presentation, OLS regression models are estimated except in the case of binary outcome variables, for which logit models are estimated. Similar results are found if we estimate ordered logits for outcomes with fewer than seven response options. In all models, we control for age, race, gender, income, marital status, education, and full-time work status. Analyses were conducted on unweighted data, though the model results are similar when weighted using the provided weight variable. We also replicated the models using the unmatched sample with nearly identical results.

10.7   Results

The full set of empirical results are reported in Table 10.1 for online panel membership and Table 10.2 for number of surveys completed.[2] Overall, our analysis finds that individuals who participate in more surveys and on more survey panels have consistently lower levels

---

[2] All reported predicted probabilities in the text are calculated holding all indicators at their mode and other variables at their means.

Table 10.1 Model results for self-reported panel membership.

	Political knowledge	Political interest	Turnout	Political activity	Ideological strength
(Intercept)	1.15*	1.91*	−2.93*	−0.77*	0.87*
	(0.23)	(0.13)	(0.65)	(0.20)	(0.20)
Panel Membership	−0.11*	−0.04*	−0.15*	−0.05*	−0.05*
	(0.02)	(0.01)	(0.05)	(0.02)	(0.02)
Age	0.02*	0.02*	0.06*	0.02*	0.01*
	(0.00)	(0.00)	(0.01)	(0.00)	(0.00)
Female	−0.39*	−0.27*	−0.50†	−0.09	−0.19*
	(0.09)	(0.05)	(0.26)	(0.08)	(0.08)
White	0.21*	0.25*	−0.07	0.16†	0.26*
	(0.09)	(0.05)	(0.26)	(0.08)	(0.08)
Income	0.05*	0.04*	0.09*	0.01	0.02
	(0.02)	(0.01)	(0.04)	(0.01)	(0.01)
Education	0.14*	0.05*	0.36*	0.12*	0.03
	(0.03)	(0.02)	(0.10)	(0.03)	(0.03)
Married	0.08	−0.01	0.56*	0.14†	0.07
	(0.09)	(0.05)	(0.25)	(0.08)	(0.08)
Work Full-Time	0.12	0.04	0.09	0.07	0.02
	(0.09)	(0.05)	(0.26)	(0.08)	(0.08)
N	840	842	691	853	843
Adj. R^2	0.21	0.33	0.18	0.13	0.07

Notes: Standard errors in parentheses.
†$p < .10$
*$p < .05$
All models OLS regression except Turnout, which is logit with Nagelkerke R^2 reported.
Panel membership range is 0–12, all higher values top-coded as 12.

of political interest and engagement. That is, we find a negative and statistically significant relationship between our professionalism measures and the political outcome variables.

Looking first at the relationship between panel membership and political knowledge in the first column in Table 10.1 finds that each additional panel membership is associated with a .11 *lower* score on the political knowledge scale (0–4 scale). The model predicts that, holding all else constant, the most active respondents – those belonging to 10+ panels – have a knowledge level nearly a full point lower than those belonging to a single online panel. It is also worth noting that the political knowledge level of respondents in this survey is substantially higher than what is found in probability samples. For example, a probability-based telephone survey from March 2011 by the Pew Research Center finds that 38% of Americans knew which party holds the majority in the House of Representatives. In contrast, the same question in the CCES survey (one component of the knowledge scale) finds that 91% of those belonging to a single online panel answered correctly, compared to 74% of those belonging to three or more panels. Both groups are more knowledgeable than the telephone sample, but the frequent survey-takers are less biased in this measure than those with less experience. This finding is also counter to the idea that respondents in web surveys might look up the answers

Table 10.2   Model results for self-reported number of surveys in past four weeks.

	Political knowledge	Political interest	Turnout	Political activity	Ideological strength
(Intercept)	0.99*	1.86*	−2.89*	−0.88*	0.77*
	(0.23)	(0.13)	(0.65)	(0.20)	(0.20)
Survey Number	−0.03*	−0.01*	−0.06*	−0.01	−0.01*
	(0.01)	(0.00)	(0.02)	(0.01)	(0.01)
Age	0.02*	0.02*	0.06*	0.02*	0.01*
	(0.00)	(0.00)	(0.01)	(0.00)	(0.00)
Female	−0.46*	−0.29*	−0.59†	−0.10	−0.21*
	(0.09)	(0.05)	(0.25)	(0.07)	(0.08)
White	0.20*	0.25*	−0.02	0.17*	0.27*
	(0.09)	(0.05)	(0.26)	(0.08)	(0.08)
Income	0.05*	0.04*	0.09*	0.01	0.02
	(0.02)	(0.01)	(0.04)	(0.01)	(0.01)
Education	0.14*	0.05*	0.38*	0.12*	0.03
	(0.03)	(0.02)	(0.10)	(0.03)	(0.03)
Married	0.04	−0.02	0.53*	0.12	0.06
	(0.09)	(0.05)	(0.25)	(0.08)	(0.08)
Work Full-Time	0.11	0.04	0.07	0.05	0.03
	(0.09)	(0.05)	(0.26)	(0.08)	(0.08)
N	843	847	692	858	847
Adj. R^2	0.20	0.33	0.18	0.12	0.07

Notes: Standard errors in parentheses.
†$p < .10$
*$p < .05$
All models use OLS regression except Turnout, which is a logit with Nagelkerke R^2 reported.
Survey number range is 0–30, all higher values top-coded as 30.

on political knowledge questions, though previous research has similarly found little evidence of this behavior (Ansolabehere & Schaffner, 2011; Prior & Lupia, 2008).

The effects are similar, but smaller, for political interest. Those who belong to more survey panels report lower levels of political interest. The model predicts that the most professionalized respondents (10+ panel memberships) are, on average, about one half a point less interested in politics on the 4-point scale than those belonging to a single panel. A simple bivariate comparison, for instance, shows that 73% of those belonging to a single online panel were interested in politics "most of the time," compared to 48% of those belonging to three or more online panels.

A similar pattern is found for more specific political behaviors, including voter turnout in the 2010 congressional election and participation in other political activities, such as attending a local political meeting, displaying a yard sign or bumper sticker, or working for a political campaign. For example, the model predicts that, holding all else constant, the most professionalized respondents are 15 percentage points less likely to vote and they participate in an average of 0.50 fewer political activities (0–3 scale) than those belonging to a single panel. Turnout is one of the few variables for which we have an especially strong external benchmark. The 2010 Current Population Survey (CPS) finds that 45.5% of eligible citizens reported voting in the 2010 congressional election. In contrast, in the CCES sample (unweighted), those belonging to three or more online panels had a self-reported turnout rate of 74%, compared to 89% among those in a single online panel. Thus, once again, we find less bias in the estimate for the professional respondents.

The final column reports results for a measure of ideological extremism. The negative and significant relationship indicates that those participating in more online survey panels are more ideologically moderate than those belonging to fewer online panels. The model predicts that, holding all else constant, the most professionalized respondents were about one half point less extreme on a 0–3 scale than those belonging to a single panel. Looking descriptively at the bivariate data, for instance, shows that 21% of those belonging to a single panel call themselves moderate compared to 31% of those belonging to 3 or more panels.

As shown in Table 10.2, we find similar results when using the number of surveys completed in the past four weeks rather than number of panel memberships as our measure of professionalism. For example, the models predict that those who answered surveys on at least a daily basis (30+ surveys in the past four weeks) were, on average, 20 percentage points less knowledgeable, 8 percentage points less interested, and 8 percentage points less ideologically extreme than those who answered no other surveys. These frequent survey-takers were also predicted to be 19 percentage points less likely to have voted in the 2010 congressional election. The relationship between number of surveys completed and number of other political activities is in the expected negative direction but does not reach the expected level of statistical significance ($p = .16$, one-tailed). It is again perhaps easiest to visualize the magnitude of these relationships with a simple descriptive look at the data. The "professional" respondents (5+ surveys in past four weeks) were less likely than nonprofessional respondents to get all four political knowledge questions correct (31% vs. 51%), less likely to say they are interested "most of the time" (53% vs. 70%), less likely to have reported voting (76% vs 95%), less likely to have participated in at least one other political activity (38% vs. 51%), and more likely to report being ideologically "moderate" (30% vs. 20%). In sum, at least for this sample, those respondents who were more professionalized–belonged to more panels and answered more surveys–where less knowledgeable than other respondents, even though the overall sample appears to be more knowledgeable and engaged than the general population.

10.8   Satisficing behavior

We next consider the possibility that professional respondents reduce data quality due to higher levels of satisficing. That is, do they speed through the survey without giving adequate cognitive attention to answering the questions in a thoughtful manner? Researchers have used a variety of different indicators of satisificing that we explore here: self-reported survey effort, interview duration, attrition after the pre-election wave, response straight-lining, percentage of skipped questions, and percentage of "don't know" responses. See Appendix 10.A for detailed information about question wording and variable construction. The results are shown in Table 10.3 for online panel membership and Table 10.4 for surveys completed. In contrast to the expectations of many (including us), we did not find consistent evidence that frequent participation in surveys or multiple survey panel participation was related to bad survey-taking behavior.

The first columns in Table 10.3 and Table 10.4 report the results for self-reported survey effort. As one of the final questions on the questionnaire, we asked each respondent how much effort they had put into answering the survey. Not surprisingly, all respondents were more likely to say "a lot" than "a little," but we also find that those belonging to more panels and completing more surveys report more effort than the less frequent survey-takers. Holding all

Table 10.3 Quality measures: self-reported panel membership.

	Survey effort	Interview duration	Attrited	Straight-line	Percent missing	Percent DK	Open junk
(Intercept)	5.73*	36.39*	−0.78	−0.15	6.69*	27.41*	0.95[†]
	(0.29)	(5.72)	(0.52)	(1.12)	(0.55)	(1.93)	(0.54)
Panel	0.06*	0.81	−0.06	0.01	0.10*	0.57*	0.10*
Membership	(0.03)	(0.50)	(0.05)	(0.10)	(0.05)	(0.17)	(0.05)
Age	−0.001	0.11	−0.01	−0.03[†]	−0.01	−0.22*	−0.04*
	(0.00)	(0.08)	(0.01)	(0.02)	(0.01)	(0.03)	(0.01)
Female	0.08	−0.82	0.32	0.74	−0.16	2.71*	0.51*
	(0.11)	(2.16)	(0.21)	(0.54)	(0.21)	(0.73)	(0.24)
White	−0.27*	−2.08	−0.62*	−1.19*	0.45*	−2.12*	−0.26
	(0.12)	(2.28)	(0.21)	(0.49)	(0.22)	(0.77)	(0.23)
Income	−0.05*	−0.25	−0.03	−0.19*	0.005	−0.33*	−0.04
	(0.02)	(0.38)	(0.04)	(0.09)	(0.04)	(0.13)	(0.04)
Education	−0.01	−0.30	−0.003	−0.22	−0.1	−0.98*	−0.24*
	(0.04)	(0.81)	(0.08)	(0.19)	(0.08)	(0.27)	(0.09)
Married	0.32*	2.48	0.19	−0.08	−0.54*	−1.87*	−0.25
	(0.12)	(2.29)	(0.22)	(0.50)	(0.22)	(0.77)	(0.23)
Work	0.07	1.42	−0.26	0.55	0.28	−1.58*	−0.54*
Full-Time	(0.12)	(2.31)	(0.22)	(0.51)	(0.22)	(0.78)	(0.25)
N	851	853	853	853	853	853	853
Adj. R^2	0.02	0.00	0.04	0.18	0.01	0.23	0.17

Notes: Standard errors in parentheses.
[†]$p < .10$
*$p < .05$
All models use OLS regression except Straightline, Open-Ended Junk, and Attrited which are logits with Nagelkerke R^2 reported.
Panel membership range is 0–12, all higher values top-coded as 12.

else constant, the model finds that those who belonged to 10+ panels report 6% more effort than those belonging to a single panel. Table 10.4 similarly shows that those taking more surveys in the past four weeks were significantly more likely to report higher levels of survey effort.

Of course, it is also possible that the more professional respondents are simply more likely to lie in response to this question. One alternative metric for survey effort could be the amount of time spent completing the interview. Previous research has found that professional respondents are more likely to speed through surveys (Toepoel et al., 2008). However, in this sample, the more professional respondents took slightly longer to complete the survey – a mean of 39 minutes among those answering fewer than five surveys a week, and 44 minutes for those taking five or more. Once we control for other factors, we find that the relationship between duration and panel memberships or surveys completed is positive, but not statistically significant. If we look just at excessive speeding (completion in less than 20 minutes – one half the average time), we likewise find that professional respondents were slightly less likely to speed through the questionnaire.

Table 10.4   Quality measures: number of completed surveys.

	Survey effort	Interview duration	Attrited	Straight-line	Percent missing	Percent DK	Open junk
(Intercept)	5.66*	37.42*	−0.34	0.53	6.63*	28.40*	1.05*
	(0.28)	(5.57)	(0.51)	(1.15)	(0.54)	(1.90)	(0.54)
Number of	0.03*	0.17	−0.10*	−0.21*	0.06*	$0.12^{\dagger}$	$0.03^{\dagger}$
Surveys	(0.01)	(0.19)	(0.03)	(0.10)	(0.02)	(0.06)	(0.02)
Age	0.0002	0.12	−0.01	$−0.03^{\dagger}$	−0.01	−0.22*	−0.04*
	(0.00)	(0.07)	(0.01)	(0.02)	(0.01)	(0.03)	(0.01)
Female	0.12	0.04	0.31	0.83	−0.19	3.14*	0.57*
	(0.11)	(2.12)	(0.20)	(0.55)	(0.20)	(0.72)	(0.23)
White	−0.30*	−1.85	−0.58*	−1.17*	$0.41^{\dagger}$	−2.14*	−0.23
	(0.11)	(2.25)	(0.21)	(0.50)	(0.22)	(0.77)	(0.23)
Income	−0.05*	−0.35	−0.05	−0.23*	0.004	−0.38*	−0.05
	(0.02)	(0.38)	(0.04)	(0.10)	(0.04)	(0.13)	(0.04)
Education	−0.01	−0.44	−0.003	−0.21	−0.09	−1.04*	−0.25*
	(0.04)	(0.80)	(0.08)	(0.19)	(0.08)	(0.27)	(0.09)
Married	0.33*	2.73	0.18*	0.1	−0.54*	−1.69*	−0.25
	(0.11)	(2.26)	(0.22)	(0.52)	(0.22)	(0.77)	(0.23)
Work	0.07	1.44	−0.26	0.7	0.27	$−1.51^{\dagger}$	−0.58*
Full-Time	(0.12)	(2.29)	(0.22)	(0.53)	(0.22)	(0.78)	(0.25)
N	856	858	858	858	858	858	858
Adj. R^2	0.03	0.00	0.07	0.22	0.02	0.22	0.17

Notes: Standard errors in parentheses.
$^{\dagger}p < .10$
$^{*}p < .05$
All models use OLS regression except Straight-line, Open-Ended Junk, and Attrited which are logits with Nagelkerke R^2 reported.
Survey number range is 0–30, all higher values top-coded as 30.

Column 3 in Tables 10.3 and 10.4 shows that more professional respondents are also less likely to attrite – that is, drop out before the post-election wave – though the difference is statistically significant only for the survey number measure. The model predicts that, holding all else constant, those answering just one survey in the last month had a 15% probability of attriting, compared to a less than 1% probability of dropping out among those answering at least 30 surveys in the last month. Likewise, professional respondents were no more likely to straight-line; that is, select the same response for an entire battery of questions.

While the measures thus far find no support for the hypothesis that professional respondents are more likely to satisfice, we do find they are more likely to skip individual questions, give junk answers to open-ended questions, and give "don't know" responses. Overall, however, many of these behaviors are fairly rare – the average percentage of "don't know" responses through the questionnaire was just 9%, the average percentage of missing responses was 6%, and 13% gave junk responses to the open-ended questions. Nonetheless, the analysis does find that professional respondents are more likely to engage in these bad behaviors. For example, column 5 of Table 10.3 and Table 10.4 shows that, holding all else constant, those belonging to more online panels or answering more surveys in the past four weeks had a higher percentage of missing responses. Likewise, column 6 shows a positive and significant relationship with the percentage of "don't know" responses. The model predicts that those with 10+ panel memberships gave an average of 5.7% "don't know" responses, compared to less than 1% among those belonging to a single panel. And in the final column, we find that more frequent survey-takers are more likely to give junk responses to two open-ended questions in the survey; that is, they gave no answer, volunteered a "don't know" or "no opinion" answer, or simply typed in gibberish.

It will likely be noted that the adjusted R^2 is very small for many of these models, suggesting we do not do a very good job predicting variation in these bad behaviors. Although this is not problematic given the purpose of our analysis – to evaluate if there was a significant relationship between these outcomes and our measures of professionalism – it does suggest that our ability to more generally predict satisficing behaviors in this survey is quite limited.[3] A more comprehensive analysis of satisficing behavior in the CCES and other online panels is a topic worthy of further study.

Indeed, while the items considered here are commonly used measures of satisficing, we want to raise the possibility of an alternative explanation for the observed patterns. It could be that the failure to give a substantive response could reflect a lack of ability rather than a lack of motivation. In other words, less politically knowledgeable professional respondents might be more likely to sincerely answer "don't know" (or skip an individual question) because they were not sure how to answer, despite giving the question adequate thought and consideration. As a test of this alternative hypothesis, we re-estimated these models including a control for political knowledge. Doing so finds that the professionalism measures are no longer statistically significant in 5 out of 6 cases. The relationship remains significant between number of surveys completed and percentage missing, but not in any other case once we control for political knowledge. This suggests that the bad behaviors could be attributable to respondent competencies rather than respondent laziness.

10.9   Discussion

Our analysis finds that higher levels of participation in surveys and online panels are associated with lower levels of political knowledge, interest, engagement, and ideological extremism in the 2010 CCES. Our analysis does not explicitly test why that is the case, but we suspect it reflects differences in the initial motivation to participate between professional and nonprofessional respondents. Frequent survey-takers may have been motivated to participate in the survey for the compensation offered, while the less frequent survey-takers were interested in the survey topic. Certainly, this hypothesis is worthy of further study.

These conclusions are based on the analysis of a single political survey so we cannot assume these patterns will hold in other nonprobability samples. For one, the CCES was an especially lengthy political survey taken in the midst of a heated midterm election. As such, it may have been more likely to attract some politically-interested respondents. Second, these results might be specific to this survey company or specific study, reflecting, say, the panel management procedures of YouGov or the academic source of the survey. Third, as with any nonprobability sample, the observed sample may differ in unknown ways from the broader target population. Thus, we cannot necessarily generalize our findings here to other nonprobability samples. Nonetheless, these findings offer a somewhat different take

---

[3] For a more thorough discussion of the appropriateness of R-squared, see King (1990).

on the potential biases in nonprobability samples. These results make clear that assessing a nonprobability sample as a whole could mask countervailing biases within the sample. In this survey, professional respondents actually reduce bias by lowering the average level of political interest, knowledge, and engagement in the sample. This means that eliminating professional respondents from this nonprobability political surveys, as some have recommended, would have resulted in a *more* biased sample along these dimensions.

In contrast to some expectations, we did not find consistent evidence that more professional respondents gave less thoughtful responses to the questionnaire. On the contrary, frequent survey-takers spent more time completing the questionnaire, were less likely to attrite, were less likely to straight-line, and reported putting more effort into answering the survey. While panel memberships and number of surveys completed were related to skipping questions, answering "don't know," or giving junk responses to open-ended questions, these relationships did not hold once political knowledge was accounted for. In some respects, then, the findings are reassuring for those concerned about professional respondents. Indeed, deleting professional respondents from a sample from an online nonprobability survey could decrease both the validity *and* reliability of the data.

In sum, the results here offer one more analysis about the consequences of having a growing class of professional respondents participating in nonprobability online panels. As the prevalence of online survey research grows, so too does the need to learn more about who is "opting-in" to these samples and why they are doing so. Our analysis suggests the problem may not be so much with the number of surveys completed, but an individual's reason for completing them. It is inherently difficult to measure motivation – in no small part because respondents might not be conscious of their motivations – but it would nonetheless be worthwhile for future research to consider how motivation to participate interacts with the survey topic to shape the content and quality of survey responses.

References

Adams, A., Atkeson, L., & Karp, J. (2012). Panel conditioning in online survey panels: Problems of increased sophistication and decreased engagement. In *Proceedings of the 2012 Annual Conference of the American Political Science Association*. New Orleans.

Ansolabehere, S. (2012, August 10). *Guide to the 2010 cooperative congressional election survey*. Cambridge, MA: Harvard University Press.

Ansolabehere, S., & Schaffner, B. F. (2011). *Re-examining the validity of different survey modes for measuring public opinion in the U.S.: Findings from a 2010 multi-mode comparison*. Amherst, MA: University of Massachusetts Press.

Baker, R., Blumberg, S., Brick, J., Couper, M., Courtright, M., Dennis, J., ... Zahs, D. (2010). American association of public opinion researchers report on online panels. *Public Opinion Quarterly*, 74, 711–781.

Baker, R., & Downes-LeGuin, T. (2007). Separating the wheat from the chaff: Ensuring data quality in internet panel samples. In *Proceedings of the Fifth International Conference of the Association of Survey Computing* (pp. 157–166). Berkeley, UK: ASC.

Bartels, L. (1999). Panel effects in the American National Election Studies. *Political Analysis*, 8, 1–20.

Battaglia, M., Zell, E., & Ching, P. (1996). Can participating in a panel sample introduce bias into trend estimates? In *Proceedings of the Section on Survey Research Methods at the 1996 Annual Conference of the American Statistical Association* (pp. 1010–1013). Alexandria, VA.

Binswanger, J., Schunk, D., & Toepoel, V. (2006). *Panel conditioning in difficult attitudinal questions*. Dutch Online Panel Study NOPVO 2006.

Casdas, D., Fine, B., & Menictas, C. (2006). Attitudinal differences: Comparing people who belong to multiple versus single panels. In *Proceedings of ESOMAR Panel Research 2006*. Orlando.

Chang, L., & Krosnick, J. (2009). National surveys via RDD telephone interviewing versus the internet comparing sample representativeness and response quality. *Public Opinion Quarterly, 73*, 641–678.

Coen, T., Lorch, J., & Piekarski, L. (2005). *The effects of survey frequency on panelists' responses.* White Paper, Survey Sampling International.

Comley, P. (2005). Understanding the online panelist. In *Proceedings of ESOMAR World Research Conference*, Barcelona.

Conrad, F., Couper, M., Tourangeau, R., Galesic, M., & Yan, T. (2005, May). Interactive feedback can improve the quality of responses in web surveys. In *Proceedings of the 2005 annual meeting of the American Association for Public Opinion Research* (pp. 3835–3840). Miami Beach.

Das, M., Toepoel, V., & van Soest, A. (2011). Nonparametric tests of panel conditioning and attrition bias in panel surveys. *Sociological Methods & Research, 40*, 32–56.

Dennis, J. (2001). Are internet panels creating professional respondents? *Marketing Research, 13*, 34–39.

De Wulf, K., & Berteloot, S. (2007). Duplication and multi-source online panel recruitment: real quality differences or idle rumours? In *Proceedings of ESOMAR World Research Conference on Panel Research* (pp. 49–62).

Dillman, D., Smyth, J., & Christian, L. (2009). *Internet, mail, and mixed-mode surveys: the total design method.* Hoboken, NJ: John Wiley & Sons.

Downes-Le Guin, T., Mechling, J., & Baker, R. (2006). Great results from ambiguous sources: Cleaning internet panel data. In *Proceedings of ESOMAR Panel Research 2006* (pp. 1–10).

Frede, S., Markowitz, L., & Coffery, S. (2006). *Heavier Responders in Online Survey Research.* White Paper, Ipsos, NPD Group, and TNS.

Fulgoni, G. M. (2005). The "professional respondent" problem in online survey panels today. In *Proceedings of the 2005 Marketing Research Association Annual Conference.*

Garland, P., Santus, D., & Uppal, R. (2012). *Survey lockouts: Are we too cautious?* White Paper, Survey Sampling International.

Gittelman, S., & Trimarchi, E. (2009). *Variance between purchasing behavior profiles in a wide spectrum of online sample sources.* White Paper, Marketing. Inc.

Gittelman, S., & Trimarchi, E. (2010). Are hyperactive respondents different? *Quirks Marketing Research Review, November.*

Goyder, J. (1987). *The silent minority: Nonrespondents on sample surveys.* Boulder, CO: Westview Press.

Groves, R., Presser, S., & Dipko, S. (2004). The role of topic interest in survey participation decisions. *Public Opinion Quarterly, 68*, 2–31.

Groves, R., Singer, E., & Corning, A. (2000). Leverage-saliency theory of survey participation: Description and an illustration. *Public Opinion Quarterly, 64*, 299–308.

Grover, R., & Vriens, M. (Eds.). (2006). *The handbook of marketing research: Uses, misuses, and future advances.* Thousand Oaks, CA: Sage.

Harlow, A. (2010). *Online surveys – possibilities, pitfalls and practicalities: The experience of the TELA evaluation.* Faculty of Education, University of Waikato, New Zealand.

Honda, N., & Motokawa, A. (2004). *Can internet surveys be used for social surveys?: Results of an experimental study.* The Japan Institute for Labour Policy and Training Research Report, No. 17.

Kapelner, A., & Chandler, D. (2010). Preventing satisficing in online surveys: A "kapcha" to ensure higher quality data. In *Proceedings of CrowdConf ACM.*

King, G. (1990). When not to use R-squared. *The Political Methodologist, 3*, 11–12.

Knapton, K., & Garlick, R. (2007). Catch me if you can. *Quirks, November*, 58–63.

Krosnick, J. (1991). Response strategies for coping with the cognitive demands of attitude measures in surveys. *Applied Cognitive Psychology*, 5, 213–236.

Krosnick, J., Nie, N., & Rivers, D. (2005). *Web survey methodologies: A comparison of survey accuracy.* American Association for Public Opinion Research, 60th Annual Conference, Miami.

Kruse, Y., Callegaro, M., Dennis, J., Subias, S., Lawrence, M., DiSogra, C., & Tompson, T. (2010). Panel conditioning and attrition in the AP-Yahoo! news election panel study. In *Proceedings of the 2010 Annual Conference of the American Association for Public Opinion Research* (pp. 5742–5756). Washington, DC: American Statistical Association.

Malhotra, N., & Krosnick, J. (2007). The effect of survey mode and sampling on inferences about political attitudes and behavior: Comparing the 2000 and 2004 ANES to Internet surveys with nonprobability samples. *Political Analysis*, 15, 286–323.

Mason, W., & Watts, D. (2010). Financial incentives and the performance of crowds. *ACM SIGKDD Explorations Newsletter*, 11, 100–108.

Mathiowetz, N., & Lair, T. (1994). Getting better? Change or error in the measurement of functional limitations. *Journal of Economic and Social Measurement*, 20, 237–262.

Matthijsse, S., Leo, E., & Hox, J. (2006). Professional respondents in online panels: A threat to data quality? In *Proceedings of ESOMAR Panel Research*.

Meurs, H., Wissen, L., & Visser, J. (1989). Measurement biases in panel data. *Transportation*, 16, 175–194.

Miller, J. (2007). *Burke panel quality research and development summary.* White Paper, Advertising Research Foundation Online Research Quality Council.

Nancarrow, C., & Cartwright, T. (2007). Online access panels and tracking research: The conditioning issue. *International Journal of Market Research (formerly Journal of the Market Research Society)*, 49, 573–594.

Oppenheimer, D., Meyvis, T., & Davidenko, N. (2009). Instructional manipulation checks: Detecting satisficing to increase statistical power. *Journal of Experimental Social Psychology*, 45, 867–872.

Paolacci, G., Chandler, J., & Ipeirotis, P. (2010). Running experiments on Amazon Mechanical Turk. *Judgment and Decision Making*, 5, 411–419.

Peruzzi, N. (2010, July 8). Online survey sample is not clean enough – clean it yourself. Retrieved from: www.researchaccess.com.

Pineau, V., Nukulkij, P., & Tang, X. (2005). Assessing panel bias in the knowledge networks panel: Updated results from 2005 research. In *Joint Statistical Meeting Proceedings* (pp. 3480–3486).

Prior, M., & Lupia, A. (2008). Money, time and political knowledge: Distinguishing quick recall and political learning skills. *American Journal of Political Science*, 52, 169–183.

Rogers, F., & Richarme, M. (2009). *The honesty of online survey respondents: Lessons learned and prescriptive remedies.* White Paper, Decision Analyst.

Schlackman, W. (1984). A discussion of the use of sensitivity panels in market research: The use of trained respondents in qualitative studies. *Journal of the Market Research Society*, 26, 191–208.

Smith, R., & Brown, H. H. (2006). Data and panel quality: Comparing metrics and assessing claims. In *Proceedings of the ESOMAR Panel Research Conference*.

Sparrow, N. (2007). Quality issues in online research. *JOURNAL OF ADVERTISING RESEARCH*, 47, 179.

Toepoel, V., Das, M., & Van Soest, A. (2008). Effects of design in web surveys: Comparing trained and fresh respondents. *Public Opinion Quarterly*, 72, 985–1007.

Toepoel, V., Das, M., & Van Soest, A. (2009). Relating question type to panel conditioning: A comparison between trained and fresh respondents. *Survey Research Methods*, 3, 73–80.

Walker, R., Pettit, R., & Rubinson, J. (2009). *The foundations of online research quality executive summary 3: Interstudy comparability and benchmark analysis.* White Paper, Advertising Research Foundation.

Waterton, J., & Lievesley, D. (1989). Evidence of conditioning effects in the British Attitudes Panel Survey. In G. Duncan, G. Kalton, D. Kasprzyk, & M. P. Singh (Eds.), *Panel surveys*. New York: Wiley.

Willems, P., van Ossenbruggen, R., & Vonk, T. (2006a). The effects of panel recruitment and management on research results. In *Proceedings of the conference of the European Society for Opinion and Marketing Research*. Barcelona.

Willems, P., Vonk, T., & van Ossenbruggen, R. (2006b). The effects of panel recruitment and management on research results: A study across 19 online panels. In *Proceedings of ESOMAR Panel Research*.

Yan, T., & Tourangeau, R. (2008). Fast times and easy questions: The effects of age, experience and question complexity on web survey response times. *Applied Cognitive Psychology, 22*, 51–68.

Appendix 10.A

10.A.1 Detailed variable information

Survey Frequency: "About how many online surveys (on all topics) have you completed in the past 4 weeks?"

Panel Memberships: "How many survey panels do you belong to?"

Political Interest: "How often are you interested in news and public affairs? Most of the time (4) Some of the time (3) Only now and then (2) Hardly at all (1)"

Turnout: "Which of the following statements best describes you? I did not vote in the election this November (0) I thought about voting this time - but didn't (0) I usually vote, but didn't this time (0) I attempted to vote but did not or could not (0) I definitely voted in the General Election on November 2 (1)"

The *Political Activity* scale is a sum of "yes" responses to three items (0–3): "During the past year, did you ... attend local political meetings (such as school board or city council)? (1); put up a political sign (such as a lawn sign or bumper sticker)? (1); work for a candidate or campaign? (1)"

The *Political Knowledge* scale is the sum of correct responses to four separate questions (0–4): "Please indicate whether youve heard of this person and if so which party he or she is affiliated with." The Governor, US Senators, and US House member for each respondent based on residency as listed.

Ideological strength is calculated by folding the standard 7-point strongly liberal to strongly conservative question. "How would you rate each of the following individuals and groups? Yourself. Very Liberal (3), Liberal (2), Somewhat Liberal (1), Moderate (0), Somewhat Conservative (1), Conservative (2), Very Conservative (3), Not Sure (0)."

Survey effort: "Finally, we are interested in your survey experience. Overall, how much effort would you say that you put into answering the question on scale that ranges from 1–7, where 1 means very little effort and 7 means a lot of effort?"

The *Percent Don't Know* measure is calculated as the percent of "don't know" responses to 81 questions asked of all respondents for which this response was not a substantive response (e.g., knowledge items were not included).

Percent Missing is calculated as the percent of 301 questions asked of all respondents that were skipped.

The *Straight-line* variable is an indicator if a respondent gave an identical response to each of 13 items in a grid that asked "How would you rate each of the following individuals and groups? very liberal, liberal, somewhat liberal, middle of the road, somewhat conservative, conservative, very conservative, not sure." The items to rate were: yourself, governor of respondents' state, Barack Obama, Democratic Party, Republican Party, Senator 1, Senator 2, U.S. Senate candidate 1 (if up for election in 2010), U.S. Senate candidate 2 (if in election), U.S. House candidate 1, U.S. House candidate 2, U.S. House Member (if retiring and candidates are both new), Tea Party Movement.

Open-ended Junk is an indicator if a respondent gave a "don't know," gibberish, or no response to either of two open-ended questions: (1) "What do you think is the most important problem facing the country today?" and (2) "What policy issue do you think is most at stake in this election?"

Interview Duration is calculated by subtracting the end time from the start time. Times that exceeded 2 hours were coded as 2 hours to try to account for respondents who might have walked away from the computer or were distracted by other activities.

Attrition is an indicator that a respondent completed the pre-election wave but failed to complete the post-election wave.

11

The impact of speeding on data quality in nonprobability and freshly recruited probability-based online panels

Robert Greszki, Marco Meyer, and Harald Schoen
University of Bamberg, Germany

11.1   Introduction

Online surveys have become a widely used tool in survey research (e.g., Baker et al., 2010; Bethlehem & Biffignandi, 2012; Tourangeau, 2004). They differ from traditional survey techniques in a number of respects. Some of them, including low costs and short field times, make them attractive. Other features are more ambivalent, including the absence of an interviewer (e.g., Chang & Krosnick, 2010; Schaeffer, Dykema, & Maynard, 2010). While the absence of an interviewer might reduce social desirability effects, it also implies that the interviewing process is uncontrolled and accordingly data quality might be low (Groves et al., 2009). In particular, measurement problems arising from inattentive respondents are likely to be quite pervasive in online surveys.

Analysts of data gleaned from online surveys thus face a threat to data quality arising from inattentive respondents especially due to the uncontrolled interviewing situation. Aiming at valid results, they have strong incentives to identify low-quality responses and to remove them from analyses. Identifying low-quality responses is not an easy task, however. To be sure, some low-quality responses might be detected using conventional indicators like item-specific

Online Panel Research: A Data Quality Perspective, First Edition.
Edited by Mario Callegaro, Reg Baker, Jelke Bethlehem, Anja S. Göritz, Jon A. Krosnick and Paul J. Lavrakas.
© 2014 John Wiley & Sons, Ltd. Published 2014 by John Wiley & Sons, Ltd.
Companion website: www.wiley.com/go/online_panel

nonresponse or "don't know" answers. But sophisticated respondents who want to get material rewards for completing surveys might not choose these responses because otherwise they risk being easily identified as inattentive. Rather, they might straight-line in item batteries, choose frequently middle categories, or any (random) answer to quickly get through the questionnaire and receive rewards. One way to identify those low-quality responses is to use response time as an indicator of data quality. Building on the fact that reading questions and processing information require time (e.g. Tourangeau, Rips, & Rasinski, 2000), particularly quick responses, the so-called speeding, might indicate minor data quality. Accordingly, removing those responses from the data might improve the quality of the data and help avoid biases in substantive results.

In this chapter, we thus explore whether removing data on the basis of response time affects substantive findings. If it turns out that response time is a valid indicator of data quality and removing "too fast" responses from the data set changes substantive findings, this might be a viable strategy to increase data quality in online surveys. Otherwise, i.e. if response time is a poor indicator of data quality or removing "speeders" does not alter findings, this strategy might be not applicable.

We will study the prevalence and impact of speeding in a probability-based and nonprobability online panel (see Chapter 1 in this volume; Couper, 2000) to explore whether the nature of the sample makes a difference. It might be argued that inattentive respondents are more prevalent in nonprobability than in probability-based samples, because the former sampling is more self-selective and might attract more persons who are likely to satisfice. At the same time, self-selection might be driven by factors that are not closely tied to satisficing, e.g., profound interest in the survey topic. It is thus an empirical question whether the samples differ in the prevalence and the impact of speeding on substantive findings.

The remainder of the chapter is structured as follows. Section 11.2 outlines the theoretical framework underlying the notion that speeding might be a symptom of low data quality. After a short description of the data, we will demonstrate that response time is a reasonable indicator of data quality and we will describe the page-specific procedures to identify too fast responses. The analyses show that the nonprobability survey is somewhat more plagued by speeding than the probability-based survey. In both, however, removing too fast responses from the data sets does not alter substantive findings in terms of marginal distributions and multivariate models. The chapter concludes by summing up key findings and discussing implications.

11.2   Theoretical framework

According to the total survey error framework (Groves et al., 2009), measurement error is one of the most serious sources of low data quality. Such "a departure from the true value of the measurement as applied to a sample unit and the value provided" (Groves et al., 2009, p. 52) can be caused by respondents as well as external factors. Responses given by inattentive respondents, who do not pay close attention or give not a sufficient level of thought to their responses, might prove invalid (Baker et al., 2010). As we study just two cases, effects of external factors such as questionnaire design or technical obstacles cannot be examined in this contribution. The empirical analysis in this chapter, rather, focuses on problems arising from inattentive respondents. Nevertheless, potential effects of external factors will be addressed if applicable and necessary.

Given the lack of control in self-administered online interviews, as compared to interviewer-administered surveys, inattentive respondents are likely to be particularly prevalent in online surveys. To identify inattentive respondents, research institutes pursue different strategies (e.g., Baker & Downes-Le Guin, 2007; Balden, 2008; Knapton & Garlick, 2007). For example, red herring questions are widely employed. These control questions are implemented in item batteries as a single item such as "mark the option 'Neither like nor dislike'" to test whether respondents have carefully read the survey questions (Berinsky, Margolis, & Sances, 2012; Miller, 2006; Miller & Baker-Prewitt, 2009). In a study of 13 US online panels, Miller (2006) reported about 5–10% respondents, who answered red herring questions wrongly. Moreover, employing red herring questions early in a survey appears to increase completion time and to decrease the likelihood of straight-lining (Miller & Baker-Prewitt, 2009). Using straight-lining as an indicator of inattentiveness, Smith and Brown (2005) identified just 1% of the respondents in 20 large surveys as inattentive. Finally, Meade and Craig (2011) utilized a battery of 11 indicators to capture inattentiveness and found that between 5–15% of the respondents in an undergraduate Internet survey lacked sufficient attention.

Moreover, respondents, though initially motivated, might become fatigued or distracted in the course of the interview (Krosnick, 1991). In line with this notion, Galesic and Bosnjak (2009) as well as Puleston and Sleep (2008) have demonstrated that respondents in online surveys appear to speed up in the course of the interview. This acceleration effect is likely to be stronger in online surveys than in interviewer-administered surveys, as the interviewer might be able to keep the respondent at the interview and positively influence the respondent to optimize only through his presence (Baker et al., 2010).

While online surveys raise specific issues concerning data quality, they might also provide a device to overcome this problem. Collecting survey data via the Internet enables the automatic detention of paradata such as response time (see Couper, 2000, 2005; Heerwegh, 2003, 2011; Kaczmirek, 2009).[1] As survey software automatically captures the time a respondent spends on a specific survey page, researchers are in a position to explore response times regarding different pages and compare within and across respondents (for techniques to measure response time in web surveys, see e.g., Fraley, 2004; Heerwegh, 2003; Kaczmirek, 2009). This information could prove helpful in identifying low-quality data – provided response time is linked to data quality.

To establish this link, we build on the response process model (Tourangeau, Rips & Rasinski, 2000; Tourangeau, 1984, 1987; Tourangeau & Rasinski, 1988; see also Kahn & Cannell, 1957) which suggests that the process of answering a survey question comprises four major steps. A respondent who fills in a web survey, first of all has to read the whole question text to comprehend a question. Reading only response options is insufficient and may lead to invalid answers. After comprehension of the question the respondent has to access the relevant information in memory before forming a judgment with this accessible information. Finally, the respondent formulates and reports an answer by clicking a response option or writing his answer in a textbox. To be sure, some steps might be skipped depending on the kind of question or the accuracy of the respondent's answer (Tourangeau et al., 2000). But it is reasonable to assume that the first and the fourth step of the response process model – comprehension and reporting an answer – are indispensable for valid survey answers. Thus, the response process, by its very nature, takes some time.

---

[1] Prior to the use of automatically generated paradata in web surveys, response times were already used in CATI surveys to measure, e.g., attitude accessibility (see Bassili, 1993, 1996; Bassili & Fletcher, 1991; Bassili & Scott, 1996; Draisma & Dijkstra, 2004; Johnson, 2004). In contrast to paradata in web surveys, response times in CATI surveys are usually gathered manually by interviewers.

Provided that the response process takes a certain period of time, extremely short response latencies of individual respondents compared to all respondents, can be considered as indicative of invalid response behavior and thus inattentiveness. In this line of reasoning, very quick answers result from a response process in which several necessary steps are skipped because respondents engage in satisficing, rather than attempting to give valid answers (Krosnick, 1991; Krosnick & Alwin, 1987; Krosnick et al., 2002; Narayan & Krosnick 1996).

This time-saving strategy might lead to different forms of behavior, depending on the questions asked. When dealing with single items, an inattentive respondent might somewhat satisfice by simply selecting the first (given) response option which she considers reasonable. Thus, steps two and three are done quite quickly or skipped completely. A stronger form of satisficing implies that a respondent does not read the question content properly. In the worst case, respondents perform only step four: the formation of a (random) answer, e.g., choosing always middle categories (Schuman & Presser, 1996; Tourangeau, Couper, & Conrad, 2004) or selecting the first response option (Malhotra, 2008). When asked to answer item batteries using a grid pattern, some respondents may straight-line, i.e., they answer the questions rapidly by clicking always the same response category in each "line" of the grid without paying attention to the substantive meaning of response options. Provided with a "don't know" category, some respondents perform all steps of the response process, requiring some time, and actually come to no answer. Others, however, might skip one or more steps of the response process, thereby strongly satisficing, and give a "don't know" (for a more sophisticated view, see Krosnick et al., 2002). Finally, in online surveys, clicking "next" without paying any attention to question content is a time-saving strategy, even if not all online surveys provide respondents with an opportunity to click "next," which is equivalent to "no answer" in interviewer-administered surveys. Clicking "next," however, does not necessarily reflect a lack of motivation because some respondents might perform all steps of the response process and give no answer after thorough consideration.

In effect, there is a, though not perfect, link between very quick responses and low data quality which is supported by evidence (Callegaro et al., 2009; Malhotra, 2008; Rossmann, 2010). Yet, we have to keep in mind that raw response time might not be an appropriate indicator because respondents might differ in the time it takes to perform the task of providing valid responses. To give just two examples, cognitive ability as well as training might make a difference. Accordingly, it might be appropriate to utilize somewhat adjusted rather than raw response time as indicator of satisficing.

Nevertheless, it is reasonable to consider too fast responses as low-quality answers that have to be removed from the data before conducting substantive analyses. This conclusion raises the question for the appropriate yardstick to distinguish too fast from regular responses. Survey analysts know from pretests how long it usually takes to complete the survey. Assuming a usual response time of, e.g., 25 minutes, experienced respondents might be able to complete the questionnaire in 15–20 minutes without any loss in data quality. If a respondent manages to complete this survey in 5 or 10 minutes, however, concerns about data quality will arise, even if a respondent is highly skilled and well trained. The same reasoning applies to the completion of individual pages within a survey. In effect, there is some objective information on reasonable response times. At the same time, the choice of a specific threshold to identify speeding is somewhat arbitrary.

Irrespective of the threshold chosen, analysts might remove too fast responses from the data before performing substantive analyses to get valid results. This raises the question whether this procedure will change the substantive results. Quite obviously, the answer to this question depends on the proportion of speeders in a sample. If just a tiny fraction of the sample gives too quick responses, removing them will hardly change substantive findings. Moreover, the distribution of too quick answers plays a role. Concerning marginal distributions, removing too quick answers will change results considerably if speeding is correlated with the variable in question. If speeders exhibit a low level of interest in politics, for example, removing speeders will increase the level of political interest in the sample. When it comes to correlations and multivariate models with such biased variables, a similar reasoning applies (see, e.g., Faas & Schoen, 2006; Schoen, 2004).

These factors are not completely independent of *how* respondents speed. If a respondent chooses a "don't know" answer or just clicks next ("no answer"), removing too-fast answers will not make a difference because those answers are treated as missing values anyway. If respondents choose any (random) answer, thereby skipping steps two and three or even steps one to three of the response process model, marginal distributions will not be affected by the exclusion of speeders, whereas correlations might be altered. The impact of satisficing by always choosing middle categories or straight-lining might not easily be identified. Moreover, it is, by and large, an empirical question whether, and in which way, removing those speeders from a data set alters substantive findings.

The prevalence and effects of speeding might, *inter alia*, depend on the nature of samples. It might be argued that nonprobability and freshly probability-based recruited samples differ in several respects that may affect also the motivation of respondents (Bethlehem & Biffignandi, 2012; Coen, Lorch, & Piekarski, 2005; Topoel, Das, & Van Soest, 2008), and, thus, may vary also in the proportion of speeders. Compared to freshly probability-based recruited respondents, nonprobability online panels are in most instances recruited via a highly self-selective process, and therefore, likely to attract other persons with other intentions in the first instance compared to the random recruitment methods. It is thus reasonable to compare nonprobability-based and probability-based samples in terms of the prevalence and effects of speeding on substantive findings. Yet, it remains an empirical question whether there are really relevant differences.

In sum, speeding through web surveys is a potential threat to data quality. What is more, it cannot be taken for granted that speeding is irrelevant to the substantive findings gleaned from online surveys. In the subsequent sections of this chapter, we will explore these questions using data from a nonprobability online survey and a freshly probability-based recruited online survey.

11.3   Data and methodology

To explore speeding in a nonprobability online survey and a freshly recruited probability-based online sample, we use data from the *German Longitudinal Election Study (GLES)* and data from the *American National Election Study (ANES)*. The 2008–2009 ANES Panel Study was fielded from January 2008 through September 2009, including 21 monthly panel waves. For our analysis, we use data from the first panel wave conducted in January 2008 in order to study "freshly recruited" respondents who have not been subject to panel conditioning and are most different from online panelists. The target population comprises a representative sample of US citizens aged 18 or older as of Election Day in November 2008. Respondents were telephone-recruited using RDD. In the first contact, a short recruitment interview was conducted and respondents were offered $10 per month to complete the surveys on the Internet. Willing respondents without a computer and Internet connection were provided with a free web appliance for the duration of the study (see DeBell, Krosnick, & Lupia, 2010).

The German election study includes a series of online surveys with a nonprobability online panel as sampling frame. Online surveys were conducted in the run-up to and after the 2009 German federal election. As substantive findings do not vary across surveys, we report the results from the survey conducted from 18–26 September 2009 (results from the other surveys are available from the authors). The dataset comprises 1153 respondents who were drawn in a quota sample, where the panelists were quoted in terms of gender, age, and education via a nonprobability online panel (see Rattinger et al., 2009). The quotas were a mix of the distribution in the target population and the current online population. The target population comprises all German citizens who were eligible to vote and thus differs considerably from the frame population of the online panel. The online panel comprised about 65000 active panelists in Germany in 2009. As defined by the panel provider, active panelists are those persons who completed the double-opt-in registration, completed the master questionnaire about basic personal information, and successfully participated in at least one survey within the last 12 months. The panel uses different on- and offline channels to recruit new panelists which include opinion portals, on-site surveys, search engines, and recruitment by telephone. To participate in the surveys, members of the online panel are offered incentives, namely 10 panel-points per minute, which, in 2009, was the equivalent of approximately 0.10 €. Having collected at least 10 Euros, panelists may choose between cash payment, shopping coupons, or a donation (GESIS, 2009a, 2009b; Respondi, 2009a, 2009b).

Both surveys were designed to be completed within 30 minutes (GLES) and 25–30 minutes (ANES), respectively. Empirically, respondents on average spent 35–36 minutes to complete either survey. Due to the right-skewed distribution of response times (Ratcliff, 1993), the median is considerably lower than the mean with 33 minutes in the GLES study and about 27 minutes in the ANES surveys. The fastest respondents, however, managed to complete the GLES survey in 2 minutes and the ANES survey in 6 minutes. Given the design of the surveys and the empirical distribution of response times, these minimal response times are indicative of speeding by inattentive respondents. Leaving aside these extreme cases, it is an empirical question whether short response times indicate low data quality. This is the question we address in the next section.

11.4   Response time as indicator of data quality

Using response time in general, and speeding in particular, as an indicator of data quality rests on the assumption that valid indicators of data quality are correlated with response time. We consider four well-known indicators that might reflect satisficing behavior which allows respondents to complete a survey particularly quickly (see codebook at: www.wiley.com/go/online_panel_research, for a detailed description of the four indicators). First, it is quite straightforward to interpret giving "no answer" as time-saving response strategy. A respondent who clicks "next" instead of giving a substantive answer simply skips virtually all steps that the response process model includes. It thus does not come as a surprise that item-nonresponse is often correlated with low data quality (e.g., de Leeuw, 1992; Groves, 1991). Second, we calculated the proportion of "don't know" answers because choosing this response option is for some respondents a well-known satisficing strategy (Krosnick, 1991; Krosnick et al., 2002). Third, choosing the middle category from five or more response options on an ordinal scale is another satisficing strategy (Kaminska, McCutcheon, & Billiet,

2010; Krosnick, Narayan, & Smith, 1996). We thus calculated the proportion of cases in which respondents chose the middle category when offered at least five response options in ordinal scales. Finally, we measured straight-lining by capturing systematic response patterns that do not reflect substantive preferences (Kaminska, McCutcheon, & Billiet, 2010; Krosnick et al., 1996). We considered item-batteries comprising at least five items with at least five response options each. In analyzing the data, we were careful not to mistake substantive responses, e.g., consistently high ratings for a politician, for straight-lining. In effect, we calculated for each respondent the proportion of pages with straight-lining as compared to the number of pages with opportunities for straight-lining (for a detailed description of the procedure, see the codebook at www.wiley.com/go/online_panel_research). To be sure, these indicators are not perfect because these kinds of response behavior might in some instances reflect substantive answers, rather than satisficing. In light of prior research (Kaminska, Goeminne, & Swyngedouw, 2006; Kaminska et al., 2010; Krosnick et al., 1996), these are reasonable choices, however.

These indicators are easily applicable to the GLES data but not to the ANES survey because the question formats differed considerably across surveys. First of all, the ANES questionnaire does not include item batteries that are displayed in grid fashion and also it rarely includes items with at least five response options. Moreover, when ANES respondents clicked "next" without choosing any response option, they did not go immediately to the next page. Instead, the question was displayed once again and the respondent was asked to provide his best answer even if he was not completely sure. Only if the respondent clicked "next" for a second time, was the question skipped and recorded as a "no answer". Thus, the warning message automatically forced respondents to spend some additional time on the respective survey page. As a consequence, this part of the analysis is confined to GLES data.

We calculated the four quality indicators by dividing the number of actual satisficing by the number of opportunities to satisfice. In order to explore whether the quality measures are correlated with response time, we regressed the overall response time spent on the survey pages that are relevant for the respective quality measure on the quality indicators, using OLS regression (Table 11.1).

Table 11.1   Response time as a function of satisficing behavior (OLS).

	Constant	B	β (r)	Adj. R^2	N
No answer	1380.4***	−22.3***	−0.21	0.05	1,153
	(16.6)	(3.0)			
Don't know	235.4***	−8.3***	−0.20	0.04	1,153
	(4.0)	(1.2)			
Middle category	291.3***	−2.0	−0.05	0.002	1,153
	(8.2)	(1.2)			
Straight-lining	372.9***	−26.8***	−0.33	0.11	1,153
	(5.1)	(2.3)			

Note: Cell entries are b-coefficients and standardized β-coefficients; standard errors in parentheses.
Significance levels:
*: $p < 0.05$;
**: $p < 0.01$;
***: $p < 0.001$.

As the results reported in Table 11.1 show, indicators of satisficing are negatively correlated with response time. Straight-lining is most strongly correlated with response time. According to the evidence, a one-unit increase in straight-lining is accompanied by a decrease of 27 seconds on the respective survey sites. The correlations of response time with "no-answers" and "don't know" answers prove also statistically significant, though they are somewhat smaller. The impact of the middle-category index on response time, however, is indistinguishable from zero. Accordingly, we might speculate that GLES respondents chose middle categories not primarily to satisfice. Notwithstanding the latter result, the evidence suggests that response time is correlated with – imperfect – measures of satisficing. It is thus reasonable to use response time as an indicator of the attention respondents paid to the survey and ultimately of data quality.

Due to data limitations, the above analysis was confined to GLES data. Yet, it is reasonable to consider the GLES results, by and large, applicable to the ANES data and to utilize response times as an indicator of data quality in both surveys. This conclusion rests on the fact that for a large number of comparable items the distributions of response times in both surveys resemble each other quite closely. To give just two examples, Figure 11.1 contains the distributions of response time in both surveys for political interest and the question of how

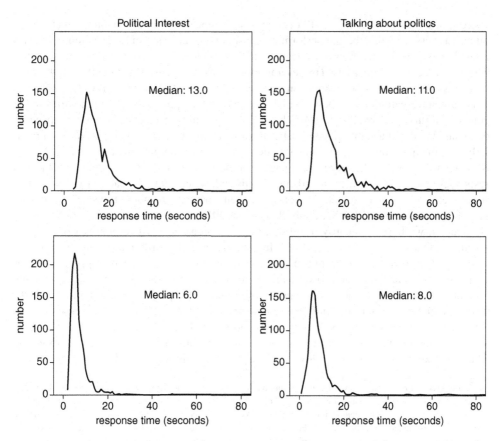

Figure 11.1   Distribution of response times for "political interest" and "talking about politics" in ANES and GLES.

often respondents talk about politics. The GLES and ANES distributions are quite similar. Moreover, the curves reach their peaks at rather short response times. The proportion of very fast respondents might be explained by the absence of an interviewer in self-administered online surveys. In interviewer-administered studies, such short response times to similar questions are very unlikely, hence the question and response options had to be read first of all by the interviewer. Given the similarities in the distributions, we conclude that responses in the ANES and the GLES surveys are subject to similar regularities. It is thus warranted to utilize response time as – proxy – indicators of data quality in both surveys.

Comparing the distributions of response times in both surveys, it turns out that the median is considerably lower in GLES rather than in ANES. This finding suggests that the former survey is somewhat more plagued by inattentive and thus speeding respondents. This pattern hints at the fact that nonprobability online surveys are conducive to larger problems arising from satisficing than probability-based surveys. Yet, the evidence is suggestive rather than conclusive because the two surveys under study differ in a number of respects, and not just in terms of sampling procedures.

11.5   How to measure "speeding"?

As response time is correlated with data quality, it appears to be reasonable to exclude too quick answers from data before performing substantive analyses. This suggestion raises two related questions. First, we have to decide whether to identify speeding case- or page-wise. The case-wise procedure, as used by several research institutes (see, e.g., GESIS, 2009b), rests on the assumption that some respondents answer all the questions too quickly. This notion is at odds with research on satisficing which suggests that respondents' attention varies in the course of a survey (Krosnick, 1991). What is more, previous research suggests that speeding is, by and large, not a stable characteristic of respondents but that respondents respond regularly to some questions, whereas they satisfice when answering other items (Greszki, Meyer, & Schoen, 2012). We therefore use a page-specific strategy that aims at detecting too fast responses rather than too fast respondents. To this end, we calculate for each survey page, i.e., for all items on a page alike, the difference between a person's response time and the median response time (see also GESIS, 2009b; Rossmann, 2010).

Second, we have to identify a threshold to distinguish too fast responses from regular responses. One can easily imagine that different criteria lead to different substantive findings (see, e.g., Meyer & Schoen, in press). Moreover, despite some reasonable upper and lower limits, the ultimate choice of a specific threshold is somewhat arbitrary. To avoid results that crucially depend upon an arbitrarily chosen criterion, we employ three different thresholds. The most inclusive criterion identifies those responses as too fast who were given more than 30% faster than the median response time. According to the second measure, answers which were given more than 40% faster than the median response time are flagged as speeding. The most exclusive measure employs the 50% criterion.[2] Being more than 50% faster than

---

[2] The ANES survey makes extensive use of automatic conditional branching on the same page of the parent item. That means, that dependent on the response choice, follow-up questions appear on the same screen below the original question. In terms of speeding, it is warranted to take this into account because survey pages appear different depending on respondents' response option. For example when a respondent is asked whether he likes or dislikes the Democratic Party, a follow-up question is displayed on the same screen that asks if he likes or dislikes the Democratic Party a little, a moderate amount or a great deal. On the contrary, if respondents choose "neither nor" on the parent item, no follow-up question is displayed. Thus, such respondents are naturally faster because they only answer one

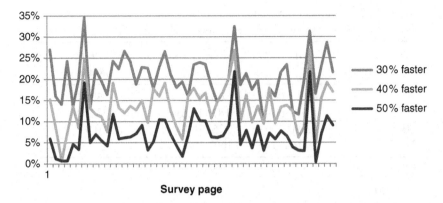

Figure 11.2   Percentage of flagged speeders for the first 50 survey pages (GLES). Note: In order to avoid biases resulting from varying numbers of observations, we included only survey pages which have been seen by 99% or more respondents.

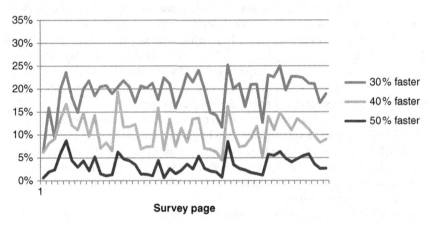

Figure 11.3   Percentage of flagged speeders for the first 50 survey pages (ANES). Note: In order to avoid biases resulting from varying numbers of observations, we included only survey pages which have been seen by 99% or more respondents.

the median response is a rather hard criterion. The latter measure might thus be considered as being capable of excluding very fast respondents and being simultaneously insensitive to "false positives" (see GESIS, 2009b; Rossmann, 2010).

Figures 11.2 and 11.3 show the percentage of flagged speeders for the first 50 survey pages according to these three criteria. Moreover, Table 11.2 reports the mean numbers of speeders across survey pages when applying the three criteria. First, the evidence demonstrates that the proportion of speeders is small to modest in the ANES and the GLES survey. Employing the most exclusive criterion results in less than 10%, while the most inclusive measure suggests that roughly one in five respondents speeds through the survey. To be sure, 20% is a considerable proportion. But even this percentage indicates rather a minority and it is likely to include a considerable proportion of "false positives."

question on that survey page. We take this into account and calculate separate page medians for those with and without follow-up questions.

Table 11.2 Number of excluded respondents and percentage of speeding flags across all survey pages (for three median-based criteria).

Percentage faster	GLES (%)	ANES (%)	Difference in percentage points
50	92 (8.0)	66 (4.1)	3.9***
40	169 (14.7)	187 (11.6)	3.1*
30	250 (21.7)	335 (20.8)	0.9
N	1144–1153	1607–1609	

Note: Significance levels:
*: $p < 0.05$;
**: $p < 0.01$;
***: $p < 0.001$. In order to avoid biases resulting from varying numbers of observations, we included only survey pages which have been seen by 99% or more respondents.

Moreover, the evidence suggests that thresholds make a difference in the proportion of respondents flagged as speeders. As the criterion for speeding becomes more exclusive, the number of speeders declines considerably. To give just an example, in the GLES data applying the 30 percent criterion, on average some 22% of the respondents are flagged as speeders. Utilizing the 40% threshold, speeders approximate 15% of all respondents, whereas the 50% criterion leads to 8% speeders. In effect, choosing a criterion might considerably affect conclusions about the pervasiveness of respondent inattentiveness.

Finally, comparing the findings from ANES and GLES, we find that for the 40% and 50% criteria, ANES data exhibit a significantly smaller proportion of speeders than the GLES survey. Yet, the differences are modest in substantive terms. Moreover, the 30% criterion yields virtually identical proportions of speeders in both surveys. So, the evidence supports the notion weakly that the freshly probability-based recruited respondents exhibit less speeding than self-selected respondents, if at all.

As already mentioned, raw response times (and the according medians) might be biased because respondents – due to differences in, e.g., cognitive ability and training – might differ in the time that they need to provide a valid response. With respect to this kind of respondent characteristics, however, the GLES and ANES surveys differ considerably. To make our conclusion more convincing and to demonstrate that response times can be employed as quality indicator in both surveys alike, we tested two rival explanations that focus on the composition of the ANES and GLES samples.

First, GLES respondents are considerably older and better educated than ANES respondents. As cognitive ageing and cognitive abilities appear to have an impact on response behavior (Malhotra, 2008; Yan & Tourangeau, 2008), we might expect that younger, more Internet-experienced, and better educated respondents handle online questionnaires more efficiently and thus answer faster. Provided the validity of these assumptions, the age and education differences between ANES and GLES samples might lead to between-samples differences in the prevalence of speeding.

In order to explore this hypothesis, we divided respondents into five age groups and three education groups and calculated page medians per group. Tables 11.3 and 11.4 first of all report how many respondents are defined as speeders according to the three criteria (as already reported in Table 11.2). Additionally, the tables report the proportion of respondents

Table 11.3 Proportion of speeding flags by age group for the overall and the group-wise speeding measure (total and for different age groups).

	GLES		ANES	
	Raw (%)	Corrected (%)	Raw (%)	Corrected (%)
50% faster (total)	8.0	7.2	4.1	4.2
18–29 years	12.1	8.6	10.9	3.6
30–39 years	11.3	7.8	8.2	1.9
40–49 years	7.1	6.4	4.2	3.1
50–59 years	5.2	7.4	2.5	3.2
60 years +	2.3	5.7	0.9	3.9
40% faster (total)	14.7	13.0	11.6	10.3
18–29 years	21.2	14.4	24.8	10.9
30–39 years	20.6	14.2	21.8	8.2
40–49 years	14.0	11.7	12.5	10.3
50–59 years	6.8	13.2	8.9	10.4
60 years +	4.4	11.4	3.3	11.1
30% faster (total)	21.7	21.2	20.8	19.5
18–29 years	30.2	22.8	37.2	19.6
30–39 years	29.2	22.4	35.3	17.1
40–49 years	21.7	19.4	23.8	19.0
50–59 years	14.9	21.5	18.1	19.8
60 years +	7.0	19.8	8.1	20.9

Note: "Raw" entries indicate the proportion of flags when speeding is measured by comparing an individual's response time to the response times in the whole sample. "Corrected" entries indicate the proportion of flags when speeding is measured by comparing an individual's response time to the response times in the respective age group.

flagged within each group, i.e., the proportion of respondents who answer too quickly as compared to their fellow respondents in the respective age or education bracket (right-hand columns). For example, applying the 50% criterion to GLES data with no group correction, we see the expected pattern that 12% of the young respondents are flagged, but only 2% of those 60 years and older. A similar pattern applies to the ANES data, although the proportions of speeders are smaller.

The within-group page medians yield a somewhat more balanced distribution of speeding flags across groups in the GLES and the ANES sample. In the GLES survey, for example, when the 50% group-specific measure is employed, the proportion of speeding flags ranges from roughly 6% among the oldest respondents to 8.6% in the 18–29 years group as compared to a 2%–12% range when the 50% overall speeding measure is utilized. Similar patterns apply to education groups. But even when the group-based speeding measures are employed, we find a higher proportion of speeders in the GLES sample than in the ANES sample. As a result, the differences in the prevalence of speeding between the two samples cannot be explained by compositional differences in terms of education and age.

Second, whereas ANES respondents were recruited "fresh" for the first panel-wave, the GLES respondents are regular members of an online panel and are more likely to have answered surveys before. As more experienced respondents are supposed to answer faster, this difference in sample composition might account for the differences in speeding

Table 11.4 Proportion of speeding flags by education group for the overall and the group-wise speeding measure (total and for different education groups).

	GLES		ANES	
	Raw (%)	Corrected (%)	Raw (%)	Corrected (%)
50% faster (total)	7.9	8.0	4.1	4.2
education low	6.6	8.6	3.3	7.3
education medium	7.7	8.1	3.6	4.1
education high	9.9	7.1	4.9	2.8
40% faster (total)	14.6	14.2	11.6	11.5
education low	11.5	14.9	8.5	15.5
education medium	14.0	14.7	10.6	11.3
education high	19.0	12.6	14.1	9.8
30% faster (total)	21.6	21.5	20.8	20.5
education low	16.8	22.1	14.8	24.0
education medium	20.7	21.5	19.2	20.5
education high	28.3	20.8	25.0	18.8

Note: "Raw" entries indicate the proportion of flags when speeding is measured by comparing an individual's response time to the response times in the whole sample. "Corrected" entries indicate the proportion of flags when speeding is measured by comparing an individual's response time to the response times in the respective education group.

between ANES and GLES. To explore this hypothesis, we once more utilized GLES data. In particular, we regressed the proportion of survey pages that a respondent completed faster than the respective median respondent on indicators of experience (e.g., number of completed surveys in the previous four weeks, duration of panel membership). Moreover, we included predictors capturing the recruitment process. As it turns out, these predictors explain just 3% of the variance in the dependent variable (Table 11.5). As concerns experience, panel membership exhibits a positive effect, suggesting that the longer a respondent is active in the panel, the more likely he or she is to have a higher amount of survey pages below the median. This effect, however, is far from being substantively relevant. Moreover, the number of surveys completed in the four weeks prior to the current survey does not exhibit a statistically significant effect. However, as respondent activity in online panels might vary over different time periods, considering only the number of completed surveys in the past four weeks might be a rather weak indicator of experience. Therefore, we also tested the number of completed surveys in larger time frames (last 12 weeks, last 12 months). These additional analyses exhibit no different results, suggesting that survey experience does not make a difference.[3]

In sum, survey experience, by and large, does not affect response time. Age and education make a difference in response time, but age and education differences between ANES and GLES do not account for the differences in speeding. Having shown that these alternative explanations do not account for the differences in speeding, we are in a better position to conclude that response time is a functionally equivalent indicator across surveys. Accordingly, we might conclude that the nature of the recruitment process – probability-based vs.

---

[3] Yet, we have to keep in mind that we have only information about experience in the GLES panel. Respondents, however, might also be members of other panels about which we have no information.

Table 11.5 Response time as a function of recruitment and respondent behavior (GLES; OLS).

	b	β
Constant	0.48***	
	(0.03)	
Number of survey completions (last 4 weeks)	−0.004	−0.04
	(0.003)	
Panel membership (weeks)	0.0004***	0.11
	(0.00)	
Self-recruitment	0.003	0.01
	(0.03)	
Recruitment through extern link	−0.083**	−0.14
	(0.03)	
Adj. R^2		0.03
N		1140

Note: Cell entries are b-coefficients and standardized β-coefficients; standard errors in parentheses; significance levels:
*: $p < 0.05$;
**: $p < 0.01$;
***: $p < 0.001$. Reference group for recruitment: friend, advertising.

nonprobability – makes a small difference in the proportion of speeders in the ANES and the GLES samples. We have to keep in mind, however, that the surveys differ also in questionnaire design for which we could not account for in our analysis. Given the evidence, we are thus cautious drawing strong conclusions about causal effects of the recruitment process.

11.6 Does speeding matter?

Having established that speeding, as measured above, occurs in the ANES and the GLES surveys, we now turn to the issue of whether it matters for substantive findings. As speeding is correlated with low-data quality, not removing "too fast responses" from the data set might lead to biased results. To address this question, we examine whether excluding page-specific speeders from the analysis alters substantive findings both in terms of marginal distributions and parameters of interest in explanatory models. The results are gleaned from analyses using uncorrected speeding flags; using age and education corrected flags does not alter the substantive findings we report in the remainder of this chapter.

We explore this question by analyzing two phenomena that are a kind of standard topics in research on public opinion and political behavior. For one thing, we analyze attitudes towards the head of government's handling of the economy (e.g., Alesina & Rosenthal, 1995; Duch & Stevenson, 2008). For another, we address electoral participation (e.g., Blais, 2000; Hansen, 1975). Given the comparative nature of our analysis, we used explanatory models that include variables which were, by and large, covered both by ANES and GLES. More sophisticated models could not be used due to a lack of appropriate data.

Starting with evaluations of the president's and chancellor's handling of the economy, Tables 11.6 and 11.7 report the marginal distributions in both surveys. In each table, the

Table 11.6 Evaluations of Merkel's handling of the economy (marginal distributions across different thresholds).

	Without exclusion (%)	50% faster than median (%)	40% faster than median (%)	30 % faster than median (%)
1 not at all suitable	18.1	17.8	17.4	17.3
2	16.0	17.1	17.4	17.3
3	26.3	24.4	25.0	25.6
4	25.7	26.9	26.8	26.6
5 very suitable	13.9	13.9	13.4	13.2
N valid	1091	972	891	836
N speeders of valid (%)	-	119 (11)	200 (18)	255 (23)
N speeders total	-	134	219	278

Note: None of the differences between the proportions is statistically significant ($p > 0.05$). Question text: "Angela Merkel has suitable concepts for stimulating the economy."

Table 11.7 Evaluations of Bush's handling of the economy (marginal distributions across different thresholds).

	Without exclusion (%)	50% faster than median (%)	40% faster than median (%)	30 % faster than median (%)
Extremely disapprove	31.2	31.5	30.5	29.9
Moderately disapprove	20.7	20.8	20.9	21.6
Slightly disapprove	4.8	4.8	4.9	5.0
Neither nor	24.9	24.3	24.8	24.2
Slightly approve	6.1	6.2	6.3	6.7
Moderately approve	8.1	8.1	8.3	8.6
Extremely approve	4.2	4.2	4.3	4.0
N valid	1610	1569	1452	1321
N speeders of valid (%)	-	41(3)	158 (10)	289(18)
N speeders total	-	43	160	292

Note: None of the differences between the proportions is statistically significant ($p > 0.05$). Question text: "Do you approve, disapprove, or neither approve nor disapprove of the way George W. Bush is handling the economy?"

left-hand column contains the results for the sample without any speeders removed. The remaining columns show the marginal distributions for the samples when speeders, as defined by the three above criteria, are excluded. The evidence clearly shows that marginal distributions do not considerably change when speeders are excluded, irrespective of the threshold. Although a somewhat larger proportion of the GLES respondents are flagged as speeders, in neither GLES nor ANES does removing speeders alter the substantive findings.

Turning to explanatory models, we model evaluations of handling of the economy as a function of party evaluations, ideological self-placement, and the respondent's perception of the national economy (all variables rescaled to run from 0–1). The results of the respective OLS regressions are reported in Tables 11.8 and 11.9. Once again, in each table the left-hand

Table 11.8 Determinants of evaluations of Merkel's handling of the economy (across different speeding thresholds; OLS).

	Baseline model	50 % faster than median		40% faster than Median		30% faster than median	
	B (s.e.)	B (s.e.)	Diff. to baseline	B (s.e.)	Diff. to baseline	B (s.e.)	Diff. to baseline
Constant	0.15*** (0.02)	0.13*** (0.03)	0.02	0.14*** (0.03)	0.01	0.14*** (0.03)	0.01
Evaluation CDU	0.66*** (0.03)	0.65*** (0.03)	0.01	0.61*** (0.03)	0.05	0.60*** (0.04)	0.06
Evaluation SPD	−0.01 (0.03)	0.02 (0.03)	−0.03	0.01 (0.03)	−0.02	0.00 (0.03)	−0.01
Ideology (high = conserv.)	0.03 (0.04)	0.06 (0.04)	−0.03	0.09* (0.04)	−0.06	0.10* (0.05)	−0.07
Economy better than 1 year ago	0.14*** (0.03)	0.16*** (0.04)	−0.02	0.18*** (0.04)	−0.04	0.18*** (0.05)	−0.04
Adjusted R^2	0.49	0.50		0.49		0.47	
N	926	806		693		605	

Note: Cell entries are b-coefficients; Standard errors in parentheses; Significance levels:
*: $p < 0.05$;
**: $p < 0.01$;
***: $p < 0.001$. Difference tests for regression coefficients for different samples were applied following Cohen et al. (2003).

column reports the baseline model with no speeding respondents excluded; only those respondents are removed who exhibit missing values on at least one variable, i.e., "don't know" or "no answer" responses, in the analysis. The right-hand columns contain the models without those respondents being excluded as speeders according to the three different criteria. We additionally included columns that report the respective coefficient differences to the baseline model. Comparing models with and without speeders is warranted because alternative strategies, e.g., including interactions between speeding indicators and substantive predictors, imply that respondents who speed on the dependent variable cannot be adequately adressed.

The evidence shows that in US model, the four independent variables exhibit statistically significant effects in the expected direction. Accordingly, approving of the Republicans and subscribing to conservative ideology, e.g., makes citizens more favorable of George W. Bush's handling of the economy. What is more, the substantive conclusions concerning the predictors of the president's handling of the economy are independent of whether speeders are excluded or not. In fact, the largest difference between coefficients in baseline model and a model with speeders removed amounts to a tiny 0.04.

Turning to GLES data, the evidence supports an identical conclusion. To be sure, there are some differences. For one thing, two predictors of economic evaluations, i.e., ideology and evaluations of the Social Democrats, prove statistically insignificant in the baseline model. For another thing, we find somewhat more sizable differences between the baseline model and the models with speeders excluded. In particular, ideology turns out to pass conventional levels of statistical significance when speeders are excluded according to the 30% or the 40%

Table 11.9   Determinants of evaluations of Bush's handling of the economy across different speeding thresholds (OLS).

	Baseline model	50 % faster than median		40% faster than median		30% faster than median	
	B (s.e.)	B (s.e.)	Diff. to baseline	B (s.e.)	Diff. to baseline	B (s.e.)	Diff. to baseline
Constant	0.14***	0.14***	0.00	0.16***	−0.02	0.16***	−0.02
	(0.03)	(0.03)		(0.03)		(0.04)	
Evaluation	−0.21***	−0.21***	0.00	−0.24***	−0.03	−0.24***	−0.03
Democrats	(0.02)	(0.03)		(0.03)		(0.03)	
Evaluation	0.28***	0.31***	−0.03	0.29***	−0.01	0.27	0.01
Republicans	(0.03)	(0.03)		(0.03)		(0.03)	
Ideology	0.12***	0.10***	0.02	0.11***	0.01	0.12**	0.00
(high=conserv.)	(0.03)	(0.03)		(0.03)		(0.04)	
Economy better	0.43***	0.41***	0.02	0.39***	0.04	0.40***	0.03
than 1 year ago	(0.03)	(0.03)		(0.04)		(0.04)	
Adjusted R^2	0.46	0.47		0.45		0.44	
N	1598	1336		1067		862	

Note: Cell entries are b-coefficients; Standard errors in parentheses; Significance levels:
*: $p < 0.05$;
**: $p < 0.01$;
***: $p < 0.001$. Difference tests for regression coefficients for different samples were applied following Cohen et al. (2003).

criterion. The differences between models and the baseline model are so small, however, that they prove neither statistically significant nor substantively relevant.

We thus conclude that both in ANES and GLES, the exclusion of speeders does not affect substantive conclusions about the distribution of attitudes toward the head of government's handling of the economy and about the determinants of these attitudes. Irrespective of whether speeders are removed, the substantive message is the same.

Next we address electoral participation. In both surveys, respondents were asked whether they intend to participate in the next federal/presidential election. Whereas the ANES survey offered two response options (yes/no), respondents had five response options in the German survey running from "definitely" to "definitely not." To ease comparison, we dichotomized the German answer scheme ("definitely," "probably" = 1; "perhaps", "probably not", "definitely not" = 0). Again, we address marginal distributions before exploring the determinants of (prospective) turnout.

The marginal distributions reported in Table 11.10 demonstrate that the exclusion of speeders does not make a difference in results on turnout intention. As the first column in Table 11.10 indicates, some 92% of the respondents were likely to vote. This percentage is much higher than the actual turnout in both elections and thus reflects sampling problems as well as measurement error (overreporting) in both surveys (see e.g., Bernstein, Chadha, & Montjoy, 2001; Silver, Anderson, & Abramson, 1986). Removing speeders from the sample does not alter the proportion of voters, irrespective of the criterion for speeding. The reason for this finding is the fact that speeders are similarly frequent among voters and nonvoters in the ANES and the GLES data.

Table 11.10   Marginal distribution of turnout intention in GLES and ANES.

	Without exclusion	50% faster than median	40% faster than median	30% faster than median
GLES	(%)	(%)	(%)	(%)
yes	91.5	92.2	92.6	92.8
no	8.5	7.8	7.4	7.2
N valid	1123	912	850	737
N speeders of valid (%)	–	211 (19%)	273 (24%)	386 (34%)
N speeders total	–	221	283	399
ANES	(%)	(%)	(%)	(%)
yes	92.7	93.0	93.1	92.9
no	7.3	7.0	6.9	7.1
N valid	1620	1588	1488	1362
N speeders of valid (%)	–	32 (2)	132 (8)	258 (16)
N speeders total	–	32	132	258

Note: None of the differences between the proportions is statistically significant ($p > 0.05$).

Turning to multivariate models, we perform logistic regression analyses with turnout intention as dependent variable. Due to data limitations, we cannot perform models with exactly identical sets of predictor variables in ANES and GLES. But the ANES and GLES models alike build on the notion that electoral turnout is driven by political involvement, which includes interest in politics, feelings of political efficacy, and citizen duty. The results of these analyses are reported in Tables 11.11 and 11.12, with the left-hand columns including the baseline models.[4]

Starting with ANES results, political interest and internal efficacy appear to be particularly powerful in shaping electoral turnout. Comparing these findings to the results from the analyses with speeder corrections, we find that the coefficients on political interest, party identification, and – to a smaller extent – internal and external efficacy are quite stable across models. The results concerning the impact of talking about politics on turnout, however, differ somewhat across models, in particular between the baseline and the 30%-criterion model. But the differences in coefficients do not pass conventional levels of statistical significance. Moreover, we calculated from the models reported in Table 11.11 predicted probabilities of turnout for different levels of the independent variable of interest, i.e., talking about politics, while setting the remaining variables in the model to their respective mean (not reported in the tables). The results suggest that moving from respondents who never talk about politics to those who talk about politics every day increases the likelihood of turnout slightly from 94%–98%. In the model with the 30%-speeder criterion, the increase is from 91%–99%, i.e.

---

[4] In an attempt to explore whether cognitive ageing and cognitive abilities do matter (Malhotra, 2008; Yan & Tourangeau 2008), we studied the distribution of age and education in terms of valid cases according to listwise deletion in the regression models (see online appendix at: www.wiley.com/go/online_panel_research). Comparing the baseline model with the 40% and 30% rule, we find patterns consistent with findings in prior studies. Accordingly, young and highly educated respondents are more likely to be flagged as speeders. For the 50% threshold, the differences to the baseline model are rather marginal. These patterns suggest, that using the 50% rule leads to the exclusion of rather "real" speeding behavior that is independent of "predispositions," such as age and education. In this vein, the 40%- and 30%-variants appear to be sensitive, but not very specific. In effect, young and highly educated respondents who do not satisfice but quickly give valid responses have been removed from the data set.

Table 11.11   Determinants of turnout intention (ANES) with different speeding thresholds (logistic regression).

	Baseline model	50% faster than median		40% faster than median		30% faster than median	
	B	B	Diff. to baseline	B	Diff. to baseline	B	Diff. to baseline
Constant	−1.48***	−1.43***	−0.05	−1.42***	−0.06	−1.55***	0.07
	(0.30)	(0.32)		(0.37)		(0.43)	
Political Interest	4.84***	4.60***	0.24	4.90***	−0.06	4.70***	0.14
	(0.59)	(0.62)		(0.73)		(0.86)	
Efficacy external	0.93	0.94	−0.01	0.08	0.85	−0.30	1.23
	(0.62)	(0.66)		(0.73)		(0.84)	
Efficacy internal	1.73**	1.59**	0.14	2.06**	−0.33	2.19**	−0.46
	(0.55)	(0.57)		(0.66)		(0.74)	
Talking Politics	1.24*	1.52*	−0.28	1.50*	−0.26	2.39**	−1.15
	(0.59)	(0.63)		(0.70)		(0.88)	
PID yes	0.65**	0.74**	−0.09	0.72**	−0.07	0.74*	−0.09
	(0.23)	(0.24)		(0.28)		(0.31)	
Nagelkerke R^2	0.36	0.34		0.33		0.35	
N	1603	1472		1196		886	

Note: Cell entries are logit coefficients; Standard errors in parentheses; Significance levels:
*: $p < 0.05$;
**: $p < 0.01$;
***: $p < 0.001$. Difference tests for logit coefficients for different samples were applied following Cohen et al. (2003).

seemingly larger but indistinguishable from the findings in the model without corrections. The same pattern, which might also reflect some kind of ceiling effect, applies to the remaining models and variables. So in ANES data, speeders do not make a difference.

Turning to the GLES baseline model in Table 11.12, citizen duty and again political interest are important determinants of turnout in Germany. When taking into account models with speeder corrections, the results concerning citizen duty and – to a smaller degree – political interest and party identification turn out to be remarkably stable across models. By contrast, point estimates for campaign interest and satisfaction with democracy appear to vary somewhat across models. Whereas the former's impact appears to decrease when excluding an increasing number of speeders, the latter's effect on turnout appears to increase. None of the differences in coefficients passes conventional levels of statistical significance, however. Calculating predicted probabilities from the regression estimates lends additional credence to the conclusion that speeder corrections do not make a difference (not reported in the tables). To give just an example, in the baseline model moving from the low to high campaign interest increases turnout from 94%–98%, in the model with the 30% correction from 97%–98%. Neither difference attains statistical significance. As a result, removing speeders from the analysis does not affect substantive conclusions.

In summary, we conclude that both in ANES and GLES, the exclusion of speeders does not affect substantive conclusions about the distribution of the intention to turnout and its predictors. Irrespective of whether speeders are removed and how they are defined, the substantive findings do not differ. As the findings on turnout parallel those on attitudes toward

Table 11.12 Determinants of turnout intention (GLES) with different speeding thresholds (logistic regression).

	Baseline model	50% faster than median		40% faster than median		30% faster than median	
	B	B	Diff. to baseline	B	Diff. to baseline	B	Diff. to baseline
Constant	−2.50***	−2.37***	0.13	−2.24***	−0.26	−2.59***	0.09
	(0.39)	(0.51)		(0.59)		(0.72)	
Political Interest	2.85***	2.71**	0.14	2.96**	−0.11	3.46***	−0.61
	(0.67)	(0.87)		(0.99)		(1.16)	
Campaign	1.38*	1.14	0.24	0.81	0.57	0.44	0.94
Interest	(0.66)	(0.80)		(0.89)		(1.01)	
Citizen duty	3.15***	3.24***	−0.09	3.11***	0.04	3.17***	−0.02
	(0.47)	(0.54)		(0.57)		(0.66)	
Satisf. w.	1.56**	1.76*	−0.20	1.72*	0.16	3.09**	−1.53
Democracy	(0.58)	(0.75)		(0.79)		(0.96)	
PID strength	0.54	0.69	−0.15	0.52	0.02	−0.03	0.57
	(0.37)	(0.47)		(0.59)		(0.60)	
Nagelkerke R^2	0.42	0.39		0.34		0.36	
N	1077	781		633		491	

Note: Cell entries are logit coefficients; Standard errors in parentheses; Significance levels:
*: $p < 0.05$;
**: $p < 0.01$;
***: $p < 0.001$. Difference tests for logit coefficients for different samples were applied following Cohen et al. (2003).

the head of government's handling of the economy, we might conclude that speeding, though existent, does not pose severe problems for scholars interested in substantive results.

11.7   Conclusion

This chapter addressed the phenomenon of speeding in web surveys as an indicator of inattentive respondents and explored the problems arising from it for data quality. Building on the response process model, we suggested that speeding implies skipping necessary steps in the process of providing a valid answer to a survey question. In this vein, response time might be used as an indicator of data quality in self-administered online surveys. Because of the absence of an interviewer, web surveys are notably encountered with the uncontrolled interview situation. Utilizing a page-wise procedure, we identified too fast responses and explored whether removing them from the data might affect substantive results in a nonprobability online panel and a freshly probability-based recruited sample.

Using data from these two surveys, we demonstrated that a considerable number of respondents "speed" through the pages of web surveys. However, our analysis demonstrates that the choice of criteria to identify speeders makes a difference in the proportion of speeders. Moving from the 30% to the 50% criterion decreases the proportion of speeders considerably. These differences suggest that the more lenient criteria are likely to identify "false positives" as speeders whereas stiffer criteria might yield some "false negatives." In substantial terms, however, the proportion of speeders is quite small unless the very lenient 30% criterion is employed. So, speeding does occur, but it is clearly a minority of our samples that exhibit speeding behavior.

We explored the prevalence and effects of speeding in a nonprobability sample and a probability-based sample. The findings did not differ systematically across samples. To be sure, relying on the more exclusive criteria the nonprobability GLES sample yielded a somewhat higher percentage of speeders than the probability-based ANES sample. But these differences were not so sizable that they led to different substantive conclusions about the prevalence of speeding. When it comes to the effect of speeding on substantive findings, both samples led to identical conclusions. Irrespective of whether studying marginal distributions or multivariate models, irrespective of which threshold is applied, excluding too fast responses does not make a considerable difference in substantive results. We may caution, however, that we have evidence from just two cases that, in addition, differ not only in terms of recruitment. These qualifications notwithstanding, we might conclude that the evidence does not support the notion that the nonprobability/probability recruitment differs considerably or is of critical importance when it comes to speeding.

Nevertheless, this finding is good news for online-based survey research. There are indeed inattentive respondents who "speed" through the pages, but speeding appears not to bias substantive findings. We might account for this pattern by two related arguments. On the one hand, obvious speeders are often already excluded from analysis because of invalid missing data (e.g., item nonresponse). On the other hand, for valid answers, speeding is not systematically linked to certain variables or parameters of interest in our analyses. In this line of reasoning, there appear to be some variables that are not vulnerable to biases resulting from speeding.

As already mentioned, this research is subject to several limitations. The most severe limitations stems from the limited number of data sets available. Future research should thus utilize more data from more diverse online surveys. This approach would permit scholars to explore the prevalence and substantive impact of speeding in different societies, in surveys on a diversity of topics, with different speeding measures, in samples that comprise professional and novice respondents who were recruited in different ways. This kind of evidence might lend additional credence to the notion that speeding, though existent, does not affect substantive results. Presented with this finding, scholars might conclude that inattentive respondents are not a real problem for online surveys. It cannot be taken for granted, however, that the evidence will not suggest that speeding poses severe problems for analyses of online data that were collected on specific topics, in particular samples or societies.

A related approach does not aim to assemble a diversity of data but identifies potential predictors of speeding and its substantive relevance in the first place and then employs a (quasi-)experimental design to gauge the impact of the factor of interest. For example, scholars might study the impact of certain features of online surveys, e.g., batteries of grid questions, on respondent motivation and speeding behavior. In order to study these questions, scholars might find it convenient to not only analyze response times but might employ also other techniques like eye-tracking that permit the scrutiny of respondent behavior in more depth. In identifying factors conducive to respondent attention, scholars might also find complex interactions between respondent characteristics and features, be they technical or substantive, of the survey in shaping respondent motivation. Here, a valuable strategy to study the role of dispositional traits in the interplay with survey features might be also to employ experimental designs in panel studies.

Identifying dispositional traits and survey features that are conducive to respondent motivation and thus presumably valid responses might be considered a first step toward strategies that do not aim at detecting speeders but at avoiding speeding at all. In this vein, scholars might consider how to design online surveys in order make or keep all respondents sufficiently attentive to give valid responses. This research agenda might lead to the conclusion that there is a single optimal strategy for all respondents and topics. Alternatively, scholars might learn that the suitability of survey features varies across topics or respondents, thereby raising additional issues of comparability. Probably, findings will also change over time as online surveys become more pervasive and respondents more acquainted to them. In any event, we consider respondent attentiveness, and speeding as an indicator thereof, as a topic that warrants scholarly attention in the future.

References

Alesina, A., & Rosenthal, H. (1995). *Partisan politics, divided government, and the economy.* New York: Cambridge University Press.

Baker, R., Blumberg, S. J., Brick, M., Couper, M. P., Courtright, M., Dennis, J. M., et al. (2010). Research synthesis: AAPOR report on online panels. *Public Opinion Quarterly, 74,* 711–781.

Baker, R., & Downes-Le Guin, T. (2007). Separating the wheat from the chaff: Ensuring data quality in internet samples. In M. Trotman, T. Burrell, L. Gerrard, K. Anderton, G. Basi, M. Couper, K. Moris, et al. (Eds.), *Proceedings of the Fifth International Conference of the Association for Survey Computing: The challenges of a changing world* (pp. 157–166). Berkeley, UK: ASC.

Balden, W. (2008). An unwanted impact. Retrieved June 28, 2013, from: http://www.quirks.com/articles/2008/20080103.aspx?searchID=765725748&sort=5&pg=1.

Bassili, J. N. (1993). Response latency versus certainty as indexes of the strength of voting intentions in a CATI survey. *Public Opinion Quarterly, 57,* 54–61.

Bassili, J. N. (1996). The how and the why of response latency measurement in telephone surveys. In N. Schwarz, & S. Sudman (Eds.), *Answering questions: Methodology for determining cognitive and communicative processes in survey research* (pp. 319–346). San Francisco, CA: Jossey-Bass.

Bassili, J. N., & Fletcher, J. F. (1991). Response-time measurement in survey research: A method for CATI and a new look at nonattitudes. *Public Opinion Quarterly, 55,* 331–346.

Bassili, J. N., & Scott, B. S. (1996). Response latency as a signal to question problems in survey research. *Public Opinion Quarterly, 60,* 390–399.

Berinsky, A. J., Margolis, M., & Sances, M. W. (2012). Separating the shirkers from the workers? Making sure respondents pay attention on internet surveys. Paper presented at the NYU CESS 5th Annual Experimental Political Science Conference, New York.

Bernstein, R., Chadha, A., & Montjoy, R. (2001). Overreporting voting: Why it happens and why it matters. *Public Opinion Quarterly, 65,* 22–44.

Bethlehem, J., & Biffignandi, S. (2012). *Handbook of web surveys.* Hoboken, NJ: John Wiley & Sons, Inc.

Blais, A. (2000). *To vote or not to vote: The merits and limits of rational choice theory.* Pittsburgh, PA: University of Pittsburgh Press.

Callegaro, M., Yang, Y., Bhola, D. S., Dillman, D. A., & Chin, T. (2009). Response latency as an indicator of optimizing in online questionnaires. *Bullétin de Méthodologie Sociologique, 103,* 5–25.

Chang, L., & Krosnick, J. A. (2010). Comparing oral interviewing with self-administered computerized questionnaires: An experiment. *Public Opinion Quarterly, 74,* 154–167.

Coen, T., Lorch, J., & Piekarski, L. (2005). The effects of survey frequency on panelists' responses. Paper presented at the Worldwide Panel Research Conference ESOMAR, Amsterdam.

Cohen, J., Cohen, P., West, S. G., & Aiken, L.S. (2003). *Applied multiple regression/correlation analysis for the behavioral science*. New York: Routledge.

Couper, M. P. (2000). Usability evaluation of computer-assisted survey instruments. *Social Science Computer Review, 18*, 384–396.

Couper, M. P. (2005). Technology trends in survey data collection. *Social Science Computer Review, 23*, 486–501.

DeBell, M., Krosnick, J. A., & Lupia, A. (2010). *Methodology report and user's guide for the 2008–2009 ANES panel study*. Palo Alto, CA and Ann Arbor, MI: Stanford University and the University of Michigan.

De Leeuw, E. D. (1992). *Data quality in mail, telephone and face to face surveys*. Amsterdam: TT-Publikaties.

Draisma, S., & Dijkstra, W. (2004). Response latency and (para)linguistic expressions as indicators of response error. In S. Presser, J. M. Rothgeb, M. P. Couper, J. T. Lessler, E. Martin, J. Martin, & E. Singer (Eds.), *Methods for testing and evaluating survey questions* (pp. 131–147). Hoboken, NJ: John Wiley & Sons, Inc.

Duch, R. M., & Stevenson, R. T. (2008). *The economic vote: How political and economic institutions condition election results*. New York: Cambridge University Press.

Faas, T., & Schoen, H. (2006). Putting a questionnaire on the web is not enough: A comparison of online and offline surveys conducted in the context of the German Federal Election 2002. *Journal of Official Statistics, 22*, 177–190.

Fraley, R. C. (2004). *How to conduct behavioral research over the internet: A beginner's guide to HTML and CGI/Perl*. New York: The Guilford Press.

Galesic, M., & Bosnjak, M. (2009). Effects of questionnaire length on participation and indicators of response quality in a web survey. *Public Opinion Quarterly, 73*, 349–360.

GESIS – Leibniz-Institute for the Social Science (2009a). *German Longitudinal Election Study 2009: Langfrist-Online-Tracking, T6: ZA5339, Version 2.0.0, Studienbeschreibung*. Retrieved August 20, 2012, from: http://info1.gesis.org/dbksearch19/SDesc2.asp?no=5339&tab=3&ll=10&notabs=&af=&nf=1&search=gles&search2=&db=E.

GESIS – Leibniz-Institute for the Social Science (2009b). *German Longitudinal Election Study 2009: Short-term Campaign Panel: ZA5305, Version 3.0.0, Study Description* (ZA5305_sd_v3-0-0.pdf). Retrieved August 20, 2012, from: http://info1.gesis.org/dbksearch19/SDesc2.asp?no=5305&tab=3&ll=10&notabs=&af=&nf=1&search=gles&search2=&db=E.

Greszki, R. P., Meyer, M., & Schoen, H. (2012). Speeding in web surveys: A serious threat to data quality? Paper presented at the annual European Political Science Association Conference EPSA, Berlin.

Groves, R. M. (1991). *Survey errors and survey costs*. New York: John Wiley & Sons, Inc.

Groves, R. M., Fowler, F. J., Couper, M. P., Lepkowski, J. M., Singer, E., & Tourangeau, R. (2009). *Survey methodology*. Hoboken, NJ: John Wiley & Sons, Inc.

Hansen, S. B. (1975). Participation, political structure, and concurrence. *The American Political Science Review, 69*, 1181–1199.

Heerwegh, D. (2003). Explaining response latencies and changing answers using client-side paradata from a web survey. *Social Science Computer Review, 21*, 360–373.

Heerwegh, D. (2011). Internet survey paradata. In M. Das, P. Ester, & L. Kaczmirek (Eds.), *Social and behavioral research and the internet* (pp. 325–348). New York: Routledge.

Johnson, M. (2004). Timepieces: Components of survey question response latencies. *Political Psychology, 25*, 679–702.

Kaczmirek, L. (2009). *Human-survey interaction: Usability and nonresponse in online surveys*. Cologne: Herbert von Halem Verlag.

Kahn, R. L., & Cannell, C. F. (1957). *The dynamics of interviewing*. New York: John Wiley & Sons, Inc.

Kaminska, O., Goeminne, B., & Swyngedouw, M. (2006). *Satisficing in early versus late responses to a mail survey.* Leuven, NL: Katholieke Universiteit Leuven.

Kaminska, O., McCutcheon, A. L., & Billiet, J. (2010). Satisficing among reluctant respondents in a cross-national context. *Public Opinion Quarterly, 74,* 956–984.

Knapton, K., & Garlick, R. (2007). *Catch me if you can.* Retrieved August 28, 2013, from: http://www.quirks.com/articles/2007/20071107.aspx?searchID=765736159&sort=5&pg=1.

Krosnick, J. A. (1991). Response strategies for coping with the cognitive demands of attitude measures in surveys. *Applied Cognitive Psychology, 5,* 213–236.

Krosnick, J. A., & Alwin, D. F. (1987). An evaluation of a cognitive theory of response-order effects in survey measurement. *Public Opinion Quarterly, 51,* 201–219.

Krosnick, J. A., Holbrook, A. L., Berent, M. K., Carson, R. T., Hanemann, W. M., Kopp, R. J., et al. (2002). The impact of "no opinion" response options on data quality: Non-attitude reduction or an invitation to satisfice? *Public Opinion Quarterly, 66,* 371–403.

Krosnick, J. A., Narayan, S., & Smith, W. R. (1996). Satisficing in surveys: Initial evidence. In M. T. Braverman, & J. K. Slater (Eds.), *Advances in survey research* (pp. 29–44). San Francisco, CA: Jossey-Bass.

Malhotra, N. (2008). Completion time and response order effects in web surveys. *Public Opinion Quarterly, 72,* 914–934.

Meade, A. W., & Craig, S.B. (2011). Identifying careless responses in survey data. Paper presented at the 26th Annual Meeting of the Society for Industrial and Organizational Psychology. Chicago, IL.

Meyer, M., & Schoen, H. (in press). Response latencies and attitude-behavior consistency in a direct democratic setting: Evidence from a subnational referendum in Germany. *Political Psychology.*

Miller, J. (2006). Research reveals alarming incidence of "undesirable" online panelists. Retrieved June 28, 2013, from: http://www.burke.com/Library/Articles/Jeff%20Miller%20RCR%20PDF.pdf.

Miller, J., & Baker-Prewitt, J. (2009). Beyond "trapping" the undesirable panelist: The use of red herrings to reduce satisficing. Retrieved June 28, 2013, from: http://www.burke.com/Library/Conference/Beyond%20Trapping%20the%20Undesirable%20Panelist_FINAL.pdf.

Narayan, S., & Krosnick, J.A. (1996). Education moderates some response effects in attitude measurement. *Public Opinion Quarterly, 60,* 58–88.

Puleston, J., & Sleep D. (2008). A look at the impact of boredom on the respondent experience. Retrieved June 28, 2013, from: http://www.quirks.com/articles/2008/20081106.aspx?searchID=765759022&sort=5&pg=1.

Ratcliff, R. (1993). Methods for dealing with reaction time outliers. *Psychological Bulletin, 114,* 510–532.

Rattinger, H., Rossteutscher, S., Schmitt-Beck, R., & Wessels, B. (2009). *German Longitudinal Election Study: Long-term Online Tracking, T6 (GLES 2009).* Cologne: GESIS Leibnitz-Institute for the Social Science.

Respondi AG (2009a). Our answers to the 26 ESOMAR questions for determining the quality of online samples and online panels. Retrieved August 20, 2012, from: http://www.respondi.com.

Respondi AG (2009b). Panel book. Retrieved August 20, 2012, from: http://www.respondi.com.

Rossmann, J. (2010). Data quality in web surveys of the German Longitudinal Election Study 2009. Paper presented at the 3rd ECPR Graduate Conference, Dublin.

Schaeffer, N. C., Dykema, J., & Maynard, D. W. (2010). Interviewers and interviewing. In P. T. Marsden, & J. D. Wright (Eds.), *Handbook of survey research* (pp. 437–470). Bingley, UK: Emerald.

Schoen, H. (2004). Online-Umfragen – schnell, billig, aber auch valide? Ein Vergleich zweier Internetbefragungen mit persönlichen Interviews zur Bundestagswahl 2002. *ZA-Information, 54,* 27–52.

Schuman, H., & Presser, S. (1996). *Question and answers in attitude surveys.* Thousand Oaks, CA: Sage.

Silver, B. D., Anderson, B. A., & Abramson, P. R. (1986). Who overreports voting? *The American Political Science Review*, *80*, 613–624.

Smith, R., & Brown, H. H. (2005). Assessing the quality of data from online panels: Moving forward with confidence. Retrieved November 14, 2012, from: http://www.harrisinteractive.com/servicebureau /pubs/HI_Quality_of_Data_White_Paper.pdf.

Topoel, V., Das, M., & Van Soest, A. (2008). Effects of design in web surveys: Comparing trained and fresh respondents. *Public Opinion Quarterly*, *72*, 985–1007.

Tourangeau, R. (1984). Cognitive science and survey methods. In T. B. Jabine, M. L. Straf, J. M. Tanur, & R. Tourangeau (Eds.), *Cognitive aspects of survey methodology: Building a bridge between disciplines* (pp. 73–100). Washington, DC: National Academy Press.

Tourangeau, R. (1987). Attitude measurement: A cognitive perspective. In H. J. Hippler, N. Schwarz, & S. Sudman (Eds.), *Social information processing and survey methodology* (pp. 149–162). New York: Springer.

Tourangeau, R. (2004). Survey research and societal change. *Annual Review of Psychology*, *55*, 775–801.

Tourangeau, R., Couper, M. P., & Conrad, F. (2004). Spacing, position, and order: Interpretive heuristics for visual features of survey questions. *Public Opinion Quarterly*, *68*, 368–393.

Tourangeau, R., & Rasinski, K. (1988). Cognitive processes underlying context effects in attitude measurement. *Psychological Bulletin*, *103*, 299–314.

Tourangeau, R., Rips, L. J., & Rasinski, K. (2000). *The psychology of survey response*. New York: Cambridge University Press.

Yan, T., & Tourangeau, R. (2008). Fast times and easy questions: The effects of age, experience and question complexity on web survey response times. *Applied Cognitive Psychology*, *22*, 51–68.

Part IV

WEIGHTING ADJUSTMENTS

Introduction to Part IV

Jelke Bethlehem[a] and Mario Callegaro[b]

[a]Statistics Netherlands, The Netherlands
[b]Google UK

IV.1 Panel problems

It is the objective of most online panels to collect and publish reliable and accurate statistical information about specific populations. If the fundamental principles of probability sampling are applied, unbiased estimates of population characteristics can be computed, and also a margin of error for these estimates can be determined.

Computing estimates and margins of error for non-probability online panels are still much-debated topics. The Office of Management and Budget (2006) defines *estimation error* for nonprobability samples as "the difference between a survey estimate and the true value of the parameter in the target population" (p. 31). For estimates for nonprobability online panels, a *credibility interval*, rather than a margin of error, is a popular metrics used. Online electoral polling is a good example of its use. "The credibility interval reflects the statistical uncertainty generated by a statistical model that relies on Bayesian statistical theory" (AAPOR, 2012, p. 1).

In daily practice, ensuring that results from a panel represent the target population is often not an easy task. There are always phenomena affecting the representativity of the outcomes of online panels. One such problem is *undercoverage*, a phenomenon in which not all members of the target population are represented in the sampling frame. This can happen, for example, in the recruitment phase of an online panel for the general population if every member of the population does not have access to the Internet. (See the Introduction to Part I of this volume.)

Another problem is *nonresponse*. This phenomenon occurs when members of the target population who have been selected for the sample do not provide the required information. Nonresponse can have several causes, the most common of which are refusal, non-contact, and inability. (See the Introduction to Part II in this volume.)

Online Panel Research: A Data Quality Perspective, First Edition.
Edited by Mario Callegaro, Reg Baker, Jelke Bethlehem, Anja S. Göritz, Jon A. Krosnick and Paul J. Lavrakas.
© 2014 John Wiley & Sons, Ltd. Published 2014 by John Wiley & Sons, Ltd.
Companion website: www.wiley.com/go/online_panel

Nonresponse can occur in the recruitment phase. It can also occur in the waves of a longitudinal study panel, in which case it usually has a monotone pattern: the group of respondents decreases with each subsequent wave. Once individuals stop responding, they are lost to the panel. This type of nonresponse is usually called *attrition*.

In the case of a cross-sectional study panel, there can also be nonresponse in both phases. Nonresponse in the second phase (for specific surveys) may have a monotone pattern (attrition), but nonresponse may also be in reaction to a specific topic of one survey. Panel members who do not like a particular topic may decide not to participate in that survey but may respond to a subsequent survey.

A third problem affecting the representativity of online panels is *self-selection* (Public Works and Government Services Canada, 2008). If panel recruitment is not based on probability sampling but is voluntary, that is, it is left to individuals themselves to become a member of the panel, the researcher has no control over the composition of the panel. As a result, the panel will typically consist only of persons who like to do surveys and/or are interested in the topics of these surveys. Therefore, self-selection panels often suffer from a substantial lack of representativity.

The problem with nonresponse, undercoverage, and self-selection is that panel members are usually different from those not in the panel. Consequently, the panel is not representative of the population, which in turn prohibits valid statistical inference. Wrong conclusions about the population would be drawn from the panel. To avoid this, the results must be corrected for this lack of representativity. Weighting adjustments make up a family of commonly used techniques designed to correct for this problem. A short overview will be given here.

IV.2   Weighting adjustments

Weighting adjustments attempt to improve the accuracy of survey estimates by using auxiliary information. *Auxiliary information* is defined as a set of variables that have been previously measured in a survey and for which information on their population distribution (or complete sample distribution) is available.

By comparing the response distribution of an auxiliary variable to its population (or complete sample) distribution, it can be determined if the sample is representative of the population (with respect to this variable). If this distribution differs considerably, one must conclude that the sample lacks representativity. To correct this, adjustment weights are computed. Weights are assigned to records of the respondents. Estimates of population characteristics are then computed by using weighted instead of unweighted values. Weighting adjustments are often used to correct surveys that are affected by nonresponse. An overview of weighting adjustments can be found in Bethlehem and Biffignandi (2012) and Särndal and Lundström (2005).

We will explore various correction techniques in the rest of this chapter. Our focus is on weighting adjustments for cross-sectional study panels, but note that these techniques are equally applicable for longitudinal study panels.

Because nonresponse can occur during recruitment phase and during the subsequent surveys of the cross-sectional study panel, it would imply that two corrections are required. A possible first approach could be to ignore the two phases of nonresponse. Weights would then be obtained by directly aligning response distributions for auxiliary variables with their population distributions. However, this is not the most effective way to conduct adjustment weighting. Weighting in two steps is preferred. In the first place, recruitment nonresponse may be a different phenomenon than survey nonresponse; therefore, it may require a different model containing different variables. In the second place, there are a lot more auxiliary variables available to correct for the survey nonresponse. For many online panels, new members take a profile survey in which they answer basic demographic questions. All these variables can be used to weight the survey data. In contrast, there are often fewer auxiliary variables available for weighting adjustments in the recruitment phase.

To summarize, weighting adjustments for an online panel is a two-step process as follows:

1. Compute weights for all panel members in such a way that the panel is representative with respect to the target population.

2. For each survey, compute weights in such a way that the survey is representative with respect to the panel.

The final weights are obtained by multiplying the recruitment weights by the survey weights.

IV.3   Effective weighting

Auxiliary variables are a vital ingredient of weighting techniques, but not every auxiliary variable is effective in terms of weighting. The set of auxiliary variables used for weighting should satisfy two conditions:

1. The auxiliary variables must be able to completely explain the response behavior of the individuals.

2. The auxiliary variables must be able to completely explain the target variables of the survey.

Auxiliary variables for the recruitment phase are often scarce; therefore, it will not be easy to find auxiliary variables that satisfy both conditions. Moreover, due to the multi-purpose nature of many online panels, it will not be clear in advance what the target variables will be.

If recruitment is based on probability sampling, the main representativity problem will be caused by nonresponse. The target population will usually be well-defined, and there will be a corresponding sampling frame, which may contain auxiliary variables. For example, members for the LISS panel (Scherpenzeel, 2008) were selected by means of a random sample from the population register. Consequently, for all individuals in the sample (whether they responded or not) a set of demographic variables were available. Moreover, the Statistics Netherlands had population distributions for many more auxiliary variables for the target population of this panel. So there were a lot of opportunities for weighting adjustments.

If panel recruitment is based on self-selection, weighting techniques can be applied as well. There may, however, be a problem with the definition of the target population. Due to self-selection, it is not always clear what the target population is. Consequently, individuals who choose to join the panel may not belong to the intended target population. If the actual target population is not clear, it will not be possible to find the proper population distributions of auxiliary variables. Moreover, for self-selection panels, there is no sampling frame; therefore, this source of auxiliary variables is not available. For these reasons, it may be difficult to make effective weighting adjustments.

Baker et al. (2010) describe weighting adjustments for online panels in more detail in an AAPOR report. Their conclusion is that researchers should avoid self-selection online panels when one of the research objectives is to accurately estimate population values. The report states that there is "no generally accepted theoretical basis from which to claim that survey results using samples from non-probability online panels are projectable to the general population."

In a recent AAPOR report on non-probability sampling, the authors discuss the performance of weighting adjustment for self-selection panels (see Baker et al., 2013) and conclude that "adjustments seems to reduce to some extent, but do not by any means eliminate coverage, nonresponse, and selection bias inherent to opt-in panels."

The next section describes a number of weighting adjustment techniques. For the sake of convenience, only weighting adjustments for the recruitment phase are described. Note that the weighting adjustments for each specific survey are similar. To keep things simple, it is assumed the all individuals in the target population have access to the Internet; therefore, there are no under coverage effects. We also assume that a simple random sample has been selected from the population.

IV.4   Weighting adjustment techniques

IV.4.1   Post-stratification

Post-stratification is a well-known and often used weighting technique (see Cochran, 1977, or Bethlehem, 2002). To perform post-stratification adjustments, categorical auxiliary variables are needed. By crossing these variables, the population and sample are divided into a number of non-overlapping strata (subpopulations).

All elements in one stratum are assigned the same weight, and this weight is equal to the population proportion in that stratum divided by the sample proportion in that stratum. Suppose that crossing the stratification variables produces L strata. The number of population elements in stratum h is denoted by N_h, for $h = 1, 2, \ldots, L$. Hence, the population size (N) is equal to $N_1 + N_2 + \ldots + N_L$. The weight w_k for an element k in stratum h is now defined by

$$w_k = \frac{N_h/N}{n_h/n},$$ (IV.1)

where n_h is the number of respondents in stratum h, and n is the sample size. If the values of the weights are taken into account, the result is the post-stratification estimator

$$\bar{y}_{ps} = \frac{1}{N} \sum_{h=1}^{L} N_h \bar{y}_h$$ (IV.2)

where $\bar{y}_h$ is the response mean in stratum h. So, the post-stratification estimator is equal to a weighted sum of response means in the strata.

It can be shown that the bias of weighted estimates is small if there is a strong relationship between the target variable and the stratification variables. The variation in the values of the target should manifest itself between strata but not within strata. In other words, strata should be homogeneous with respect to the target variables. In nonresponse correction terminology, this type of missing data comes down to the missing at random (MAR) assumption.

The bias of the estimator will also be small if the variation of the response probabilities is small within strata. This implies that there must be strong relationships between the auxiliary variables and the response probability.

In conclusion, application of post-stratification will successfully reduce the bias of the estimator if proper auxiliary variables can be found. Such variables should satisfy the following four conditions:

1. They have to be measured in the survey.

2. Their population distribution must be known.

3. They must be strongly correlated with all target variables.

4. They must be strongly correlated with the response behavior.

Unfortunately, such variables are not very often available, or there is only a weak correlation. A reference survey may be an option in this situation.

IV.5   Generalized regression estimation

Post-stratification is a rather simple and straightforward weighting technique. More advanced weighting adjustment techniques are described in Bethlehem (2002) and Särndal & Lundström (2005). One such technique is *generalized regression estimation*, also known as *linear weighting*.

Generalized regression estimation assumes there is a set of auxiliary variables $X_1, X_2, \ldots$ X_p that can be used to predict the values of a target variable Y. The generalized regression estimator is defined by

$$\bar{y}_{GR} = \bar{y} + (\bar{X} - \bar{x})'b, \tag{IV.3}$$

where $\bar{y}$ is the sample mean of the target variable. $\bar{X}$ is the vector of population means of the auxiliary variables, and $\bar{x}$ is the vector of sample means of the auxiliary variables. Furthermore, b is the (estimated) vector of regression coefficients. This estimator reduces the bias if the underlying regression model fits the data well.

Post-stratification is a special case of generalized regression estimation. If the stratification is represented by a set of dummy variables, where each dummy variable denotes a specific stratum, expression IV.3 reduces to expression (Iv.2).

By rewriting expression (Iv.3), it can be shown that generalized regression estimation is a form of weighting adjustment (see, for example, Bethlehem & Biffignandi, 2012). The value of a weight for a specific respondent is determined by the values of the corresponding auxiliary variables.

Generalized regression estimation can be applied in situations other than post-stratification. For example, post-stratification by age, class, and sex requires the population distribution of the crossing of age, class, by sex to be known. If just the marginal population distributions of age, class, and sex separately are known, post-stratification cannot be applied. Only one variable can be used. However, generalized regression estimation makes it possible to specify a regression model that contains both marginal distributions. In this way, more information is used, and this will generally lead to better estimates.

Generalized regression estimation has the disadvantage that some correction weights may turn out to be negative. Such weights are not wrong but simply a consequence of the underlying theory. Usually, negative weights indicate that the regression model does not fit the data too well. Some analysis packages are able to work with weights, but they do not accept negative weights. This may be a reason not to apply generalized regression estimation.

It should be noted that generalized regression estimation will only be effective in substantially reducing the bias if the MAR assumption applies to the set of auxiliary variables used.

IV.6 Raking ratio estimation

Correction weights produced by generalized regression estimation are the sum of a number of weight coefficients. It is also possible to compute correction weights in a different way, namely, as the product of a number of weight factors. This weighting technique is usually called *raking ratio estimation, iterative proportional fitting*, or *multiplicative weighting*.

Multiplicative weighting can be applied in the same situations as generalized regression estimation as long as only qualitative auxiliary variables are used. Correction weights are the result of an iterative procedure. They are the product of factors contributed by all cross-classifications (of stratification variables). To compute weight factors, the following process has to be carried out:

1. Introduce a weight factor for each stratum in each cross-classification term. Set the initial values of all factors to 1.

2. Adjust the weight factors for the first cross-classification term so that the weighted sample becomes representative with respect to the auxiliary variables included in this cross-classification.

3. Adjust the weight factors for the next cross-classification term so that the weighted sample is representative for the variables involved. Generally, this will distort representativeness with respect to the other cross-classification terms in the model.

4. Repeat this adjustment process until all cross-classification terms have been dealt with.

5. Repeat steps 2, 3, and 4 until the weight factors no longer change.

The advantage of using multiplicative weighting is that computed weights are always positive. The disadvantage is that there is no clear model underlying the approach. Moreover, there is no simple and straightforward way to compute standard errors of weighted estimates. In contrast, a generalized regression estimation is based on a regression model that allows for computing standard errors.

IV.7 Weighting adjustment with a reference survey

In the previous section, it was shown that correction techniques are effective provided that auxiliary variables have a strong correlation with the target variables of the survey and with the response behavior. If such variables are not available, one might consider conducting a *reference survey*. A reference survey is based on a probability sample and a data collection mode that leads to high response rates and little bias, e.g., CAPI (computer-assisted personal interviewing) with laptops or CATI (computer-assisted telephone interviewing). CAPI and CATI surveys tend to have high response rates. They can be used to produce accurate estimates of population distributions of auxiliary variables. These estimated distributions can be used as benchmarks in weighting adjustment techniques.

The reference survey approach has been used by several market research organizations (see Börsch-Supan et al., 2004 and Duffy et al., 2005) to reduce the bias caused by respondents' self-selection process.

An interesting aspect of the reference survey approach is that any variable can be used for adjustment weighting as long as it is measured both in the reference survey and in the online panel. For example, some market research organizations use "webographics" or "psychographic" variables that divide the population into "mentality groups." (See Schonlau et al. (2004) for more details about the use of such variables.)

It should be noted that use of estimated population distribution will increase the variance of the estimators. The increase in variance depends on the sample size of the reference survey: the smaller the sample size the larger the variance. Therefore, using a reference survey may reduce the bias at the cost of increasing the variance.

IV.8   Propensity weighting

Propensity weighting is used by several market research organizations to correct for a possible bias in their web surveys. Examples can be found in Börsch-Supan et al. (2004) and Duffy et al. (2005). The original idea behind propensity weighting goes back to Rosenbaum and Rubin (1983, 1984), who developed a technique for comparing two populations. The technique is used to attempt to make the two populations comparable by simultaneously controlling for all variables that were thought to explain the differences. In the case of an online panel, there are also two populations: those who participate in the online panel (if asked), and those who do not participate.

Propensity scores are obtained by modeling a variable that indicates whether or not someone participates in the survey. Usually a logistic regression model is used where the indicator variable is the dependent variable and attitudinal variables are the explanatory variables. These attitudinal variables are assumed to explain why someone participates or not. Fitting the logistic regression model involves estimating the probability (propensity score) of participating, given the values of the explanatory variables.

The propensity score $\rho(X)$ is the conditional probability that a person with observed characteristics X responds, i.e.,

$$\rho(X) = P(r = 1|X)$$

It is assumed that within the strata defined by the values of the observed characteristics X_k, all persons have the same response probability. This is the MAR assumption. The propensity score is often modeled using a logit model:

$$\log\left(\frac{\rho(X_k)}{1 - \rho(X_k)}\right) = \alpha + \beta'X_k$$

Once response propensities have been estimated, they can be used to reduce a possible response bias. There are two general approaches: *response propensity weighting* and *response propensity stratification*.

Response propensity weighting is based on the principle of Horvitz and Thompson (1952) that an unbiased estimator always can be constructed if the selection probabilities are known. In the case of nonresponse, selection depends on both the sample selection mechanism and the response mechanism. The idea is to adapt the Horvitz–Thompson estimator by including the (estimated) response probabilities.

There are more advanced estimators than the Horvitz–Thompson estimator. One example is the generalized regression estimator. This estimator also can be improved by including response propensities. (For more details, see Bethlehem, Cobben, & Schouten, 2011.)

Response propensity stratification takes advantage of the fact that estimates will not be biased if all response probabilities are equal. In this case, selection problems will only lead to fewer observations, but the composition of the sample is not affected. The goal is to divide the sample in strata in such a way that all elements within a stratum have (approximately) the same response probabilities. Consequently, unbiased estimates can be computed within strata. Next, stratum estimates are combined into a population estimate.

IV.9   The chapters in Part IV

In Chapter 12, Stephanie Steinmetz, Kea Tijdens, Annamaria Bianchi, and Silvia Biffignandi explore the possibility of improving a self-selection online panel by applying weighting adjustments. They use a sample from the WageIndicator survey for their exploration. This is a continuous self-selection web survey that is conducted in 75 different countries with the objective of collecting labor-related information. For the analysis in this chapter, only the Dutch version of the survey is used.

Steinmetz et al. compare the results of their survey with the results of a different survey taken from the LISS panel. This is an online panel whose sample was selected from the Dutch population register. They use the panels in two ways. First, they use the LISS panel as a benchmark for assessing the quality of the WageIndicator survey. In their comparison of estimates for each survey, they conclude that the estimates for the WageIndicator survey are substantially biased. They note also that LISS panel estimates are not unbiased. Second, they use the LISS panel as a source of auxiliary variables for weighting adjustments. In fact, the LISS panel is used as a reference survey. The authors analyze the effects of various forms of propensity weighting.

Steinmetz et al. demonstrate that propensity weighting can help to reduce the bias, but they also conclude that the effect is rather limited and depends on the type of propensity weighting that is applied.

In Chapter 13, Weiyu Zhang takes a completely different approach to correction. Instead of applying some kind of weighting adjustment technique, she attempts to solve the nonresponse problem by imputing the values of the variables for the missing persons. Her focus is on nonresponse in specific surveys taken from an online panel. This panel already contains many survey variables for both respondents and nonrespondents, which can be used as auxiliary variables in a correction procedure. Zhang investigates an approach using a regression model where the missing values of the target variables for the nonrespondents are estimated by means of imputation.

After imputation, estimates based on the complete (imputed) sample can be compared with estimates based on just the respondents. It becomes clear that indeed uncorrected estimates may be biased. The author warns that often the explanatory power of the regression models is low, which may affect the accuracy of the imputed variables.

References

AAPOR. (2012, October 8). *AAPOR Statement*: Understanding a "credibility interval" and how it differs from the "margin of sampling error" in a public opinion poll. Retrieved July 1, 2013 from: http://www.aapor.org/Understanding_a_credibility_interval_and_how_it_differs_from_the_margin_of_sampling_error_in_a_publi.htm.

Baker, R., Blumberg, S.J., Brick, J.M., Couper, M.P., Courtright, M., Dennis, et al., (2010). Research Synthesis: AAPOR Report on Online Panels. *Public Opinion Quarterly*, *74*, 711–781.

Baker, R., Brick, J.M., Bates, N.A., Battaglia, M., Couper, M.P., Dever, J.A., Gile, K.J. & Tourangeau, R. (2013). Report on the AAPOR Task Force on Non-probability Sampling.

Bethlehem, J.G. (2002). Weighting nonresponse adjustments based on auxiliary information. In R. M., Groves, D. A., Dillman, J. L. Eltinge, & Little, R. J .A. (Eds.), *Survey nonresponse* (pp. 275–288), New York : John Wiley & Sons.

Bethlehem, J. G., & Biffignandi, S. (2012), *Handbook of web surveys*. Hoboken, NJ : John Wiley & Sons.

Bethlehem, J. G., Cobben, F., & Schouten, B. (2011), *Handbook of nonresponse in household surveys*. Hoboken, NJ: John Wiley & Sons.

Börsch-Supan, A., Elsner, D., Faßbender, H., Kiefer, R., McFadden, D., & Winter, J. (2004), *Correcting the participation bias in an online survey*. Report, University of Munich, Munich, Germany.

Cochran, W. G. (1977). *Sampling techniques* (3rd ed). New York: John Wiley & Sons.

Duffy, B, Smith, K., Terhanian, G., & Bremer, J (2005), Comparing data from online and face-to-face surveys. *International Journal of Market Research*, *47*, 615–639.

Horvitz, D. G., & Thompson, D. J. (1952), A generalization of sampling without replacement from a finite universe. *Journal of the American Statistical Association*, *47*, 663–685.

Office of Management and Budget. (2006). *Standards and guidelines for statistical surveys*. Retrieved July 1, 2013 from: http://www.whitehouse.gov/sites/default/files/omb/inforeg/statpolicy/standards _stat_surveys .pdf.

Public Works and Government Services Canada. (2008). *The advisory panel on online public opinion survey quality: Final report June 4, 2008*. Ottawa: Public Works and Government Services Canada. Retrieved July 1, 2013 from: http://www.tpsgc-pwgsc.gc.ca/rop-por/rapports-reports /comiteenligne-panelonline/tdm-toc-eng.html.

Rosenbaum, P. R., & Rubin, D. B. (1983), The central role of the propensity score in observational studies for causal effects, *Biometrika*, *70*, 41–55.

Rosenbaum, P., & Rubin, D. (1984). Reducing bias in observational studies using sub classification on the propensity score. *Journal of the American Statistical Association*, *79*, 516–524.

Särndal, C. E., & Lundström, S. (2005). *Estimation in surveys with nonresponse*. Chichester: John Wiley & Sons.

Scherpenzeel, A. (2008), An online panel as a platform for multi-disciplinary research. In I. Stoop, & M. Wittenberg (Eds.), *Access panels and online research, panacea or pitfall?* (pp.101–106) Amsterdam: Aksant.

Schonlau, M., Zapert, K., Payne Simon, L., Haynes Sanstad, K., Marcus, S., Adams, J. Kan, H., Turber, R. & Berry, S. (2004), A comparison between responses from propensity-weighted web survey and an identical RDD survey. *Social Science Computer Review*, *22*, 128–138.

12

Improving web survey quality

Potentials and constraints of propensity score adjustments

Stephanie Steinmetz[a], Annamaria Bianchi[b], Kea Tijdens[a], and Silvia Biffignandi[b]

[a]University of Amsterdam, The Netherlands

[b]University of Bergamo, Italy

12.1 Introduction[1]

The increasing popularity of web surveys has triggered a heated debate about their quality and reliability for scientific use (Bethlehem, 2010; Couper, 2000; Fricker & Schonlau, 2002; Groves, 2004; Taylor, 2005; Tuten, Urban, & Bosnjak, 2002). Arguments in favor of web surveys emphasize cost benefits, fast data collection, ease of processing results, flexibility of questionnaire design, and the potential to reach respondents across national borders. The most obvious drawback of web surveys is that they may not be representative of the population of interest. Following Couper (2000), two main types of web surveys can be distinguished: *Probability-based* web surveys, such as intercept and email requests, have the advantage of a proper sample frame which allows the drawing of a probability-based random sample from a population in which every individual has a known probability of being selected.[2] While

---

[1] The authors would like to acknowledge networking support by (WEBDATANET COST Action IS1004 as well as the support from the ex 60% Biffignandi grant, University of Bergamo).

[2] For probability-based web surveys, all members of the target population are known (via contact or email addresses).

Online Panel Research: A Data Quality Perspective, First Edition.
Edited by Mario Callegaro, Reg Baker, Jelke Bethlehem, Anja S. Göritz, Jon A. Krosnick and Paul J. Lavrakas.
© 2014 John Wiley & Sons, Ltd. Published 2014 by John Wiley & Sons, Ltd.
Companion website: www.wiley.com/go/online_panel

those surveys are also affected by typical survey errors, they can be analyzed using standard inference procedures allowing for the generalization of results across the target population. In contrast, *nonprobability* web surveys, such as entertainment and volunteer web surveys, are problematic because not every individual has a known probability of being selected, being exposed to the invitation, and accepting the invitation. Due to that, the resulting sample is problematic as the answers of volunteers may differ from non-volunteers.

To deal with these problems and to improve the quality of estimates based on a volunteer sample, different weighting techniques have been considered. Particularly, propensity score adjustment (PSA) has been proposed to statistically surmount problems of nonprobability web surveys. However, as its application can produce rather diverse results, depending on the used variables, there is no certainty as to whether the representativeness of volunteer web surveys can be improved through PSA (Bethlehem, 2009; Lee & Valliant, 2009; Lensvelt-Mulders, Lugtig, & Hubregtse, 2009; Loosveldt & Sonck, 2008; Schonlau, van Soest, Kapteyn, & Couper, 2009; Valliant & Dever, 2011). Moreover, it is not clear to what extent specific behavioral, attitudinal or lifestyle (so-called webographic questions) increase the efficiency of PSA with regard to various outcomes. Therefore, the implications of this method, particularly for (web) survey methodology, still need to be studied much more extensively.

Against this background, the present chapter focuses on nonprobability web surveys. The aim is to explore and evaluate in more detail the efficiency of different weighting techniques, in particular PSA, in adjusting biases arising from volunteer samples. The empirical application is based on the Dutch sample of the WageIndicator Survey[3] which is a multi-country, continuous volunteer web survey devoted to the collection of labor-related variables. In the analysis, the target variable is the monthly gross wage. The sample is compared with a probability-based web sample from the LISS panel (Longitudinal Internet Studies for the Social Sciences)[4] which is also used as a reference survey in the PSA application.

The chapter is divided into five sections. After a short introduction and exploration of the aims of the chapter, Section 12.2 provides an overview of findings concerning the specific problems of nonprobability web surveys and the performance of different weighting applications correcting for biases. Section 12.3 introduces the data sets and explains the applied PSA procedure. In Section 12.4, first, the biases are described comparing means of central auxiliary variables from the two used data sets with population data. This is also an essential step for selecting relevant variables for the PSA. Second, the performance of PSA is tested by comparing unweighted and weighted estimates from the volunteer web survey with the population data. Finally, in Section 12.5, the findings and the sensitivity of the results are discussed.

12.2   Survey quality and sources of error in nonprobability web surveys

When the primary purpose of a survey is to gather information about the general population, the information is useless unless it is accurate and representative. However, what does representativeness mean? As indicated by Kruskal and Mosteller (1979), the term itself is rather vague and requires a clear explanation. In this chapter, the term is used in a rather broad sense and refers to the aim that sample data gain external validity in relationship to the target population they are meant to represent. The concept of external validity is strongly (but

[3] The WageIndicator Survey is administered by the WageIndicator Foundation (Amsterdam, The Netherlands).
[4] The LISS panel is part of the MESS project (Measurement and Experimentation in the Social Sciences) and it is administered by CentERdata (Tilburg University, The Netherlands).

not exclusively) related to generalization. If a sample is adequate (i.e., probability-based), it will be representative of the desired population, and only then can findings be generalized to the population of interest. However, in this study, generalization is not approached via representative probability-based data, but by modeling the estimates using some population benchmarks. In this respect, it means that suitable weighting techniques are applied and survey results are compared to the external benchmarks. Due to that, the deficiency of the nonprobability web survey should be corrected, and the aim of making the results projectable to the general population should be reached.

As indicated before, in the case of nonprobability web surveys, representativeness and therefore the possibility to generalize have been questioned. As stated in the AAPOR report on online panels (Baker et al., 2010), it is even recommended that researchers should avoid nonprobability online panels when one of the research objectives is to accurately estimate population values. Recently, Yeager et al. (2011) concluded, based on an experimental study that, in comparison to nonprobability samples, probability samples, even those without especially high response rates, still yield the most accurate results. This, however, does not automatically imply that nonprobability samples have no value or always provide invalid findings. Studies based on nonprobability samples are mainly intended to assess whether two variables are related to each other along the lines of theoretical anticipation. Therefore, the continued use of nonprobability samples seems quite reasonable if the goal is not to document the strength of an association in a population, but rather to reject the null hypothesis that two variables are completely unrelated to each other throughout the population (Yeager et al., 2011).

The lack of representativeness in nonprobability volunteer web surveys is due to sampling bias (defined as the difference between a statistically expected and the corresponding true population value) which may be caused by three different types of problems. The first problem is related to the use of the Internet as the solely data collection mode and affects every web-based survey to the same extent. Due to the fact that not everybody in the population has Internet access web survey estimates refer in principle only to Internet users which could be different from non Internet users.[5] The extent of this problem varies across countries and depends on the Internet penetration. The second problem is related to the fact that even if only the Internet population is considered (or this population is considered as a good proxy for the whole population), volunteer web surveys often use open website invitations to get survey participants. As a consequence, no representative sample is *a priori* generated. Moreover, the probability of being exposed to such an invitation is unknown, as the probability of coming across the invitation depends on the visitors' interest in the website topic and its marketing strategy. A third problem is related to the fact that the probability of accepting the invitation, once being exposed to it, is also unknown. While the nonresponse problem is not unique to web surveys, it is quite severe in this case (see Lozar Manfreda et al., 2008), as web survey response rates tend to be lower when compared to other modes (Lynn, 2008; Shih & Fan, 2008). Moreover, the described nonresponse can arise for different reasons, such as inefficiency of response-stimulating efforts (incentives, follow-up contacts), technical difficulties (slow, unreliable connections, low-end browsers), personal problems in using a computer, and privacy and confidentiality concerns (Bosnjak & Tuten, 2003; Göritz, 2006; Heerwegh, 2005; Kaczmirek, 2008; Vehovar, Batagelj, Manfreda, & Zaletel, 2002). In particular, for volunteer web surveys there is often no way to assess the potential magnitude of the nonresponse bias, since there is generally no information on those who choose not to opt in. Due to the described problems the resulting sample is problematic as the answers of volunteers may be

---

[5] In the report of the AAPOR task-force on nonprobability sampling (Baker et al., 2013), this problem is also labeled an "exclusion bias" rather than a "coverage bias."

different from non-volunteers. This has been confirmed by previous research showing that the differences lie in socio-demographics but also in time availability, web skills, or altruism to contribute to the project (Couper, Kapteyn, Schonlau, & Winter, 2007; Fricker, 2008; Malhotra & Krosnick, 2007). In sum, it seems that, considering the absence of an adequate sampling frame and the applied self-selection recruitment methods, the basic methodological condition for generalizing for the whole population can hardly be met (Horvitz & Thompson, 1952).

To reduce the bias resulting from the outlined inferential problems, and improve the quality of web surveys, researchers employ different correction techniques, such as post-stratification weighting, linear weighting, multiplicative weighting, and more generally calibration, to correct predominantly for socio-demographic differences[6] between the (web) sample and the population under consideration (for an overview, see Bethlehem & Biffignandi, 2012; Bethlehem, 2009; Bethlehem & Stoop, 2007). A comparison of estimates based on alternative weighting schemes (Horvitz-Thompson, calibration based on Euclidean distance, and propensity scores estimates) advised that, especially for online collected data (experiment based on LISS panel data) all the proposed methods are very sensitive to the choice of the auxiliary variables. By considering late respondents and their updated socio-demographic information, however, the study shows improved estimates are expected (Bianchi & Biffignandi, 2013). As variables of interest often do not show a sufficiently strong relationship with socio-demographic/other background auxiliary variables, it has been emphasized that such weights can correct for proportionality but not necessarily for representativeness (Loosveldt & Sonck, 2008; Steinmetz, Tijdens, & de Pedraza, 2009; Steinmetz, Raess, Tijdens, & de Pedraza, 2013). For example, weighting does not solve the problem that Internet users may differ substantially from non-users in some of their attitudes (Schonlau et al., 2004; Bandilla, Bosnjak, & Altdorfer, 2003). As a consequence, PSA has been proposed to statistically overcome problems inherent in web survey data (e.g., Terhanian, Bremer, Smith, & Thomas, 2000; Lee, 2006; Rosenbaum & Rubin, 1983; 1984; Valliant & Dever, 2011). A core feature of PSA is that the imbalances between web respondents and the general population are not only captured by common auxiliary variables but also by the aforementioned "webographic" variables relating to the likelihood of web survey participation. Webographic variables measure general attitudes or behaviors that are hypothesized to differ between the web sample and the general population (Schonlau et al., 2007). There are several ways of using propensity scores in estimations. For instance, while Harris Interactive uses a kind of post-stratified estimator based on propensity score (Terhanian et al., 2000), the firm YouGovPolimetrix uses a variation of matching (Valliant & Dever, 2011).

Even though it has been emphasized that to generalize web survey results for the whole population, weights are necessary (Duffy, Smith, Terhanian, & Bremer, 2005), the implications of the different adjustment procedures, in particular PSA, are still under discussion. Commercial market research agencies seem to have "successfully" applied this correction technique to their volunteer web surveys (Terhanian et al., 2000; Terhanian & Bremer 2012). However, their application for scientific surveys has produced rather controversial results until now. Hence, it is not certain whether the representativeness of (volunteer) web surveys can be improved. Critics assume that no simple weighting factor or adjustment strategy can make on- and offline samples comparable (e.g., Malhotra & Krosnick, 2007; Vehovar & Manfreda, 1999).

---

[6] It should be underlined that, depending on the availability, not only socio-demographic but also other auxiliary variables have been used for weighting purposes.

12.3   Data, bias description, and PSA

12.3.1   Data

The analysis is based on the 2009q4 release (collected between October and December 2009) of the Dutch WageIndicator Survey (in Dutch, Loonwijzer, here abbreviated as LW) and the LISS panel (Longitudinal Internet Studies for the Social Sciences). The WageIndicator Survey is a continuous volunteer web survey running now in 75 countries (see http://www.wageindicator.org). Since 2001, it has collected information on a wide range of subjects, including basic socio-demographics, wages and other work-related topics. Most importantly, the data set also includes variables which can be considered as webographic variables. To increase the transparency of labor markets, a freely accessible salary check for different occupations has been developed. After having explored the salary check, web-visitors are encouraged to complete the international comparable questionnaire on work and wages with a lottery prize incentive. The WageIndicator Survey has been quite successful in gathering large samples over the last years. Although in some countries the number of observations of the WageIndicator Survey is larger than in national Labor Force Surveys, the question remains whether these samples are representative of the general population.

The LISS panel is a probability-based online panel in the Netherlands and consists of 5000 households, comprising 8000 individuals. The panel was drawn from the population register in collaboration with Statistics Netherlands (*Centraal Bureau voor de Statistiek, CBS*). Even though the questionnaire is completed online, all people in the sample were recruited in traditional ways (by letter, followed by telephone call and/or house visit) with an invitation to participate in the panel (for more details about the recruitment, see Scherpenzeel & Bethlehem, 2011). Households that could not otherwise participate were provided with a computer and an Internet connection.[7] In October 2009, the LISS panel members were asked to answer the LW questionnaire. The response rate was 60.5% (the monthly response of participants varies between 50%–80%). The LISS panel survey is used as a reference survey for the PSA application. This entails two main advantages: first, it provides a proper probability-based reference survey stemming from the same questionnaire (Heckman, LaLonde, & Smith, 1999). Second, survey mode effects deriving from using different modes (i.e., online, face-to-face or telephone interviews) can be excluded as both surveys are completed as a self-filling questionnaire on the computer. Several authors (Bandilla et al., 2003; Dillman, 2006; Klassen & Jacobs, 2001) have underlined that in the current age of increased survey alternatives, different survey modes often produce different results. However, that obviously does not rule out the fact that both surveys could be affected by the same (web) mode effect.

A specific feature of this chapter is to test whether the use of PSA and webographic variables is a better tool to correct for observed biases in the volunteer LW sample. Therefore, it is crucial to identify potentially relevant covariates affecting both web participation and wages (as the outcome of interest). Based on earlier research (Steinmetz, Tijdens, & de Pedraza, 2009; Steinmetz, Raess, Tijdens, & de Pedraza, 2013), for the current analyses eight additional survey questions were considered covering two categories of webographics: two questions related to the quality of life and six questions related to the quality of working life. As indicated, it is assumed that these questions are able to explain web participation but also, to some

---

[7] Panel members complete online questionnaires of about 20–30 minutes in total every month. They are paid for each completed questionnaire. One member in the household provides the household data and updates this information at regular intervals.

extent, wage differentials. Table 12.A.1 in Appendix 12.A shows the descriptive statistics for these variables. The strongest differences between the LW sample and the (unweighted) LISS sample can be observed for the variables LIFESUCC and JOBADV. On average, it seems that people in the LW consider it very important in their life to be successful in their job and to have career chances. These two variables are very likely to be the main characteristics distinguishing the visitors to the WageIndicator website from the visitors to other websites.

With respect to the target variable –the monthly gross wage – in both surveys the same wage question was asked (for more detail, see Tijdens, van Zijl, Hughie-Williams, van Klaveren, & Steinmetz, 2010) ruling out question wording effects. The wage variable was then adapted to the definition given by the population information presented by Berkhout and Salverda (2012). Moreover, in order to compare the samples to the available population information, they are restricted to employees aged between 15–64 and living in the Netherlands, working at least 3.1 hours per week. The sample also excludes employees in the agriculture and fishery sector as well as temporary agency workers, on-call work and apprentices. Additionally, unreliable values and outliers in the wage variable are excluded (hourly gross wages are restricted to vary between Euros 0.87–347.2). For the application of PSA, moreover, all missing values are eliminated. This finally results in an overall sample for the weighting analysis of 2706 respondents (N = 1693, LW sample; N = 1063, LISS sample).

12.3.2   Distribution comparison of core variables

Before adjustment techniques can be implemented, it is important to evaluate the bias by comparing both web samples with available population information. The information used in this chapter is from a study by Berkhout and Salverda (2012) providing relevant statistics in their appendix.

Starting with a bivariate comparison of the distribution of socio-demographic and employment-related covariates with population distributions, it is obvious that the LW as well as the (unweighted) LISS[8] samples are affected by typical biases (Loosveldt & Sonck, 2008; De Pedraza, Tijdens, & Muñoz de Bustillo, 2007; Van Der Laan, 2009). The differences (see Figure 12.1 and Table 12.A.2 in Appendix 12.A) reveal that in the LW sample men, people between 25–44 years of age as well as full-timers are overrepresented in comparison to the target population. A slight overrepresentation can also be observed for highly educated people, people working in sectors related to industries and services (NACE codes G–N) and in scientific occupations.

The comparison of the unweighted LISS sample with the population information reveals a somehow different bias pattern. In the LISS sample, particularly men, people between 45 and 64, highly educated people as well as people working full-time and people with a permanent contract are overrepresented. A slight overrepresentation can also be observed for people employed in the service sector (NACE codes O–U) and in high-level and scientific occupations. This confirms previous findings that the probability-based LISS panel is also affected by biases and therefore less representative of the population of interest (Knoef & de Vos, 2009; Van Der Laan, 2009; de Vos, 2010). In order to compare the distributions in more detail *Average Relative Differences*, focusing on the magnitude of the differences, are computed for the

---

[8] At this stage the unweighted LISS sample is used since this is the data provided by CentERdata. Due to the nature of the LISS panel, CentERdata does not provide any standard sample weights (for more details, see http://www.lissdata.nl/lissdata/About_the_Panel/Sample%20Weigths).

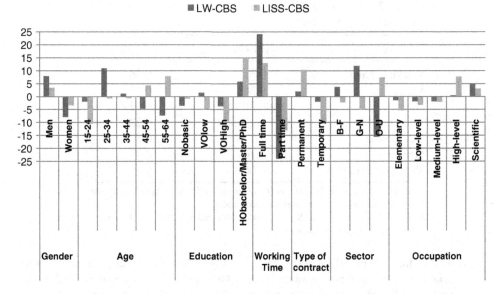

■ LW-CBS   ■ LISS-CBS

Figure 12.1   Differences between the distribution of covariates from the unweighted LW and LISS and the population (CBS), 2009q4.

Notes: The term difference refers to the percentage difference between the LW/LISS and CBS data. A positive value indicates that the percentage in the web samples (LW or LISS) in comparison to the population information (CBS) is higher (overrepresentation), whereas a negative value indicates that the percentage is lower in the web samples. Education: nobasic refers to the Dutch educational levels which can be classified as nor or basic education (ISCED 0 and 1); VOlow/high refers to lower and higher secondary education (ISCED 2 and 3); HObachelorMasterPhD refers to higher education (ISCED 5 and 6). Sector: B–F = Mining, Manufacturing and Constructions, G–N = Wholesale and retail trade, Transport and Services, O–U = Public administration, Education and other Services.

Sources: Dutch LW and LISS panel, 2009q4; CBS data for 2009q4 (see Berkhout & Salverda, 2012); own calculations.

samples of the LW and the LISS.[9] The *Average Relative Difference* for the covariate x (with p categories) is defined by $p^{-1}\sum_{j=1}^{p}|d_j|$, where d_j is the relative difference at category j.[10] It can assume any value between 0 and $+\infty$.

Table 12.1 confirms the findings of Figure 12.1 that both samples are affected by different biases. The LW shows a strong deviation from the population information with respect to working time, sector, but also age and education. The LISS sample is affected by biases related to working time, age, type of contract and type of occupation. Whether these variables should be included in a weighting model for the LW or not depends strongly on whether they are

---

[9] Alternatively, a Chi-square test could have been used. However, due to the large number of observations, every relationship, how small it may be, will turn out to be significant. Therefore, a Chi-square test is hardly informative.

[10] Suppose that a covariate x has p categories. For each category j, let $Perc_j^{LW}$ ($Perc_j^{POP}$) denote the corresponding percentage for individuals in the LW (Population). The *relative difference* for category j is defined by

$$d_j = \frac{Perc_j^{LW} - Perc_j^{POP}}{Perc_j^{POP}}$$

Table 12.1   Average Relative Differences between CBS and the
LW and the LISS, 2009.

Variable	CBS-LW	CBS-LISS
Working Time	0.61	0.46
Sector	0.30	0.15
Age	0.28	0.32
Education	0.27	0.28
Occupation	0.17	0.30
Gender	0.16	0.07
Type of contract	0.06	0.31

Note: The shaded fields indicate the main/strongest differences between the population information (CBS) and the two web samples.
Sources: Dutch LW and LISS panel, 2009; CBS data for 2009 (see Berkhout & Salverda, 2012); own calculations.

also related to the variable of interest (gross monthly wage). The bivariate analysis of the target variable and the auxiliary variables indicate (see Table 12.A.3 in Appendix 12.A) that categories of people which are overrepresented or underrepresented in the samples show also differences in mean gross monthly wages. This is expected to cause biased estimates.

The analysis also demonstrates that, in order to use the LISS panel as a reference survey for the PSA application it has to be corrected beforehand. This exploits a number of variables in the LISS to correct the LW. Moreover, it enables the calculation of better estimates as they are based on a larger set of auxiliary variables. In this context, several weights have been tested and final weights have been computed in order to calibrate the LISS sample to population counts of the target population.[11] The final set of weights chosen for the subsequent analysis is based on the ranking ratio method calibrating on marginal distributions of the variables working time, age (in five classes), type of contract, occupation (in five classes), and education (in four classes). This set of weights (scaled to have an average value of 1) has a coefficient of variation (CV) of 88.9%, ranges from 0.28–10.02 and has a Max.–Min. ratio of 36.2. These measures provide an idea of the variability of the weights, i.e., the importance given to different observations in the sample. In general, the variability of the weights should be not too large, since it may increase the variance of the corresponding weighted estimates. Meng, Duan, Chen, and Algeria (2009) indicate that the ratio from the largest to the smallest weight should not exceed 10 and is unacceptable beyond 100. In the present analysis the variability of the computed weights is considered to be acceptable.

12.3.3   Propensity score adjustment and weight specification

For the above-described biases the implementation of PSA weights assumes that the significant differences between the weighted LISS and the LW are rendered insignificant or even decrease. Consequently, the adjusted estimates of the LW should not differ significantly from the LISS survey estimates.

A propensity score is the conditional probability that a person will be in one condition rather than in another (e.g., being in the web or reference survey) when considering a set of observed covariates (X_i) used to predict that person's condition (Rosenbaum & Rubin, 1983).

---

[11] An overview of the various techniques and tested weighting variables is presented in Table in Appendix 12.A.

Like all probabilities, a propensity score ranges from 0–1. It is a very convenient method as the propensity score is a single number summarizing a person's score on all the observed covariates and weighing the importance of each background characteristic according to its ability to predict the treatment assignment. The propensity score of person i is defined as

$$\pi(x_i) = P(I_i = 1|x_i)$$

where I_i is an indicator variable for membership in the web survey, and x_i contains information on covariates which are collected in both surveys.

For the application of PSA a probability-based reference survey is needed in which each member of the population has a known probability of selection. The reference survey must contain the required covariates. As already mentioned, the weighted LISS panel is used as reference survey to estimate the propensity score model. The volunteer base weights are set to 1. In this way the fitted parameters estimate those in the model that would be estimated if all persons of the target population were in the sample. The weighted estimates of propensities will refer to the probability of volunteering within the target population (Valliant & Dever, 2011). After merging the two samples by using variables common to both data sets, an indicator variable (I_i) is defined, indicating whether the respondent belongs to the volunteer or reference survey. The volunteer sample is then adjusted to the reference sample by estimating the probability of each respondent to participate in the web survey using the selected set of covariates. The most commonly used method for computing propensity scores is the logistic regression, with the observed selected covariates as predictors and the dummy coded treatment assignment (web participation) as dependent variable.[12]

After calculating the propensity scores, they can be used in different ways for estimation (matching, stratification, covariance adjustment, or weighting on the estimated propensity score). Based on the study of Valliant and Dever (2011), three options are tested in this chapter.

In the following, denote $\hat{\pi}_i$ the estimated propensity for person i, y_i the value observed for the volunteer sample unit i, and s_V the volunteer sample with n_V units. Option A, also called the *individual propensity weights* (denoted as PS.A), uses weights equal to $1/\hat{\pi}_i$. The estimated mean of the analysis variable Y is

$$\hat{\bar{y}}_A = \frac{\sum_{s_V} y_i/\hat{\pi}_i}{\sum_{s_V} 1/\hat{\pi}_i}.$$

Option B, called *average propensity weights* (denoted as PS.B), sorts the combined data by the predicted propensity scores and divides it into G subclasses, where each subclass has about the same number of units. Based on that, an average propensity $\bar{\hat{\pi}}_g$ within subclass g ($g=1, \ldots, G$) can be computed. Finally, the weight adjustment $1/\bar{\hat{\pi}}_g$ is used for every unit in subclass g. Following Cochran (1968), five subclasses ($G = 5$) are used. The estimated mean can be written as a combination of unweighted class means as

$$\hat{\bar{y}}_B = \sum_{g=1}^{G} \hat{p}_{V_g} \hat{\bar{y}}_{V_g},$$

---

[12] Rosenbaum and Rubin (1984) suggested constructing one model that uses all the predictors for respondents who have completed data. For respondents with missing data, one or more additional models should be constructed in which only variables with complete data are predictors (more than one model if more than one group is identified with different patterns of missing data).

where $\widehat{p}_{V_g} = (n_{V_g}/\widehat{\overline{\pi}}_g)/\sum_{g=1}^{G}(n_{V_g}/\widehat{\overline{\pi}}_g), \widehat{\overline{y}}_{V_g} = \sum_{s_{V_g}} y_i/n_{V_g}$ is the unweighted mean of the volunteer sample in subclass g, and s_{V_g} is the set of volunteer sample units in subclass g with n_{V_g} units.

The final option C, called *propensity post-stratified weights* (denoted as PS.C), forms sub-classes as in option B and creates a type of post-stratified estimator. More specifically, the weight assigned to all units in subclass g of the volunteer panel is

$$\frac{\widehat{N}_g^R/\widehat{N}^R}{n_{V_g}/n_V},$$

where $\widehat{N}_g^R$ is the estimated population count of units based on the reference survey in subclass g and $\widehat{N}^R = \sum_g \widehat{N}_g^R$. These weights make the distribution of the volunteer sample equal to the reference survey target population in terms of propensity scores . The estimated mean can be written as a combination of unweighted subclass means as

$$\widehat{\overline{y}}_C = \sum_{g=1}^{G} \widehat{p}_{R_g}\widehat{\overline{y}}_{V_g},$$

where $\widehat{p}_{R_g} = \widehat{N}_{R_g}/\widehat{N}_R$. This estimator appears to be similar to the one used by Harris Interactive (see Terhanian et al., 2000).

As indicated by Valliant and Dever (2011) using ungrouped estimated propensities can produce highly variable weights (PS.A), which can inflate variances of estimated means. On the other hand, using class adjustments (as in the case of PS.B and PS.C) can substantially reduce the variability of the weights.

For the computation of the propensity scores four different logistic models are used (see Table 12.A.5 in Appendix 12.A). Model 1 is based only on socio-demographic covariates. Model 2 adds work-related covariates. To test the effect of webographic variables, Model 3 and Model 4 are defined including the former covariates together with the two categories of webo-graphic variables: attitudes about what is important in life (Model 3), plus attitudes about what is important in a job (Model 4). For the estimation of the propensity scores, different covariate selection strategies can be used, for example, via a stepwise regression excluding variables that are not significant in explaining the treatment (the significance level for removing a variable is 0.05) (e.g., Berk & Newton, 1985). An alternative can also be a one-step covariate selection based on theoretical and/or logical relevance (e.g., Duncan & Stasny, 2001). Over-all, there are no clear-cut criteria for selecting variables for propensity score model building. Moreover, Drake (1993) has shown in her simulation study that a misspecification of a PSA model, such as mistakenly adding a quadratic term or dropping a covariate, does not seriously affect the outcome. In this chapter, however, only statistically significant covariates ($p < 0.05$) are included in the calculation of the propensity scores.

Table 12.A.5 in Appendix 12.A shows the coefficient estimates for the four logistic models. The Nagelkerke pseudo R^2, the total χ^2, and the degrees of freedom of the models are also reported. The results of Model 4 show, that being young, medium and highly educated, having a full-time job and a permanent contract significantly explain web participation. Moreover, being employed in sectors related to services, health, education and public administration (NACE classes O–U), having a scientific or high-level occupation, and using the Internet frequently increase the chances of web participation. With respect to the webographic variables, people who indicate that it is important in life to be successful in the job, and that it is less important to have an income from work are more likely to participate in the LW. This also holds for people saying that it is important to have career chances in the job and a job that allows them to help others, as well as for people indicating that it is less important to have an interesting job or a high income.

For each of the four models, the three above-defined types of propensity weights are computed (PS.A, PS.B and PS.C) resulting in a total set of 12 weights. The type of model is denoted after the type of weight. For instance, the set of individual propensity weights with propensity scores based on Model 1 is denoted PS.A.1. In order to allow for a more extensive comparison of the efficiency of PSA, also a set of calibration weights is computed (Deville & Särndal, 1992). They are based on the ranking ratio method calibrating on the marginal distributions for working time, sector (NACE), age (CAL.1), + education (CAL.2), + occupation (CAL.3), + gender (CAL.4). This method is preferred to post-stratification as it allows calibration on marginals, and to linear weighting as it guarantees positive weights. A summary description of the designated calibration and propensity weights is presented in Table 12.2. For each weight the coefficient of variation (CV), minimum and maximum values (based on weights scaled to have an average value of 1), and their ratio are reported. The results clearly show that the individual propensity weights (PS.A) have a much higher variability in comparison to the others. This variability is too high as it will give too much importance to a very little number of observations. Therefore, the use of this set of weights for estimation can already be considered problematic.

The performance of the defined weights is tested by looking at differences in means comparing the unadjusted and adjusted LW results with CBS data. Adjustments for several

Table 12.2 Description of weights based on the logistic models.

Weights	CV	Min	Max	Ratio
CAL.1	72.77	0.43	4.77	11.03
CAL.2	84.27	0.37	13.30	36.33
CAL.3	85.79	0.23	11.98	51.44
CAL.4	88.28	0.18	11.75	66.95
PS.A.1	53.95	0.39	5.16	13.12
PS.B.1	48.18	0.49	2.01	4.08
PS.C.1	45.78	0.0001	0.0005	3.55
PS.A.2	92.44	0.13	14.27	**110.70**
PS.B.2	80.82	0.36	2.90	8.16
PS.C.2	87.01	0.0001	0.0008	9.04
PS.A.3	102.81	0.08	11.93	**143.14**
PS.B.3	92.57	0.31	3.28	10.61
PS.C.3	97.43	0.0001	0.0009	11.89
PS.A 4	135.51	0.02	19.80	**980.66**
PS.B.4	107.44	0.22	3.82	17.02
PS.C.4	106.92	0.0001	0.001	15.33

Sources: Dutch LW and LISS panel, 2009; own calculations.

outcomes are examined: first, the extent to which the designed weights are capable of correcting the fundamental differences for the covariates described in Figure 12.1. Furthermore, it is analyzed in how far PSA helps to improve the estimation of the mean wages (also by different groups).

12.4   Results

12.4.1   Applying PSA: The comparison of wages

Starting with the evaluation of the different weights with respect to the target variable, Table 12.3 shows the estimates for the mean monthly gross wages (using different sets of weights for the LW) as well as the corresponding standard errors and the percentage relative bias, defined by $\frac{\hat{\bar{y}}-\bar{Y}}{\bar{Y}} \times 100$, where $\bar{Y}$ is the population benchmark, and $\hat{\bar{y}}$ is an estimate (according to the different sets of weights). Following Lee and Valliant (2009), the standard errors are computed using a jackknife estimator. In the is context of volunteer samples, however, variance estimates cannot be interpreted as repeated sampling variances.

Table 12.3   Estimates for the monthly gross wage in Euro (S.D. in Euro) and percentage relative bias from LW and LISS (comparison to population information CBS, 2009).

Weights	Mean wage	Bias %
Population	**2492**	
LISS weighted	2465 (53)	−1.08
LW unweighted	3259 (125)	30.79
LW CAL.1	3025 (138)	21.38
LW CAL.2	2982 (145)	19.66
LW CAL.3	2954 (149)	18.55
LW CAL.4	3034 (174)	21.75
LW PS.A.1	3133 (136)	25.70
LW PS.B.1	3162 (131)	26.88
LW PS.C.1	3148 (129)	26.32
LW PS.A.2	2926 (135)	17.42
LW PS.B.2	2961 (138)	18.84
LW PS.C.2	2947 (142)	18.27
LW PS.A.3	3065 (199)	22.98
LW PS.B.3	3006 (166)	20.61
LW PS.C.3	2997 (170)	20.28
LW PS.A.4	2910 (155)	16.76
LW PS.B.4	3066 (193)	23.04
LW PS.C.4	3074 (192)	23.37

Sources: Dutch LW and LISS, 2009; CBS data for 2009 (see Berkhout & Salverda, 2012); own calculations.

They should be considered as reflecting the variance with respect to an underlying model which describes the volunteering mechanism.

In order to allow a comparison, Table 12.3 also reports the estimate from the weighted LISS sample. The results clearly show that the mean gross monthly (weighted) wage estimate from the LISS sample is quite accurate. The percentage relative bias is only −1.08%. For the LW, Table 12.3 reveals that weighting helps reducing the initial bias. The calibration estimates are quite similar and show a percentage relative bias between 18.5%–21.8%. The most efficient calibration weight in terms of bias reduction is the one calibrating on working time, sector (NACE), age, education, and occupation (LW CAL.3). Turning to propensity scores, it seems that, generally those based on Model 2 (including socio-demographic and work-related covariates) perform better than the others. Moreover, it seems that the inclusion of webographic variables does not improve the estimates. When looking at the application of weights by working time and education (see Table 12.A.6 in Appendix 12.A) an improvement in the estimates can be observed. Also here, propensity weights based on Model 2 (PS.A.2) seem to have a positive effect on the wage estimates of both different working time and educational groups. However, it is obvious that the improvements are not homogeneous for all groups.

In sum, it can be concluded that the corrected probability-based LISS sample still provides more accurate estimates for the target variable than the corrected volunteer LW sample (Yeager et al., 2011). However, it seems that the availability of an accurate reference survey, which can be used for the calculation of propensity weights, is capable to reduce biases in volunteer samples and therefore increase their representativeness. In this context, however, the analysis has shown that the inclusion of additional webographic variables does not increase the efficiency of weights when compared to, for instance, the used calibration weights.

12.4.2 Applying PSA: The comparison of socio-demographic and wage-related covariates

In addition to the central question whether the developed weights are capable to adjust the distribution of the target variable, it is also important to examine in how far their adjustment efficiency holds for other central covariates (socio-demographics and wage-related). In this context, only the propensity weights based on Model 2 (PS.A.2, PS.B.2, and PS.C.2) are considered as they have shown the best performances in the previous analysis. Calibration weights are not taken into account, since by definition they perfectly calibrate to the population.

Table 12.A.7 in Appendix 12.A shows the *Average Relative Differences* between the propensity-adjusted LW and the population. The comparison shows that all the selected propensity weights are able to adjust the distribution of covariates between the volunteer sample and the population, and therefore significantly reduce the differences between them. However, PS.A.2 seems to be the most effective one for balancing covariates, as the difference for at least three covariates (education, age, and type of contract) are rendered insignificant. From the previous section, it is also clear that the variability of this set of weights is too high to be used directly for the estimates. One possible solution for using this type of propensity weight could be to trim or winsorize the most extreme weights in order to reduce their variance. Censoring would introduce bias into the estimates, but the bias incurred should be more than compensated for by the variance reduction. In this respect, it should also be noted that the general calibration framework (Deville & Särndal, 1992) also allows for the computation of weights under side-conditions (as they should be between certain boundaries).

12.5   Potentials and constraints of PSA to improve nonprobability web survey quality: Conclusion

This chapter has addressed the question of how trustworthy results stemming from volunteer (nonprobability-based) web surveys can be. Can they be generalized for the whole population? Is PSA capable of remedying shortcomings?

In this context, the example of the Netherlands has shown that the selected LW sample deviates significantly from the available population information with regard to core socio-demographic and work-related covariates (in particular, working time and sector). The descriptive analysis, moreover, has revealed differences between the two samples and the selected webographic questions, indicating that in particular the importance of having career chances as well as being successful in the job are crucial for participation in a volunteer web survey. As the topic of the WageIndicator website is predominantly related to information about wages, these findings are reasonable. They show that web-visitors and survey participants are obviously searching for better wages and career opportunities. Also with respect to the main variable of interest – the mean monthly gross wage – a significant difference between the two web samples and the population information could be observed overall as well as differentiated between groups. As to the bias description, it is interesting to note that in comparison to population information, both unweighted samples (the LW and the LISS) are affected by different biases. This finding implies that representativeness should be questioned not only for nonprobability-based but also for probability-based web surveys.

Following the stricter recommendation by Valliant and Dever (2011) to use an adjusted reference survey for the PSA applications, weights have been computed for the LISS sample based on the ranking ratio method. The weighted LISS panel has been further used for the computation of propensity weights. Besides the described bias problem, it has been argued that the LISS panel seems to be the most appropriate reference survey for this analysis. The advantages are threefold: first, the LISS allows the use of the same questionnaire and the same mode, preventing at least biases streaming from different modes. Second, while a thoroughly set-up CAPI survey might appeal as a better choice for a reference survey, the increased risks of mode effects as well as the higher costs of such a survey have to be taken into account. Finally, the third advantage is that the longitudinal nature of the LISS eases updates and future analysis of the LW without the need to set up additional expensive CAPI surveys.

Based on four logistic regression models considering different sets of covariates, three types of propensity weights have been computed, resulting in a total set of 12 weights. The weights have been applied to adjust the mean monthly gross wage as well as central socio-demographic and work-related variables. The findings for the overall mean gross monthly wage have revealed that the use of propensity weights can help to reduce biases in volunteer samples. However, the effect is rather limited and depends on the type of propensity weight. In this context, the results confirm findings of Valliant and Dever (2011) that ungrouped estimated propensities produce rather highly variable weights (PS.A), whereas class adjustments (as in the case of PS.B and PS.C) reduce the variability of weights, and therefore limit the inflation of the variances of estimated means. Moreover, the analysis has demonstrated that, contrary to the expectations, the inclusion of additional webographic variables does not increase the efficiency of the weights. When turning to the effect of the computed weights on the mean income for different working time and educational groups, generally an estimated improvement could be observed. However, these improvements have not been homogeneous for all groups. Also here earlier findings are confirmed that it is rather difficult to derive a weight which corrects the observed biases for all covariates in the expected direction (Lee, 2006; Loosveldt & Sonck, 2008; Schonlau et al., 2007; Steinmetz, Tijdens, & de Pedraza, 2009; Valliant & Dever, 2011). Part of that might be explained by the fact, that the validity of the inference from nonprobability samples rests on the appropriateness of the underlying model assumptions as well as on how deviations from those assumptions affect the specific estimates (Baker et al., 2013). In the present study, there are still problems in the estimates due to the underlying assumptions. For instance, it could be expected that the estimates improve by adding other relevant variables in the model which have not be included (such as *paradata*, see Groves & Lyberg 2010).

Besides the mean income adjustment also the efficiency of selected propensity weights to correct biases of other relevant covariates has been examined. The comparison has revealed that the selected propensity weights (PS.A.2, PS.B.2, and PS.C.2) are able to adjust the distribution of covariates between the volunteer sample and the population and therefore significantly reduce the differences between them. However, due to the fact that these types of weights show too much variability (in particular PS.A.2), their direct use for estimation is problematic and demands further steps (such as trimming or winsorizing) to reduce the variance.

In addition, this study shows that the application of PSA depends on a variety of factors which have to be taken into account: first, one core aspect is the selection of a *proper reference survey*. As indicated before, even a probability-based reference survey can be affected by biases which have to be corrected before the PSA application. Alternatively, calibration weights seem to efficiently adjust a volunteer sample to the target population. However, even though it is desirable to use population information for the adjustment directly, the availability of such sources depends on several aspects, such as the topic of interest, anonymity restrictions, etc. A second important criterion for the application of PSA is the selection of *meaningful covariates*, in particular, webographic variables, related to Internet participation and the outcome of interest. While previous studies have underlined the importance of the inclusion of webographic variables in the computation of propensity weights (e.g., Schonlau et al., 2007), in the present analysis, the addition of them was rather ineffective. However, it would be misleading to conclude that webographic variables are useless. The inefficiency might be due to a limited selection in the present study or missing additional webographics, or both. Finally, the question whether the developed propensity weights could also be applied in a cross-national setting still needs to be explored. Problems might arise from cross-national differences in bias patterns (Steinmetz, Raess, Tijdens, & de Pedraza, 2013) or from difficulties in finding or conducting an adequate reference survey including the relevant covariates, having the same mode, and being conducted in the same time period.

Against this background, it can be concluded that with the availability of an accurate probability-based reference survey, the application of PSA can help reduce biases in volunteer samples. However, the effectiveness of the adjustment depends heavily on the pre-conditions described above. If no reference survey is available and has to be conducted, researchers should carefully evaluate beforehand whether the expense is worthwhile and which measures they have to undertake. With respect to the inclusion of webographic variables, at least for the target variables wages, the present analysis did not lead to the expected improvements. Even though some of the propensity weights effectively reduced the bias between the samples, the problem of increased variances has to be considered. Therefore, researchers should first test the efficiency of "simple" adjustment methods before proceeding with PSA. Finally, and to answer the question whether nonprobability-based web survey results can be generalized, it

seems that the presented adjustment approaches could offer improvements for a better generalization. In particular, when considering the advantages of volunteer web surveys, such as reduced costs, flexibility, worldwide coverage, etc. critics simply stressing the impossibility of generalizings findings of nonprobability-based web surveys should be perceived with greater reflection. As indicated at the beginning, the ability to generalize is also dependent on the purpose of a study. In this context, researchers using nonprobability surveys should be transparent regarding the method they use to evaluate the quality of the survey and critically reflect on the capability of the developed weights to improve representativeness.

Generally, the study has demonstrated that, when speaking about representativeness of surveys, researchers should also be aware that with a decline in response rates even probability-based (web) samples face the problem of self-selection (Brick, 2011; Groves, 2006). Persons who are willing to participate in a survey may always differ from those who are not participating. Figure 12.2 supports this argument for the Dutch sample, by comparing the results[13] used from the WageIndicator data with those arising from selected "representative" surveys frequently used in sociological research.[14] In all data sets only the working population aged 15–64 (employees and self-employed) has been selected.

Using the Dutch Labor Force Survey (black bar) as a benchmark for representativeness, this comparison shows that biases can be observed not only for the "unrepresentative" Dutch WageIndicator data (light gray bar) but also for each of the other "representative" surveys. With regard to most of the variable classifications also, here the comparison of the different data sets indicates that it would be exaggerated to speak about a fundamental bias in case of the WageIndicator data set. A calculated deviation score over all cells even shows that the deviation of the Dutch Labor Force Survey is smallest for the Dutch WageIndicator (a score of 8% in comparison with more than 11% for the other surveys). Against this background, it seems appropriate to recall the comment by Couper and Miller (2008) who recommend treating survey quality not as an absolute value, but evaluating it relative to other features such as the design and the stated goals of the survey. This is also in line with the notion of "fit for purpose" (Biemer & Lyberg, 2003) as a basic definition of survey quality which defines three key dimensions: accuracy, timeliness, and accessibility.

With reference to a future research agenda, the present chapter can be seen as an application and evaluation of different PSA models on real data and for a set of selected variables. However, the evaluation should assess the appropriateness of the assumptions of such models under various circumstances and for different estimates. Moreover, a rather under-researched area in nonprobability samples is variance estimation. More research in this area is essential as variance estimates are needed for inferences and for evaluation of the estimates. In addition, further sources of bias – such as repeated survey participation and high rates of satisficing – should be considered as they might have an impact on survey estimates.

Finally, the present chapter has also shed light on the difficulty to quantify the quality of a nonprobability survey. Even though established techniques for probability surveys are being used, and new techniques have been devised, the situations in which nonprobability surveys may be most appropriate still need more investigation. In this context, it is advisable to explore in more detail the proposal of the AAPOR task-force on nonprobability sampling (Baker et al., 2013) to use the Total Survey Error Model as a single framework to develop well-defined measures to assess the quality of nonprobability samples. This might also lead to further insights

[13] Results here refer to the distribution over 24 cells (2genders*2workinghours*2agegroups* 3educationgroups).
[14] The Labor Force Survey, the European Working Conditions Survey and the European Social Survey.

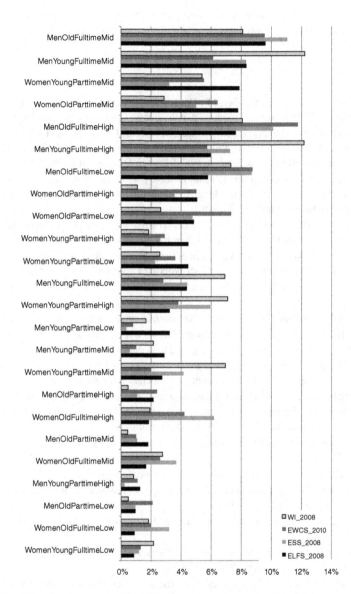

Figure 12.2 Distribution over 24 cells (2genders*2workingtime*2agegroups*3education groups) in four probability and nonprobability surveys.

Notes: The two working time groups include full-time (>=30 hours per week) vs. part-time (< 30 hours per week), the two age groups include young (16–39 years) vs. old (40–65 years), and three educational groups refer to low education (ISCED 1–2), middle education (ISCED 3–4), and high education (ISCED 5–6); the online appendix at: www .wiley.com/go/online_panel_research details how the national educational categories are classified according to ISCED.

Sources: Dutch samples (workers in dependent employment) of the European Social Survey 2008, the Labor Force Survey 2008, the five-yearly European Working Conditions Survey 2010, and the WageIndicator web survey 2008.

into underexplored topics such as ways of mixing probability and nonprobability web surveys, and the use of nonprobability web surveys in a longitudinal context.

References

Baker, R., Blumberg, S., Brick, J., Couper, M., Courtright, M., Dennis, J., Dillman, D., et al. (2010). Research synthesis: AAPOR report on online panels. *Public Opinion Quarterly*, *74*, 711–781.

Baker, R., Brick, M., Bates, N., Battaglia, M., Couper, M., Dever, J., Gile, K., et al. (2013, May). Report of the AAPOR task-force on nonprobability sampling. Retrieved from: http://www.aapor.org/AM/Template.cfm?Section=Reports1&Template=/CM/ContentDisplay.cfm&ContentID=5963.

Bandilla, W., Bosnjak, M., & Altdorfer, P. (2003). Survey administration effects? A comparison of web-based and traditional written self-administered surveys using the ISSP environment module. *Social Science Computer Review*, *21*, 235–243.

Berk, R., & Newton, P. (1985). Does arrest really deter wife battering? An effort to replicate the findings of the Minneapolis Spouse Abuse Experiment. *American Sociological Review*, *50*, 253–262.

Berkhout, E., & Salverda, W. (2012). Development of the public–private wage differential in the Netherlands, 1979–2009. SEO Discussion Paper/AIAS Working Paper. University of Amsterdam.

Bethlehem, J. (2009). *Applied Survey Methods: A Statistical Perspective*. Hoboken, NJ: Wiley-Interscience.

Bethlehem, J. (2010). Selection bias in web surveys. *International Statistical Review*, *78*, 161–188.

Bethlehem, J. & Biffignandi, S., (2012). *Handbook of Web Surveys*. Hoboken, United States: Wiley.

Bethlehem, J., & Stoop, I. (2007). Online panels: a paradigm theft? In M. Trotman et al. (Eds.), *Proceedings of the Fifth International Conference of the Association for Survey Computing: The challenges of a changing world* (pp. 113–131). Southampton: Association for Survey Computing.

Bianchi, A., & Biffignandi, S. (2013). Web Panel Representativeness. In P. Giudici, S. Ingrassia, & M. Vichi (Eds.), *Statistical Models for Data Analysis. Series: Studies in Classification, Data Analysis, and Knowledge Organization*, XV (pp. 37–44). Cham: Springer.

Bosnjak, M., & Tuten, T. (2003). Prepaid and promised incentives in web surveys. *Social Science Computer Review*, *21*, 208–217.

Brick, J. (2011). The future of survey sampling. *Public Opinion Quarterly*, *75*, 872–888.

Cochran, W. G. (1968). The effectiveness of adjustment by subclassification in removing bias in observational studies. *Biometrics*, *24*, 295–313.

Couper, M. (2000). Web surveys: a review of issues and approaches. *Public Opinion Quarterly*, *64*, 464–481.

Couper, M., Kapteyn, A., Schonlau, M., & Winter, J. (2007). Noncoverage and nonresponse in an Internet survey. *Social Science Research*, *36*, 131–148.

Couper, M., & Miller, P. (2008). Web survey methods: Introduction. *Public Opinion Quarterly*, *72*, 831–835.

De Pedraza, P., Tijdens, K., & Muñoz de Bustillo, R. (2007). Simple bias, weights and efficiency of weights in a continuous voluntary web survey. AIAS Working Paper WP 07–58, Amsterdam Institute for Advanced Labour Studies.

Deville, J., & Särndal, C. (1992). Calibration estimation in survey sampling. *Journal of the American Statistical Association*, *87*, 376–382.

de Vos, K. (2010). *Representativeness of the LISS-panel 2008, 2009, 2010*. Tilburg: CentERdata, University of Tilburg.

Dillman, D. (2006). Why choice of survey mode makes a difference. *Public Health Reports*, *121*, 11–13.

Drake, C. (1993). Effects of misspecification of the propensity score on estimators of treatment effect. *Biometrics*, *49*, 1231–1236.

Duffy, B., Smith, K., Terhanian, G., & Bremer, J. (2005). Comparing data from online and face-to-face surveys. *International Journal of Market Research, 47*, 615–639.

Duncan, K., & Stasny, E. (2001). Using propensity scores to control coverage bias in telephone surveys. *Survey Methodology, 27*, 121–130.

Fricker, R. (2008). Sampling methods for web and E-mail surveys. In N. Fielding, R. Lee, & G. Blank (Eds.), *The SAGE Handbook of online research methods* (pp. 195–216). London: Sage.

Fricker, R., & Schonlau, M. (2002). Advantages and disadvantages of Internet research surveys: Evidence from the literature. *Field Methods, 14*, 347–367.

Göritz, A. (2006). Cash lotteries as incentives in online panels. *Social Science Computer Review, 24*, 445–459.

Groves, R. (2004). *Survey errors and survey costs*. Hoboken, NJ: Wiley-Interscience.

Groves, R. (2006). Nonresponse rates and nonresponse bias in household surveys. *Public Opinion Quarterly, 70*, 646–675.

Groves, R., & Lyberg, L. (2010). Total survey error: Past, present, and future. *Public Opinion Quarterly, 74*, 849–879.

Heckman, J., LaLonde, R., & Smith, J. (1999). The economics and econometrics of Active Labour Market Programs. In O. Ashenfelter, & D. Card (Eds.), *Handbook of Labor Economics* (pp. 1865–2097, Vol. 3a). Amsterdam: Elsevier.

Heerwegh, D. (2005). Effects of personal salutations in e-mail invitations to participate in a web survey. *Public Opinion Quarterly, 69*, 588–598.

Horvitz, D., & Thompson, D. (1952). A generalization of sampling without replacement from finite universe. *Journal of the American Statistical Association, 47*, 663–685.

Klassen, R., & Jacobs, J. (2001). Experimental comparison of web, electronic and mail survey technologies in operations management. *Journal of Operations Management, 19*, 713–728.

Knoef, M., & de Vos, K. (2009). *The representativeness of LISS, an online probability panel*. Tilburg: CentERdata, University of Tilburg.

Kruskal, W., & Mosteller, F. (1979). Representative sampling, II: Scientific literature, excluding statistics. *International Statistical Review, 47*, 113–127.

Lee, S. (2006). Propensity Score Adjustment as a weighting scheme for volunteer panel web surveys. *Journal of Official Statistics, 22*, 329–349.

Lee, S., & Vaillant, R. (2009). Estimation for volunteer panel web surveys using propensity score adjustment and calibration adjustment. *Sociological Methods Research, 37*, 319–343.

Lensvelt-Mulders, G., Lugtig, P., & Hubregtse, M. (2009). Separating selection bias and non-coverage in Internet panels using propensity matching. Retrieved September 2009 from: www.surveypractice.org.

Loosveldt, G., & Sonck, N. (2008). An evaluation of the weighting procedures for online access panel surveys. *Survey Research Methods, 2*, 93–105.

Lozar Manfreda, K., Bosnjak, M., Berzelak, J., Haas, I., & Vehovar, V. (2008). Web surveys versus other survey modes: A meta-analysis comparing response rates. *International Journal of Market Research, 50*, 79–104.

Lynn, P. (2008). The problem of nonresponse. In E. de Leeuw, J. Hox, & D. Dillman (Eds.), *International Handbook of Survey Methodology* (pp. 35–55). Hillsdale, NJ: Lawrence Erlbaum Associates.

Malhotra, N., & Krosnick, J. (2007). The effect of survey mode and sampling on inferences about political attitudes and behaviour: Comparing the 2000 and 2004 ANES to Internet surveys with non-probability samples. *Political Analysis, 15*, 286–324.

Meng, X., Duan, N., Chen, C., & Algeria, M. (2009). Power-shrinkage: An alternative for dealing with excessive weights. *Joint Statistical Meetings*.

Rosenbaum, P., & Rubin, D. (1983). The central role of the propensity score in observational studies for causal effects. *Biometrika, 70*, 41–55.

Rosenbaum, P., & Rubin, D. (1984). Reducing bias in observational studies using subclassification on the propensity score. *Journal of the American Statistical Association, 79,* 516–524.

Scherpenzeel, A., & Bethlehem, J. (2011). How representative are online-panels? Problems of coverage and selection and possible solutions. In M. Das, P. Ester, & L. Kaczmirek (Eds.), *Social and behavioral research and the Internet: Advances in applied methods and research strategies* (pp. 105–132). New York: Routledge.

Schonlau, M., van Soest, A., & Kapteyn, A. (2007). Are "Webographic" or attitudinal questions useful for adjusting estimates from web surveys using propensity scoring? *Survey Research Methods, 1,* 155–163.

Schonlau, M., van Soest, A., Kapteyn, A., & Couper, M. (2009). Selection bias in web surveys and the use of propensity scores. *Sociological Methods Research, 37,* 291–318.

Schonlau, M., Zapert, K., Simon, L., Sanstad, K., Marcus, S., Adams, J., et al. (2004). A comparison between a propensity weighted web survey and an identical RDD survey. *Social Science Computer Review, 22,* 128–138.

Shih, T.–H., & Fan, X. (2008). Comparing response rates from web and mail surveys: A meta-analysis. *Field Methods, 20,* 249–271.

Steinmetz, S., Raess, D., Tijdens, K., & de Pedraza, P. (2013). Measuring Wages Worldwide: Exploring the potentials and Constraints of Volunteer Web Surveys. In N. Sappleton (Ed.), *Advancing Research Methods with New Technologies* (pp. 100–119). Hershey, PA: IGI Global.

Steinmetz, S., Tijdens, K., & de Pedraza, P. (2009). Comparing different weighting procedures for volunteer online panels: Lessons to be learned from German and Dutch WageIndicator data. AIAS Working Paper WP 09–76. Amsterdam Institute for Advanced Labour Studies.

Taylor, H. (2005). Does Internet research "work"? Comparing online survey results with telephone surveys. *International Journal of Market Research, 42,* 51–63.

Terhanian, G., & Bremer, J. (2012). A smarter way to select respondents for surveys? *International Journal of Market Research, 54,* 751–780.

Terhanian, G., Bremer, J., Smith, R., & Thomas, R. (2000). Correcting data from online survey for the effects of non-random selection and non-random assignment. Research Paper: Rochester, NY: Harris Interactive.

Tijdens, K., van Zijl, S., Hughie-Williams, M., van Klaveren. M., & Steinmetz, S. (2010). Codebook and explanatory note on the WageIndicator dataset: A worldwide, continuous, multilingual web-survey on work and wages with paper supplements. AIAS Working Paper WP 10–102. Amsterdam Institute for Advanced Labour Studies.

Tuten, T., Urban, D., & Bosnjak, M. (2002). Internet surveys and data quality: A review. In B. Batinic, U. Reips, & M. Bosnjak (Eds.), *Online social science* (pp. 7–26). Seattle, WA: Hogrefe and Huber Publishers.

Valliant, R., & Dever, J. (2011). Estimating propensity adjustments for volunteer web surveys. *Sociological Methods, 40,* 105–137.

Van der Laan, J., (2009). Representativity of the LISS panel. CBS Discussion paper (09041). The Hague: Statistics Netherlands (CBS).

Vehovar, V., Batagelj, Z., Manfreda, K., & Zaletel, M. (2002). Nonresponse in web surveys. In R. Groves, D. Dillman, J. Eltinge, & R. Little (Eds.), *Survey Nonresponse* (pp. 229–242). New York: John Wiley & Sons, Inc..

Vehovar, V., & Manfreda, K. (1999). Web surveys: Can the weighting solve the problem? In: *Proceedings of the Survey Research Methods Section, American Statistical Association (1999)*(pp. 962–967). Retrieved from: http://www.amstat.org/sections/srms/Proceedings/y1999f.html.

Yeager, D., Krosnick, J., Chang, L., Javitz, H., Levendusky, M., Simpser, A., & Wang, R. (2011). Comparing the accuracy of RDD telephone surveys and Internet surveys conducted with probability and nonprobability samples. *Public Opinion Quarterly, 75,* 709–747.

Appendix 12.A

Table 12.A.1 Descriptive summary of webographics, LW and LISS, 2009.

Variable	Description	LW Mean	S.D.	LISS Mean	S.D.
LIFESUCC	Be successful in the job	4.2	0.79	3.8	0.73
LIFEWAGE	Have an income from work	4.3	0.77	4.2	0.67
JOBHIGH	A high income	4.5	0.73	4.4	0.66
JOBADV	Good opportunities for advancement	4.1	0.87	3.7	0.84
JOBINT	An interesting job	4.5	0.75	4.4	0.63
JOBINDEP	A job that allows working independently	4.3	0.81	4.2	0.75
JOBHELP	A job that allows helping others	3.9	0.92	3.8	0.87
JOBUSE	A job that is useful to society	3.6	0.99	3.7	0.90

Notes: The shaded fields indicate for which covariates the unweighted LW sample is closer to population information than the unweighted LISS sample; for education and sector specifications, see Table 12.A.2.
Sources: Dutch LW and LISS panel, 2009; own calculations.

Table 12.A.2 Distribution of covariates from the LW, (unweighted) LISS, and the population (CBS), 2009.

Variable	CBS	LW	LISS
Gender			
Men	52.0	59.9	55.4
Women	48.0	40.1	44.6
Age			
15–24	13.7	11.7	3.1
25–34	22.1	33.0	21.3
35–44	25.9	27.9	25.2
45–54	24.8	20.5	29.1
55–64	13.5	6.9	21.4
Education			
Nobasic	5.0	1.4	4.1
VOlow	19.5	21.0	14.7
VOHigh	43.4	39.7	33.9
HObachelor/Master/PhD	32.1	37.9	47.3
Working Time			
Full-time	51.8	75.9	64.8
Part-time	48.2	24.1	35.2
Type of contract			
Permanent	78.2	80.2	88.6
Temporary	21.8	19.8	11.4

(continued overleaf)

Table 12.A.2   (*continued*)

Variable	CBS	LW	LISS
Sector			
B–F	19.6	23.4	17.3
G–N	44.1	55.9	39.0
O–U	36.3	20.8	43.7
Occupation			
Elementary	9.0	7.5	3.6
Low-level	24.1	22.2	20.8
Medium-level	36.7	34.8	34.6
High-level	21.1	21.6	28.8
Scientific	9.1	13.9	12.2

Notes: Education: nobasic refers to the Dutch educational levels which can be classified as nor or basic education (ISCED 0 and 1); VOlow/high refers to lower and higher secondary education (ISCED 2 and 3); HObachelorMasterPhD refers to higher education (ISCED 5 and 6).

Sector: B–F = Mining, Manufacturing and Constructions, G–N = Wholesale and retail trade, Transport and Services O–U = Public administration, Education and other Services

Sources: Dutch LW and LISS panel, 2009; CBS data for 2009 (see Berkhout & Salverda, 2012); own calculations.

Table 12.A.3   Unweighted estimates for the monthly gross wage in Euros from LW, LISS and the population (CBS), 2009.

Variable	CBS	LW	LISS
Total	2492	3259	3048
Gender			
Men	3168	3779	3674
Women	1760	2484	2271
Age			
15–24	712	1807	1752
25–34	2338	2911	2756
35–44	2838	3644	2955
45–54	3023	3872	3267
55–64	3047	4015	3339
Education			
No basic	1422	2534	2605
VO low	1544	2536	2195
VO high	2168	2805	2618
HObachelor/Master/PhD	3696	4165	3659
Working Time			
Full-time	3528	3549	3568
Part-time	1587	2326	2102
Type of contract			
Permanent	2805	3461	3152

(continued overleaf)

Table 12.A.3   (*continued*)

Variable	CBS	LW	LISS
Temporary	1339	2441	2236
Sector			
B–F	3133	4107	3568
G–N	2293	2974	2982
O–U	2509	3072	2901
Occupation			
Elementary	1058	1932	2206
Low-level	1552	2417	2242
Medium-level	2502	3059	2870
High-level	3571	4606	3590
Scientific	4719	3729	3893

Notes: The shaded fields indicate for which covariates the unweighted LW sample is closer to population information than the unweighted LISS sample; for education and sector specifications, see Table 12.A.2. *Sources:* Dutch LW and LISS panel, 2009; CBS data for 2009 (see Berkhout & Salverda, 2012); own calculations.

Table 12.A.4   Weights tested for adjusting the LISS panel (method, variables used in the model, and main characteristics of the weights scaled to have an average of 1).

	Method	Variables	CV	Min.	Max.	Ratio
W1	Ranking Ratio	Working time	27.27	0.80	1.37	1.71
W2	Linear	Working time + Age	69.45	0.36	4.81	13.20
W3	Ranking Ratio	Working time + Age	71.99	0.47	6.15	13.10
W4	Linear	Working time + Age + Type of contract	71.91	0.36	5.14	14.40
W5	Ranking Ratio	Working time + Age + Type of contract	75.70	0.47	7.52	16.10
W6	Linear	Working time + Age + Type of contract + Occupation	79.44	0.12	5.20	43.90
W7	Ranking Ratio	Working time + Age + Type of contract + Occupation	83.26	0.33	8.56	25.60
W8	Linear	Working time + Age + Type of contract + Occupation + Education	82.97	−0.05	5.40	106.70
W9	Ranking Ratio	Working time + Age + Type of contract + Occupation + Education	88.90	0.28	10.02	36.20
W10	Linear	Working time + Age + Type of contract + Occupation + Education + NACE	83.98	−0.13	5.50	41.30
W11	Ranking Ratio	Working time + Age + Type of contract + Occupation + Education + NACE	90.99	0.25	11.09	44.60

Sources: Dutch LW and LISS panel, 2009; own calculations.

Table 12.A.5 Estimation of web-participation, step-wise inclusion of relevant variables.

Variable	Category	M1	M2	M3	M4
Intercept		−8.33***	−8.80***	−9.04***	−8.66***
Age	55–64	−0.66***	−0.44***	−0.47***	−0.41***
(Ref. 15–24)	45–54	−0.06	−0.03	−0.01	0.04
	35–44	0.16***	0.15**	0.15**	0.21***
	25–34	0.55***	0.28***	0.27***	0.25***
Gender (Ref. Man)	Woman	−0.58***	–	–	–
Education (Ref. nobasic)	HOBachelor/	0.34***	0.36***	0.31***	0.34***
	Master/PhD				
	VOhigh	0.11*	0.09	0.11	0.13*
	VOlow	0.46***	0.45***	0.46***	0.41***
Working Time (Ref. Part-time)	Full-time		0.43***	0.42***	0.38***
Public sector (Ref. Private)	Public		−1.84***	−1.94***	−2.10***
Nace (Ref. B-F)	O-U		0.55***	0.62***	0.66***
	G-N		−0.28***	−0.32***	−0.37***
Occupation (Ref. Elementary)	Scientific		0.32***	0.25***	0.31***
	High-level		0.10	0.11*	0.11*
	Medium-level		−0.01	0.01	0.04
	Low-level		−0.04	−0.03	−0.02
Type of Contract	Permanent		0.16*	0.21**	0.21**
(Ref. Limited duration)					
Internet use freq. (Ref. No)	Yes		0.50**	0.83***	1.03***
Important in life LIFESUCC	Very important			0.67***	0.37***
(Ref. Neither nor)					
	Not at all important			−0.45***	−0.17
LIFEWAGE	Very important			−0.60***	−0.73***
(Ref. Neither nor)					
	Not at all important			0.82***	1.02***
Important in job JOBADV	Very important				0.73***
(Ref. Neither nor)					
	Not at all important				−0.66***
JOBHIGH (Ref. Neither nor)	Very important				−0.26**
	Not at all important				0.69***
JOBHELP (Ref. Neither nor)	Very important				0.26***
	Not at all important				−0.11
JOBINT (Ref. Neither nor)	Very important				−0.60***
	Not at all important				0.61**
Pseudo R^2		0.13	0.30	0.35	0.41
Wald Chi^2		348.8	803.6	993.0	1150.0
Degrees of freedom		8	17	19	29

Sources: Dutch LW and LISS panel, 2009; CBS data for 2009 (see Berkhout & Salverda, 2012); own calculations; *$p < 0.05$; **$p < 0.01$; ***$p < 0.001$

Notes: The estimations are based on the sample where all relevant variables have been included. This leads to a final N of 2706; for education and sector specifications, see Table 12.A.2.

Table 12.A.6 Estimates for the monthly gross wage in Euro and percentage relative biases (comparison with CBS, 2009).

Weighting method	Working time Part Mean	Bias (%)	Full Mean	Bias (%)	Nobasic Mean	Bias (%)	Education Volow Mean	Bias(%)	Vohigh Mean	Bias (%)	HOBAMAPhD Mean	Bias (%)
Population	1587		3528		1422		1544		2168		3696	
LISS Weighted	1859	17	3064	-15	2122	49	1880	22	2263	4	3147	-15
LW Unweighted	2326	47	3549	1	2534	78	2536	64	2805	29	4165	13
LW CAL.1	2340	47	3653	4	2536	78	2293	48	2721	26	3821	3
LW CAL.2	2333	47	3583	2	2551	79	2293	49	2725	26	3814	3
LW CAL.3	2317	46	3553	1	2487	75	2252	46	2700	25	3797	3
LW CAL.4	2451	68	3585	8	2531	88	2349	63	2787	38	3860	11
LW PS.A.1	2268	43	3527	0	3009	112	2349	52	2798	29	4087	11
LW PS.B.1	2306	45	3536	0	2534	78	2309	50	2823	30	4169	13
LW PS.C.1	2299	45	3515	0	2534	78	2258	46	2829	31	4176	13
LW PS.A.2	2230	41	3513	0	2295	61	2373	54	2716	25	3611	-2
LW PS.B.2	2271	43	3484	-1	2634	85	2366	53	2766	28	3583	-3
LW PS.C.2	2270	43	3491	-1	2638	86	2357	53	2764	28	3547	-4
LW PS.A.3	2388	51	3635	3	1988	40	2543	65	2858	32	3788	2
LW PS.B.3	2294	45	3592	2	2111	48	2637	71	2733	26	3664	-1
LW PS.C.3	2300	45	3590	2	2057	45	2654	72	2725	26	3644	-1
LW PS.A.4	2133	34	3598	2	2190	54	2280	48	2640	22	3765	2
LW PS.B.4	2300	45	3671	4	2813	98	2297	49	2807	29	3901	6
LW PS.C.4	2299	45	3675	4	2831	99	2285	48	2816	30	3916	6

Sources: Dutch LW and LISS panel, 2009; CBS data for 2009 (see Berkhout & Salverda, 2012); own calculations.

Table 12.A.7 Average Relative Differences for the weighted LW (comparison with CBS data, 2009).

Variable	CBS-LW	LW PS.A.2	LW PS.B.2	LW PS.C.2
Working Time	0.61	0.32	0.32	0.31
Sector	0.30	0.09	0.08	0.10
Education	0.28	0.05	0.10	0.09
Age	0.27	0.07	0.08	0.08
Gender	0.16	0.14	0.10	0.12
Type of contract	0.06	0.01	0.03	0.03

Notes: The shaded fields indicate for which covariates the unweighted LW sample is closer to population information than the unweighted LISS sample; for education and sector specifications, see Table 12.A.2.
Sources: Dutch LW and LISS panel, 2009; CBS data for 2009 (see Berkhout & Salverda, 2012); own calculations.

13

Estimating the effects of nonresponses in online panels through imputation

Weiyu Zhang

National University of Singapore, Singapore

13.1   Introduction

Since the early stages of public opinion research, nonresponse has been identified as an important threat to the degree to which our sample can represent the population we are interested
in (Smith, 2002: 27–28), both in demographic and opinion terms. Nonresponse means that no
information is gathered about the selected units either on the entire questionnaire or on items
of interest (Dillman Eltinge, Groves, & Little, 2002). Missing data on the entire questionnaire
is called unit nonresponse while missing data on particular items in the questionnaire is called
item nonresponse. In survey research, unit nonresponse is often caused by noncontacts, sample units that could not be reached; refusal, units that refuse to participate in the research;
and lack of ability, units that are unable to participate in the research because of either language problems or physical/mental problems (Dillman et al., 2002). The reasons for refusal
are multi-faceted (Zhang, 2008), including lack of resources (e.g., the respondents do not
have time to answer the questionnaire), an unsupportive environment (e.g., crying babies are
demanding the respondent's attention constantly), unmotivated respondents (e.g., the respondents are not motivated to complete a long questionnaire or the respondents don't like the topic
of the survey), uncomfortable interactions between respondents and interviewers (e.g., the
respondent does not like the way the interviewer asks questions), worries regarding intrusion
of privacy (e.g., the respondents are afraid of leaking private information during the survey),
and more (Lavrakas, 1993).

Online Panel Research: A Data Quality Perspective, First Edition.
Edited by Mario Callegaro, Reg Baker, Jelke Bethlehem, Anja S. Göritz, Jon A. Krosnick and Paul J. Lavrakas.
© 2014 John Wiley & Sons, Ltd. Published 2014 by John Wiley & Sons, Ltd.
Companion website: www.wiley.com/go/online_panel

Item nonresponse often takes the form of missing values on particular items. In other words, while we have some information about the surveyed respondents, we are missing data for the variable(s) of interest (Berinsky, 2004). Item nonresponse includes "don't knows" and other missing values for unspecified reasons (Zhang & Cao, 2008). There are various ways to handle "don't knows" in survey questions: (1) explicitly mention this option; (2) implicitly mention the option; (3) do not offer the option; (4) use a filter question. For instance, opinion/policy preference variables nowadays increasingly include "don't know" options in order to enhance the quality of measures (Berinsky, 2004). However, how to treat "don't know" answers in analyses is still open to debate (see Luskin & Bullock, 2011, for an example).

Researchers have documented the trend of declining response rate over the years (Brick & Williams, 2013; Singer, 2006). In addition to the problem that nonresponses increase field-work costs, the nonresponse rate becomes an analytical concern when it introduces error or bias into the survey results. Groves (2006) argues that nonresponse rate does not always lead to nonresponse bias. A key parameter determining the relationship between nonresponse rate and nonresponse bias is how strongly correlated the variable of interest is with response propensity, the likelihood of responding. Nonresponse bias in this chapter is hypothesized based on a combination of Groves' (2006) two causality models of nonresponse bias. Figure 13.1 illustrates the theoretical model of nonresponse bias used in this chapter. Y means the variable of interest in this chapter, i.e., an opinion measure. P is the response propensity to answer the question Y. Z refers to a set of causes that influence both Y and P. The Z→P relationship exists because response propensity could be explained by individual resources: Socioeconomic status, gender, race, urbanicity, and children at home were found to have ubiquitous main effects on response propensity (Groves, 2006). Previous research has shown that these resource variables can affect opinion placements as well (e.g., Berinsky & Tucker, 2006), which means the Z→Y relationship is highly plausible. Therefore, response propensity correlates with opinion variables because they share a series of common causes Z, i.e., demographics. Bias would thus exist due to this common-cause covariance.

One way to estimate nonresponse bias is through imputation. Essentially, the imputation process is one that uses what we know (e.g., basic personal information obtained from online panelists) to estimate what we do not know due to nonresponses. After we obtain the imputed values of nonrespondents, we are able to compare the imputed distribution to the observed distribution on variables of interest. Such a comparison helps us to understand how biased our measured opinions might be, due to nonresponses. There are a variety of ways to carry out imputation. Allison (2001, pp. 11–29) states that four imputation approaches are often used in estimating missing values. First, *marginal mean imputation* substitutes missing values

Z = A series of common causes
Y = The variable of interest
P = Response propensity

Figure 13.1   The theoretical model of nonresponse bias.

using marginal means of that variable. This method is known to produce biased variances and covariances. Second, *model-based imputation* regresses the variable of interest, Y, on all potential independent variables for those cases with complete data, and, then, using the estimated equation, to generate predicted values for those cases with missing data on Y. This method is approximately unbiased in large samples, if the model used is correct. However, often we are not sure about the accuracy of the models simply because the relevant explanatory variables are not available. Third, *maximum likelihood* uses the expectation-maximization algorithm to produce estimates of the means and the covariance matrix, and then inputs these estimates into regression models to get consistent estimates of the parameters of the variable of interest. The last method is *multiple imputation*, which runs multiple rounds of imputations that include a random component, and uses a formula to incorporate the multiple estimates to produce the final estimates. The latter two methods are deemed more reliable but less efficient due to the multiple steps and lack of software support.

Online panels (Baker et al., 2010), which maintain a pool of respondents who are invited to participate in research through electronic means, face unique opportunities as well as challenges with regard to nonresponses and their imputations. On one hand, participants in online panels have often already been successfully contacted through phone calls (or other means) and some of their personal information has already been obtained. This limited yet important personal information allows analysts to make basic analyses regarding who did not respond to the requests to fill out online questionnaires, and so on. In addition, a unique benefit of online panels is that panelists are not restrained to answering questionnaires only. They can be randomly assigned to online experiment conditions through different versions of questionnaires. They can even be recruited to participate in other online activities such as real-time chats. On the other hand, the estimate of the effects of nonresponse becomes further complicated as the very first step of recruiting people into online panels already has incurred nonresponse. This type of two phases of self-selection poses an additional threat to the representativeness of online panels. In addition, the nonresponse rate increases as the burden on panelists becomes heavier, when they are asked to do more than simply filling out a questionnaire.

Based on both the theoretical model of nonresponse and the unique process of online panels, it is expected that nonresponse bias may exist due to the common causes shared between response propensity and opinion placements. Therefore, the two correlations are first hypothesized:

> *H1: Demographics will be correlated with response propensity.*

> *H2: Demographics will be correlated with opinion placements.*

After identifying the possible common causes, it becomes necessary to understand how serious the nonresponse biases might be. There are three possibilities of this inquiry: First, indeed, there is little nonresponse bias; second, there is a nonresponse bias but due to the imperfect models, such bias cannot be detected; third, there is indeed a nonresponse bias. Using the limited yet important personal information as explanatory variables, regression models are built and predicted values based on such models are calculated as imputations of missing responses. After the imputation models have been operated, the differences between measured opinions and imputed opinions that include all potential respondents can be seen. A research question is asked to verify the differences:

> *RQ1: What are the differences between imputed opinions and measured opinions?*

13.2   Method

13.2.1   The Dataset

The analyses in this chapter try to estimate the problem of opinion misrepresentation due to nonresponses in online panels. Data were taken from the Electronic Dialogue 2000 project (ED2K), a multi-wave panel project lasting roughly one year. The project used a sample of American citizens aged 18 and older, drawn from a probability-based online panel (Callegaro & DiSogra, 2008) of survey respondents maintained by Knowledge Networks, Inc. of Menlo Park, California. The online panel included a large number of households (in the tens of thousands) who have agreed to accept free WebTV equipment and service in exchange for completing periodic surveys online. The Knowledge Networks Panel Sample began with a list-assisted RDD sample provided by Survey Sampling, Inc. (SSI). Samples were acquired approximately once a month to ensure that they were drawn from up-to-date databases. Numbers in the SSI sample were then matched against a database of numbers known to be in the WebTV network. These numbers were then contacted, and households were asked to participate as members of the Knowledge Network panel. In exchange for completing questionnaires (approximately 40 minutes of cumulative survey time per household per month), panelists received WebTV equipment and access free of charge. It needs to be noted that nowadays Knowledge Networks recruits panelists via Address Based Sample and non-Internet households are given a laptop PC with an Internet connection in exchange for participating in the studies (Knowledge Networks, 2013).

In February 2000, a random sample of respondents (N = 2327) was selected from the Knowledge Networks panel and invited to take part in the ED2K project. A screening survey asked about the willingness to join the project and a limited number of basic demographics such as number of children at home was obtained. The response rate to the screening survey calculated as the AAPOR RR5 (AAPOR, 2008) was 45%. A baseline survey extensively assessed respondents' opinions, policy preferences, and a variety of other variables among those who said they were willing to join. The response rate to the baseline survey calculated as the AAPOR RR5 (AAPOR 2008) was 40%. The surveys were fielded from February 28–March 23, 2000. The core of the project consisted of groups of citizens who engaged in a series of real-time electronic discussions about issues facing the unfolding 2000 presidential campaign. The multiple rounds of discussions ran for roughly a year. Each discussion lasted about an hour. The analyses in this study are based on the screening and the baseline survey and did not concern attrition (see Callegaro & DiSogra, 2008).

13.2.2   Imputation analyses

During the analyses, three steps were undertaken (see Althaus, 2003): First, regressions were used to test the relationship between all available variables (i.e., demographics) and the response propensity variables. This first step is to show the Z→P relationship (see Figure 13.1). Second, all opinion and policy preference questions were recoded into dummies: "1" means supporting while "0" means non-supporting in order to relax the assumption of linearity of linear regressions. It also helps to make the comparison of measured opinions and imputed opinions explicit because means of these opinion variables now refer to percentage of respondents who support certain opinions or policies. Logistic regressions were used to test the relationship between all available variables (i.e., demographics) and opinion placements. This second step is to show the Z→Y relationship (see Figure 13.1). By completing the first two steps, we are able to see whether there is the potential of nonresponse bias in this study. Third, imputations were done by saving predicted probabilities for everybody who was contacted to participate (N = 2327) based on the logistic regressions. In other words, this chapter uses model-based imputation as the imputation method (see Allison, 2001). Then the means of the imputed opinions were taken and compared to those who actually answered the questions.

13.3 Measurements

13.3.1 Demographics

Education was measured as years of education ($M = 13.30$, $SD = 1.84$). Similar to education, a continuous version of *age* was used in analyses ($M = 42.19$, $SD = 15.17$). Some 50% of the respondents were female. The variable was a dummy one, with "1" referring to *male* and "0" to female. 78% of the respondents were Whites, 8% Blacks, 7% Hispanic, 3% Asian, 1% American Indian, and 3% "others" or "don't know." The race variable was recoded into a dummy one, with "1" referring to *Whites* and "0" to non-Whites. *Number of children at home* was a continuous variable ($M = .29$, $SD = .78$). Whether one was *employed* (75%) or *retired* (5%) was coded as dummies. The information on whether one was a *student* was obtained (3%), and so were region (18% live in the *Northeast*) and the status of being *parents* (15%). These measure were already available when Knowledge Networks recruited the respondents to join its regular panels.

13.3.2 Response propensity

There are three measures of response propensity. *Non-acceptance* refers to online panel members who did not accept the invitation to join the project when asked in the screening survey (either refusal or non-response to our invitation). Out of 2327 people who were invited, 1054 accepted the invitation. The remaining 1273 people were thus not included in the baseline survey and considered as unit nonresponse, or missing on the entire questionnaire. Of the 1054 people who accepted the invitation, 938 completed the baseline survey. This *non-completion* is considered a unit nonresponse, too. The last measure, *missing opinion*, was derived from the policy favorability items by coding missing cases as those who did not answer the item in the baseline survey. "Don't know" options were not offered. The measure included 15 items (Range = 0–15, $M = .16$, $SD = .93$). Higher scores on this measure reflected a lower tendency to hold or report opinions about various issues. This measure is considered as item nonresponse, missing on the particular items.

13.3.3 Opinion items

A large number of policy preferences and candidate evaluations were analyzed, and they could be grouped into several categories. The baseline survey first asked respondents about their perceived problems facing the country. Then questions were asked to elicit things that they thought the federal government in Washington should do. The baseline survey also asked respondents how much either effort or money they thought the federal government should put into addressing certain issues. Finally, a series of evaluation questions were asked with regards to the two presidential candidates competing that year. Exact wording of all the questions used in the analyses can be found in Table 13.3.

13.4 Findings

The first set of analyses used regressions to test the relationship between demographics and response propensity (see Table 13.1). Table 13.1 shows that demographics do have significant

Table 13.1 Effects of demographics on response propensity based on logistic and OLS regression models.

	Non-acceptance (B)	Non-completion (B)	Missing opinion (Beta)
(Constant)	5.660***	2.799	
Education	.011	.034	−.061
Male	−.026	.254	−.051
Age	.008	−.014	.077+
Whites	−.417*	.541	−.074*
Employed	−7.459***	1.397+	−.046
Retried	22.654	.370	.005
Student	22.301	−.720+	.012
Children under 18	14.602	−.334	−.028
Parent	−9.475	.820	.026
Northeast	−.354+	.853	.017
N	2,245	938	911
R-Square	.31	.01	.004

+$p < .10$,
*$p < .05$,
**$p < .01$,
***$p < .001$.

Note: Non-acceptance and non-completion were tested using Logistic regression models. The R-squares reported are Cox & Snell R-squares. Missing opinion was tested using an OLS regression model. The R-square reported is adjusted R-square.

impact on all the three measures of response propensity. It is found that demographics have the highest power to explain non-acceptance, with an R-square of .31. People who were not Whites and who were unemployed tended not to accept the invitation to join the project. There was a marginally significant effect ($p < .10$) of being located in Northeastern America, as those who reside outside of the Northeastern region tended not to accept the invitation. It could be because that political decision-making centers (e.g., Washington, DC) are located in the Northeast and people in this area are particularly aware of policy issues. However, the prediction power of such demographics to explain noncompletion and missing opinion is very small (R-squares not higher than .01). It basically suggests that these variables are not the best explanatory variables to model non-completion and missing opinion. But among those variables that are available for analyses, being White again shows a significant effect on missing opinion, which indicates that people who were not White tended not to report their opinions. A few marginally significant coefficients were found, too. People who were employed tended not to complete the baseline survey whereas students tended to complete it. People who were older also tended not to report their opinions. It is thus summarized that hypothesis 1 is supported because demographics do show some significant correlations with response propensity while the strongest relationship was found in explaining the acceptance of the invitation to join the project.

The second set of analyses includes a series of logistic regression models on 38 policy preference and candidate evaluation items. All the 38 models show some significant effects of the demographics included. Table 13.2 is an example of the logistic regression models used to test such relationships. The question asked the respondents whether they think the federal

Table 13.2 Effects of demographics on policy favorability based on logistic regression models, an example.

Please check which of the following you think the federal government in Washington should do. (1 = checked, 0 = not checked): Make sure public school students can pray as part of some official school activity.

	B
Constant	.840
Year of education	−.169***
Male	.109
Age	.025***
Income	−.196
Employed	.441*
Retried	−.467
Student	−.075
Children under 18	.275*
Parent	−.304
Northeast	−.339+
N	912
Cox & Snell R-square	.06

$^{+}p < .10$,
$^{*}p < .05$,
$^{**}p < .01$,
$^{***}p < .001$.

government in Washington should make sure public school students can pray as part of some official school activity (1 = yes, 0 = no). The model shows that the higher education one has, the less likely is one to support such a policy. In contrast, age, being employed, and number of children at home all positively related to supporting this policy. There was a weak negative correlation between residing in the Northeast and supporting the policy. Therefore, hypothesis 2 is supported too.

The last set of analyses imputes the opinions of nonrespondents and compares the collective distribution of opinions of all potential respondents to those measured. The differences between imputed and measured opinions show how the general percentage of support might have been different if all potential respondents had given an answer. Table 13.3 shows that 5 out of 38 items displayed no difference between imputed and measured opinion distributions. Among the rest, most differences have a small range from 1% to 2%. However, in a few cases, there are relatively large differences that range from 4% to 5%. These differences were further submitted to paired sample t-tests to see whether they were statistically significant. Basically, the larger differences were significant but not the smaller differences. For instance, when asking whether one thinks the federal government should try to reduce the income differences between rich and poor Americans, the measured responses show a 48% support whereas the imputed shows a 53% support. This finding illustrates the problem that might be associated with nonresponses because it indicates that due to nonresponses, the majority opinion might seem to be a minority opinion. For contentious issues that have highly divided opinions (e.g., close elections), this type of difference could be crucial. In general, the answer to RQ1 could

Table 13.3 Differences between the imputed opinions and the measured opinions.

	Nonresponses imputed	Responses measured	Differences
How much a problem do you think each of the following is in our country today? (1 = serious problem and extremely serious, 0 = not so serious and not a problem at all)			
Amount of poverty in the United States	80	79	1
Number of criminals who are not punished	83	83	0
Amount of money Americans pay in taxes	73	73	0
Amount of illegal drug use	86	87	−1
Number of Americans losing jobs to foreign competition	66	64	2
Number of immigrants coming into the US	56	56	0
Number of people who can't afford health insurance	89	87	2
Please check which of the following you think the federal government in Washington should do. (1 = checked, 0 = not checked)			
Give tax credits of vouchers to help parents send their children to private schools.	37	40	−3+
Limit the amount of money that can be given to political parties.	75	74	0
Make it harder for a woman to get an abortion.	31	28	3+
Make all Americans pay the same percentage of their income in taxes.	50	52	−2
Make sure public school students can pray as part of some official school activity.	45	43	2
Try to reduce the income differences between rich and poor Americans.	53	48	5*
Use American military forces to stop civil wars in other countries.	11	9	2
How much effort do you think the federal government should put into each of the following? (1 = should do more, 0 = do nothing at all)			
Trying to stop discrimination against homosexuals.	31	32	−1
Eliminating many of the regulations that businesses have to follow.	37	38	−1
Restricting the kinds of guns that people can buy.	61	62	−1
Protecting patients' rights in the health care system.	77	76	1

Table 13.3   (*continued*)

How much money do you think the federal government should spend on each of the following? (1 = spend more money, 0 = no money at all)			
Maintaining a strong military defense.	45	47	−2
Providing health care for people who don't have it.	71	68	3+
Social Security benefits.	56	55	1
Programs designed to reduce the flow of illegal drugs.	54	52	2
How well does each of the following traits describe George W. Bush? (1 = very well, 0 = not at all)			
Really cares about people like me.	50	49	1
Honest	57	59	−2
Inspiring	50	48	2
Knowledgeable	64	68	−4*
How well does each of the following traits describe Al Gore? (1 = very well, 0 = not at all)			
Really cares about people like me	47	46	1
Honest	48	48	0
Inspiring	37	32	5**
Knowledgeable	72	71	1
Has George W. Bush, because of something he said or did, ever made you feel? (1 = yes, 0 = no)			
Proud	55	51	4*
Anxious	56	55	1
Enthusiastic	55	52	3+
Worried	58	59	−1
Has Al Gore, because of something he said or did, ever made you feel? (1 = yes, 0 = no)			
Proud	46	44	2
Anxious	56	55	1
Enthusiastic	44	42	2
Worried	63	62	1

+*p* < .10,
**p < .05,
**p* < .01,
***p* < .001.

Note: Each of the second and third columns presents the imputed opinions among nonrespondents and the measured opinions among respondents. The differences are listed in the fourth column. The differences were calculated by subtracting the percentage of support among respondents from the percentage of support among all people invited. Positive values mean that the imputed all generally support the issue more than respondents whereas negative values mean that the imputed all generally support the issue less than respondents. For instance, −5 means that the proportion of the imputed all who support the issue is 5% lower than the proportion of respondents who support the issue. These differences were submitted to paired sample *t*-tests to see whether they are significant.

be that nonresponse biases are not widespread according to the comparisons between imputed and measured opinions. But there are rare cases which do seem to indicate a critical bias.

13.5   Discussion and conclusion

This chapter shows that imputation could be a method to estimate the nonresponse biases that may exist in online panels. First, the theoretical model of nonresponse bias was supported because the common-cause pattern was found in the dataset. In other words, response propensity and opinion items that are of interest appeared to share common causes, including mostly demographic variables. Second, imputation analyses show that though most of the differences between imputed and measured opinions do not indicate serious discrepancies, there were a few cases where the differences were statistically significant and seemed to be critical.

However, this chapter has produced evidence that is far from conclusive for a few reasons. The most important reason is with regard to the imputation method. Due to the fact that only limited information was obtained from the initial recruitment of the online panelists, most of the imputation models are short of explanatory power, seen in the low model fit statistics (most of which are not higher than .10). The accuracy of imputed opinions thus becomes questionable when the model fits are low. One solution to the problem is to conduct additional studies to understand why people hold certain opinions (e.g., how socioeconomic status may influence one's view on health care policies; how psychological factors such as sociability may influence one's tendency to reveal opinions) and ask these questions at the very beginning of calling respondents to join the online panels. In other words, even if these people refuse to answer later requests to complete questionnaires, we can estimate their opinions based on these explanatory factors. Unfortunately, it could be seen as infeasible to the practitioners if adding questions to the recruitment survey might further lower the response rate in the first phase.

Another way to investigate the potential threat of the problem is to understand how low model fits may influence the accuracy of imputed opinions. Regression models are used in this chapter for two reasons: (1) to show that there is the potential of nonresponse bias due to the common-cause pattern; and (2) to impute missing values. So what is the potential threat when explanatory variables are few and model fits are low, in such a situation? For the first reason, as long as there are some common causes (no matter how many such causes are identified), we are able to say that the common-cause pattern exists and nonresponse bias is probable. For the second reason, the imputed values might not be very different from the measured values in terms of means because the model fits are low. The actual nonresponse biases might be even much larger than what we see here if better model fits can be obtained.

A second limitation of this dataset is that "don't know" options were not included. If a respondent really does not have an opinion on an issue, imputing his/her opinions would be misleading. "Don't know" options offer the respondents an opportunity to declare their lack of opinion on a matter. The imputation method is only appropriate when the respondents do have an opinion and their answers are missing due to reasons like negligence or self-censorship. It is suggested that future research should include "don't knows" and impute opinions that are missing for other reasons than "don't knows," as Luskin and Bullock (2011) found that discouraging "don't knows" does not help much in encouraging correct responses to political knowledge questions.

This chapter ends with an attempt to provide suggestions regarding questions to be included in the entry panel survey. According to the empirical findings, I first suggest that basic socioeconomic status variables (e.g., gender, income, age, race, education) must always be included. These variables not only serve as basic criteria against which we can evaluate the representativeness of the sample but also can be used in imputations. My second suggestion is to consider the future use of online panels when deciding what to ask in the entry survey. For example, if online panelists are expected to participate in activities that are more time-consuming than completing online surveys, questions such as how much free time one has, occupation, family obligations (e.g., marital status, number of children at home), social obligations (e.g., volunteer work, organizational membership) should be asked to understand whether refusals are simply due to time constraints. A third suggestion is to understand the nature of online surveys. If the surveys are for commercial purposes, some measures regarding consumption habits might be of help. If the surveys are highly academic, it does not hurt to include those well-established social and political measures (e.g., media exposure, political participation, social networks). An example could be a study of Internet users' psychology, which then justifies the inclusion of psychographic variables in the entry survey. However, there is always the concern about the length of the survey. The tradeoff between response rate and survey length urges researchers to clarify their purposes of online panels before compiling the panels.

Acknowledgement

This research is based on a project supported by grants to Vincent Price and Joseph N. Cappella, both of professors at Annenberg School for Communication, University of Pennsylvania. Views expressed are those of the author alone and do not necessarily reflect opinions of the sponsoring agencies and the principle investigators.

References

AAPOR (American Association for Public Opinion Research). (2008). *Standard definitions: Final dispositions of case codes and outcome rates for surveys* (5th ed.). Lenexa, KS: AAPOR.

Allison, P. (2001). *Missing data*. Thousand Oaks, CA: Sage.

Althaus, S. L. (2003). *Collective preferences in democratic politics: Opinion surveys and the will of the people*. Cambridge: Cambridge University Press.

Baker, R., Blumberg, S. J., Brick, J. M., Couper, M. P., Courtright, M., Dennis, J. M., Dillman, D. A., et al. (2010). Research synthesis: AAPOR report on online panels. *Public Opinion Quarterly*, *74*, 711–781.

Berinsky, A. J. (2004). *Silent voices: Public opinion and political participation in America*. Princeton, NJ: Princeton University Press.

Berinsky, A. J., & Tucker, J. A. (2006). "Don't knows" and public opinion towards economic reform: Evidence from Russia. *Communist and Post-Communist Studies*, *36*, 73–99.

Brick, J. M., & Williams, D. (2013). Explaining rising nonresponse rates in cross-sectional surveys. *The ANNALS of the American Academy of Political and Social Science*, *645*, 36–59.

Callegaro, M., & DiSogra, C. (2008). Computing response metrics for online panels. *Public Opinion Quarterly*, *72*, 1008–1031.

Dillman, D. A., Eltinge, J. L., Groves, R. M., & Little, R. J. A. (2002). Survey nonresponse in design, data collection, and analysis. In R. M. Groves, D. A. Dillman, J. L. Eltinge, & R. J. A. Little (Eds.), *Survey nonresponse* (pp. 3–26). New York: John Wiley & Sons, Inc.

Groves, R. M. (2006). Nonresponse rates and nonresponse bias in household surveys. *Public Opinion Quarterly, 70*, 646–675.

Knowledge Networks. (2013). Knowledge Panel design summary. Retrieved from: http://www.knowledgenetworks.com/knpanel/KNPanel-Design-Summary.html.

Lavrakas, P. (1993). *Telephone survey methods: Sampling, selection, and supervision.* Newbury Park, CA: Sage.

Luskin, R. C., & Bullock, J. G. (2011). "Don't know" means "don't know": DK responses and the public's level of political knowledge. *The Journal of Politics, 73*, 547–557.

Singer, E. (2006). Introduction: Nonresponse bias in household surveys. *Public Opinion Quarterly, 70*, 637–645.

Smith, T. W. (2002). Developing nonresponse standards. In R. M. Groves, D. A. Dillman, J. L. Eltinge, & R. J. A. Little (Eds.), *Survey nonresponse* (pp. 27–40). New York: John Wiley & Sons, Inc.

Zhang, W. (2008). Noncooperation rate. In P. Lavrakas (Ed.) *Encyclopedia of survey research methods.* London: Sage.

Zhang, W. & Cao, X. (2008). Nonresponse rates. In P. Lavrakas (Ed.), *Encyclopedia of survey research methods.* London: Sage.

Part V

NONRESPONSE AND MEASUREMENT ERROR

Introduction to Part V

Anja S. Göritz[a] and Jon A. Krosnick[b]
[a]*University of Freiburg, Germany*
[b]*Stanford University, USA*

V.1   Nonresponse and measurement error

Part V of the book deals with a possible error tradeoff in online panels. One error is non-response. Panel managers try to lower nonresponse through special efforts directed towards hard-to-reach and reluctant sample members such as repeated contact and refusal conversion attempts. Another error is measurement error. Panel managers try to combat measurement error by optimally designing the questionnaire in terms of attractiveness, clarity, usability, and low respondent burden.

But are nonresponse and measurement error independent of each other? On the one hand, some of the techniques that combat nonresponse such as offering a reward might also lower measurement error. On the other hand, response-enhancing techniques directed towards hard-to-reach and reluctant sample members might bring participants into a study who are less motivated and/or less skilled, which might result in increased measurement error. Two chapters look at these two errors and especially at their relationship, thus achieving a much needed view on total survey error and deriving balanced recommendations for both researchers and practitioners.

In Chapter 14, Neil Malhotra, Joanne Miller, and Justin Wedeking compare cross-sectional online panel surveys to traditional surveys with regard to nonresponse and measurement error. In Chapter 15, Caroline Roberts, Nick Allum, and Patrick Sturgis work on two longitudinal online panel surveys, which allows these authors to examine panel conditioning as a type of measurement error that might strike with repeated survey participation.

In spite of addressing the same topic and using similar independent and dependent variables the two chapters have a small overlap in underlying data. Yet both chapters reach similar conclusions. From having two chapters that complement each other so nicely we can be more confident in the stability of their conclusions.

Online Panel Research: A Data Quality Perspective, First Edition.
Edited by Mario Callegaro, Reg Baker, Jelke Bethlehem, Anja S. Göritz, Jon A. Krosnick and Paul J. Lavrakas.
© 2014 John Wiley & Sons, Ltd. Published 2014 by John Wiley & Sons, Ltd.
Companion website: www.wiley.com/go/online_panel

14

The relationship between nonresponse strategies and measurement error

Comparing online panel surveys to traditional surveys

Neil Malhotra[a], Joanne M. Miller[b], and Justin Wedeking[c]

[a]*Stanford University, USA*
[b]*University of Minnesota, USA*
[c]*University of Kentucky, USA*

14.1   Introduction

As recent societal, political, and technological changes have contributed to declining response rates in face-to-face and telephone surveys (Brick & Williams, 2013; Curtin, Presser, & Singer, 2005; de Leeuw & de Heer, 2002; Groves & Couper, 1998; Singer, 2006), one of the most daunting challenges facing survey researchers is nonresponse. For instance, technological advances such as caller identification and cellular phones have made it difficult to reach sampled individuals and people have been increasingly unwilling to participate in interviews when contacted (Groves, 1989). Many researchers have become alarmed at these trends because inferential sample statistics assume 100% response (Brehm, 1993; Groves, 2006; Groves & Couper, 1998). Consequently, there has been great emphasis on boosting response rates via

Online Panel Research: A Data Quality Perspective, First Edition.
Edited by Mario Callegaro, Reg Baker, Jelke Bethlehem, Anja S. Göritz, Jon A. Krosnick and Paul J. Lavrakas.
© 2014 John Wiley & Sons, Ltd. Published 2014 by John Wiley & Sons, Ltd.
Companion website: www.wiley.com/go/online_panel

aggressive callback and refusal conversion strategies (Curtin et al., 2005), along with incorporating various interviewer and respondent incentives (Fowler, 1993; Singer, 2002). Similarly, administrators of Internet surveys have sought to increase the completion rate (i.e., the percentage of invited panelists who complete the online questionnaire) via reminders and incentives in order to boost cumulative response rates (Callegaro & DiSogra, 2008; Göritz & Crutzen, 2012). Although they effectively decrease nonresponse (albeit not completely), some of these procedures can increase the cost of administration. Additionally, they may compromise data quality by increasing random or systematic measurement error if hard-to-reach respondents are not fully engaged with the survey, and instead treat it like a coerced task.

This chapter engages this last possibility by exploring the implications of strategies aimed at increasing response rates for random and systematic measurement error. Specifically, we examine the impact of attempts to capture hard-to-reach respondents on various expressions of satisficing that increase systematic measurement error (don't know/no opinion responses, midpoint selection, nondifferentiation, and selecting the first reasonable response) and random measurement error (mental coin flipping). We compare the satisficing behavior of reluctant respondents across three survey modes: Internet, phone, and face-to-face. As we review in more detail below, nonresponse strategies may be more problematic for Internet surveys where there is not a live interviewer present to engage and motivate the respondent. On the other hand, because the task of completing a survey over the web is often separated from initial recruitment, the effect of reluctance on satisficing may be muted. The direction of moderation by mode is therefore theoretically unclear, requiring empirical investigation.

There is, of course, a compelling rationale for implementing nonresponse strategies (see Brehm, 1993, for a comprehensive review of the impact of nonresponse on scientific research). Namely, the assumption is that these methods will increase representativeness by boosting response rates (depending on the size of the response rate increase and the characteristics of the respondents added to the sample). In this chapter, we examine a research question posed by Groves et al.: "When efforts to interview reluctant respondents succeed, do they provide responses more tainted by measurement error?" (2004a: 195).[1] In addition, we assess whether the answer to this question varies by mode of administration. This is a crucial part of what Weisberg (2005) terms the "total survey error approach," which focuses on the trade-offs between survey errors and costs. Thus, our study connects two separate lines of research (representing two distinct sources of survey error) that have important implications for one another: methods to increase representativeness and methods to decrease measurement error.

14.2   Previous research and theoretical overview

Previous research on nonresponse has generally studied the effects of increased response rates on the demographic representativeness of the sample and its effects on the distribution of responses to survey questions (e.g., Brehm, 1993; Keeter, Miller, Kohut, Groves, & Presser, 2000; Keeter, Kennedy, Dimock, Best & Craighill, 2006; Teitler, Reichman, & Sprachman, 2003; Heerwegh, Abts, & Loosveldt, 2007; Peytchev Baxter, & Carley-Baxter, 2009;

[1] Some scholars (e.g., Groves et al., 2004a) refer to such respondents as "reluctant." We use this term to maintain consistency with the literature even though we do not have concrete evidence that they are, in fact, reluctant responders. They may lack the motivation to be conscientious but they may also (or instead) lack the ability to be conscientious. Additionally, nonresponse strategies (such as refusal conversions and multiple callback attempts) do not directly assess a respondent's stated willingness to participate in the survey. We also refer to reluctant respondents as "difficult-to-obtain."

Curtin et al., 2000; Merkle & Edelman, 2002; Groves, 2006; Groves & Peytcheva, 2008; Groves, Presser, & Dipko, 2004b; Pew Center, 2012). The research reported in this chapter explores a separate possible effect of strategies aimed at increasing response rates – that systematic and/or random measurement error may increase as difficult-to-obtain respondents are introduced into the sample. Specifically, regardless of the effect on the *distributions* of survey variables, the *quality* of the data may be compromised as response rates increase. It is possible that nonresponse strategies may bring respondents into the sample who are more likely than their easy-to-obtain counterparts to engage in less-than-optimal question answering strategies, thus increasing one type of measurement error (either systematic or random) that is due to respondents (as opposed to measurement error that can come from other sources, such as the interviewer, the mode of interview, or the instrument). This form of measurement error is often referred to as "response bias" (Groves, 1989).

The hypothesis that data quality may be compromised as response rates increase is derived from theories that break down the question answering process into its component parts. According to Tourangeau (1984; see also Tourangeau & Rasinski, 1988), there are four steps respondents must complete in order to answer a survey question optimally: (1) determine the intent of the question; (2) search memory for relevant information; (3) integrate information into a summary judgment; and (4) translate that judgment into the response alternatives provided. Performing all of these steps less thoroughly, or skipping one or more of the steps altogether, is known as satisficing (Krosnick, 1991).

There are a number of factors that determine whether a person will choose to satisfice rather than optimize, including task difficulty, respondent ability, and respondent motivation (Krosnick, 1991). Aspects of the response task that make it more difficult, such as length of the interview, unclear questions, unclear response alternatives, or long lists of response alternatives, may increase the likelihood that respondents will satisfice (e.g., Krosnick & Alwin, 1987). Cognitive sophistication, domain-relevant expertise, and the degree to which respondents have crystallized domain-relevant attitudes are all indicators of respondent ability that may impact satisficing. And aspects of respondent motivation such as the degree to which respondents enjoy engaging in effortful cognitive tasks (i.e., need for cognition, Cacioppo & Petty, 1982, 1984) and the degree to which the topic of the survey or survey questions is personally relevant, important, or interesting also impact satisficing.

There are a variety of response strategies that satisficers can use to get through a survey without expending a substantial amount of cognitive effort. Weak satisficers (who complete each of the four response steps, but less thoroughly than optimizers) may select the first reasonable alternative from a list (thus evidencing response order effects in favor of options presented early in a list), or may agree with any assertion an interview makes (thus evidencing a confirmatory bias). Strong satisficers (who skip one or more of the steps altogether) may: (1) fail to differentiate their responses to a list of questions with the same response options, (2) endorse the status quo instead of endorsing change, (3) say "don't know" or opt out of a question rather than stating an opinion (hereafter, DK/NO); (4) select a midpoint response that reflects a non-opinion; or (5) choose among response alternatives randomly (known as mental coin-flipping; see Krosnick, 1991, for more details on these strategies).

We argue that methods used to increase response and completion rates – doing refusal conversions and making a high number of contact attempts (via phone, email, or face-to-face) to hard-to-reach respondents – will bring respondents into the sample who have less motiva-tion (and possibly less ability) to optimize. Respondents who initially refuse to participate in a survey may have done so because they feel that they do not have the requisite knowledge/

ability to complete the survey (which may or may not correspond to actual ability). Alternatively, they may resent the fact that they lost control of a situation and were coaxed into participating. According to Reactance Theory (Bem, 1966; Bem & Bem, 1981), people do not like to think that their freedom to do what they want is being violated. To reassert their freedom, reluctant respondents may try to actively sabotage the survey by providing non-sensible answers, or, more likely, they will satisfice by providing "good enough" answers without much cognitive engagement. Alternatively, they may agree to participate simply to get the interviewer to stop contacting them. If their goal in assenting to be interviewed is to prevent further bothersome contacts, they may satisfice to complete the interview as quickly as possible. Yet another possibility is that they may be more susceptible to memory effects that introduce measurement error. Groves and Couper (1998) suggest that even when people have sufficient motivation to agree to an interview (perhaps to avoid the unpleasant task of saying "no" and continued interactions), they may have insufficient motivation to complete the more effortful task of optimizing their survey responses.

Respondents recruited via numerous contact attempts or email reminders may similarly lack sufficient motivation, or, alternatively (or in combination), they may lack sufficient ability (due to time constraints) to optimize. The goal of multiple contact attempts is to finally "catch" hard-to-reach respondents at an opportune moment. These respondents, though they have agreed to complete the interview when they are finally "caught," might very well still be too busy at the time of the interview to thoroughly engage in the survey. In sum, reluctant respondents may lack the motivation and/or ability to optimize, and may therefore satisfice by using one or more of the low-effort strategies described above.

Consistent with our reasoning, some studies have suggested that measurement error increases as survey organizations engage in more strenuous efforts to increase response rates. In an early analysis, Cannell and Fowler (1963) found that respondents brought into the sample through such strategies were more likely to misreport factual information. Other studies have similarly raised the concern of whether successfully supplementing the respondent pool via persuasive efforts may introduce measurement error (e.g., Biemer, 2001; Groves & Couper, 1998).

Recent research that has examined the impact of strategies aimed at reducing nonresponse on measurement error has considered the following types of error: item nonresponse, reports of round values, classification errors, inconsistencies between interviews, differences between survey and administrative records data, nondifferentiation, no opinion or "don't know" responses, straight-lining, and a count of the number of extreme plus middle responses. The extant literature is quite inconsistent. Some studies have found that nonresponse strategies increase measurement error (e.g., Fricker & Tourangeau, 2010; Kreuter, Muller & Trappmann, 2010), others find no relationship (e.g., Göritz & Crutzen, 2012; Kaminska, McCutcheon, & Billiet, 2010; Sakshaug, Yan, & Tourangeau, 2010), and still others present mixed results (e.g., Olson, 2006; Olson & Kennedy, 2006). In an attempt to provide coherence to this area of research, a meta-analysis by Olson et al. (2008) found mixed evidence for the hypothesis that adding "high recruitment effort" respondents would lead to worse data quality, noting it tended to occur for individual measures rather than the entire general questionnaire and for behavioral items rather than attitudinal items. Thus, Olson, Feng, and Witt (2008) suggest this means the relationship is item-specific. This chapter builds on the extant literature by examining how mode of interview may moderate the potential relationship between reluctance and satisficing. As explained below, in one of our datasets, respondents were randomly assigned to mode so as to avoid confounding sampling technique with mode.

Given the past research, we suspect that strategies aimed at increasing response rates by conducting refusal conversions and making multiple contact attempts will have a negative effect on data quality because they add respondents who are more likely to satisfice. However, the effects may not be uniform across all types of satisficing, especially when interview mode is taken into account.

14.3   Does interview mode moderate the relationship between nonresponse strategies and data quality?

On balance, research suggests that the level of satisficing in a survey depends, in part, on the interview mode, and that different types of satisficing are more prevalent depending on mode. For example, in their comparison of the response quality of survey data obtained via the telephone and the Internet,[2] Chang and Krosnick (2009) theorize that satisficing may be lower in the telephone mode due to the presence of an interviewer who may create a sense of accountability. Further, telephone interviewing does not require literacy/reading comprehension among respondents and therefore may increase the quality of responses. On the other hand, the Internet mode's lack of an interviewer may reduce socially desirable responding. In addition, the Internet mode allows respondents to complete the survey at their own pace, and enables them to re-read questions and response options. This reduces the need for respondents to hold items in working memory, thus reducing cognitive burden and possibly increasing response quality as a result. Chang and Krosnick (2009) find that respondents in the telephone mode evidenced higher midpoint responding, greater socially desirable responses, and greater error variance, and less concurrent and predictive validity than the Internet mode.

Consistent with these findings, studies that randomly assign respondents to complete the same survey via computer or with an interviewer (see Chang & Krosnick, 2009, for a more comprehensive review) find data quality to be higher in the computer mode (e.g., Chang and Krosnick, 2010; Rogers et al., 2005). Holbrook, Green, and Krosnick (2003) found that the telephone mode showed more evidence of satisficing (as measured by "don't know" responding, nondifferentiation, and acquiescence) than the face-to-face mode, indicating that the physical presence of an interviewer does more to increase respondent accountability than an interviewer's voice on the phone. Other research indicates greater socially desirable responding when the survey is administered via face-to-face interview compared to audio-CASI (Newman et al., 2002), and that there is more "don't know" responding in the Internet mode compared to face-to-face, especially when the "don't know" response is explicitly shown on the computer screen (Smith, 2003; cf. Dennis, Li, & Hadfield, 2007).

How might nonresponse strategies interact with mode to affect data quality? On the one hand, difficult-to-obtain respondents may lack sufficient motivation (and/or ability) so as to eliminate any apparent data quality advantage of one mode over another, in which case we would see no interaction between mode and nonresponse strategies. On the other hand, mode differences may exacerbate the underlying differences between reluctant and willing

---

[2] Chang and Krosnick (2009) compare telephone responses obtained via probability sampling with Internet responses obtained via probability and nonprobability sampling. Here and elsewhere we focus only on the findings from comparisons of probability sample surveys so as not to confound the effect of mode with the effect of sampling strategy (accordingly, our study, described below, also focuses only on mode differences among probability sample surveys).

respondents. The direction of such an effect depends, in part, on the (as yet unknown) characteristics of difficult-to-reach respondents, who may be less motivated or have less ability than their easy-to-reach counterparts. If the presence of an interviewer increases accountability enough to motivate potentially unmotivated difficult-to-reach respondents, then we might expect to see lower data quality among reluctant respondents in the Internet mode (compared to the face-to-face and telephone modes). However, if the self-pacing nature of the Internet mode is enough to reduce any ability differences between the two types of respondents, then we might expect to see lower data quality in the face-to-face and telephone modes. In addition, reluctance is more proximal in the case of telephone interviewing; an interviewer usually commences with the interview immediately after "catching" or "convincing" a reluctant respondent. On the other hand, a reluctant respondent who agrees to join an Internet panel or complete a face-to-face interview is more distal from the interview, potentially allowing the effects of reluctance to wear off and allowing the respondent to feel more comfortable with her decision to participate.

Alternatively, it may be the case that the reluctance-mode interaction depends on the type of satisficing examined (with reluctance exacerbating mode effects that have been found in previous research). For example, consistent with past research, we may expect to see greater midpoint responding among reluctant respondents in the modes with an interviewer present (via telephone or face-to-face) and greater "don't know" responding among reluctant respondents in the Internet mode compared to telephone or face-to-face. This is because respondents interviewed over the telephone may select midpoints to provide a façade of compliance or engagement that is unnecessary in the Internet setting. Our analyses are primarily designed to test the effect of nonresponse strategies on satisficing in online panels. We also examine whether mode (Internet vs. phone and Internet vs. face-to-face) interacts with nonresponse strategies to affect data quality, as well as the sign of the interaction across a variety of forms of strong (nondifferentiation, DK/NO responding, midpoint responding, and mental coin-flipping) and weak (choosing the first reasonable response) satisficing.

14.4   Data

We analyze three data sources, each with a unique set of advantages and disadvantages. Via the three studies, we hope to triangulate a reliable answer to the question of whether reluctant respondents are more likely to satisfice, and whether this relationship is stronger for some modes than others.

14.4.1   Study 1: 2002 GfK/Knowledge Networks study

We analyze the *Survey on Civic Attitudes and Behaviors after 9/11*, collected by GfK/Knowledge Networks (hereafter, GfK/KN) between January and March of 2002 (see Dennis et al., 2005, for another analysis of this dataset). A probability sample of respondents was randomly assigned to complete the same survey via telephone or Internet. The dataset is ideal for our purposes because we can examine mode differences that are not confounded by sampling technique.[3] GfK/KN uses list-assisted RDD sampling to recruit participants for its Internet panel (see Pineau & Dennis, 2004). People who had previously

---

[3] Response rates did differ across modes. Accordingly, we also assess the robustness of our results controlling for a set of demographic variables: age, education, gender, and race (see online appendix 4 at: www.wiley.com/go/online_panel_research).

agreed to be part of the panel were randomly assigned to receive an email request to complete the survey on the web or a phone request to complete the survey on the phone with an interviewer. A third group of respondents who had previously declined to participate in the KN panel were contacted and asked to complete the survey over the phone. We only examine the two groups of "panel acceptors" so as to eliminate any potential confounding factors and restrict the sample to respondents from a common sampling frame.

The sample size for the telephone portion was 300 (of 477 initially contacted, yielding a completion rate of 62.9%); these cases were administered by RTI International. The sample size for the Internet portion was 2979 (of 3627 initially contacted, yielding a completion rate of 82.1%; see Dennis, Chatt, Li, Motta-Stanko, & Pulliam, 2005).

Study 1 enabled us to assess the relationship between reluctance and several measures of satisficing (DK/NO responses, midpoint selection, nondifferentiation, mental coin flipping). However, this dataset did not contain any experimental manipulations of response order, which is less than ideal since experimental manipulations enable the researcher to distinguish satisficing from genuine responses. Additionally, in Study 1 we are only able to compare Internet to telephone as there were no cases conducted face-to-face. Finally, Study 1 was conducted in 2002 when Internet interviewing was still a relatively new technology, and the results may not speak to the more advanced state of Internet interviewing and panel management today.

14.4.2   Study 2: 2012 GfK/KN study

To determine whether the findings from the *Survey on Civic Attitudes and Behaviors after 9/11* replicate in a more recent dataset, we also analyzed data from a survey conducted by GfK/KN between November 29, 2012 and December 12, 2012 as part of a Timeshare Experiments in the Social Sciences (TESS) module. The sample size was 1567 (from 2459 KnowledgePanel members invited), and the completion rate was 63.7%. The panel was recruited in a similar manner as the Internet cases in the 2002 study, except that addressed-based sampling (ABS) techniques were also used. The topic area of the survey was assessing the level of political polarization across several issue areas (e.g., taxes, trade, immigration, campaign finance).

Because Study 2 was a standalone data collection effort, there are no telephone cases. However, the questionnaire is sufficiently extensive to examine the same satisficing measures as in Study 1, along with the same measures of reluctance. It therefore provides a useful replication of the Internet cases from the first study.

14.4.3   Study 3: American National Election Studies

We leverage two datasets from the American National Election Studies (ANES; the 2008 Panel Study and the 2008 Time Series Study), representing two interview modes (face-to-face and Internet). These datasets address two limitations of Studies 1 and 2. First, they contain numerous experimental manipulations of response order. These experiments allow us to examine an additional form of satisficing that we cannot examine with the Gfk/KN data – selecting the first reasonable response – in a way that disentangles satisficing from genuine response. The 2008 ANES studies are unique in terms of the sheer number of response order manipulations available to analyze, especially given that many items employed unipolar rating scales. Therefore, unlike studies that rely on a few experiments, we can assess whether findings are consistent and common or rare and infrequent. Second, the ANES conducted many cases in person, allowing us to compare the Internet mode to face-to-face administration. We first analyzed the 2008 ANES Panel Study. We focus solely on the first (January 2008) wave of the study, which was conducted over the Internet by GfK/KN. Respondents were recruited into the panel in late 2007 over the phone via random digit dialing (RDD). If a respondent agreed to participate in the panel as a result of the phone request, he or she was sent an email notification with a link to the web survey up to a few weeks later. Some 1623 respondents were interviewed in the first wave. The AAPOR RR5 response rate was 75.0%. Additional methodological details and full questionnaires can be found at www.electionstudies.org.

We also analyzed the main 2008 ANES Time Series Study. The pre-election wave was conducted face-to-face on a nationally representative sample of 2322 U.S. adults between September 2, 2008, and November 3, 2008. The target population was English- or Spanish-speaking U.S. residents of voting age in the lower 48 states, with an oversample of African Americans and Latinos. Respondents were chosen through a five-stage procedure and data collection was done face-to-face by RTI International using Computer Assisted Personal Interviewing technology. Respondents were asked questions in four main ways: (1) orally by the interviewer; (2) orally by the interviewer with the use of a visual aid, a booklet for respondents; (3) orally self-administered by computer (through headphones); and (4) visually self-administered by computer. A fresh cross-section of respondents was recruited for the study. The AAPOR RR5 response rate was 78.2%. Additional methodological details and full questionnaires can be found at www.electionstudies.org.

One limitation of the ANES studies is that the Internet and face-to-face respondents were not from a common sample frame. Additionally, while there are some questions with identical wordings that were answered by both groups, all respondents did not complete the same questionnaire, precluding us from replicating any of the analyses from Study 1, which all rely on a common questionnaire. Nonetheless, the same organization (ANES) designed both studies and all respondents were initially recruited via RDD, similar to the Gfk/KN study. Finally, it is important to note that the Internet respondents in Study 3 were not existing members of a panel, but were instead recruited for a fresh panel. Consequently, whereas in Studies 1 and 2 we examine the reluctance to participate in a given (web) survey, in Study 3 we are looking at the reluctance to join the panel itself.

14.5   Measures

14.5.1   Studies 1 and 2 dependent variables: Measures of satisficing

14.5.1.1   DK/NO responses

For our DK/NO measure we relied on recent scholarship for guidance. Holbrook, Green, and Krosnick (2003) constructed an index of no-opinion responding by counting the number of times respondents gave a "don't know" response to nine questions in the 2000 NES that explicitly offered the "don't know" option. We adapted Holbrook et al.'s (2003) strategy to examine whether reluctant respondents are more likely to satisfice and whether the effect differs by mode.

Our index for Study 1 (the GfK/KN phone and Internet surveys of the common sampling frame, which used the same questionnaire) is constructed from 10 questions in which the Internet version did not offer an explicit DK option. We used this subset of questions because an explicit DK option was never offered to telephone respondents. While volunteered DKs are technically not measures of satisficing in Krosnick's (1991) framework, they represent a lack of engagement with the questionnaire, similar to the selection of explicitly offered DKs. The dependent variable in the analyses is simply the count of the number of questions for which the respondent volunteered the DK response to the interviewer (in the phone mode) or skipped the question (in the Internet mode). The question wordings for all items used in the analyses of the GfK/KN dataset are presented in online appendix 1. In Study 2, the DK/NO index was constructed from the complete set of 15 questions asked in the question-naire; an explicit DK option was never offered. Strictly speaking, our operationalization of DK responding is slightly different in the phone vs. Internet modes (there are of course many differences between these two modes, including oral vs. visual presentation and the presence vs. absence of an interviewer). Conceptually, however, the measures are similar; in both cases the respondent volunteers a "don't know" response or a refusal to answer the question.

14.5.1.2 Midpoint selection

In Study 1, we identified 19 questions in which a midpoint response option was used in a rating scale that expressed a form of ambivalence or non-opinion (e.g., "neither agree nor disagree") or a status quo response (e.g., "keep the same as it is now"). The dependent variable is simply the count of questions in which the midpoint was selected. All 15 questions in Study 2 had midpoints.

14.5.1.3 Nondifferentiation

For nondifferentiation we adopt a measure from Anand et al. (2005) that gauges the dispersion of a respondent's answers using the standard deviation because, unlike the computed variance, the standard deviation does not give disproportionate weight to outliers. Specifically, in Study 1, we focused on five sets of questions in which respondents were asked to perform the same cognitive task over and over again. We calculated the standard deviation of the responses, with higher numbers reflecting greater differentiation and less satisficing. We then standardized each of these nondifferentiation scales to lie between 0 and 1, and then computed an average to construct a single measure of nondifferentiation for the interview. We adopted a similar technique for Study 2 except there was a single scale with nine items.

14.5.1.4 Mental coin flipping

We use seven criterion variables from Study 1 to assess our expectation that difficult-to-obtain respondents are more likely to choose response options randomly (another indicator of strong satisficing), thus attenuating theoretically expected relationships: (1) approval of Bush's per-formance as President; (2) approval of Bush's performance on the issue of terrorism; (3) a feeling thermometer of Bush; (4) a feeling thermometer of Gore; (5) a question assessing whether respondents think it would be better if Christians had more influence on public issues; (6) attitudes towards environmental regulation of business; and (7) attitudes toward govern-ment's role in ensuring a high standard of living. We examine the relationships between the aforementioned criterion variables and party identification, a variable that has been shown to be correlated with these types of attitudes time and time again. The end result of these analyses is a difference-in-differences (DID) estimate that assesses whether the correlation between party identification and the criterion item is stronger among willing versus unwilling respondents.

 We chose these variables for both theoretical and methodological reasons. This group of variables and relationships is prevalent in the public opinion and political behavior literatures. In addition, we have a strong expectation that these variables are highly correlated, and thus should provide a strong test for our attenuation analyses. While considering the theoretical reasons, we jointly considered what would be attractive methodological characteristics. Among the measures, there are three different types of scales: (1) four- and five-point ordered scales (Bush overall and terrorism approval; views on evangelical Christians); (2) seven-point bipolar scales (environmental attitude and attitude toward government intervention; and (3) numerical continuous-interval level scales (feeling thermometers). This diversity enables us to test the generalizability of our findings. We have reason to believe that if we are to find a pattern of attenuation among this diverse collection of variables, it suggests we would likely find attenuation among relationships between other variables that are constructed using one of these common response scales (categorical, ordinal, and interval).

In Study 2, we again used party identification as the main independent variable. The criterion items were positions on four policy items: capital gains tax rates, illegal immigration, free trade, and public financing of campaigns.

14.5.2   Study 3 dependent variable: Measure of satisficing

14.5.2.1   Selecting the first reasonable response

In the 2008 ANES Panel, there were 47 response order manipulations administered to either the entire sample or a randomly selected portion of the sample (we excluded response-order manipulations for items into which respondents self-selected. In the 2008 ANES Time Series, there were 103 response order manipulations administered to either the entire sample or a randomly-selected portion of the sample. As noted above, Studies 1 and 2 did not contain any response order manipulations.

To examine whether reluctant respondents were more likely to satisfice by choosing the first reasonable alternative (and whether this relationship differs across survey modes) we estimate the following difference-in-differences (DID):

$$(\overline{Y_{rh}} - \overline{Y_{fh}}) - (\overline{Y_{re}} - \overline{Y_{fe}}) \tag{14.1}$$

where $\overline{Y_{rh}}$ represents the average value of the response option (measured on a Likert-type rating scale) for respondents who were "hard" to obtain in the "reverse" order condition and $\overline{Y_{fh}}$ represents the average value of the response option for respondents who were "hard" to obtain in the "forward" order condition. $\overline{Y_{re}}$ and $\overline{Y_{fe}}$ represent corresponding values for respondents who were "easy" to obtain. In the "forward" order condition, the response options coded with lower numerical values are presented on top, whereas in the "reverse" order condition, those same response options are coded identically but are presented last. Hence, a positive DID estimate (as represented by 14.1) indicates that the order effect (our measure of survey satisficing) is higher among hard-to-obtain respondents as compared to easy-to-obtain respondents. If the DID estimate is positive and significantly different from zero (as ascertained by a t-test), then it suggests that using nonresponse strategies has the tradeoff of introducing systematic measurement error into the data.

The DID estimate is equivalent to the coefficient estimate on the interaction term from a linear regression predicting the item response with: (1) the order dummy; (2) the measure of whether the respondent was "hard-to-obtain"; and (3) the interaction between the two. We did not include demographic covariates because we wanted to maximize the chance we observed significant DID estimates and did not want to include variables that might be correlated with the survey administration measures. The sizes of the effects are similar when controlling for a host of demographics (see online appendix 4).

14.5.3   Studies 1 and 2 independent variables: Nonresponse strategies

Recall that for Study 1, a subset of respondents who had already agreed to be part of the GfK/KN Internet panel was randomly assigned to be asked to complete the same questionnaire via the Internet or telephone (information about the initial panel intake survey was not available). For the phone respondents, the nonresponse strategy we use is the number of call attempts it took to reach the respondent (we do not have information on refusal conversions). The distribution of the callback variable ranges from 1–19 and is highly positively skewed (26% were contacted with just 1 call and over half received less than 4 calls). As a result, we also take the natural log of the number of calls, as well as a dummy variable for whether the person required a callback of any kind or not.

For the Internet respondents in both Study 1 and Study 2 (recall that Study 2 was administered only via the Internet), the nonresponse strategy we analyze is a dummy variable indicating whether or not the respondent received an email reminder, which was sent to all respondents two days after the initial email request if they had not yet completed the survey. Some 52.5% and 41% of respondents received an email reminder in Studies 1 and 2, respectively. Additionally, GfK/KN sent additional reminders after the two-day period to the remaining respondents who did not complete the survey. Although we do not have information in Study 1 on how many reminders respondents received after the first one, we do know how long it took them to complete the survey after the field date, which is a rough proxy of both willingness to complete the survey and the number of necessary email reminders. This variable ranges from 0 days (indicating that the respondent completed the survey on the same day that it was fielded) to 42 days, and is highly positively skewed. As such, we take the natural log of the number of days after the field date it took for the respondent to complete the survey. For Study 2, the variable ranges from 1 day to 13 days. For Study 2, GfK/KN provided us with the number of reminders respondents received, which we also analyze as an independent variable. Some 59% of respondents received zero reminders, 18.6% received one reminder, 12.4% received two reminders, and the remaining 10.0% received three reminders.

Additionally, for Internet respondents in Study 1, we analyze a measure of general propensity to positively respond to survey invitations. We obtained from GfK/KN the number of surveys that respondents had been invited to complete prior to the *Survey on Civic Attitudes and Behaviors after 9/11*, and the number of surveys completed. This enables us to compute an overall response rate to survey invitations. We only examine respondents who were invited to participate in at least 10 surveys so as to not include potentially skewed data based on a small number of invitations. The continuous variable ranges from 0–1 to indicate the proportion of surveys the respondent completed (mean=.88, or 88%). We use a similar measure for Study 2; the mean proportion of surveys completed was 78%.

14.5.4   Study 3 independent variable

14.5.4.1   The 2008 ANES Panel

All adults chosen to be part of the sample were offered $10 a month to complete surveys over a 21-month period. A unique feature of this dataset is that there were many opportunities for respondents to opt out of the survey. For example, after joining the panel, respondents might not have responded to email invitations to participate in the survey. Therefore, it is possible that even if a respondent was difficult to recruit into the panel initially, he or she may have been more willing to participate conditional on accepting an email invitation. Given that our analyses focus on just the first wave of the 21-wave study, the nonresponse strategy of interest is initial panel recruitment attempts. For the initial RDD recruitment, the dataset contained 47 initial refusals. Initial refusers were assigned to "refusal conversion specialists" (see the 2008–2009 ANES Panel Study User Guide for more details). We construct a dummy variable to indicate whether respondents were obtained via a refusal conversion.

With regard to number of callbacks (which ranged from 1–50; mean: 5.8; SD: 6.5), we construct two variables. First, we examine the natural log (to reduce skewness) of the number of calls it took to contact the respondent to complete the interview (coded as a continuous variable; for the Panel Study all the callback attempts were made via telephone). Second, given that a low number of callbacks may be due to inefficiencies on the part of the survey organization, we dichotomized the callback variable at the median to capture differences in the number of callbacks that are more likely to indicate a busy, or hard-to-reach, respondent. Finally, we create a combined measure of reluctance using both the refusal conversion and callback attempts measures, coded "1" if the respondent was *either* a refusal conversion or required a high number of calls.

14.5.4.2 The 2008 ANES Time Series

All adults chosen to be part of the sample received an advanced mailing with a $25 incentive that was increased to $50 on October 7th. Interviewers were given various incentives ranging from $5 for completing a household screener interview to $30 for both the screener and pre-election interview. We analyze two nonresponse strategies used in the data collection: refusal conversions and the number of call attempts to reach the respondent (the 2008 Time Series Study does not distinguish phone calls from house visits, so we use "calls" as a catchall term to refer to both contact attempts).

For refusal conversions, if a respondent initially refused, he/she was sent a specially tailored letter to address the reasons for refusing. The dataset contained 366 refusal conversions. Similar to the ANES Panel Study, we construct a dummy variable to indicate whether respondents were obtained via a refusal conversion attempt or not (16% were refusal conversions).

For callbacks, the range of call attempts was 5–48 (mean: 13.4; SD: 5.1). We constructed the same callback variables as with the ANES Panel Study: (1) a continuous variable indicating the log number of calls it took to contact the respondent to complete the interview; (2) a dummy variable constructed by splitting the continuous callbacks variable at the median; and (3) the indicator combining refusal conversion and the "high calls" measure.

Descriptive statistics for the independent variables (nonresponse strategies) and dependent variables (satisficing metrics) for all three studies are presented in Table 14.1.

14.6 Results

14.6.1 Internet mode

To examine the impact of respondent reluctance on satisficing in the Internet mode, we focus on the Internet respondents from Study 1, all respondents in Study 2 (which was administered only via the Internet), and the ANES Panel Study component of Study 3.

Table 14.1   Descriptive statistics.

	Study 1: 2002 KN Dataset							
	Telephone Sample (N = 300)				Internet Sample (N = 2979)			
	Mean	SD	Min	Max	Mean	SD	Min	Max
Independent Variables:								
Call Attempts	4.44	3.87	1.00	19.00	–	–	–	–
Log Call Attempts	1.13	0.86	0.00	2.94	–	–	–	–
Multiple Calls (Dummy)	0.74	0.44	0.00	1.00	–	–	–	–
Reminder Email	–	–	–	–	0.53	0.50	0.00	1.00
Log Days to Completion	–	–	–	–	1.44	0.72	0.00	3.76
Participation Rate[1]	–	–	–	–	0.88	0.09	0.18	1.00
Dependent Variables:								
Midpoints Selected	1.77	1.50	0.00	8.00	2.56	2.20	0.00	17.00
DK/NO Responses	0.73	1.33	0.00	8.00	1.40	2.07	0.00	10.00
Nondifferentiation[2]	0.35	0.11	0.12	0.85	0.36	0.10	0.05	0.70

	Study 2: 2012 KN Dataset							
	No Telephone Sample				Internet Sample (N = 1567)			
					Mean	SD	Min	Max
Independent Variables:								
Reminder Email	–	–	–	–	0.41	0.49	0.00	1.00
No. of Reminder Emails	–	–	–	–	0.73	1.02	0.00	3.00
Log Days to Completion	–	–	–	–	0.78	0.87	0.00	2.56
Participation Rate[1]	–	–	–	–	0.78	0.16	0.08	1.00
Dependent Variables:								
Midpoints Selected	–	–	–	–	3.60	2.67	0.00	15.00
DK/NO Responses	–	–	–	–	0.18	0.79	0.00	9.00
Nondifferentiation[2]	–	–	–	–	1.84	0.66	0.00	3.44

	Study 3: ANES Dataset							
	Face-to-Face Sample (N = 2322)				Internet Panel Sample (N = 1623)			
	Mean	SD	Min	Max	Mean	SD	Min	Max
Independent Variables:								
Refusal	.158	.364	0	1	.026	.159	0	1
Log Call Attempts	2.53	.343	1.61	3.87	1.23	.954	0	3.56
High Calls (Dummy)	.469	.499	0	1	.467	.499	0	1
Combined	.509	.500	0	1	.470	.499	0	1

[1]Participation rate variable only calculated for respondents who received at least 10 previous invitations from KN to participate in Internet surveys: 2002 KN dataset (N = 2021), 2012 KN dataset (N = 1557).
[2]Sample sizes for non-differentiation scale are: 2002 KN dataset telephone sample (N = 271), 2002 KN dataset Internet sample (N = 2565), 2012 KN dataset (N = 1567).

14.6.1.1   Midpoint selection

For Study 1, we first estimated Poisson regression models predicting the number of questions (out of 19) for which respondents selected the middle response category. We employ Poisson regression because the dependent variable is the discrete count of the number of midpoints selected. As the first three columns of Table 14.2 show, reluctant Internet respondents were not more likely to select the middle response category than non-reluctant Internet respondents. This null result holds across all three indicators of reluctance (reminder email, log days to completion, and participation rate). We replicated this analysis for the 15 questions in Study 2. As with the Internet respondents in Study 1, there was no relationship between three of the reluctance measures (sending a reminder email, number of reminders, log number of days to completion) and midpoint selection in Study 2 (see the first three columns of Table 14.3). However, we did observe that respondents with a lower participation rate (i.e., responded to fewer survey invitations overall) were more likely to satisfice by selecting the middle response option ($b = -0.20$, $p = .01$).

14.6.1.2   DK/NO responses

Whereas reluctance was (with one exception) not associated with midpoint selection among Internet respondents, it was associated with DK/NO responding. We again employed Poisson regression to model the count of the number of DK/NO responses. As shown in the middle three columns of Table 14.2, Study 1 respondents who were more difficult to obtain in the Internet survey were more likely to express a non-opinion. Specifically, Internet respondents who completed the survey only after being sent a reminder email were significantly more likely to not answer questions ($b = 0.14$, $p < .01$). Being sent a reminder email is associated with an increased expected count of about 15%. We find similar effects for the other two measure of reluctance. The log of days to completion was significantly related to DK/NO responding ($b = 0.10$, $p < .01$, see column 5 of Table 14.2). Internet respondents who responded to invitations at higher rates, indicating a general willingness to participate, were less likely to provide DK/NO responses ($b = -.1.21$, $p < .01$, see column 6 of Table 14.2). Going from 0% participation to 100% participation decreases the expected count by about 70%.

These findings are replicated in Study 2 (see the middle four columns of Table 14.3). Specifically, three of the four indicators of reluctance were positively and significantly related to DK/NO responding–being sent a reminder email, the number of reminders sent, and the log of the days to taking the survey. Moreover, the DK/NO effect sizes in Study 2 are larger than those of Study 1.

14.6.1.3   Nondifferentiation

In Study 1, we find little relation between nondifferentiation and satisficing in the Internet survey, although we did observe a marginally significant effect for the participation rate measure (see the final three columns of Table 14.2). As the outcome variable is continuous, we employ OLS regression. Moving from 0%–100% participation increased nondifferentiation by about 4.8% ($p = .084$) – an effect that does not quite reach conventional levels of statistical significance. However, there is no relationship between nondifferentiation and being sent a reminder email or time taken to respond to the survey. Moreover, the participation rate finding

Table 14.2 Study 1 results (Internet cases).

	Selection of midpoints		DK/NO Responding			Nondifferentiation		
Reminder email	0.01 (0.02)	—	0.14** (0.03)	—	—	−0.003 (0.004)	—	—
Log days to completion	0.00 (0.02)	—	—	0.10** (0.02)	—	—	−.002 (0.003)	—
Participation rate	—	−0.02 (0.16)	—	—	−1.21** (0.19)	—	—	0.048+ (0.028)
Constant	0.93** (0.02)	0.96** (0.14)	0.26** (0.02)	0.19** (0.03)	1.32** (0.17)	0.359** (0.003)	0.361** (0.004)	0.314** (0.025)
N	2979	2021	2979	2979	2021	2565	2565	1750
Log lik. / R²	−6398.3	−4368.0	−6036.8	−6036.1	−3913.4	.00	.00	.00

Notes: "Selection of midpoints" is number of times the midpoint response was selected for 19 questions in which a midpoint was used. "DK/NO Responding" is number of times a DK/NO response was provided for 10 questions in which an explicit DK/NO option was offered. "Nondifferentiation" is nondifferential scale constructed by averaging standardized standard deviations of responses to five indices. Columns (1)–(6) present coefficient estimates from OLS regressions; columns (7)–(9) present coefficient estimates from Poisson regressions. Standard errors in parentheses.

$**p<.01$;
$*p<.05$;
$+p<.10$ (two-tailed).

Table 14.3 Study 2 results.

	Selection of midpoints				DK/NO responding				Nondifferentiation			
Reminder email	0.024 (0.027)	—	—	—	0.42** (0.12)	—	—	—	0.00 (0.03)	—	—	—
Number of reminders	—	0.024 (0.013)	—	—	—	0.24** (0.05)	—	—	—	0.00 (0.02)	—	—
Log days to completion	—	—	0.22 (.015)	—	—	—	0.24** (0.07)	—	—	—	0.00 (0.02)	—
Participation rate	—	—	—	−0.20* (0.08)	—	—	—	−0.02 (0.36)	—	—	—	0.08 (0.10)
Constant	1.27** (0.02)	1.26** (0.02)	1.26** (0.02)	1.44** (0.06)	−1.89** (0.08)	−1.90** (0.08)	−1.90** (0.09)	−1.67** (0.29)	1.84** (0.02)	1.84** (0.02)	1.84** (0.02)	1.78** (0.08)
N	1567	1567	1567	1557	1567	1567	1567	1557	1567	1567	1567	1557
Log lik. / R^2	−3735.8	−3734.5	−3735.1	−3713.9	−942.8	−938.9	−942.8	−947.2	.00	.00	.00	.00

Notes: "Selection of midpoints" is number of times the midpoint response was selected for 15 questions in which a midpoint was used. "DK/NO Responding" is number of times a DK/NO response was provided for 15 questions in which an explicit DK/NO option was not offered. "Nondifferentiation" is nondifferential scale is the standard deviation of responses to a nine-item scale. Columns (1)–(6) present coefficient estimates from Poisson regressions; columns (7)–(9) present coefficient estimates from OLS regressions. Standard errors in parentheses.

**p<.01;

*p<.05;

+*p*<.10 (two-tailed).

does not replicate in Study 2 (see the last four columns of Table 14.3), nor were any of the other indicators of reluctance in Study 2 associated with nondifferentiation.

14.6.1.4 Mental coin flipping

We found absolutely no relationship between reluctance and mental coin flipping among the Study 1 Internet cases or in Study 2, across the different tests we performed. As mentioned above, we assessed whether the relationship between party identification and various theoretically related criteria (e.g., feeling thermometer rating of President Bush) were significantly lower for reluctant respondents. Detailed results are presented in Tables 2a and 2b of online appendix 2.

14.6.1.5 Selecting the first reasonable response

We used the ANES Panel Study to explore the relationship between reluctance and a fifth type of satisficing–selecting the first reasonable response – in the Internet mode. To do so, we examine the difference-in-differences (DID) estimates of the response order effects. However, before describing the overall results, it is instructive to describe a statistically significant DID estimate in detail. In the 2008 Panel Study, respondents were asked: "Compared to 2001, would you say the nation's crime rate is now much better, somewhat better, about the same, somewhat worse, or much worse?" on a five-point scale ranging from "much better" (1) to "much worse" (5). Among respondents who were *not* refusal conversions, the difference-in-means between the "reverse" condition and the "forward" condition was .05 points (3.42 − 3.37). Among respondents who were refusal conversions, the difference-in-means between the "reverse" condition and the "forward" condition was .65 points (3.78 − 3.13). Hence, the DID estimate was .60 units (.65 − .05), meaning that the primacy effect was significantly larger among refusal conversions ($p = .02$).

The top panel of Table 14.4 displays a summary of the DID estimates for the 2008 Internet (ANES Panel) analyses (the full set of individual estimates is presented in online appendix 3. We first examined the full set of 47 items included in the Internet portion. As shown in the top row of Table 14.4, response order effects are slightly more pronounced among reluctant respondents surveyed via the Internet (given that we have well-specified theoretical reasons to assume a positive DID estimate, we apply the standard $p < .05$ statistical significance threshold using one-tailed tests). Across the four measures of reluctance, the percentage of items that had significant difference-in-differences estimates was between 4.3%–8.5%. This is a bit more than what we would expect to see by chance alone.

We next examined the 19 items which exhibited significant primacy effects in the full sample of items, determined by calculating a difference-in-means in response outcomes between respondents in the "reverse" condition and respondents in the "forward" condition. As shown in the second row of Table 14.4, significant, positive DID estimates appeared in 15.8%–21.1% of cases in the Internet survey for the callbacks and combined measure (no significant DID estimates were observed for the refusal conversion measure). Given that we would expect to observe significant effects by chance 5% of the time, these findings suggest that this type of systematic measurement error (a result of a strong form of satisficing) is more prevalent among difficult-to-obtain respondents surveyed via the Internet (particularly in the case of callback attempts).

Table 14.4 Percentage of statistically significant difference-in-difference estimates for response order analyses (Study 3).

	Measure of reluctance			
	Refusal conversion	Log number of calls	High calls	Combined measure
2008 ANES Panel (Web)				
All items (47 items) (%)	4.3	6.4	6.4	8.5
Items with significant order effects (19 items)	0.0	15.8	15.8	21.1
2008 ANES Time Series (face-to-face)				
All items (102 items) (%)	2.9	3.9	4.9	2.9
Items with significant order effects (48 items)	0.0	4.2	4.2	4.2

Note: Cell entries represent percentages. "Combined measure" indicates respondents coded "1" for either refusal conversion or high calls.

14.6.2   Internet vs. telephone

To compare the effects of nonresponse strategies on satisficing between the Internet and telephone modes, we make use of Study 1, the GfK/KN dataset in which we were able to compare respondents from a common sampling frame that were randomly assigned to either complete the same questionnaire over the web or over the telephone. Before presenting the main results, it is important to note that satisficing behavior was generally higher in the Internet survey across all measures except for nondifferentiation (see Table 14.1). However, the question of interest in this analysis is whether the relationship between reluctance and satisficing depends on mode. Recall that in the Internet mode, there was no effect of reluctance on midpoint selection, nondifferentiation (save for the participation rate indicator of reluctance; a finding that does not replicate in Study 2), or mental coin flipping. However, there was a consistent, positive, statistically or marginally statistically significant relationship between nonresponse strategy and DK/NO responding (across all but one indicator of reluctance in both Studies 1 and 2).

14.6.2.1   Midpoint selection

As shown in the first two columns in Table 14.5, whereas reluctance was not associated with likelihood of selecting the middle response option in web surveys, reluctant telephone respondents were, in fact, more likely to select the middle response category than non-reluctant telephone respondents. Requiring additional call attempts was marginally significantly associated with the selection of more midpoints ($b = 0.09$, $p = 0.08$; see column 1). Similarly, respondents who required a callback were significantly more likely to select midpoints ($b = 0.25$, $p = 0.02$; see column 2) than respondents who were reached on the first attempt. The expected count of number of midpoints selected increases by about 29% if a respondent required multiple callback attempts.

Table 14.5   Study 1 results (telephone cases).

	Selection of midpoints		DK/NO responding		Nondifferentiation	
Log call attempts	0.09^+	–	−0.03	–	−0.01	–
	(0.05)		(0.08)		(0.01)	
Multiple call attempts	–	0.25^*	–	0.03	–	0.01
		(0.11)		(0.15)		(0.01)
Constant	0.47^{**}	0.38^{**}	$−0.28^*$	$−0.33^*$	0.36^{**}	0.34^{**}
	(0.07)	(0.09)	(0.11)	(0.13)	(0.01)	(0.01)
N	300	300	300	300	271	271
Log lik. / R^2	−515.4	−513.9	−407.3	−407.4	.00	.00

Note: Dependent variables defined as in Table 14.2. Columns (1)–(6) present coefficient estimates from Poisson regressions; columns (7)–(9) present coefficient estimates from OLS regressions. Standard errors in parentheses.
$^{**}p<.01$;
$^*p<.05$;
$^+p<.10$ (two-tailed).

14.6.2.2   DK/NO responses

We observed the opposite effect with DK/NO responses. Whereas reluctance was consistently, positively associated with DK/NO responses in the Internet mode, there was no relationship between reluctance and satisficing for the telephone cases (see the middle two columns of Table 14.5). We discuss these divergent results in the Conclusion.

14.6.2.3   Nondifferentiation and mental coin flipping

As shown in the last two columns in Table 14.5, similar to the Internet mode, we found no relationship between either of the two reluctance measures and nondifferentiation in the telephone surveys. Finally, as with the Internet respondents, we found no relationship between reluctance and mental coin flipping among the respondents randomly assigned to be surveyed via the telephone in Study 1. Detailed results are presented in Table 2c of online appendix 2.

14.6.3   Internet vs. face-to-face

To examine the effects of nonresponse strategies on satisficing between the Internet and face-to-face modes, we compared the results from the response order analyses using the 2008 ANES Panel Study (conducted over the Internet) and the 2008 ANES Time Series Study (conducted face-to-face). Recall that in the Internet mode, there was a relationship between reluctance (particularly the number of callbacks required) and primacy effects.

14.6.3.1   Selecting the first reasonable response

The bottom panel of Table 14.4 presents the analyses for the face-to-face mode. Examining the full set of items, we find little evidence that primacy effects are concentrated or higher among difficult-to-obtain respondents surveyed face-to-face. The percentage of DID estimates that are significantly significant ranges between 2.9%–4.9% across the reluctance measures.

Examining the 48 items which exhibited significant primacy effects, significant, positive DID estimates appeared in only 0.0%–4.2% of cases in the face-to-face survey. These findings suggest that this type of systematic measurement error (a result of a strong form of satisficing) is more prevalent among difficult-to-obtain respondents surveyed via the Internet compared to face-to-face.

In sum, to the extent that any difference is present between the Internet and face-to-face modes, there is a slight tendency for difficult-to-reach respondents to evidence more primacy effects when surveyed via the Internet than face-to-face.

14.7   Discussion and conclusion

Survey researchers in industry, government, and academia have all been concerned with the trend of declining response rates in all survey modes (see CNSTAT 2013, for a recent comprehensive review). Technological and social changes have only exacerbated these concerns, challenging the paradigm of representative sampling. Not surprisingly, these worries have generated a great deal of valuable research assessing the impact of nonresponse rates on bias and measurement error, and potential strategies to improve response rates. However, the effort to increase response rates might be counterproductive because difficult-to-obtain respondents may not be fully engaged with the survey.

Leveraging several datasets and some novel methodological approaches to tackle the question, this chapter found evidence that satisficing is sometimes greater for "hard-to-reach" respondents. Specifically, our results suggest that the presence of an interviewer reduces the negative impact of reluctant respondents on data quality. Across the four indicators of satisficing in Study 1, the only one for which difficult-to-reach respondents were more likely to evidence satisficing in the phone mode was middle responding (more on this finding below). With regard to Study 3, we found that reluctant respondents in the face-to-face mode (when the interviewer is physically present and thus accountability pressures are presumably even stronger) evidenced very few response order effects (i.e., very few instances of selecting the first reasonable response). In contrast, in the self-administered Internet mode, we found evidence of greater satisficing among reluctant respondents for one of the indicators in Studies 1 and 2 (DK/NO responding) and in the one indicator (selecting the first reasonable response) we examined in Study 3.

The most unambiguous pattern among our results is the difference in how reluctant respondents choose to satisfice in the phone versus the Internet mode (Studies 1 and 2). Specifically, difficult-to-reach respondents are significantly more likely to engage in DK/NO responding than their easy-to-reach counterparts in the Internet mode and are significantly more likely to evidence middle responding in the phone mode.

What explains the symmetric results from the analyses predicting midpoint selection and those predicting DK/NO responding? As explained earlier, satisficing respondents in a telephone survey will be more likely to select midpoints, since that provides a façade of compliance. This difference is likely due to the fact that the presence of an interviewer in the phone mode makes it less appealing for respondents to offer that she does not know, or does not have an opinion. She may worry about irritating a live interviewer if she continuous to offer no-opinion responses, or may be embarrassed to represent to an interviewer that she is not knowledgeable or does not have opinions on important social and political issues. On the other hand, in the absence of a live interviewer, an Internet respondent does not have to worry about maintaining a façade of compliance or competence. He can simply not respond to questions without fear of embarrassment or of upsetting an interviewer. In fact, the Internet mode makes this form of satisficing quite easy (all the respondent needs to do is "click through" the question to move on to the next one) and therefore a more appealing option for reluctant (perhaps unmotivated) respondents. Hence, the relationship between reluctance and satisficing is expressed in different ways across mode: midpoint selection in telephone interviewing and DK/NO responding in web interviewing. In other words, the problem of reluctant respondents introducing measurement error is common to both modes, but is manifested differently. Whereas researchers might be quick to conclude that there are no systematic explanations of the interaction between mode and nonresponse strategies on satisficing, we strongly emphasize that the satisficing effects in our analyses were theoretically predictable. In short, satisficing is still an issue that must be addressed with surveys, regardless of mode.

A few caveats are in order. First, we caution that the bulk (though not all) of our items measure political attitudes on Likert scales. Based on both the response format and subject matter, these types of items may be most subject to comprehension errors. Thus, we must be circumspect in generalizing our findings to other types of questions and topics commonly employed by public opinion researchers. Second, we only analyze surveys from three survey houses (and the mode comparisons differed between the survey houses). We look forward to additional research that investigates this question using data from a diverse set of survey sponsors and survey houses. Finally, our measure of reluctance for the Internet vs. face-to-face comparison was different than our measure of reluctance for the Internet vs. phone comparison. Ideally, we would have liked to compare similar measures, but we still feel our results are beneficial to survey researchers because each measure provides insight into how or whether hard-to-reach respondents introduce measurement error into surveys. This speaks to the theoretical tradeoff mentioned at the beginning of the chapter. Future research could replicate the excellent design of the GfK/KN study but include response order manipulations as well as experiments to assess acquiescence bias, another form of satisficing not considered here.

In sum, research on the relationship between response rates and measurement error is in its infancy. Clearly, additional study is needed to more fully understand the costs and benefits of achieving higher response rates, particularly with panel surveys. Nonetheless, it is comforting to note that efforts to ameliorate nonresponse (i.e., doing refusal conversions, sending reminder emails, and making a high number of attempts to contact hard-to-reach respondents) do not *always* introduce additional complications in the form of increasing the prevalence of satisficing in survey items. However, for some manifestations of satisficing, we did observe that the relationship between reluctance and satisficing was present in both the Internet and telephone modes, suggesting caution for survey researchers who are implementing nonresponse strategies. Overall, the Internet mode performed well compared to more traditional approaches to conducting surveys. Yet, given that we found suggestive evidence that some forms of satisficing were more present among reluctant respondents in the Internet mode, these new forms of survey interviewing may present additional challenges going forward.

References

Anand, S., Krosnick, J.A., Mulligan, K., Smith, W.R., Green, M.C., & Bizer, G.Y. (2005). Task difficulty, participant motivations, and nondifferentiation: A test of satisficing theory. Paper presented at the Annual Meeting of the American Association for Public Opinion Research.

Bem, J. W. (1966). *A theory of psychological reactance*. New York: Academic Press.

Bem, S. S., & Bem, J. W. (1981). *Psychological reactance: A theory of freedom and control*. New York: Academic Press.

Biemer, P. P. (2001). Nonresponse bias and measurement bias in a comparison of face to face and telephone interviewing. *Journal of Official Statistics*, *17*, 295–320.

Brehm, J. (1993). *The phantom respondents*. Ann Arbor, MI: University of Michigan Press.

Brick, J. M., & Williams, D. (2013). Explaining rising nonresponse rates in cross-sectional surveys. *The Annals of the American Academy of Political and Social Science*, *645*, 36–59.

Cacioppo, J. T., & Petty, R. E. (1982). The need for cognition. *Journal of Personality and Social psychology*, *42*, 116–131.

Cacioppo, J. T., & Petty, R. E. (1984). The need for cognition: Relationship to attitudinal processes. In R. McGlynn, J. Maddux, C. Stotlenberg, & J. Harvey (Eds.), *Social perception in clinical and counseling psychology* (pp. 113–140). Lubbock, TX: Texas Tech Press.

Callegaro, M., & DiSogra, C. (2008). Computing response metrics for online panels. *Public Opinion Quarterly*, *72*, 1008–1032.

Cannell, C. F., & Fowler, F. J. (1963). Comparison of a self-enumerative procedure and a personal interview: A validity study. *Public Opinion Quarterly*, *27*, 250–64.

Chang, L., &. Krosnick, J. A. (2009). National surveys via RDD telephone interviewing versus the internet: Comparing sample representativeness and response quality. *Public Opinion Quarterly*, *73*, 641–678.

Chang, L., & Krosnick, J.A. (2010). Comparing oral interviewing with self-administered computerized questionnaires: an experiment. *Public Opinion Quarterly*, *74*, 154–167.

CNSTAT. (2013). *Nonresponse in social science surveys: A research agenda*. Washington, DC: National Academies Press.

Curtin, R., Presser, S., & Singer, E. (2000). The effects of response rate changes on the index of consumer sentiment. *Public Opinion Quarterly*, *64*, 413–428.

Curtin, R., Presser, S., & Singer, E. (2005). Changes in telephone survey nonresponse over the past quarter century. *Public Opinion Quarterly*, *69*, 87–98.

de Leeuw, E., & de Heer, W. (2002). Trends in household survey nonresponse: A longitudinal and international perspective. In R. M. Groves, D. A. Dillman, J. L. Eltinge, & R. J. A. Little (Eds.), *Survey nonresponse* (pp. 41–54). New York: Wiley.

Dennis, J. M., Chatt, C., Li, R., Motta-Stanko, A., & Pulliam, P. (2005). Data collection mode effects controlling for sample origins in a panel survey: Telephone versus Internet. Unpublished manuscript.

Dennis, J. M., Li, R., & Hadfield, J. (2007). Results of a within-panel survey experiment of data collection mode effects using the general social survey's national priority battery. Paper presented at the 2007 Annual Meeting of the American Association for Public Opinion Research, Anaheim, CA.

Fowler, F. J. (1993). *Survey research methods*. Newbury Park, CA: Sage Publications.

Fricker, S., & Tourangeau, R. (2010). Examining the relationship between nonresponse propensity and data quality in two national household surveys. *Public Opinion Quarterly*, *74*, 934–955.

Göritz, A. S., & Crutzen, R. (2012). Reminders in web-based data collection: increasing response at the price of retention? *American Journal of Evaluation*, *33*, 240–250.

Groves, R. M. (1989). *Survey errors and survey costs*. New York: John Wiley & Sons, Inc.

Groves, R. M. (2006). Nonresponse rates and nonresponse error in household surveys. *Public Opinion Quarterly*, *70*, 646–675.

Groves, R. M., & Couper, M. P. (1998). *Nonresponse in household interview surveys*. New York: John Wiley & Sons, Inc.

Groves, R. M., Fowler, Jr.,, F. J., Couper, M. P., Lepkowski, J. M., Singer, E., & Tourangeau, R. (2004a). *Survey methodology*. Hoboken, NJ: John Wiley & Sons, Inc.

Groves, R. M., & Peytcheva, E. (2008). The impact of nonresponse rates on nonresponse bias. *Public Opinion Quarterly*, *72*, 167–189.

Groves, R. M., Presser, S., & Dipko, S. (2004b). The role of topic interest in survey participation decisions. *Public Opinion Quarterly*, *68*, 2–31.

Heerwegh, D., Abts, K., & Loosveldt, G. (2007). Minimizing survey refusal and noncontact rates: Do our efforts pay off? *Survey Research Methods*, *1*, 3–10.

Holbrook, A. L., Green, M. C., & Krosnick, J. A. (2003). Telephone versus face-to-face interviewing of national probability samples with long questionnaires: Comparisons of respondent satisficing and social desirability response bias. *Public Opinion Quarterly*, *67*, 79–125.

Kaminska, O., McCutcheon, A. L., & Billiet, J. (2010). Satisficing among reluctant respondents in a cross-national context. *Public Opinion Quarterly*, *74*, 956–984.

Keeter, S., Kennedy, C., Dimock, M., Best, J., & Craighill, P. (2006). Gauging the impact of growing nonresponse on estimates from a national RDD telephone survey. *Public Opinion Quarterly*, *70*, 759–779.

Keeter, S., Miller, C., Kohut, A., Groves, R. M., & Presser, S. (2000). Consequences of reducing nonresponse in a national telephone survey. *Public Opinion Quarterly*, *64*, 125–148.

Kreuter, F., Muller, G., & Trappmann, M. (2010). Nonresponse and measurement error in employment research: Making use of administrative data. *Public Opinion Quarterly*, *74*, 880–906.

Krosnick, J. A. (1991). Response strategies for coping with the cognitive demands of attitude measures in surveys. *Applied Cognitive Psychology*, *5*, 213–236.

Krosnick, J. A., & Alwin, D. F. (1987). An evaluation of a cognitive theory of response order effects in survey measurement. *Public Opinion Quarterly*, *51*, 201–219.

Merkle, D., & Edelman, M. (2002). Nonresponse in exit polls: A comprehensive analysis. In R. M. Groves, D. A. Dillman, J. L. Eltinge, & R. J. A. Little (Eds.), *Survey nonresponse* (pp. 243–257). New York: John Wiley & Sons, Inc.

Newman, J. C., Des Jarlais, D. C., Turner, C. F., Gribble, J., Cooley, P., & Paone, D. (2002). The differential effects of face-to-face and computer interview modes. *American Journal of Public Health*, *92*, 294–297.

Olson, K. (2006). Survey participation, nonresponse bias, measurement error bias, and total bias. *Public Opinion Quarterly*, *70*, 737–758.

Olson, K., Feng, C., & Witt, L. (2008). When do nonresponse follow-ups improve or reduce data quality? A meta-analysis and review of the existing literature. Paper presented at the International Total Survey Error Workshop, Research Triangle Park, NC June 1–4. Retrieved from: 2008.http://www.niss.org/sites/default/files/OlsonTSEWorkshopNRMEReview080108.pdf (accessed May 22, 2013).

Olson, K., & Kennedy, C. (2006). Examination of the relationship between nonresponse and measurement error in a validation study of alumni. In *Proceedings of the American Statistical Association*, Survey Research Methodology Section.

Pew Center. (2012). Assessing the representativeness of public opinion surveys, May 15, 2010. Pew Research Center for the People & the Press. Retrieved from: http://www.people-press.org/2012/05/15/assessing-the-representativeness-of-public-opinion-surveys/ (accessed May 22, 2013).

Peytchev, A., Baxter, R. K., & Carley-Baxter, L. R. (2009). Not all survey effort is equal: Reduction of nonresponse bias and nonresponse error. *Public Opinion Quarterly*, *73*, 785–806.

Pineau, V. J., & Dennis, M. J. (2004). Methodology for probability-based recruitment for a web-enabled panel. Retrieved from: http://www.knowledgenetworks.com/ganp/docs/Knowledge-Networks-Methodology.pdf.

Rogers, S. M., Willis, G., Al-Tayyib, A., Villarroel, M. A., Turner, C. F., Ganapathi, L., et al. (2005). Audio computer-assisted interviewing to measure HIV risk behaviors in a clinic population. *Sexually Transmitted Infections*, *81*, 501–507.

Sakshaug, J. W., Yan, T., & Tourangeau, R. (2010). Nonresponse error, measurement error, and mode of data collection: Tradeoffs in a multi-mode survey of sensitive and non-sensitive items. *Public Opinion Quarterly*, *74*, 907–933.

Singer, E. (2002). The use of incentives to reduce nonresponse in household surveys. In R. M. Groves, D.A. Dillman, J. L. Eltinge, & R. J. A. Little (Eds.), *Survey nonresponse* (pp. 163–178). New York: John Wiley & Sons, Inc.

Singer, E. (2006). Introduction: Nonresponse bias in household surveys. *Public Opinion Quarterly, 70,* 637–645.

Smith, T. W. (2003). An experimental comparison of knowledge networks and the GSS. *International Journal of Public Opinion Research, 15,* 167–179.

Teitler, J. O., Reichman, N. E., & Sprachman, S. (2003). Costs and benefits of improving response rates for a hard-to-reach population. *Public Opinion Quarterly, 67,* 126–138.

Tourangeau, R. (1984). Cognitive sciences and survey methods. In T. Jabine, M. Straf, J. Tanur, & R. Tourangeau (Eds.), *Cognitive aspects of survey methodology: Building a bridge between disciplines* (pp. 73–100). Washington, DC: National Academy Press.

Tourangeau, R. & Rasinski, K. A. (1988). Cognitive processes underlying context effects in attitude measurement. *Psychological Bulletin, 103,* 299–314.

Weisberg, H. F. (2005). *The total survey error approach: A guide to the new science of survey research.* Chicago: University of Chicago Press.

15

Nonresponse and measurement error in an online panel

Does additional effort to recruit reluctant respondents result in poorer quality data?

Caroline Roberts[a], Nick Allum[b], and Patrick Sturgis[c]

[a]*University of Lausanne, Switzerland*
[b]*University of Essex, UK*
[c]*University of Southampton, UK*

15.1 Introduction

Over the past two decades or so, in the face of almost universally falling response rates (e.g., Brick & Williams, 2013; Groves, 2011), much research attention has been focused on discovering the best ways to minimize nonresponse to household surveys. Such methods include using repeated call attempts to sample members who are hard to contact, and changing the standard recruitment protocol to try to persuade more reluctant sample members to participate (e.g., as in "refusal conversion" interviews). Underlying these now standard fieldwork practices is the assumption that such efforts to improve response rates will reduce bias in survey estimates which arises as a result of underrepresentation of population subgroups in the achieved sample. While the generality of this assumption has been called into question by recent empirical findings (see Groves & Peytcheva, 2008), methodologists have also begun to pay closer attention to other sources of nonsampling error and, in particular, to how different sources of error may be causally related to one another (Biemer, 2010; Groves &

Online Panel Research: A Data Quality Perspective, First Edition.
Edited by Mario Callegaro, Reg Baker, Jelke Bethlehem, Anja S. Göritz, Jon A. Krosnick and Paul J. Lavrakas.
© 2014 John Wiley & Sons, Ltd. Published 2014 by John Wiley & Sons, Ltd.
Companion website: www.wiley.com/go/online_panel

Lyberg, 2010). One example of how apparently independent errors may in fact be inter-related concerns nonresponse bias and measurement error. It has increasingly been observed that efforts to recruit hard-to-contact or reluctant sample members may have the undesired conse-quence of bringing respondents into the sample who have little impact on nonresponse bias but are more likely to provide inaccurate responses (Groves & Couper, 1998b). Thus, while response enhancement procedures may serve to increase response rates and thereby (poten-tially) reduce nonresponse bias, they may inadvertently result in an increase in total survey error due to the introduction of higher levels of measurement error from less motivated respon-dents.

The majority of the research evidence on this issue is based on cross-sectional survey designs. However, the question of how these errors relate to one another and, importantly, to survey costs is perhaps even more germane in the context of longitudinal survey research, where the evidence base is considerably thinner. The issue of error tradeoffs and costs is particularly important for longitudinal surveys because the quality of panel data depends not only on achieving an adequate representation of the population at the recruitment stage, but also on ensuring the cooperation of sample members throughout the duration of the study. Moreover, it is likely that the nature of measurement error is somewhat different in a longi-tudinal context as a result of practice, or so-called "panel conditioning" effects (Warren & Halpern-Manners, 2012). It is, therefore, particularly important to examine whether invest-ment in recruiting the most reluctant or hardest-to-contact respondents is an efficient strategy with regard to total errors and costs.

Another important dimension to the question of how survey errors might be related to one another is whether the existing research evidence, based as it is primarily on face-to-face and telephone modes, generalizes to surveys conducted online. In surveys of the general population conducted over the Internet, the requirement to minimize nonresponse errors at the recruitment stage is compounded by a need to minimize noncoverage by supplying sam-ple members who do not have Internet access with the equipment and services they need to complete surveys online. And, as with the contrast between cross-sectional and longitudinal designs, it is known that errors of measurement are rather different in the online context than in conventional modes (Dillman, 2007). If efforts to reduce coverage errors also (inadvertently) augment errors of measurement, for example, because respondents without Internet access are less competent in the use of computers and web-based survey instruments, then it is necessary to question their value.

These questions are the focus of the present chapter. Using data from the 2008 American National Election Studies (ANES) Internet Panel Survey, we investigate the relationship between the amount of fieldwork effort required to recruit panel members and a range of measurement error indicators. Before describing our analytical approach, the data we analyze and our findings, we first review what is currently known about the relation between nonresponse and measurement error, and consider some particular challenges associated with longitudinal studies. The chapter concludes with a discussion of our findings and their implications for the design of future online panel studies based on probability-based samples.

15.2   Understanding the relation between nonresponse and measurement error

Most studies to date which have examined the relationship between nonresponse error and measurement error have focused on whether measurement quality, variously defined, varies as a function of response propensity (Olson, 2013). This approach is based on methods that compare estimates on key variables over subsets of respondents who vary in the amount or type of fieldwork effort that was required to recruit them (e.g., Hox, de Leeuw, & Chang, 2012). In this approach, fieldwork effort has been defined by indicators based on survey process data or "paradata" (Couper, 1998; Kreuter, 2013). Paradata include records of contact procedures used during the survey (e.g., call records generated by automated CATI programs), keystroke or other data available from an online survey (e.g., browser type), or interviewer observations about sampled addresses in a face-to-face survey (Kreuter & Casas-Cordero, 2010). The analysis of such data has become increasingly popular in the study of nonresponse because of the information they provide about both nonrespondents and respondents, making them suitable for nonresponse bias adjustment, as well as for evaluating the effectiveness of different fieldwork practices.

In level of effort analyses, increased "effort" is typically defined by variables such as the number of telephone calls or visits made by a personal interviewer, the use of refusal conversion methods, or an increase in the level of incentive. The underlying assumption in this approach is that sample members who respond only after several attempts to make contact, or who are persuaded to participate via extraordinary efforts are similar to those who never respond. There is a plausible logic to this because the hardest-to-reach and the hardest-to-persuade respondents would indeed never have taken part had fieldwork efforts been terminated sooner. In addition to its use in understanding nonresponse bias, "level of effort" has also been used to investigate the relationship between nonresponse error and measurement error, whereby indicators of measurement error are compared across groups defined by level of fieldwork effort. If measurement error is found to vary by level or type of effort (and varying the type of effort is generally more effective at reducing nonresponse bias (Peytchev, Baxter & Carley-Baxter, 2009)), then it can be concluded that the two error types are inter-dependent.

Olson (2013) identifies seven different mechanisms, which might give rise to an association between the level of effort required to recruit a respondent and the quality of the answers they provide. Specifically, respondents who participate in a survey after additional recruitment efforts may give answers of worse quality due to: (1) reduced motivation; (2) as a reaction against a sense of harassment following repeated callbacks; or (3) a lack of interest in the survey topic or sponsor. These mechanisms, according to Olson, all lead to poorer quality data in comparison to respondents who required lower levels of fieldwork effort (though one might better regard them as different aspects of motivation more generally). However, respondents recruited through additional effort may give *better* quality data if: (4) they become persuaded of the importance of the study as a result of repeated follow-ups; or if (5) they perceive themselves to be a "good respondent" after agreeing to participate as a result of repeated efforts (e.g., Jobber, Allen, & Oakland, 1985). Finally, data quality may either improve or worsen as a result of additional field efforts if (6) they lead to changes in sample characteristics which are themselves associated with data quality, such as education and age (or in the case of web surveys, computer literacy, see Göritz & Crutzen, 2012); or (7) the methods used to recruit more reluctant or hard-to-reach respondents, such as increases in incentive value or a switch in data collection mode, have a direct effect on data quality (Olson, 2013).

A feature shared by a number of these proposed mechanisms is that measurement error may be related to response propensity as a result of a common underlying cause (or causes) (Brunton-Smith & Sturgis, 2011; Groves, 2006; Olson, 2007). According to the "common cause" model, variables which influence a sample member's probability of responding also influence the quality of a respondent's answers. Several studies have identified "motivation" (e.g., Bollinger & David, 2001; Cannell & Fowler, 1963) and cognitive ability (generally measured by level of education) (Kaminska, McCutcheon, & Billiet, 2010; Narayan & Krosnick, 1996; Olson, 2007) as potential common causes of both error types. Other putative common causes include topic interest (e.g., Martin, 1994) and identification with the survey sponsor (Tourangeau, Groves, Kennedy, & Yan, 2009), though these are perhaps better conceived as proximal causes of motivation. Both motivation and ability seem likely to be linked to survey topic because respondents who are interested in and knowledgeable about a topic will generally be more motivated and able to answer questions about it. Moreover, the topic can act as an important leverage in decisions to participate where it is often made salient in survey introductions (Groves, Singer, & Corning, 2000). These variables are also known to be associated with a range of indicators of response quality, particularly those related to satisficing strategies pursued during questionnaire completion (Krosnick, 1991). The notion of a common cause is appealing because if the shared cause can be identified, its detrimental effect on data quality, at least as far as bias is concerned, can be mitigated by the inclusion of common cause variables as controls in multivariate analyses or in the derivation of weights (Fricker & Tourangeau, 2010).

The common cause model implies that the relationship between measurement error and response propensity will be variable-specific, will vary as a function of the type of nonresponse, and will vary according to the indicator of measurement error used (Fricker & Tourangeau, 2010; Sakshaug, Yan, & Tourangeau, 2010; Tourangeau, Groves, & Redline, 2010). This indeed appears to be the case and is reflected in the somewhat inconsistent findings which have emerged across studies which have looked at the relation between response propensity and measurement error (Fricker & Tourangeau, 2010; Olson, 2013). Different measures of fieldwork effort have been used as indicators of response propensity, including the number of follow-up attempts (e.g., telephone calls or in-person visits in interviewer surveys); whether a respondent participated as a result of a refusal conversion; the number of days since the start of fieldwork; and combinations of all of these (Olson, 2013). These approaches also differ in their treatment of different categories of nonresponse. Whereas the number of follow-up attempts and the length of time in the field may confound contact with cooperation, the use of refusal conversion indicators explicitly distinguishes less cooperative respondents from those who have been hard to contact. This distinction is important because respondents who refuse to participate tend to share more characteristics with eventual nonrespondents than those who were harder to contact (Groves et al., 2006), suggesting that differences in response quality would be more likely to be observed for refusal conversions than for noncontacts, if motivation were a common causal factor.

Olson (2013) concludes that there is "reasonable" evidence across existing studies that item nonresponse rates increase on "some, but not all, items" (p. 135) for households that have received repeated follow-ups, refusal conversion attempts, and which took longer in the field before being given a final disposition code. However, the results for response accuracy, involving comparisons between survey self-reports and external records (e.g., Lin & Schaeffer, 1995; Olson, 2006; Peytchev et al., 2009) were less consistent. Similarly, while some differences have been observed on indirect measurement error indicators, such as acquiescence (e.g., Yan, Tourangeau, & Arens, 2004), middle or extreme responses (e.g., Kaminska et al., 2010; Malhotra, Miller, & Wedeking, Chapter 14 in this volume), non-differentiation (e.g., Yan, Tourangeau, & Arens, 2004), primacy and recency (Malhotra, Miller, & Wedeking, 2011), and internal consistency (Fricker & Tourangeau, 2010), no uniform direction in differences was evident.

These studies focusing on measurement errors in attitudinal data have also investigated possible common cause variables and found equally mixed results. For example, Yan, Tourangeau, and Arens (2004) proposed that social desirability concerns might underpin both reluctance to participate in a particular survey as well as the decision to refuse answers to certain questions, but their findings did not support this hypothesis. Fricker and Tourangeau (2010) considered time-stress, social capital and perceptions of survey burden as candidate common cause variables. They concluded that efforts to reduce nonresponse can lead to poorer quality data even after controlling for these factors. However, Kaminska et al. (2010) found that the positive association between reluctance to participate and satisficing response styles disappeared once a control for the common cause, cognitive ability, was introduced.

15.3   Response propensity and measurement error in panel surveys

While a large number of studies examining the relationship between nonresponse and measurement error have focused on cross-sectional surveys, fewer have attempted to analyze how these forms of survey error might be related in a longitudinal context. Bollinger and David (2001) used validation data to assess the discrepancy between self-reported participation in a food-stamp scheme and administrative records. They hypothesized that a latent common cause – propensity to cooperate – would predict both missed panel interviews and response errors, and found that this was indeed the case. Respondents who eventually dropped out of the panel altogether also gave less accurate answers in the waves in which they did take part. Tourangeau et al. (2009) varied several features of the survey introduction (including the description of the topic, sponsor, and length of the survey) in a two-wave web panel. They found that the manipulations had little or no effect on response propensity and careless responding (measured by item nonresponse, straight-lining in grids, and fast completion times), though they did find some evidence of bias on key survey estimates.

Other panel studies have found a link between data quality in one wave and the propensity to respond to subsequent waves. For example, Loosveldt, Pickery, and Billiet (2002) observed a relationship between several measures of item nonresponse in the first wave of a panel and unit nonresponse in the second wave. They attributed this effect to shared factors responsible for both types of nonresponse – notably, perceptions of threat associated with certain question types, and consequently, negative experiences of the initial interview. Yan and Curtin (2010) also looked at how item nonresponse in the first wave of a survey related to the propensity to respond to a subsequent wave, and similarly found that respondents with higher rates of item missing data were significantly less likely to respond to a re-interview request. For nonprobability-based Internet panel studies, there is evidence that participation in previous panel waves predicts participation in future waves (Göritz & Wolff, 2007; Göritz, Wolff, & Goldstein, 2008), but these studies did not look specifically at how response propensity relates to response quality.

A particular advantage of using panel data to investigate correlated survey errors is that information gathered at previous waves can provide important leverage on subsequent attrition and measurement error which is not, by definition, available for cross-sectional studies. This not only permits the specification of more sophisticated models of response propensity across panel waves, but also provides an array of candidate common cause variables to examine. Panel surveys also give rise to measurement errors which are not found in the cross-sectional context, notably "conditioning" (or learning) effects. Panel conditioning occurs when respondents' answers in later waves of a longitudinal study are affected by their participation in prior waves (Sturgis, Allum & Brunton-Smith, 2008). For example, people may change their behaviors and attitudes as a direct result of answering questions about them, or the quality of their answers may improve or reduce as a result of gaining experience of how a survey questionnaire is administered (Warren & Halpern-Manners, 2012). Improvements in response quality may occur when respondents become familiar with certain question formats and make fewer errors. By contrast, a reduction in data quality may occur if respondents learn strategies to reduce the cognitive burden of the survey task, such as ways to avoid subsequent follow-up questions or to answer questions more quickly (e.g., Duan, Alegria, Canino, McGuire, & Takeuchi, 2007). Evidence of panel conditioning is often undermined by methodological confounds in research designed to investigate it (Warren & Halpern-Manners, 2012), but evidence from a number of studies suggests that online panels may be susceptible to conditioning effects (e.g., Das, Toepoel, & van Soest, 2011; Dennis, 2001; Kruse et al., 2010; Toepoel, Das, & van Soest, 2009). It is important, therefore, to consider how these types of errors, which arise only in the longitudinal context, are related to response propensity and, by extension, to underlying common causes.

15.4   The present study

In the remainder of the chapter, we present a study designed to assess how efforts to reduce nonresponse at the recruitment stage of an Internet panel survey are related to subsequent cooperativeness in the panel, and the quality of responses provided. First, we compare respondents on a variety of indicators of cooperativeness and recruitment effort. We then estimate regression models predicting the likelihood of attriting from the panel using socio-demographic covariates and recruitment effort paradata. We then compare respondents on a variety of indirect indicators of response quality at wave 1 across levels of "attrition propensity" and test whether any observed associations are "explained" by candidate common cause variables that were measured at the time of recruitment.

We then extend this analysis to consider the longer-term relationship between response propensity and measurement error, by taking into account the potential negative impact that panel participation may have on respondents' execution of the response process, and respondents' propensity to engage in satisficing behaviors as a result of becoming more familiar with the invitations to satisfice posed by the questionnaire design (Krosnick et al., 2002). To address this question with the data we analyze here, we look at differences between reluctant and cooperative panelists from the original and a refresher sample: (1) to look for evidence of panel conditioning; and (2) to assess whether conditioning varies as a function of nonresponse propensity. Specifically, we address the following sets of related research questions:

1. What is the relation between recruitment effort and cooperation in the panel? To what extent does effort to recruit participants to the panel relate to their later cooperation in the panel? Do we find evidence that respondents who are recruited with greater field-work effort are less committed to participation in survey waves?

2. Do more reluctant participants contribute data of poorer quality? Do the least cooperative panelists at the recruitment stage also go on to give poor quality answers in later waves? Does giving poor quality answers in the recruitment interview relate to later nonparticipation in the panel surveys?

3. If there is a relation between cooperativeness at the recruitment stage and response quality in survey waves, can we identify common correlates of response propensity and data quality?

4. What happens over the course of time? Is there evidence of learning and panel conditioning? If so, how does this vary in relation to panel cooperativeness?

15.5 Data

The data we use are from the 2008–2009 American National Election Studies (ANES) Internet Panel Survey, a purpose-designed panel study consisting of 21 monthly surveys completed online. The panel was based on a probability sample of U.S. citizens age 18 or older as of Election Day (4th November) 2008 (see DeBell, Krosnick, & Lupia, 2010, for details). Recruitment for the panel was carried out by telephone using list-assisted random-digit-dialing (RDD) methods. Two cohorts of participants were recruited to the study. The first was recruited in Autumn 2007, and started completing monthly surveys from January 2008. The second cohort was recruited during the summer of 2008 and started completing surveys in September 2008. In this study, we focus mainly on data from the first cohort of respondents, examining data from the second cohort only later when addressing the question of panel conditioning. All monthly panel surveys were conducted via the Internet, and respondents living in households without a computer and Internet access were offered a free web appliance ("MSN TV 2") and free Internet service to enable them to access the surveys via their television set. The MSN TV 2 device (an Internet and media player manufactured by Microsoft) allows the user to surf the Internet and send emails via a standard dial-up connection, using a hand-held remote and a wireless keyboard connected to their television. Respondents receiving the device were also provided with full technical support (DeBell et al., 2010). Data collection for the panel was carried out by Knowledge Networks (KN).

The sample for the first recruitment cohort consisted of 12,809 landline telephone numbers and complete interviews were obtained at 2,371 (18.4%) telephone numbers. The fieldwork period for recruitment was around 4 months, during which time up to 50 call attempts were made to each number. Two methods were used to try to persuade members of the sample who had not been successfully recruited at the end of the call attempts to join the panel. First, at telephone numbers where someone initially refused to take part, specialist refusal conversion efforts were made by experienced interviewers at the National Opinion Research Center (after sending a letter containing an additional $5 incentive). A total of 84 recruitment interviews were completed as a result of these refusal conversion efforts (0.7%). Second, an "Internet-only" recruitment effort was used by KN to further attempt recruitment for telephone numbers at which telephone recruitment efforts had been exhausted (DeBell et al., 2010). This involved sending letters to all households where the telephone number could be matched to addresses and inviting residents to complete a recruitment questionnaire online.[1] A total of 64 (0.5%) cohort 1 recruitment interviews were completed online.

The sample for the second recruitment cohort consisted of 10,720 landline telephone numbers and complete interviews were obtained at 1,834 (17.1%) telephone numbers. The fieldwork period lasted around 3.5 months, with again, up to 50 call attempts made to each number. No specialist refusal conversion interviews were carried out for the second cohort, but

---

[1] Note that invitations to participate in wave 1 were also made to households in cohort 1 that were not successfully recruited to the panel. We do not include these cases in the analysis we present here.

55 (0.5%) respondents completed the recruitment interview online. The minimum response rate for cohorts one and two combined (AAPOR Response Rate 1) for the recruitment survey was 26% and the estimated response rate based on assumptions about the rate of eligibility among telephone numbers with unknown eligibility (AAPOR Response Rate 3) was 42% (see DeBell et al., 2010; pp. 44–51). In total, 380 (16.1%) recruited panelists in cohort 1 and 294 (16.0%) in cohort 2 required an MSN-TV2 device.

All participants successfully completing a recruitment interview were invited to complete an online "profile" survey, designed to collect basic background information about the panelists and to familiarize them with the format of the surveys and questions, and the process of completing questionnaires online. Profile surveys are standard practice in panel recruitment, and are an important basis on which the success of recruitment can be assessed, as not all participants expressing an intention to participate will go on to actively participate in the panel (DiSogra, Callegaro, & Hendarwan, 2010). In the ANES panel, email invitations to complete monthly surveys were also sent to those who had not (yet) completed the profile survey, and respondents had the opportunity to complete the background questions at the end of the main questionnaire (DeBell et al., 2010). Reminder emails were sent to respondents who did not answer the monthly survey within a few days. In all communications with prospective and recruited participants, the study was referred to as the Knowledge Networks Monthly Special Topics Panel, to avoid respondents associating it with the ANES, and to try to minimize nonresponse bias associated with interest in electoral politics. The advance letter for the recruitment survey offered an unconditional $2 incentive, plus an offer of $10 on completion of the telephone interview. The standard incentive for each monthly survey completed was $10 (some purpose-designed experiments were carried out to vary the incentive for a limited number of panel members in later waves of the study, but we do not address this here). Net attrition during the first year of the panel was relatively low, however, there was a sizeable initial loss between the telephone recruitment survey and the completion of the online surveys, constituting around 31% of the recruited sample from the two cohorts combined (DeBell et al., 2010; pp. 52–72). Table 15.1 shows the number of completed interviews at each panel wave.

In terms of study content, a total of ten survey waves were either wholly or partly devoted to ANES topics predominantly about political attitudes and electoral behavior. The remaining 11 surveys addressed a variety of non-political topics unrelated to the ANES. For the purpose of this chapter, we focus our analysis on data from the first waves of data collection completed by respondents in each cohort (wave 1 for cohort 1, and wave 9 for cohort 2), which were both designated ANES waves.

15.6   Analytical strategy

Our analytic approach is to compare response propensity and response style over groups defined by the level of effort that was required to recruit panel members to the study (Curtin, Presser, & Singer, 2000). We use call record data from the telephone recruitment survey of the 2008–2009 ANES Internet Panel Survey to derive groups based on different indicators of recruitment effort. These are: (1) the number of calls needed from an initial successful contact with the household to the completion of the recruitment interview (1–3 vs. 4 or more); (2) whether the respondent or a household member refused to participate during the call attempts (refused once or more vs. never refused); and (3) whether the respondent was recruited after a special change in fieldwork protocol, i.e., an increase in the standard recruitment effort. As noted earlier, two such "extraordinary" fieldwork efforts were made – specialist refusal

Table 15.1   Number of completed interviews at each panel wave.

Stage	Total in cohort 1	Percent of total	Total in cohort 2	Percent of total
Completed recruitment interviews	2360	100.0	1834	100.0
Standard telephone	2212	93.7	1779	97.0
Refusal conversion	84	3.6	-	-
Internet	64	2.7	55	3.0
Profile	1599	67.8	1293	70.5
Wave 1 (January 2008)[1]	**1577**	**66.8**	-	-
Wave 2	**1438**	**60.9**	-	-
Wave 3	1466	62.1	-	-
Wave 4	1343	56.9	-	-
Wave 5	1148	48.6	-	-
Wave 6 (June 2008)	**1406**	**59.6**	-	-
Wave 7	1108	46.9	-	-
Wave 8	1396	59.2	-	-
Wave 9	**1466**	**62.1**	**1098**	**59.9**
Wave 10	**1487**	**63.0**	**1118**	**61.0**
Wave 11 (November 2008)	**1482**	**62.8**	**1158**	**63.1**
Wave 12	1433	60.7	1102	60.1
Wave 13 (January 2009)	1453	61.6	1090	59.4
Wave 14	1447	61.3	1104	60.2
Wave 15	1420	60.2	1071	58.4
Wave 16	1367	57.9	1039	56.7
Wave 17	**1367**	**57.9**	**1007**	54.9
Wave 18	1328	56.3	1002	54.6
Wave 19	**1303**	**55.2**	990	**54.0**
Wave 20	**1304**	**55.3**	969	52.8
Wave 21 (September 2009)	**1267**	**53.7**	932	50.8

[1] Waves in bold print were designated ANES waves.

conversion interviews for a sample of refusing respondents; and recruitment by Internet for households where an address match could be made with the telephone number (after the maximum of 50 call attempts had been exceeded). Because the numbers recruited by each of these procedures were small, the groups were combined, so we compare those that were recruited after a protocol change with those recruited using the standard telephone recruitment.

We begin by comparing the "level of effort" groups on five different indicators of "cooperativeness" in the panel: (1) whether or not the respondent completed the profile survey after completing the recruitment interview; (2) the total number of monthly surveys completed; (3) whether the respondent completed all the monthly surveys; (4) whether the respondent did not complete the last two monthly surveys (one way of defining attrition from the panel); and (5) whether the respondent ever completed a monthly survey (i.e., if they dropped out after the recruitment interview and never completed a wave). We then compare the groups on background variables measured at the time of the recruitment survey, in order to understand the characteristics of more and less cooperative respondents and estimate a series of regression models predicting the probability of the respondent being "cooperative" according to the definitions set out above. As covariates we include sex, age in years, age squared, and ethnicity, and the survey process data denoting level of fieldwork effort. The aim of these analyses is to examine the relationship between the degree of effort required to recruit panel members and their later cooperativeness in the panel, while controlling for other available variables that might influence participation decisions. Note that respondents' level of education (often used as a proxy for cognitive ability) would also be an appropriate control variable to include here, as it has been shown to be correlated with willingness to participate in surveys (e.g., Freese & Branigan, 2012; Groves & Couper, 1998a, p. 128). However, in order to maximize the amount of auxiliary data available about nonrespondents who stopped participating after the recruitment stage, we restrict our analysis to background variables measured in the recruitment survey (respondents were first asked about their educational background in the profile survey).

Having examined the relationship between recruitment effort and cooperation, we proceed to a consideration of how reluctance to participate at the recruitment stage is related to the quality of responses provided. If the hardest-to-recruit respondents possess less ability and/or motivation to answer the survey questions, we should expect this to be reflected in the quality of the data they provide. This expectation is based on the theory of survey satisficing (Krosnick, 1991), which proposes that the prevalence of certain response styles will vary as a function of the difficulty of the response task, the ability of the respondent to answer the questions, and the respondent's motivation to make the necessary effort to provide an optimal response (Krosnick, 1991; pp. 220–225). Given that the same factors may also influence the decision to participate in a survey in the first place, respondents who participate in a survey with considerable reluctance may lack the motivation needed to respond to questions accurately.

We use multiple indirect indicators of measurement error in our analysis (described further below). Using these response quality indicators, we compare levels of data quality across different participants according to their cooperativeness. To begin with, we look at whether those who were hardest to recruit also gave data of poorer quality during the recruitment interview, by estimating OLS regression equations predicting scores on three indicators of response quality, while controlling for the same recruitment survey variables included in the equations predicting panel cooperativeness. Then, we look at response quality among respondents to wave 1 by cooperativeness, this time measured by the predicted scores from one of the logistic regression equations estimated previously. Specifically, we use the scores estimated by the equation predicting the probability of a respondent refusing to complete the last two monthly surveys, which included the socio-demographic covariates from the recruitment survey and paradata relating to fieldwork effort. From this nonresponse/attrition propensity model, we use the predicted probability of attrition to partition the sample into quintile groups (following the method used by Fricker & Tourangeau, 2010). Based on this definition of cooperativeness, we compare mean scores on our indicators of measurement error across the attrition propensity quintiles. Our prediction is that there will be a greater prevalence of response effects among the most reluctant respondents.

To investigate common causes of attrition propensity and response quality, we estimate the association between two measures of respondent cognition that were included in the recruitment survey and subsequent cooperativeness and our indicators of measurement error. The first of these is the respondent's "need to evaluate" (Jarvis & Petty, 1996) – a measure of the psychological need to form opinions about things.[2] Being low in need to evaluate has been found to negatively influence survey enjoyment (and presumably, as a consequence, willingness to participate), and positively influence the propensity to give "don't know" responses – a form of satisficing (Bizer et al., 2000). The second potential common cause variable that we examine is interest in computers,[3] which we hypothesize may simultaneously influence respondents' willingness to join an Internet panel, as well as the quality of their answers. This is because it is likely to relate to previous experience of using computers, and therefore, the ease and comfort of completing a survey questionnaire online (Dillman, 2007). To test the hypothesis that any observed relation between measurement error and attrition propensity in the panel is explained by these candidate common cause variables, we use the method of Fricker and Tourangeau (2010). This is to estimate OLS regression equations to predict measures of response quality, using the attrition propensity scores as the independent variable at the first step, then introduce the candidate common-cause variables (independently) at a subsequent step. If the covariance between attrition propensity and the measurement error indicator is the result of one of these common-cause variables, then we expect the covariance to reduce significantly after controlling for that variable (Fricker & Tourangeau, 2010, p. 944).

Finally, we turn to the question of how the quality of measurement may develop over time. As we discussed earlier, longitudinal surveys possess an additional dimension of potential measurement error which arises because participating in earlier waves of a panel may influence how respondents answer in later waves. In particular, we are interested in whether panel members' later response behavior is affected by learning that particular question formats provide opportunities to take short cuts by offering response alternatives which do not lead to follow-up or "branching" questions. Our hypothesis is that respondents who are less committed to the panel will be more likely to develop shortcutting strategies, and that this will manifest in more frequent selection of the "neither/nor" midpoint across batteries of items measured on the same scale.

To test the hypothesis that this response strategy is learned through prior experience of completing monthly surveys, we compare respondents from the first recruitment cohort who took part in wave 9 of the panel with answers from respondents from the second recruitment cohort for whom wave 9 was the first monthly survey. If the selection of "neither/nor" midpoints to avoid follow-ups on branching questions among less cooperative panelists increases over time, we expect less cooperative respondents in cohort 1 to have higher rates of midpoint responding than the less cooperative respondents in cohort 2. We test this by estimating OLS regression equations predicting midpoint responding on branching items, using recruitment cohort, attrition propensity scores, and their interaction term as independent variables.

15.6.1 Measures and indicators of response quality

A significant limitation of research into the relation between nonresponse and measurement error is that there are typically no external records of true values for survey variables that can provide direct indicators of measurement error (see Olson, 2006, and Olson & Kennedy, 2006, for examples of studies that use such data). This is especially so for measures of subjective phenomena, such as attitudes, values and beliefs where the notion of a true value

---

[2] Respondents were asked: "Some people have opinions on almost everything; other people have opinions about just some things, and still other people have very few opinions. What about you? Would you say you have opinions about just about everything, about many things, about some things, or about very few things?"

[3] Respondents were asked: "How interested are you in computers? Not interested at all, slightly interested, moderately interested, very interested, or extremely interested?"

is problematic. In the present study, we adopt a standard approach, drawing on the theory of satisficing (Krosnick, 1991; Krosnick & Alwin, 1987), which suggests that certain types of response behavior are due to respondents applying low levels of effort in answering survey questions. If this is true, then we can use indicators of these response effects as proxies for measurement error. We use a range of such indicators, basing our approach on numerous studies that have evaluated differences in the prevalence of response effects across samples of respondents with different characteristics, or exposed to different questionnaire formats (see Roberts, Gilbert & Allum, 2011, for a review). For the first part of our analysis which focuses on response quality in the recruitment questionnaire, we derived three indicators: (1) the aggregate item nonresponse rate for all items applicable to the target panel recruited across the whole of the recruitment questionnaire;[4] (2) the count of no opinion responses to four attitude measures (note that our general indicator of item nonresponse included no opinion reporting as well as all other categories of item nonresponse including refusals, no answer, and other miscellaneous reasons for item missing data); (3) the count of identical answers to the same four attitude measures – i.e., to fail to differentiate the content of the items and to seemingly offer the same response irrespective of item content (called "non-differentiation" by Krosnick (1991, p. 227), also referred to as straight-lining). For the latter two indicators, we distinguish between respondents who *ever* answered "don't know" to the attitude measures, and those who gave identical answers to all four items.

In the second part of our analysis, we examine response quality among respondents who participated in the first wave of the panel. We create indicators of response quality of two types: (1), taking shortcuts, and (2), response effects in attitudinal variables, which cover the standard indicators of satisficing.

15.6.2   Taking shortcuts

Item nonresponse rate. Item nonresponse is commonly used as an indicator of poor data quality (De Leeuw & van der Zouwen, 1988), because by definition it reduces the amount of information available for analysis, reducing precision, and potentially introducing bias. Our hypothesis is that less cooperative respondents (i.e., those with the highest propensity to drop out of the panel) will have higher rates of item nonresponse when they do participate in a survey wave. As for the recruitment survey, we computed an overall rate of item nonresponse based on the number of missing answers given to applicable items in the wave 1 questionnaire. For the purpose of this analysis, we combine all reasons for missing data (no answer, refusal and "don't know," as well as possible technical causes) and divide the total number of missing answers by the total number of items that were applicable to the respondent (which varied as a function of whether the panelist had previously completed the profile questionnaire, as well as according to answers given to questions in the wave 1 survey). This score, therefore,

---

[4] Note that because the first half of the recruitment interview consisted of questions intended to establish the eligibility of household members, and to randomly select a household member, the target panel recruit in households with more than one eligible member may not have been the household reference person responding to the first part of the recruitment interview. For this reason, their item nonresponse rate is calculated based on the number of items applicable to them, which would have been considerably fewer than the number of applicable items for respondents completing the household selection procedure as well as the remainder of the questionnaires. While it is appropriate that the item nonresponse rate is based on the number of applicable items, it is also true that comparing item nonresponse rates across these two types of respondent may not be strictly appropriate, given that overall they would have received many fewer items. Thus, the burden would have been lower, and with it, perhaps also the tendency to skip items.

represents the proportion (ranging from 0–1) of applicable items skipped, where 0 indicates that no items were skipped, and 1 indicates all applicable items were skipped.

The amount of information given. When a survey question leaves it open to respondents to decide how much information to report, respondents taking shortcuts during questionnaire completion are likely to provide less information than respondents who are optimizing. To test whether the least cooperative respondents provide less informative data than the most cooperative, we looked at answers to a "check all that apply" question designed to measure candidate approval. Respondents were asked to indicate which of 17 listed candidates they would vote for if they were able to vote for as many candidates as they would like. We took a count of the number of responses (out of the possible total of 17) given by each respondent, and compare means across our groups of interest. Scores were recoded to range from 0 (lowest count of candidates reported) to 1 (highest count of candidates reported).

15.6.3   Response effects in attitudinal variables

Non-differentiation between items presented in a battery. Non-differentiation refers to a response effect in which respondents select the same scale point to rate a number of items presented together in a battery. Krosnick (1991) identified non-differentiation as a form of weak satisficing (p. 219), based on the idea that when processing effort is reduced, respondents looking to take shortcuts may be tempted to use the same scale point on which they rated the initial items in the battery. The ANES Internet Panel generally avoided this question format either by presenting only a small number of items per screen, or by using a branching/unfolding response scale format in which respondents were first required to rate the overall valence, or direction of their attitude, before reporting its strength. This was true for two batteries of items in the wave 1 monthly survey, one of which assessed respondents' attitudes towards particular policies (containing a total of 9 items), and the other of which was intended to assess respondents' liking of possible candidates for the presidential election (containing a total of 20 items). Although the branching format is intended to minimize the risk of non-differentiation, the repetitive nature of these batteries may encourage satisficing, and we assess this by counting the maximum number of times an identical answer was given to each item in the battery and dividing it by the total number of answers given (i.e., discounting any items left unanswered). A third battery of items (11 in total) to be rated on the same response scale was included in wave 1, which did not use the branching format, and which was presented in the more conventional grid format. These items were designed to measure respondents' perceptions about the condition of the country. We similarly computed a measure of the proportion of items for which the same answer was selected in this battery. Finally, we computed an aggregate measure of non-differentiation in the wave 1 survey by calculating the mean of these three measures, and recoding these scores to range between 0 (indicating the lowest observed rate of non-differentiation) and 1 (highest rate of non-differentiation).

Preference for scale midpoints. The three batteries of items described above also formed the basis of a measure of midpoint responding, which, though it has not been found to be consistent with satisficing theory (e.g., unlike other response effects associated with satisficing, it has been shown to be more common among respondents with higher levels of education, see O'Muircheartaigh, Krosnick, & Helic, 1999; Sturgis, Roberts, & Smith, 2014), is frequently used as an indicator of poor data quality in surveys because it reduces the amount of meaningful information available for analysis. The debate about the precise mechanism underlying respondents' tendency to select a middle alternative when it is made available hinges on the meaning attributed by respondents to the middle category, which in turn, depends on how it is labeled. Early experiments by Schuman and Presser (1981) concluded that when the midpoint represents an option of preserving the status quo, compared with advocating policy change, respondents tend to be more likely to report their wish to "keep things the same" than when the midpoint is omitted. We test this effect here by looking at whether less cooperative respondents were more likely to endorse the status quo when evaluating the condition of their country than the more cooperative respondents by "over-preferring" the middle alternative, which was labeled "about the same."

When midpoints are labeled "neither/nor," then the ambiguity of their meaning leads to even greater ambiguity in the interpretation of whether respondents who select the middle alternative are communicating a genuinely ambivalent attitude, are simply satisficing by selecting an "easy to select, easy to defend" answer (Krosnick, Judd, & Wittenbrink, 1995), or using it as a more socially acceptable way to conceal that they do not know (Sturgis et al., 2012). We do not discuss these issues further here. However, we are interested in over-preference for middle alternatives in the ANES Internet panel for other reasons. In the batteries of items using branching response formats, the first part of the question asked respondents to indicate the overall direction of their attitude (whether they liked or disliked a candidate, or neither liked nor disliked them; or whether they approved or disapproved of a policy, or neither approved nor disapproved). Indicating attitude valence at the first stage then revealed a follow-up question on the screen below the first item to ask about the extent of liking/disliking or approval/disapproval, thus doubling the number of applicable questions to answer for anybody not selecting the neither/nor midpoint. Respondents looking to take shortcuts would no doubt quickly learn that selecting the middle alternative was the quickest way to advance through the questionnaire to avoid the follow-up questions associated with the unfolding format. Given this feature of the survey's design, we decided to look at preference for neither/nor middle alternatives as an indicator of satisficing, and to see whether respondents deemed to have less commitment to the panel would be more likely to select the midpoint as a way to reduce the number of items they would have to answer. We computed a simple count of the number of middle alternatives selected in the two batteries of branching items and divided each one by the number of items in the battery to yield the proportion of items for which a midpoint was selected. We then computed an aggregate measure of midpoint responding on branching questions in wave 1, using the mean proportion of middle alternative answers given across each of the two batteries. This score was then recoded to range between 0 (indicating the lowest rate of midpoint responding) and 1 (indicating the highest rate).

15.7   Results

15.7.1   The relation between recruitment efforts and panel cooperation

The respondents to the recruitment survey who were either harder to reach (requiring a greater number of call attempts before completing the CATI interview) or more reluctant to participate (refusing to take part on at least one call attempt) later showed a lower level of cooperation in the panel. This finding is presented in Table 15.2, which compares estimates across the groups described earlier on various indicators of cooperation (completing the profile survey; the mean number of monthly survey waves completed; completing less than half of the monthly surveys, completing more than half of the monthly surveys, completing all the monthly surveys; skipping the last two monthly surveys; and never answering a monthly survey). Statistically

Table 15.2 Panel cooperativeness by number of calls, initial refusals, and protocol change at the recruitment stage (cohort 1).

	Calls 1–3	Calls 4+	Never refused	Refused	Standard recruitment	Protocol change	All recruited respondents
	N = 1225	N = 1135	N = 1851	N = 509	N = 2212	N = 148	N = 2360
	% (SE)	% (SE)	% (SE)	% (SE)	% (SE)	% (SE)	% (SE)
Completed the profile survey (%)	73.9 (.01)	61.1 (.01)***	71.5 (.01)	54.2 (.02)***	71.8 (.01)	6.8 (.02)***	67.8 (.01)
Number of complete waves (mean)	13.8 (.25)	12.1 (.27)***	13.5 (.21)	11.2 (.41)***	13.1 (.19)	10.4 (.72)***	13.0 (.19)
Completed profile + 1–10 waves (%)	8.7 (.01)	10.2 (.01)	9.8 (.01)	7.9 (.02)	9.9 (.01)	1.4 (.01)**	9.4 (.01)
Completed profile + 11–20 waves (%)	35.2 (.01)	28.2 (.01)***	33.1 (.01)	27.3 (.02)*	33.6 (.01)	4.7 (.02)***	31.8 (.01)
Completed all waves (%)	28.2 (.01)	20.4 (.01)***	26.4 (.01)	17.5 (.02)***	26.1 (.01)	0.7 (.01)***	24.5 (.01)
Skipped last two waves (%)	39.9 (.01)	45.9 (.01)**	40.8 (.01)	49.9 (.02)***	42.3 (.01)	50.7 (.04)*	41.8 (.01)
Answered no waves (%)	21.1 (.01)	25.0 (.01)*	20.7 (.01)	31.2 (.02)***	22.3 (.01)	32.4 (.04)**	23.0 (.01)[†]

*** p < .001,
** p < .01,
* p < .05,
[†] p < .10.

significant differences were found between comparison groups on all seven of the indicators. Those who required greater recruitment effort completed fewer monthly surveys on average, were less likely to complete the profile survey, were less likely to complete all or more than half of the monthly surveys, and were more likely to skip the last two monthly surveys, or to never complete a monthly survey.

The only exception to this pattern concerned the likelihood of completing *less* than half of the surveys, where there was no difference between groups on the basis of the number of calls needed to recruit the participant or having refused to participate, but those recruited following a change in protocol were significantly less likely than those recruited with the standard protocol to complete less than half of the surveys. In other words, reluctance to participate at recruitment persists, and reluctant participants are more likely to become uncooperative panelists.

Next, we investigate which factors contribute to cooperation and non-cooperation in the panel following the initial recruitment interview, by fitting logistic regression models predicting varying degrees of cooperativeness (defined previously as completing the profile survey, completing all survey waves, skipping the last two waves, skipping the last five waves, and never completing a monthly survey). Using socio-demographic data reported at recruitment, together with call record data, we analyze the characteristics of the cooperative and uncooperative respondents, and the relation between fieldwork effort and cooperation. Estimated coefficients from the logistic regression models are shown in Table 15.3.

Respondent sex is not a significant predictor of panel cooperativeness, whereas age has a statistically significant nonlinear relation with participation in the panel. The odds of cooperating rise with age until up to around 47, when the odds begin to decline with age. Race and ethnicity are both important factors relating to cooperation in the panel. Compared to White panelists, both Black (non-Hispanic) and Hispanic panelists were significantly less likely to complete the profile survey after recruitment, and less likely to complete all the survey waves. Consistent with this, respondents from these and other non-White (non-Hispanic) racial and ethnic groups were also more likely to stop participating in the panel before the last two or five monthly surveys. Black and Hispanic respondents to the recruitment survey were also significantly more likely to never complete any of the monthly surveys.

Turning to the survey process variables, the number of telephone calls needed to make initial contact in the recruitment was positively associated with the likelihood of completing the profile survey and completing all survey waves, and negatively associated with ceasing to participate in the panel or never completing a wave, but these counter-intuitive findings were not statistically significant. However, consistent with our expectations, the number of call attempts needed to complete the recruitment interview following an initial contact was negatively and significantly associated with cooperation. The more calls required to complete the recruitment interview, the less likely respondents were to complete the profile survey or all of the monthly surveys. Consistent with this, the more calls required to complete the recruitment interview, the more likely respondents were to never complete a monthly survey. There was no relation between the number of call attempts and likelihood of panel attrition, however. By contrast, refusing to participate on any call attempt was positively and significantly associated with skipping the last monthly surveys of the panel and with never completing a monthly survey. Respondents to the recruitment survey who were recruited after a change in survey protocol (refusal conversion and internet recruitment) were significantly less likely to go on to complete the profile survey, or to complete all the survey waves, than respondents

Table 15.3 Coefficients from logistic regression equations predicting panel cooperativeness.

Included	Completed profile survey B	Completed all waves B	Skipped last two waves B	Skipped last five waves B	Never completed a wave B
Sex (Male)	−.11	−.02	.00	.08	−.03
Age in years	.06***	.10***	−.09***	−.07***	−.06***
Age squared	−.06***	−.09***	.08***	.07***	.07***
Race/ ethnicity (Reference: White)					
Black	−1.01***	−.67***	.53***	.68***	.96***
Hispanic	−.60**	−1.18***	.32*	.48*	.71***
Other race/ ethnicity	−.31	−.46‡	.52*	.52*	.38
Number of calls to first contact	.03	.01	−.02	−.01	−.03'
Number of calls to complete interview[1]	−.06***	−.03*	.02	.02	.03*
Ever refused	−.12	−.18	.25*	.28*	.38**
Recruited after protocol change[2]	−3.37***	−3.68***	.21	.15	.36
Required MSN-TV	−.93***	−1.33***	1.32***	1.24***	1.2***
Constant	.40	−3.42***	1.56***	.69‡	−.63
N	2360	2360	2360	2360	2360
Nagelkerke-R^2	.25	.15	.12	.12	.15

[1]From first successful contact.
[2]Refusal conversion interviews or by Internet.
***$p < .001$,
**$p < .01$,
*$p < .05$,
‡$p < .10$.

recruited using the standard telephone protocol. This variable was not related to attrition, however. Finally, we found that respondents without Internet access were more likely to drop out of the panel following recruitment and less likely to complete all the monthly surveys (including the profile survey).

15.7.2   The relation between panel cooperation and response quality

Comparing cohort 1 panelists on measures of response quality in the recruitment survey finds that the most cooperative respondents provided answers that were less troubled by response effects than less cooperative respondents. This was evidenced in lower rates of item non-response (Table available in online appendix at www.wiley.com/go/online_panel_research). There was also limited evidence of lower rates of no-opinion reporting. However, contrary to expectations, there was some suggestion that rates of non-differentiation were higher among the most cooperative compared with the least cooperative. To understand the factors contributing to this variation in response quality, we fitted OLS regression models to predict the rate of item nonresponse, no-opinion reporting and non-differentiation in the recruitment survey, using socio-demographic data reported at recruitment and the survey process data analyzed earlier. Coefficients from the equations are shown in Table 15.4.

Focusing only on the equation for item nonresponse, we find, once again, a significant quadratic relation between age and the likelihood of giving fewer answers. Being of Black or Hispanic racial or ethnic origin was significantly and positively related to item nonresponse, and, as predicted by our hypothesis, respondents who were hardest to recruit (requiring a greater number of call attempts to make contact and complete the recruitment interview, or requiring an alternative to the standard telephone recruitment protocol, or requiring an MSN TV device) had higher rates of item nonresponse than those requiring less recruitment effort. However, these findings do not extend to no-opinion reporting and non-differentiation.

Next, we compare response quality across nonresponse propensity quintiles using indicators measured at recruitment (as above) and at wave 1 (item nonresponse, non-differentiation, amount of information given in response to a check-all-that-apply response format, and preference for midpoints). Unlike item nonresponse at the recruitment stage, the item

Table 15.4 Unstandardized coefficients from OLS regression equations predicting rates of item-nonresponse, no opinion reporting, and non-differentiation in the recruitment survey.

Included	Item non-response	No opinion reporting	Non-differentiation
Constant	.02**	.01	.43***
Sex (Male)	−.00	−.00	.01*
Age in years	−.00**	.00	.00*
Age squared	.00**	.00*	−.00*
Race/ ethnicity (Reference: White)			
Black	.01**	.00	−.00
Hispanic	.01*	.00	−.01
Other race/ ethnicity	.01*	.00	.01
Number of calls to first contact	.00***	.00‡	−.00
Number of calls to complete interview[1]	.00*	.00‡	.00
Ever refused (vs. never refused)	−.00	.01**	−.01
Recruited after protocol change[2]	.16***	−.01‡	.01
Required MSN-TV	.04***	.00	−.01
N	2360	2360	2360
R squared	.56	.01	.01
$F_{(11, 2348)}$	268.91***	2.72**	1.22

[1]From first successful contact.
[2]Refusal conversion interviews or by Internet.
***$p <.001$,
**$p <.01$,
*$p <.05$,
‡$p <.10$.

nonresponse rate among wave 1 respondents shows only a very weak relationship with attrition propensity ($F_{4,1572} = 2.20$; $p = .07$; no significant differences between attrition propensity quintiles according to post-hoc Scheffé tests for multiple comparisons). However, we do find significant differences across attrition propensity quintiles on our indicators of "amount of information" ($F_{4, 1553} = 4.79$; $p < 0.01$; quintile 4 statistically different from quintiles 2 and 3); non-differentiation ($F_{4, 1572} = 11.16$; $p < .001$); and preference for midpoints ($F_{4, 1572} = 10.29$; $p < .001$; in both cases, quintiles 1 and 2 were statistically different from quintiles 4 and 5). As can be seen in Figure 15.1, these effects were all in the expected directions, with more item nonresponse, non-differentiation and midpoint responding, and fewer responses to the check-all-that-apply question format in higher attrition propensity quintiles.

15.7.3   Common causes of attrition propensity and response quality

We found no evidence that either "common cause" variable simultaneously affects attrition propensity and the propensity to give answers containing more measurement error. Interest in computers, when added as a covariate to the model predicting the likelihood of skipping the last two waves of the survey, was significantly and negatively related to attrition (those who expressed more interest in computers at recruitment were less likely to skip the last two surveys; see the Table shown in the online appendix at www.wiley.com/go/online_panel_research). By contrast, no such effect was found for "need to evaluate," when the effects of the demographic variables measured at recruitment and the recruitment effort variables were included in the equation. However, only need to evaluate was found to significantly predict response quality at wave 1 (on the indicators of non-differentiation, preference for midpoints, and the amount of information), when controlling for attrition propensity (see Table shown in the online appendix at www.wiley.com/go/online_panel_research). No such effects were found for interest in computers. The significant effect of attrition propensity on response quality remained unchanged when controlling for need to evaluate, so contrary to expectations, we cannot conclude that this variable acts as a common cause of nonresponse and measurement error.

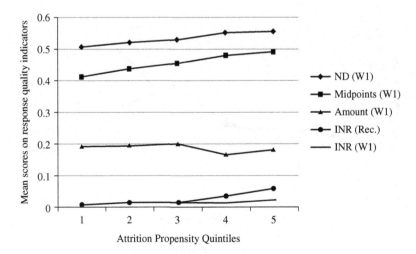

Figure 15.1   Significant relations between response quality and attrition propensity. *Note*: ND (W1): Non-differentiation rate in wave 1; Midpoints (W1): Preference for "neither/nor" midpoints in wave 1; Amount (W1): Amount of information in response to a check-all-that-apply question format; INR (Rec.): Item non-response rate (Recruitment survey); INR (W1): Item non-response rate (wave 1).

15.7.4   Panel conditioning, cooperation and response propensity

Using wave 9 data, we fitted an OLS regression model with a dummy variable indicating cohort, our attrition propensity variable, and the product term interaction of both as predictors of midpoint responding. We found that the interaction between cohort and attrition propensity was statistically significant (see Table 15.5) and suggested that the tendency for midpoint responding is greater for those with a higher propensity to attrit in the original sample than in the refreshment cohort (see Figure 15.2). In other words, those most likely to leave the panel based on the effort required to recruit them, demographics, and dropout behavior will provide data of declining quality the longer they actually stay in the panel.

Table 15.5   Unstandardized coefficients from OLS regression equations predicting rate of midpoint responding at wave 9.

Included	Main effect B	Interaction effect B
Constant	.24***	.27***
Attrition propensity (AP)	.15***	.09*
Recruitment cohort (RC)	.00	−.05‡
Interaction (AP*RC)		.13*
N	2697	2697
R squared	.10	.11
$F_{(2, 2694)/(3,2693)}$	13.26***	10.39***

***$p <.001$,
**$p<.01$,
*$p <.05$,
‡$p<.10$.

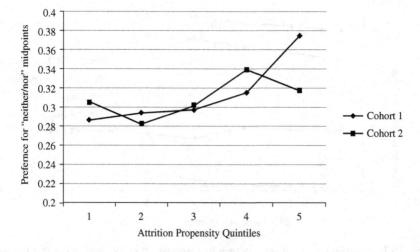

Figure 15.2   Rate of midpoint responding by attrition propensity quintile and recruitment cohort.

15.8   Discussion and conclusion

We began this chapter by noting the growing evidence that efforts to recruit hard-to-contact or reluctant sample members may have the undesired consequence of bringing respondents into the sample who are more likely to provide inaccurate responses (Groves & Couper, 1998b). In the present research, we have contributed to understanding in this area by assessing whether this phenomenon pertains in the context of online panel surveys. We also touched on some of the particularities that panels, rather than cross sections, raise.

We first asked whether respondents who were recruited with greater fieldwork effort were less committed to participation in the panel at subsequent occasions. On this, the answer is an unequivocal yes. Hard-to-recruit respondents completed fewer surveys overall, and were more likely to drop out of the panel altogether. We find this pattern across a range of indicators of recruitment effort, including number of calls and number of refusals. We also find that those recruited after a protocol change directed at the conversion of reluctant and hard-to-reach respondents completed three surveys fewer than those who had the standard protocol applied. These fieldwork effort variables remained statistically significant, and of approximately the same magnitude, as predictors of cooperation after adjusting for other known correlates of survey cooperation, namely race, age, and gender (Groves, 1989; Groves & Couper, 1998b).

A known predictor of participation in surveys is cognitive ability, typically indicated in surveys by level of education (Groves & Couper, 1998a, p. 128), and this variable holds particular interest in the present study as it is also known to moderate certain types of response effect associated with satisficing (Narayan & Krosnick, 1996). Unfortunately, we were not able to include this variable in our analysis as it was not measured at the time of recruitment in the 2008–2009 ANES panel, and our interest was in understanding the variables that predict attrition from the panel, particularly after initial recruitment efforts. To investigate the potential usefulness of this variable in understanding later cooperativeness in the panel, we re-estimated the logistic regression equations shown in Table 15.3 using only those participants who completed the profile survey (at which point they were asked about their educational background). In fact, for these respondents, education was not a significant predictor of later cooperation in the panel, though including it as a covariate did affect the strength of the associations between some of the other covariates on the dependent variables (notably, the significant effect of race/ethnicity was canceled out). Given the potential for education to act as a common cause of both nonresponse and measurement error (Kaminska et al., 2010), future online panels would benefit from including a simple measure in the recruitment questionnaire, to allow further investigation of the factors that might influence the covariance between these two error sources.

We next investigated whether respondents who are prone to drop out of the panel or complete fewer waves tend to give poorer quality data earlier on in the recruitment phase. This turned out to be true for item nonresponse, but not for our other measures of response quality. In particular, we looked at the relationship between survey process variables, along with demographic factors, and response quality. We found a similar pattern: item nonresponse was significantly more prevalent for hard-to-recruit respondents, adjusting for race, age and gender, but other response quality measures were mostly unrelated. Turning to the relation between overall attrition propensity and data quality, we found that respondents with a higher propensity to attrit also provided lower quality data, and this was evident in higher rates of midpoint responding and non-differentiation, and fewer responses to a check-all-that-apply question format for this group in the wave 1 monthly survey.

We investigated two variables measured in the recruitment interview as potential common causes of nonresponse and measurement error: the respondent's interest in computers and their "need to evaluate", both of which we hypothesized could influence willingness to participate in the panel, and at the same time, response quality. While interest in computers did indeed influence later cooperativeness in the panel, we found no relation with response quality when controlling for attrition propensity. By contrast, need to evaluate related independently to response quality and attrition propensity, but did not explain the relation between the two.

Finally, we tested the hypothesis that respondents who had been in the panel longer would learn to use midpoints more often as a way of avoiding having to answer follow-up questions. This expectation was supported, but only for those respondents with a greater potential to attrit. In other words, there is some evidence that reluctant or potentially uncooperative panel members answer questions less carefully as time goes on.

What are the practical implications of these findings for online panel studies? Our evidence comes from only one panel study, the 2008–2009 ANES Internet Panel Survey, which was conducted carefully and with considerable resources at the disposal of the investigators. Many online panels have been, and will, in the future, be less well resourced. So our findings likely underestimate the scale of problems that we have uncovered with our analysis. The key implication of our results is that there is likely to be a tradeoff between gaining more precision at the expense of more measurement error when considering which resources should be expended on obtaining cooperation from reluctant or otherwise hard-to-reach participants. This is in line with findings in other survey modes (Groves & Couper, 1998b; Olson, 2013; Peytchev et al., 2009). One of the largest differences in data quality we found was between those who had been recruited using a special protocol designed to capture hard-to-recruit responders and those subjected to the standard protocol. While every survey situation will be different, we suggest that it will be as important to question the utility of additional resource-intensive efforts for conversion in online panels as it is in other settings. The likely reduction in total survey error needs to be carefully evaluated in light of the mechanisms revealed in this study. The potential loss of quality and efficiency in a panel is of an order of magnitude greater than in a cross-section. We find that reluctant responders are more likely to provide poorer data while remaining in the panel but are also more likely to drop out. So while it may be sensible to predict the losses and gains from increased recruitment efforts, it may also pay to focus on appropriate compensatory adjustments through weighting or other model-based approaches, that are less resource-hungry. The advantage of online panels is that there is the opportunity at the outset to collect a great deal of information at relatively low marginal cost that will help to make these adjustments.

Finally, it is worth pointing out that the ANES Internet Panel was based on a probability sample design. Most online panels are not. In an era of falling response rates and uncertainty about the relationship between response rate and nonresponse bias, some argue for a paradigm shift to model-based rather than design-based inference through matching or other methods (Rivers & Bailey, 2009). The advantage of such an approach is that it can obviate the necessity for expensive conversion efforts. However, we assume that all potential respondents, whether to opt-in or probability panels, have an underlying propensity to take part and that this is likely correlated with the quality of the data provided. This being so, it remains important to understand these mechanisms so that they can at a minimum be modeled and adjustments based on them made.

The American National Election Studies are widely considered to be the gold standard of election studies internationally (Krosnick & Lupia, 2012; p. 13), in part, due to their commitment to best practice in survey data collection methodology. As such, the 2008–2009 Internet Panel Study provided an exemplary source of data to work with for our particular research needs. We were able to take advantage of a rich array of high quality, and well-documented paradata relating to the telephone recruitment phase of the Internet panel, as well as a number of candidate common cause variables measured at the time of recruitment. Such data are not routinely collected, but as interest in survey paradata increases, and researchers become more aware of its potential value for investigating total survey error (see Kreuter, 2013), they are more often being made available for methodological research. While recent years have seen a proliferation of studies using survey process data of these kinds to investigate the relation between nonresponse and measurement error (Olson, 2013), there are still comparatively few studies that have focused specifically on online panel surveys, and much that can still be learned about how to optimize the design of such studies, and the optimal balance to be gained between costs and errors. This concerns investment in nonresponse bias reduction and its potential impact on measurement error, but also the costs involved in correcting coverage error associated with the failure to include households/individuals without Internet access, and the impact of the methods used on both nonresponse and measurement error. Carefully designed survey process data will continue to play an essential role in such investigations. With new probability-based online panels being established in several countries, those responsible would do well to ensure the availability of such data so that new research can inform improvements in the design of future Internet panel surveys.

References

Biemer, P. (2010). Total survey error: Design, implementation and evaluation. *Public Opinion Quarterly, 74,* 817–848.

Bizer, G. Y., Krosnick, J. A., Petty, R. E., Rucker, D. D., & Wheeler, S. C. (2000). Need for cognition and need to evaluate in the 1998 National Election Survey Pilot Study. *ANES Pilot Study Report,* No. nes008997.

Bollinger, C. R., & David, M. H. (2001). Estimation with response error and nonresponse: Food stamp participation in the SIPP. *Journal of Business and Economic Statistics, 19,* 295–320.

Brick, J. M., & Williams, D. (2013). Explaining rising nonresponse rates in cross-sectional surveys. *The ANNALS of the American Academy of Political and Social Science, 645,* 36–59.

Brunton-Smith, I., & Sturgis, P. (2011). Do neighborhoods generate fear of crime?: An empirical test using the British Crime Survey. *Criminology, 49,* 331–369.

Cannell, C. F., & Fowler, F. J. (1963). Comparison of a self-enumerative procedure and a personal interview: A validity study. *Public Opinion Quarterly, 27,* 250–264.

Couper, M. (1998). Measuring survey quality in a CASIC environment. In *Proceedings of the Section on Survey Research Methods of the American Statistical Association.*

Curtin, R., Presser, S., & Singer, E. (2000). The effects of response rate changes on the index of consumer sentiment. *Public Opinion Quarterly, 64,* 413–428.

Das, M., Toepoel, V., & van Soest, A. (2011). Nonparametric tests of panel conditioning and attrition bias in panel surveys. *Sociological Methods & Research, 40,* 32–56.

DeBell, M., Krosnick, J. A., & Lupia, A. (2010). *Methodology report and user's guide for the 2008–2009 ANES Panel Study.* Palo Alto, CA, and Ann Arbor, MI: Stanford University and the University of Michigan.

de Leeuw, E., & van der Zouwen, J. (1988). Data quality in telephone and face-to- face surveys: A comparative analysis. In R. M. Groves, P. P. Biemer, L. E. Lyberg, J. T. Massey, W. L. Nicholls II,, & J. Waksberg (Eds.), *Telephone survey methodology* (pp. 283–299). New York: JohnWiley & Sons, Inc.

Dennis, J.M. (2001). Are internet panels creating professional respondents? A study of panel effects. *Marketing Research, 13*, 34–38.

Dillman, D. A. (2007). *Mail and internet surveys: The tailored design method*. Hoboken, NJ: John Wiley & Sons, Inc.

DiSogra, C., Callegaro, M., & Hendarwan, E. (2010). Recruiting probability-based web panel members using an address-based sample frame: Results from a pilot study conducted by Knowledge Networks. In *Proceedings of the annual meeting of the American Statistical Association* (pp. 5270–5283). Presented at the 64th Annual conference of the American Association for Public Opinion Research, Hollywood, FL: AMSTAT.

Duan, D., Alegria, M., Canino, G., McGuire, T., & Takeuchi, D. (2007). Survey conditioning in self-reported mental health service use: Randomized comparison of alternative instrument formats. *Health Services Research, 42*, 890–907.

Freese, J., & Branigan, A. (2012). Cognitive skills and survey nonresponse: Evidence from two longitudinal studies in the United States. *EurAmerica, 42*, 221–247.

Fricker, S., & Tourangeau, R. (2010). Examining the relationship between nonresponse propensity and data quality in two national household surveys. *Public Opinion Quarterly, 74*, 934–955.

Göritz, A. S., & Crutzen, R. (2012). Reminders in web-based data collection: Increasing response rates at the price of retention? *American Journal of Evaluation, 33*, 240–250.

Göritz, A. S., & Wolff, H.-G. (2007). Lotteries as incentives in longitudinal Web studies. *Social Science Computer Review, 25*, 99–110.

Göritz, A. S., Wolff, H.-G., & Goldstein, D. G. (2008). Individual payments as a longer-term incentive in online panels. *Behavior Research Methods, 40*, 1144–1149.

Groves, R. M. (1989). *Survey errors and survey costs*. New York: John Wiley & Sons, Inc.

Groves, R. M. (2006). Nonresponse rates and nonresponse bias in household surveys. *Public Opinion Quarterly, 70*, 646–675.

Groves, R. M. (2011). Three eras of survey research. *Public Opinion Quarterly, 75*, 861–871.

Groves, R. M., & Couper, M. P. (1998a). *Nonresponse in household interview surveys*. New York: John Wiley and Sons, Inc.

Groves, R. M., & Couper, M. P. (1998b). How survey design features affect participation. In R. M. Groves, & M. P. Couper (Eds.), *Nonresponse in household interview surveys* (pp. 25–46). New York: John Wiley and Sons, Inc.

Groves, R. M., Couper, M. P., Presser, S., Singer, E., Tourangeau, R., Piani Acosta, G., & Nelson, L. (2006). Experiments in producing nonresponse bias. *Public Opinion Quarterly, 70*, 720–736.

Groves, R. M., & Lyberg, L. (2010). Total survey error: Past, present, and future. *Public Opinion Quarterly, 74*, 849–879.

Groves, R. M., & Peytcheva, E. (2008). The impact of nonresponse rates on nonresponse bias. *Public Opinion Quarterly, 72*, 1–23.

Groves, R. M., Singer, E., & Corning, A. D. (2000). Leverage-salience theory of survey participation: Description and an illustration. *Public Opinion Quarterly, 64*, 299–308.

Hox, J. J., de Leeuw, E. D., & Chang, H. T. (2012). Nonresponse versus measurement error: Are reluctant respondents worth pursuing? *Bulletin of Sociological Methodology/Bullétin de Méthodologie Sociologique, 113*, 5–19.

Jarvis, W. B., & Petty, R. E. (1996). The need to evaluate. *Journal of Personality and Social Psychology, 70*, 172–194.

Jobber, D., Allen, N., & Oakland, J. (1985). The impact of telephone notification strategies on response to an industrial mail survey. *International Journal of Marketing Research, 4*, 291–296.

Kaminska, O., McCutcheon, A., & Billiet, J. (2010). Satisficing among reluctant respondents in a cross-national context. *Public Opinion Quarterly, 74*, 956–984.

Kreuter, F. (2013) *Improving surveys with paradata: Analytic uses of process information*. New York: John Wiley & Sons, Inc.

Kreuter, F., & Casas-Cordero, C. (2010). Paradata. In German Data Forum (RatSWD) (Ed.), *Building on progress. Expanding the research infrastructure for the social, economic and behavioral sciences*. Vol. 1 (pp. 509–529). Opladen: Budrich UniPress.

Krosnick, J. A. (1991). Response strategies for coping with the demands of attitude measures in surveys. *Applied Cognitive Psychology, 5*, 214–236.

Krosnick, J. A., & Alwin, D. F. (1987). An evaluation of a cognitive theory of response order effects in survey measurement. *Public Opinion Quarterly, 51*, 201–219.

Krosnick, J. A., Holbrook, A. L., Berent, M. K., Carson, R. T., Hanemann, W. M., Kopp, R. J., Mitchell, R.C., et al. (2002). The impact of "no opinion" response options on data quality: Non-attitude reduction or an invitation to satisfice? *Public Opinion Quarterly, 66*, 371–403.

Krosnick, J. A., Judd, C. M., & Wittenbrink, B. (1995). The measurement of attitudes. In D. Albarracín, B. T. Johnson, & M. P. Zanna (Eds.), *The handbook of attitudes* (pp. 21–78). Mahwah, NJ: Lawrence Erlbaum Associates.

Krosnick, J. A., & Lupia, A. (2012). The American National Election Studies and the importance of new ideas. In J. A. Aldrich, & K. M. McGraw (Eds.), *Improving public opinion surveys: Interdisciplinary innovation and the American National Election Studies*. Princeton, NJ: Princeton University Press.

Kruse, Y., Callegaro, M., Dennis, M. J., DiSogra, C., Subias, T., Lawrence, M., & Tompson, T. (2010). Panel conditioning and attrition in the AP-Yahoo news election panel study. In *Proceedings of the Joint Statistical Meeting, American Association for Public Opinion Research Conference* (pp. 5742–5756). Washington, DC: AMSTAT.

Lin, I-F., & Schaeffer, N. C. (1995). Using survey participants to estimate the impact of nonparticipation. *Public Opinion Quarterly, 59*, 236–258.

Loosveldt, G., Pickery, J., & Billiet, J. (2002). Item non-response as a predictor of unit non-response in a panel. *Journal of Official Statistics, 18*, 545–557.

Malhotra, N., Miller, J. M., & Wedeking, J. (2011). Nonresponse strategies and measurement error. Unpublished manuscript. Retrieved from: http://www.uky.edu/~jpwede2/Nonresponsepaper.pdf (accessed 14 February 2013).

Martin, C. L. (1994). The impact of topic interest on mail survey response behavior. *Journal of the Market Research Society, 36*, 327–338.

Narayan, S., & Krosnick, J. A. (1996). Education moderates some response effects in attitude measurement. *Public Opinion Quarterly, 60*, 58–88.

Olson, K. (2006). Survey participation, nonresponse bias, measurement error bias, and total bias. *Public Opinion Quarterly, 70*, 737–758.

Olson, K. (2007). An investigation of the nonresponse-measurement error nexus. PhD dissertation. University of Michigan.

Olson, K. (2013). Do non-response follow-ups improve or reduce data quality?: A review of the existing literature. *Journal of the Royal Statistical Society, Series A – Statistics in Society, 176*, 129–145.

Olson, K., & Kennedy, C. (2006). Examination of the relationship between nonresponse and measurement error in a validation study of alumni. In *Proceedings of the Survey Research Methods Section of the American Statistical Association*, 4181–4188.

O'Muircheartaigh, C., Krosnick, J. A., & Helic, A. (1999). Middle alternatives, acquiescence, and the quality of questionnaire data. Paper presented at the American Association for Public Opinion Research annual meeting, St. Petersburg, FL.

Peytchev, A., Baxter, R. K., & Carley-Baxtor, L. R. (2009). Not all survey effort is equal: Reduction of nonresponse bias and nonresponse error. *Public Opinion Quarterly, 73*, 785–806.

Rivers, D., & Bailey, D. (2009). Inference from matched samples in the 2008 US national elections. In *Proceedings of the Joint Statistical Meetings*, 627–639.

Roberts, C., Gilbert, E., & Allum, N. (2011). Research based on Satisficing Theory: A systematic review of methods and results. Paper presented at the American Association for Public Opinion Research, Phoenix, AZ, May 12–15.

Sakshaug, J., Yan, T., & Tourangeau, R. (2010). Nonresponse error, measurement error, and mode of data collection: Tradeoffs in a multi-mode survey of sensitive and non-sensitive items, *Public Opinion Quarterly, 74*, 907–933.

Schuman, H., & Presser, S. (1981). *Questions and answers in attitude surveys: Experiments on question form, wording, and context*. New York: Academic Press.

Sturgis, P., Allum, N., & Brunton-Smith, I. (2008). Attitudes over time: The psychology of panel conditioning. In P. Lynn (Ed.), *Methodology of longitudinal surveys* (pp. 113–126), Chichester: John Wiley & Sons, Ltd.

Sturgis, P., Roberts, C., & Smith, P. (2014). Middle alternatives revisited: How the neither/nor response acts as a "face-saving" way of saying "I don't know". *Sociological Methods and Research, 43*, 15–38 (Published online 27 September 2012).

Toepoel, V., Das, M., & van Soest. A. (2009). Relating question type to panel conditioning: Comparing trained and fresh respondents. *Survey Research Methods, 2*, 73–80.

Tourangeau, R., Groves, R.M., Kennedy, C., & Yan, T. (2009). The presentation of a web survey, nonresponse and measurement error among members of web panel, *Journal of Official Statistics, 25*, 299–321.

Tourangeau, R., Groves, R. M., & Redline, C. D. (2010). Sensitive topics and reluctant respondents: Demonstrating a link between nonresponse bias and measurement error, *Public Opinion Quarterly, 74*, 413–432.

Warren, J.R., & Halpern-Manners, A. (2012). Panel conditioning in longitudinal social science surveys. *Sociological Methods & Research, 41*, 491–534.

Yan, T., & Curtin, R. (2010). The relation between unit nonresponse and item nonresponse: A response continuum perspective. *International Journal of Public Opinion Research, 22*, 535–551.

Yan, T., Tourangeau, R., & Arens, Z. (2004). When less is more: Are reluctant respondents poor reporters? *Proceedings of the Survey Research Methods Section, American Statistical Association*, 4632–4651.

Part VI

SPECIAL DOMAINS

Introduction to Part VI

Reg Baker[a] and Anja S. Göritz[b]

[a]*Market Strategies International, USA*
[b]*University of Freiburg, Germany*

VI.1   Special domains

Each of the other parts of this book is focused on a single theme, with chapters that address that theme via a variety of research methods and approaches. Part VI is different in that its two chapters take up very different topics. If there is a shared theme between them, it is how the rapid pace of technological change is constantly forcing us to rethink our methods, to look over the horizon and be ready when the next big thing makes what we are doing today no longer enough.

Arguably the biggest technological change of the last decade is the emergence of mobile and, over the last few years, the dramatic growth in smartphones. While researchers have tended to focus on some of the unique opportunities that mobile offers, such as "in-the-moment" research and location tracking, survey respondents are beginning to take matters into their own hands, so to speak, and use their smartphones rather than a desktop or laptop computer to respond to online survey requests. Most of these surveys are designed for administration on the large screen of a desktop PC or even a laptop, rather than the small screen of a smartphone.

In Chapter 16, Frank Drewes looks at how these unintentional mobile respondents may be impacting online panel data collection. He analyzes the results from four experiments to gauge the frequency of this behavior and, most importantly, how it may be impacting data quality. He finds that respondents on smartphones tend to break off more often and that those respondents who stay until the end are demographically different; but at least in the reviewed studies the proportion of respondents who complete by smartphone is not yet large enough to cause problems in the collected data. However, other studies (Jue, 2012; Kinesis, 2013; Peterson et al., 2013; Tsvelik, 2013) suggest that the proportion of respondents at least starting surveys

Online Panel Research: A Data Quality Perspective, First Edition.
Edited by Mario Callegaro, Reg Baker, Jelke Bethlehem, Anja S. Göritz, Jon A. Krosnick and Paul J. Lavrakas.
© 2014 John Wiley & Sons, Ltd. Published 2014 by John Wiley & Sons, Ltd.
Companion website: www.wiley.com/go/online_panel

using a mobile device is increasing rapidly, now perhaps as much as 25% to 30%. Designing surveys that can be completed on smartphones is definitely one of the biggest challenges the research industry faces in the next few years.

These changes in how people respond to surveys reflect much broader changes in how people are using mobile devices – advanced feature phones, smartphones, and tablets – to access the Internet. A multi-device Internet browsing experience that includes a PC and one or more mobile devices is rapidly becoming the norm. Measuring the whole of an individual's Internet access is becoming more and more challenging.

In Chapter 17, Philip Napoli, Paul J. Lavrakas, and Mario Callegaro take up the topic of Internet audience measurement. Historically, traditional panels were the mainstay of audience measurement, providing the metrics to drive advertising for the television and radio industries. With the rise of the Internet and the need for a new set of metrics specific to the web, online panels seemed like the logical choice. Unlike other online panels, these Internet rating panels collect passive data from members who install tracking software on their devices that monitors their Internet behavior. But as Napoli and his colleagues show, these online panels largely have failed to provide the full range of metrics that content providers and advertisers require. The evolution of online audience measurement is now headed in a direction that looks more like big data – a three component model that integrates data collected from a panel, data from server logs, and direct measures from Internet service providers (ISPs).

References

Jue, A. (2012). Participation of mobile users in traditional online studies – Q4'12 Update. Retrieved from https://www.decipherinc.com/n/press-room/read/participation-of-mobile-users-in-traditional-online -studies-q412-update.

Kinesis (2013). Kinesis Survey Technologies' mobile survey traffic rates show notable increase across US, Europe. Retrieved from http://www.kinesissurvey.com/2013/10/09/latest-mobile-survey -traffic-stats-q3-2013.

Peterson, G., Mechling, J., LaFrance, J., Swinehart, J. & Ham, G. (2013). Solving the unintentional mobile challenge. Retrieved from https://c.ymcdn.com/sites/www.casro.org/resource/collection /0A81BA94-3332-4135-97F6-6BE6F6CEF475/Paper_-_Gregg_Peterson_-_Market_Strategies _International.pdf.

Tsvelik, M. (2013). Device diversity. Research World. Retrieved from http://rwconnect.esomar.org /device-diversity

16

An empirical test of the impact of smartphones on panel-based online data collection

Frank Drewes
Harris Interactive AG, Germany

16.1 Introduction

In less than 15 years, self-administered interviews in online panels have become an established data collection method. Online interviews have been the most common data collection method in Germany since 2010. In the following year, 36% of all interviews conducted there were online, 34% per phone, 24% were computer-aided personnel interviews, and 6% were self-administered postal interviews (Wiegand, 2011). Nowadays data collection via online panels is not only seen as being fast and cheap. The technical progress in the past few years in terms of connection speed, data transfer capacity, graphical computing power and display size makes it possible to expose respondents online to highly complex stimuli such as high-resolution photos and videos, to use innovative user interfaces, and to utilize complex research tools such as discrete choice modeling.

Starting back in the late 1990s, the first online panel providers did not have a lot of variety in terms of technical devices, operating systems, and web browsers on the respondents' side. The main challenge was ensuring the use of a survey on a few web browsers with their particularities and limitations. The technical device and the operating system could be taken for granted: According to browser statistics reported for the United States in 2001, 84% of the browsers were adjusted to a screen resolution of 800x600 or higher, 84% of them

Online Panel Research: A Data Quality Perspective, First Edition.
Edited by Mario Callegaro, Reg Baker, Jelke Bethlehem, Anja S. Göritz, Jon A. Krosnick and Paul J. Lavrakas.
© 2014 John Wiley & Sons, Ltd. Published 2014 by John Wiley & Sons, Ltd.
Companion website: www.wiley.com/go/online_panel

were versions of Microsoft's Internet Explorer, and more than 90% of them ran on a MS Windows-based operating system (Couper, 2001).

The situation has dramatically changed in recent years. The global sales of mobile devices such as smartphones and tablet PCs have overtaken the sales of desktop PCs and notebooks (Canalys, 2012). According to a survey representative of the resident population in Germany, conducted by Ipsos on behalf of Google in 2012, 29% of the respondents privately own a smartphone. In a subsequent online survey of private smartphone owners who use the Internet in general, 53% of them stated that they access the Internet via their smartphone at least once a day (Google, 2012).

These developments raise a number of challenges for panel-based online data collection: It is to be expected that members of online panels increasingly will try to access online surveys via smartphones and other mobile devices. The inherent technical limitations of these devices compared to full-scale personal computers raise concerns regarding the compatibility of proven survey and research designs and the comparability of data collected on different devices. On the other hand, strict exclusion of mobile devices would decrease the accessibility of surveys and might cause systematic sample biases.

There are few empirical studies addressing the impact of smartphones on panel-based online data collection. Some authors give face-valid advice regarding a "mobile-friendly" design of conventional web surveys. Based on the work of Tarkus (2009), Zahariev et al. (2009), Pferdekaemper (2010) and Luck (2011), Callegaro (2013) recommends short surveys, removal of all non-essential content, avoidance of grids and multimedia content and use of basic question types. It remains unclear to what extent these measures which severely limit the survey design options address and solve real threats to data quality.

Other authors report their empirical results in a way that makes it difficult to interpret their findings and replicate their research designs. Baker-Prewitt et al. (2013) conducted an online experiment and compared the data quality of an online survey of 10-minutes length taken on four different devices (PC, tablet, smartphone, and smartphone with a mobile-friendly survey design). The authors report that respondents in the unadapted smartphone condition tended to have higher dropout rates, a greater tendency to straight-line (i.e. choosing the same response option regardless of the question) in a grid question-response format, and to respond incorrectly to a trap question. The interpretability of these results is limited by a strong self-selection bias induced by mode-dependent differences in survey dropout: In the unadapted smartphone condition, the dropout rate was more than three times as high as in the PC condition (18% vs. 5%). Further biases despite randomized assignment of the respondents to the four experimental conditions were likely induced by unreported differences in participation rate.

Peterson et al. (2013) designed an experiment with eight different survey participation modes, five of them via smartphone, two of them via PC (the eighth mode was participation via a smartphone application with respondents from a different sample source). Potential participants were pre-screened for smartphone ownership and subsequently assigned randomly to one of the seven experimental conditions. Survey access against the instructed participation mode was rejected. Again the results show a strong self-selection bias: The participation rate in the PC conditions was double as high as in the mobile conditions (12% vs. 6%). Apart from this bias, the authors "removed an average of 5% of respondents from the final dataset based on standard criteria" without further explanation or documentation.

Bailey et al. (2012) conducted an experiment comparing participation via PC or laptop and participation via a smartphone survey application. The authors report very high response and completion rates which did not significantly differ between mobile and conventional participation condition. The interpretability of these results is decreased by a self-selection bias: 23% of the respondents in the online condition accessed the survey via smartphone against instruction. Median completion time for this unintended third participation way was 8.5 minutes compared to 5.5 to 6 minutes in the two regular experimental conditions.

Peterson (2012) reviewed 17 conventional online surveys in an ex-post-facto design. Mobile survey-start rates varied between 1%–30% (an overall rate is not reported). It took mobile respondents 25%–50% longer to complete the surveys, and their dropout rate was double as high. In projects where the sample source was provided by the client, mobile participants responded more quickly to invitations. Despite the incomparability of mobile and conventional respondents that was inherent in the research design, the authors found no differences in the response rates to optional open-ended questions, and mobile respondents did not break off more often on complex questions.

Most of the empirical papers addressing the issue of unintentional mobile respondents and their impact on survey quality focus on possible differences in response behavior. This chapter recognizes systematic sampling biases as a second potential threat to the validity of panel-based online survey results. It summarizes the results of four studies conducted in Germany between 2011 and 2013 by Harris Interactive AG with the following objectives:

- Assessing ownership and use of mobile devices among panel members compared to the general population.

- Monitoring the prevalence of smartphone access in conventional surveys.

- Assessing the use and attitudes of panel members regarding
  — mobile devices in general;
  — survey participation via mobile devices.

- Comparing "conventional" vs. "mobile" survey participations depending on interview length on a variety of dependent survey quality measures.

Based on these results, the chapter discusses how much the unintentional mobile respondents threaten the survey quality in online panels and proposes actions to be taken by panel providers.

16.2 Method

All data are collected in the German section of the online panel of Harris Interactive AG. The German panel section has approximately 90000 members. The members are recruited via online advertising banners on a large variety of websites. The completion of a registration survey (approx. length: 10 min.) collecting sociodemographic core data is mandatory. Subsequent participation in surveys is rewarded by bonus points. 100 bonus points (equivalent to a cumulated participation time of 250 min. on average) can be exchanged against an Amazon voucher worth 25 Euros. The retention period after the survey participation is at least six weeks. The annual dropout rate is approximately 20%. The chapter reports the results of four independent empirical studies.

- *Study 1: Mandatory registration survey/voluntary re-registration survey.* Since 2011, household ownership of smartphones has been part of both the mandatory registration survey for prospective panel members and the voluntary re-registration survey for existing panel members. Participation in these surveys is not rewarded. The length of the registration survey is approximately 10 minutes; the re-registration survey is divided into several modules with a length of 5–10 minutes each. The participation rate for the re-registration survey is approximately 50%. The reported results are from 2011 and 2012.

- *Study 2: Observation of survey access.* For a period of four weeks in December 2012 and January 2013, Harris Interactive monitored access to the surveys without an explicit instruction to use a certain device. Smartphones were identified by browser ID string and display resolution.

- *Study 3: Smartphone-related usage behavior and attitudes.* The Harris Interactive AG interviewed a panel-representative sample of N =1059 smartphone owners in January 2013. The average survey length was approximately 15 minutes. Survey topics were smartphone-related usage behavior and attitudes:

   — general usage behavior:

      - frequency, situations, locations, networks, web access via browser and specialized applications, web content;

   — general attitudes, preferences regarding web access via browser vs. specialized applications;

   — participation via smartphone on online surveys:

      - experiences, preferences (survey length, browser vs. survey application);

   — socio-demographics.

- *Study 4: Experimental test of the impact of survey participation via smartphone on the quality of survey results.* Harris Interactive screened more than 8000 smartphone owners in its panel for their willingness to participate in a survey via smartphone in December 2012. Panel members who agreed in principle were randomly assigned to one of six cells of an experimental design with two factors: "participation mode" and "survey length." These two factors were manipulated in inviting approximately 2500 of the pre-screened panelists to a subsequent survey in January 2013: The invitation asked respondents either to participate via smartphone or via desktop PC/notebook in a survey of 5, 10 or 20 minutes length (Table 16.1).

Table 16.1 Experimental design.

Survey length	Participation mode	
	Smartphone	Desktop PC /notebook
5 minutes	N = 300	N = 300
10 minutes	N = 300	N = 300
20 minutes	N = 300	N = 300

Each questionnaire version consisted of a number of consumer-related behavioral and attitudinal questions with at least one instance of the following question–response formats:

1. Open-ended question.

2. Multiple response question, at least 10 response options.

3. Brand funnel – unaided awareness, aided awareness, relevant set, first choice.

4. Statement battery, at least ten statements, agreement scale, grid format.

The questionnaire for the long survey condition (20 min.) additionally featured a choice-based conjoint exercise with 12 tasks.

Two groups of dependent survey quality measures were collected:

1. Survey participation:

    1. compliance rate (access via instructed device);

    2. latency time (from invitation to first survey access);

    3. dropout rate (as proportion of respondents who started the survey without finishing it);

    4. total participation length (from invitation to completed interview including breaks);

    5. effective participation length (from invitation to completed interview without breaks).

2. Response quality:

    1. open-ended question: number of characters;

    2. multiple response question: number of responses;

    3. brand funnel: consistency of responses;

    4. statement battery: response patterns, consistency of responses, correctness of response to trap question;

    5. choice-based conjoint: goodness of fit between predicted and observed preferences.

Results are descriptively reported due to a strong self-selection bias (cf. the results in Section 16.3) which prevented the use of inference-statistical techniques.

16.3   Results

16.3.1   Study 1: Observation of survey access

A total of 87604 members of the German Harris Interactive online panel answered questions regarding ownership of technical devices on a household level during their panel registration respectively two-yearly re-registration. From these 87604 panel members, 47.9% live in a household owning at least one smartphone. Stated household ownership is higher among male and younger panel members. The most often used operating system is Android with 22.3% followed by iOS (Apple) with 19.6%, Windows Mobile with 6.8%, BlackBerry with 5.6%

Table 16.2 Smartphone ownership by household.

	Total	Gender		Age			
		Male	Female	29 years or younger	30–39 years	40–49 years	50 years or older
Base	87,604	36960	50644	25353	22218	21005	19028
Yes (net) (%)	47.9	56.0	42.0	56.0	52.9	46.8	32.5
Yes (by operating system)							
Android OS (%)	22.3	26.8	19.1	29.3	23.5	20.4	13.9
iOS (iPhone) (%)	19.6	22.3	17.7	22.8	22.3	19.2	12.7
Windows Mobile (%)	6.8	8.5	5.6	6.6	7.4	7.7	5.4
BlackBerry (%)	5.6	6.5	4.9	5.1	7.1	6.1	3.8
Symbian OS (%)	5.5	8.6	3.3	6.8	6.3	5.4	3.2
No (%)	52.1	44.0	58.0	44.0	47.1	53.2	67.5

and Symbian OS with 5.5%. Each of these households owns smartphones with 1.2 different operating systems on average (Table 16.2).

16.3.2   Study 2: Monitoring of mobile survey access

From surveys conducted in the German Harris Interactive online panel in December 2012 and January 2013, 14 surveys were randomly selected. In total, 13904 respondents participated in these surveys. Sample sizes of the individual surveys ranged from 107 to 2868 participants. The survey topics ranged from consumer packaged goods and consumer durables to financial services, the objectives included package, concept and product tests, advertising tests, U&A and customer satisfaction surveys.

In total, 513 survey participations via mobile devices were registered. The proportion of mobile participants per survey ranged from 1.6% (N = 192) to 10.3% (N = 107) with an overall rate of 3.7%. Respondents participating via mobile devices were younger and more often female compared to respondents taking the survey in the conventional way (Table 16.3).

16.3.3   Study 3: Smartphone-related usage behavior and attitudes

A total of approximately 4000 members of the German section of the Harris Interactive AG online panel were randomly invited to the usage and attitude survey. Of the 1796 members

Table 16.3 Survey access via mobile device according to browser string and display solution.

	Total	Gender		Age			
		Male	Female	29 years or younger	30–39 years	40–49 years	50 years or older
Base	13904	5973	7931	1995	3650	4145	4,114
Yes (%)			4.5	6.2	5.0	3.0	1.8
No (%)	3.7	2.6	95.5	93.8	95.0	97.0	98.2

Table 16.4 Private Internet access via smartphone.

	Total	Gender		Age			
		Male	Female	29 years or younger	30–39 years	40–49 years	50 years or older
Base	1796	783	1013	396	473	427	496
Yes (%)	64	70	59	77	74	62	46
No (%)	36	30	41	23	26	38	54

who accepted the invitation, 64% privately accessed the Internet via smartphone and therefore were qualified to participate in the survey. Males and younger panel members were more often qualified than females and older members (Table 16.4).

A total of 1059 respondents completed the survey. Of these 1059 respondents, 20% became member of the panel in 2012, 21% in 2011 or 2010 and 21% between 2006 and 2009. 9% of them have been taking part in the panel since 2005 or earlier. During their panel membership, 37% of the respondents have participated in 20 or more surveys, 31% in 11–19 surveys and 30% in 10 surveys or less.

59% of the participants privately use a smartphone with Android OS, 26% an Apple iPhone, 6% a smartphone with an operating system by Microsoft. Smartphones with Black-Berry OS or Symbian OS are used by 5% each, 12% employ a smartphone falling in neither of the categories. On average, each respondent uses 1.1 different smartphone operating systems. 29% of the respondents also have access to a tablet PC (Table 16.5).

71% of the interviewees privately access the Internet via smartphone several times a day compared to 80% doing it via a computer (including tablet PC); 14% visit the web with a tablet PC several times per day compared to 35% doing it with a desktop PC. The proportion of respondents with several means of Internet access via smartphone per day varies between 82% for respondents who are 34 years or younger and 58% for participants who are 45 years or older. On average, each respondent accesses the Internet several times per day from 1.7 different devices (cf. Table 16.5).

76% of all respondents access the Internet at home via wireless LAN, 21% at home via cellular network; 74% access the Internet outside their home via cellular network, 24% via wireless LAN. Younger respondents use all access occasions modes more often than older respondents (Table 16.6).

A total of 15% of the participants privately have a cell phone contract with an Internet data flat-rate and unlimited high-speed volume; 56% have signed a contract with a flat-rate and limited high-speed volume; 27% of the respondents have no contract including a flat-rate. Respondents without data flat-rate are more often female and older. 61% of the respondents think that their private Internet usage via smartphone will increase somewhat or strongly. Male respondents and respondents aged between 35 and 44 years state this more often than other respondents.

Only 12% of the interviewees had participated in at least one online survey via smart-phone; 9% followed an invitation via mail from an online panel, 3% were invited on-site in a smartphone web browser, 2% via a specialized mobile application of an online panel and 2% while using a non-survey smartphone application (Table 16.7).

However, 61% of these respondents had abandoned survey participation via smartphone web browser at least once for technical problems.

Table 16.5 Private use of technical devices: in general/several times Internet access per day.

		Total	Gender		Age		
			Male	Female	34 years or younger	35–44 years	45 years or older
Base		1059	513	546	475	278	306
Computer (net) (%)	In general	100	100	100	100	100	100
	Internet	80	84	76	80	80	78
Notebook (%)	In general	77	75	78	81	75	71
	Internet	45	43	48	52	44	36
Desktop PC (%)	In general	64	74	55	56	66	77
	Internet	35	46	25	29	33	46
Tablet PC (%)	In general	29	35	23	29	31	28
	Internet	14	18	11	15	13	14
Netbook (%)	In general	18	20	15	17	20	15
	Internet	5	5	5	5	6	4
Smartphones (net) (%)	In general	100	100	100	100	100	100
	Internet	71	70	71	82	64	58
Smartphone Android OS (%)	In general	59	58	59	63	55	56
	Internet	42	41	42	51	35	34
Apple iPhone (%)	In general	26	26	27	28	30	20
	Internet	21	21	21	25	21	15
Smartphone Microsoft OS (%)	In general	6	6	5	4	7	8
	Internet	3	3	3	2	4	3
BlackBerry (%)	In general	5	5	4	4	5	5
	Internet	2	2	1	2	2	2
Smartphone Symbian OS (%)	In general	5	7	2	3	6	5
	Internet	2	3	1	1	2	2
Other Smartphones (%)	In general	12	11	12	9	9	19
	Internet	5	4	6	4	3	7
Do not know (%)	In general	0	0	0	0	0	0
	Internet	7	6	8	5	8	11

When asked about aspects of their general smartphone usage behavior, 52% of the interviewees stated that they regularly check private mail accounts via their smartphone; 48% think that they spend much more time online due their smartphones, and 15% say that they usually access the Internet via smartphone. Regarding the smartphone usability of web content, 57% of the respondents think it to be generally low; 50% prefer web content that is optimized for mobile devices when using their smartphone. On the other hand, 40% state that they have no problems filling in web forms via smartphones; 48% say that that accessing web

Table 16.6   Private Internet access via smartphone by location.

	Total	Gender		Age		
		Male	Female	34 years or younger	35–44 years	45 years or older
Base	1058	513	546	475	278	306
At home via wireless LAN (%)	76	78	75	84	74	68
At home via cellphone network (%)	21	21	21	25	21	16
Mobile via wireless LAN (%)	24	24	24	29	21	21
Mobile via cellphone network (%)	74	77	72	79	77	65
Do not know (%)	3	3	4	2	1	7

Table 16.7   Survey participation via smartphone in the past.

	Total	Gender		Age		
		Male	Female	34 years or younger	35–44 years	45 years or older
Base	1059	613	546	475	278	306
Yes (net) (%)	12	14	11	16	13	7
Yes (by invitation)						
Invited by online panel via mail (%)	9	10	8	11	10	5
Invited by online panel via survey application (%)	2	2	2	3	2	0
Invited on-site while using smartphone web browser (%)	3	4	2	4	2	1
Invited while using a non-survey application (%)	2	3	1	3	1	0
No (%)	87	86	89	84	87	93
Do not know (%)	1	0	1	1	0	1

content via specialized smartphone applications is generally more comfortable than accessing the same content via a web browser. Nearly two-thirds of the respondents are open-minded about survey participation via smartphones, but the same proportion of respondents would limit it to surveys which are no longer than 10 minutes, and 60% would like to have the choice between smartphone and PC participation for each survey. At the same time, 43% think that filling in questionnaires via smartphones is too uncomfortable (Top-2-Box proportions in a 5-agreement scale, cf. Table 16.8).

16.3.4   Study 4: Experimental test of the impact of survey participation via smartphone on the quality of survey results

Respondents' compliance with the instructed way to participate (computer vs. smartphone) was tested, based on the browser ID string and display solution, and 97% of the respondents who were assigned to the computer participation condition followed their instructions. The

Table 16.8   Smartphone use behavior: Top-2-Box 5-point-agreement scale.

	Total	Gender		Age		
		Male	Female	34 years or younger	35–44 years	45 years or older
Base	1059	513	546	475	278	306
General smartphone use behavior						
I regularly check my private mail account via my smartphone (%)	52	51	53	65	53	30
All in all I spend much more time online due to my smartphone (%)	48	48	48	60	46	32
When visiting the Internet for private reasons I usually use the smartphone	15	13	17	23	13	6
Smartphone usability of web content						
Many websites are not user-friendly when accessed via smartphones (%)	57	58	55	55	55	60
I prefer websites optimized for mobile devices when accessing the Internet via smartphone (%)	50	51	49	55	48	43
Apps usually are more comfortable than smartphone browsers (%)	48	48	48	56	49	35
I have no problems filling-in online forms via smartphone (%)	40	42	39	48	38	30
Mobile survey participation						
I would only participate via smartphone in surveys which are not longer than 10 minutes (%)	63	61	64	61	66	62
I am open-minded about survey participation via smartphone (%)	62	64	61	76	63	41
I would like having the choice between smartphone and PC participation for each survey (%)	60	59	60	60	60	60
Survey participation via smartphone is too uncomfortable (%)	43	41	45	34	41	59

compliance rate in the smartphone participation condition was much lower: only 64% used a smartphone as instructed. The compliance rate significantly varied with gender and age of the respondents but not with the announced survey length (Table 16.9). Dropout rates were strongly dependent on survey length and instructed participation mode: In the conventional participation group, dropout rates ranged from <1% (intermediate survey length) to 4% (long survey length). In the mobile participation group, dropout rates varied from 3%–16%.

Due to these two sources of self-selection biases, an inference statistical analysis of the impact of participation mode on survey quality was discarded. Results are descriptively reported for the three groups (computer participation as instructed (N = 1042), smartphone

Table 16.9 Compliance in instructed survey access.

	Survey length condition			Gender		Age				
	Short (5 min.)	Intermediate (10 min.)	Long (20 min.)	Male	Female	24 years or younger	35–44 years	45–54 years	55 years or older	25–34 years
Base	631	683	654	954	1014	339	671	502	336	120
Computer as instructed (%)	51	53	55	54	52	53	53	49	57	56
Smartphone against computer instruction (%)	1	1	1	1	1	2	2	0	0	1
Tablet PC against computer instruction (%)	1	1	0	1	1	1	1	0	0	3
Smartphone as instructed (%)	28	30	29	27	31	33	32	31	23	13
Computer against smartphone instruction (%)	17	15	14	16	14	11	12	18	19	26
Tablet PC against smartphone instruction (%)	1	1	1	1	1	1	1	2	1	2

participation as instructed (N = 571), and smartphone instruction but computer participation (N = 303)) which account for 97% of all respondents in the survey.

The average latency period between survey invitation and first survey access was much shorter in the smartphone participation group. On average, their first survey access took place 17h 6 min. after invitation, compared to 22h 44min. (computer as instructed) and 23h 45 min. (computer against instruction) (Table 16.10).

It took the smartphone participants considerably more time to complete the survey in all survey length conditions. The difference in net survey duration against both computer participation groups varied between more than two minutes (short survey length condition) and more than six minutes (long survey length condition). Across all survey length conditions, the total time between the first survey access and the completion of the survey was highest for respondents who participated via computer against instruction (126 min.) followed by smartphone participants (64 min.) and regular computer participants (43 min.). The standard error of mean is much higher in the computer-against-instruction group as well. The total time that smartphone participants spent on completing the survey strongly varied with survey duration conditions. In the short and intermediate survey length condition, they completed the survey faster than the respondents who participated via the computer whereas in the long survey condition it took them the longest amount of time (Table 16.11).

In general, the response quality of smartphone participants was at least as high as the response quality of the participants who used a computer as instructed. In a multiple choice question for online activities via a smartphone, the smartphone participants marked 5.3 out of 10 compared to 5.0 ticked off by regular computer participants.

In two open-ended questions (unaided awareness for smartphone manufacturers and unaided awareness for mobile Internet providers), smartphone participants mentioned 9.6 brands compared to 9.8 mentions by regular computer participants. In an aided brand awareness and relevant brand set question, smartphone participants marked a total of 21.7 brand list entries compared to 22.0 brands selected by regular computer participants. In a test for inconsistencies in two sets of four brand funnel questions, smartphone participants showed 0.12 inconsistencies (a brand mentioned unaided that is not marked in the aided awareness question, a brand in the relevant set the respondent is not aware of, a brand selected as first choice that is not part of the relevant set), on average showing a higher response quality than regular computer participants with 0.17 (Table 16.12).

When asked for the exact name of their smartphone model, the length of the string typed in by smartphone participants was slightly shorter (10.7 vs. 11.6 characters). In two open-ended survey evaluation questions smartphone participants wrote somewhat less about things they liked (28.3 vs. 31.5 characters) but more than double as long about things they disliked (40.3 vs. 19.1 characters) compared to regular computer participants. Both questions were answered considerably briefer in the 20-minutes survey duration condition (Table 16.13).

In a grid question with nine statements and a 5-point-agreement scale, 2.8% of the smartphone respondents selected the same scale point regardless of the statement compared to 2.2% of the regular computer participants. Some 5.4% of the smartphone participants failed to answer correctly when asked to mark a certain scale point compared to 5.8% of the regular computer respondents. The statement battery included two versions of the same statement worded in opposite directions: 18.4% of the smartphone respondents and 17.4% of the regular computer respondents indicated agreement or disagreement respectively to both statement versions (Table 16.14).

Table 16.10 Latency period in hours from survey invitation to first survey access.

	Survey access			Survey length condition			Computer as instructed			Smartphone as instructed			Computer against instruction		
	Computer as instructed	Smartphone as instructed	Computer against instruction	Short (5 min.)	Intermediate (10 min.)	Long (20 min.)	Short	Intermediate	Long	Short	Intermediate	Long	Short	Intermediate	Long
Base	851	503	251	499	576	530	250	310	291	159	182	162	90	84	77
Mean	22.7	17.1	23.8	21.2	20.8	21.4	22.9	23.3	22.0	15.5	17.6	18.0	26.5	18.4	26.5
Standard deviation	23.4	18.2	26.3	24.2	22.3	21.2	25.4	250	19.6	16.2	19.0	19.1	30.3	16.6	29.2
Standard error	0.8	0.8	1.7	1.1	0.9	0.9	1.6	1.4	1.1	1.3	1.4	1.5	3.2	1.8	3.3

Table 16.11 Participation time in minutes.

	Survey access			Survey length condition			Computer as instructed			Smartphone as instructed			Computer against instruction		
	Computer as instructed	Smartphone as instructed	Computer against instruction	Short (5 min.)	Intermediate (10 min.)	Long (20 min.)	Short	Inter-mediate	Long	Short	Inter-mediate	Long	Short	Inter-mediate	Long
Base	1,040	570	303	610	667	636	323	360	357	177	206	187	110	101	92
Net participation time (from first access to completion without breaks)															
Mean	10.9	14.7	10.5	7.8	8.6	19.6	7.2	8.0	17.3	9.3	10.2	24.8	6.9	7.6	18.0
Standard deviation	6.2	9.1	7.0	2.5	2.8	8.4	2.1	2.5	6.5	2.3	2.6	9.4	2.5	2.7	7.9
Standard error	0.2	0.4	0.4	0.1	0.1	0.3	0.1	0.1	0.3	0.2	0.2	0.7	0.2	0.3	0.8
Total participation time (from first access to completion including breaks)															
Base	1,040	570	303	610	667	636	323	360	357	177	206	187	110	101	92
Mean	42.9	64.3	125.8	46.6	43.4	97.5	34.2	29.8	64.0	16.6	25.1	152.7	131.3	129.1	115.5
Standard deviation	333.5	324.1	593.6	419.6	324.0	409.4	406.5	308.1	287.6	27.1	109.9	543.6	696.6	570.0	480.6
Standard error	10.4	13.6	34.1	17.0	12.5	16.2	22.6	16.2	15.2	2.0	7.7	39.8	66.4	56.7	50.1

Table 16.12 Detail and consistency of responses to brand funnel questions.

	Survey access			Survey length condition			Computer as instructed			Smartphone as instructed			Computer against instruction		
	Computer as instructed	Smartphone as instructed	Computer against instruction	Short (5 min.)	Intermediate (10 min.)	Long (20 min.)	Short	Inter-mediate	Long	Short	Inter-mediate	Long	Short	Inter-mediate	Long
Base	1042	571	303	611	668	637	324	361	357	177	206	188	110	101	92
Unaided brand awareness (open-ended) Number of brand mentions															
Mean	9.8	9.6	8.7	9.6	9.7	9.5	9.9	9.9	9.6	9.8	9.6	9.6	8.3	9.2	8.5
Standard deviation	3.6	3.7	3.4	3.7	3.6	3.6	3.5	3.6	3.7	3.9	3.6	3.6	3.5	3.5	3.1
Standard error	0.1	0.2	0.2	0.1	0.1	0.1	0.2	0.2	0.2	0.3	0.3	0.3	0.3	0.3	0.3
Aided brand awareness, relevant set (multiple choice) - Number of brand mentions															
Mean	22.0	21.7	21.2	22.1	21.7	21.7	22.4	21.8	21.9	21.9	21.8	21.5	21.4	21.1	21.1
Standard deviation	4.4	4.3	4.7	4.3	4.4	4.5	4.3	4.4	4.6	4.3	4.2	4.3	4.5	4.7	4.9
Standard error	0.1	0.2	0.3	0.2	0.2	0.2	0.2	0.2	0.2	0.3	0.3	0.3	0.4	0.5	0.5
Consistency of response to brand funnel question (awareness, relevant set, first choice) - Number of inconsistencies															
Mean	0.17	0.12	0.21	0.15	0.14	0.18	0.15	0.14	0.20	0.12	0.11	0.12	0.21	0.21	0.22
Standard deviation	0.55	0.39	0.61	0.49	0.48	0.58	0.49	0.50	0.64	0.38	0.37	0.41	0.64	0.57	0.63
Standard error	0.02	0.02	0.04	0.02	0.02	0.02	0.03	0.03	0.03	0.03	0.03	0.03	0.06	0.06	0.07

Table 16.13 Open-ended questions: number of characters.

	Survey access			Survey length condition			Computer as instructed			Smartphone as instructed			Computer against instruction		
	Computer as instructed	Smartphone as instructed	Computer against instruction	Short (5 min.)	Intermediate (10 min.)	Long (20 min.)	Short	Inter-mediate	Long	Short	Inter-mediate	Long	Short	Inter-mediate	Long
Base	1042	571	303	611	668	637	324	361	357	177	206	188	110	101	92
Description smartphone model															
Mean	11.6	10.7	11.1	11.2	11.1	11.4	11.7	11.8	11.2	10.9	9.9	11.3	10.3	11.0	12.2
Standard deviation	6.2	5.6	6.3	5.9	6.4	5.8	6.1	6.8	5.7	5.6	5.6	5.5	5.6	6.4	6.0
Standard error	0.2	0.2	0.4	0.2	0.2	0.2	0.3	0.4	0.3	0.4	0.4	0.4	0.5	0.6	0.7
Survey likes															
Mean	31.5	28.3	22.7	31.0	32.0	24.4	34.5	33.3	26.9	28.5	33.1	22.8	25.0	24.9	17.6
Standard deviation	38.5	39.3	28.3	38.0	40.0	33.5	38.8	39.9	36.3	40.1	44.5	31.1	30.7	28.3	24.7
Standard error	1.2	1.6	1.6	1.5	1.5	1.3	2.2	2.1	1.9	3.0	3.1	2.3	2.9	2.8	2.6
Survey dislikes															
Mean	19.1	40.3	18.1	23.6	19.2	33.2	18.0	14.3	25.0	38.5	30.8	52.2	16.5	12.7	25.9
Standard deviation	41.0	56.0	36.0	45.5	39.4	52.4	40.5	38.5	43.2	55.5	43.9	65.6	35.0	25.9	44.7
Standard error	1.3	2.3	2.1	1.8	1.5	2.1	2.3	2.0	2.3	4.2	3.1	4.8	3.3	2.6	4.7

Table 16.14 Grid statements by agreement scale: proportion of conspicuous response behavior.

	Survey access			Survey length condition			Computer as instructed			Smartphone as instructed			Computer against instruction		
	Computer as instructed	Smartphone as instructed	Computer against instruction	Short (5 min.)	Intermediate (10 min.)	Long (20 min.)	Short	Inter-mediate	Long	Short	Inter-mediate	Long	Short	Inter-mediate	Long
Base	1042	571	303	611	668	637	324	361	357	177	206	188	110	101	92
Same scale point regardless of item (%)	2.2	2.8	3.0	2.6	1.5	3.5	1.5	1.4	3.6	4.0	1.9	2.7	3.6	1.0	4.3
Trap Question answered incorrectly *(%)	5.8	5.4	6.6	4.6	5.2	7.5	4.3	5.0	7.8	4.5	4.9	6.9	5.5	6.9	7.6
Item with reversed polarity answered inconsistently (%)	17.4%	18.4	16.8	18.7	16.2	18.1	19.4	15.2	17.6	18.6	18.0	18.6	16.4	15.8	18.5

Table 16.15   Choice-based conjoint: fit between predicted and observed preferences (root likelihood, Sawtooth software).

Survey access	Computer as instructed	Smartphone as instructed	Computer against instruction
Base	357	188	92
Mean	0.78	0.78	0.78
Standard deviation	0.09	0.09	0.09
Standard error	0.00	0.01	0.01

The questionnaire in the 20-min.-survey condition included a second statement battery with 13 statements. In this statement battery, smartphone respondents showed a higher quality level than regular computer participants: 1.1% chose the same scale point regardless of the item compared to 2.2%, 6.4% failed to react correctly to the trap question compared to 8.7% and 18.6% compared to 21.3% indicated agreement or disagreement respectively to two versions of the same statement worded in opposite directions.

Respondents in the long survey condition participated in a choice-based conjoint exercise for mobile telephone contracts with eight features and a total of 39 feature levels. Respondents were asked 12 times to indicate their preferences from among varying combinations of four contract options. The observed preference patterns were analyzed with Sawtooth's Hierarchical Bayes (HB) estimation software (Sawtooth Software, 2005). Sawtooth's HB software iteratively estimates coefficients of a logit model of preferences and tests its accuracy against observed preferences. The root likelihood coefficient indicates the goodness of fit of the mathematical preference model and varies between 1, indicating a perfect fit, and 1 divided by the number of alternatives per task, indicating a poor fit not outperforming random prediction. In all three participations modes (smartphone as instructed, computer as instructed, and computer against instruction), the goodness of fit was .78 indicating a sufficient model fit three times as high as random prediction (.25) (cf. Table 16.15).

Apart from the conjoint part, respondents who participated via computer against instruction showed a lower data quality than participants following the smartphone instruction in nearly all aspects. In two of the three open-ended questions, their answers were considerably shorter; they marked less response options in multiple choice questions and showed more inconsistencies in their brand funnel responses. On average, it took them more time to get from first survey access to survey completion but the actual time they spent on the survey was shorter.

The survey-length condition showed no consistent impact on data quality. In the long questionnaire version, the proportion of respondents who chose the same point on the agreement scale regardless of the statement was 3.5% compared to 2.7% in the short and 1.5% in the intermediate survey length. Apart from that, respondents in the long survey condition answered as well as respondents in the two shorter conditions.

In a concluding open-ended question asking for aspects that respondents did not like about the survey, 39% of the smartphone respondents complained about inadequacies of the survey when accessed via smartphone 16% explicitly criticized the size and the resolution of the displays continuously enforcing zooming actions, 15% mentioned difficulties related with data entry and 2% problems related with data transfer. In the long survey duration condition, participation via smartphone was criticized by nearly half of the respondents.

16.4   Discussion and conclusion

The majority of the members of the German section of the Harris Interactive AG online panel own a smartphone. Many of these smartphone owners report intensive online usage behavior including retrieval of emails. But up to now only 4% of the respondents have accessed a conventional panel survey per smartphone on average. Therefore the potential threat that smartphone participants pose to the representativeness of samples seems to be rather small despite the expectedly low latency between invitation and survey access.

The low proportion of spontaneous mobile survey access is not especially surprising in view of the limitations of smartphones in terms of display size, display resolution, and user interface. For many panel members it seems still way too difficult and uncomfortable to participate via smartphone in conventional online surveys. Given that some smartphone models already have reached the limits of practicability in terms of size and weight, it seems unlikely that participation via mobile devices will become a major issue for online panel providers in the next years, given that the response quality is continuously monitored and secured in a way that is both transparent and fair for the panel members. In fact, the variability in the reported proportions of unintended mobile responses may be partly caused by differences between panels concerning the quality management.

But even if a larger proportion of respondents decided to participate in surveys via smartphones, it would not inevitably decrease the quality of results. There seem no longer to be major technical hurdles for mobile participation in a conventional HTML web survey. The truly crucial factor is the willingness of respondents to spend considerably more time on a survey when participating per smartphone. Therefore, it is recommended to continuously monitor the level of mobile survey access in online panels. Potential biases due to the low latency between invitation and participation can be avoided by implementation of a mandatory waiting period. Apart from that, any interview should be checked against a number of different quality criteria regardless of the participation mode.

References

Bailey, J. & Wells, T. (2012). One giant leap: Understanding the rapid evolution of smartphones for market research. Retrieved June 28, 2013, from: http://www.mrmw.net/news-blogs/305-one-giant-leap-understanding-the-rapid-evolution-of-smartphones-for-market-research.

Baker-Prewitt, J., & Miller, J. (2013). What happens to data quality when respondents use a mobile device for a survey designed for a PC? Retrieved June 28, 2013, from: https://c.ymcdn.com/sites/www.casro.org/resource/collection/0A81BA94-3332-4135-97F6-6BE6F6CEF475/Paper_-_Jamie_Baker-Prewitt_-_Burke.pdf.

Callegaro, M. (2013). From mixed-mode to multiple devices. Web surveys, smartphone surveys and apps: Has the respondent gone ahead of us in answering surveys? *International Journal of Market Research, 55,* 107–110.

Canalys (2012). Smart phones overtake client PCs in 2011. Retrieved June 28, 2013, from: http://www.canalys.com/static/press_release/2012/canalys-press-release-030212-smart-phones-overtake-client-pcs-2011_0.pdf.

Couper, M. P. (2001). Web survey research: Challenges and opportunities. Retrieved June 28, 2013, from: http://www.amstat.org/sections/srms/proceedings/y2001/proceed/00639.pdf.

Google (2012). Our mobile planet: Germany. Retrieved January 28th, 2014, from: http://ssl.gstatic.com/think/docs/our-mobile-planet-germany_research-studies.pdf.

Luck, K. (2011). Involve while you evolve: How to make mobile research work for everyone. *Quirk's Marketing Research Review, 25,* 52–58.

Peterson, G. (2012). What we can learn from unintentional mobile respondents. *CASRO Journal 2012-13,* 32–35.

Peterson, G., Mechling, J., LaFrance, J., Swinehart, J., & Ham, G. (2013). Solving the unintentional mobile challenge. Retrieved June 28, 2013, from: https://c.ymcdn.com/sites/www.casro .org/resource/collection/0A81BA94-3332-4135-97F6-6BE6F6CEF475/Paper_-_Gregg_Peterson _-_Market_Strategies_International.pdf.

Pferdekaemper, T. (2010). On-the-go and in-the-moment: Mobile research offers speed, immediacy. *Quirk's Marketing Research Review, 24,* 52–59.

Sawtooth Software (2005). CBC/HB technical paper. Retrieved June 28, 2013, from: www.sawtoothsoftware.com.

Tarkus, A. (2009). Usability of mobile surveys. In E. Maxl, N. Döring, & A. Wallisch (Eds.), *Mobile market research* (pp. 134–160). Cologne: Herbert von Halem.

Wiegand, E. (2011). Marktforschung in Deutschland 2011. In *Jahresbericht 2011* (pp. 15–19). Frankfurt am Main.Arbeitskreis Deutscher Markt- und Sozialforschungsinstitute e. V. (ADM).

Zahariev, M., Ferneyhough, C., Ryan, C., & Bishop, S. (2009). Best practices in mobile research. Paper presented at the ESOMAR Online Research Conference, Chicago.

17

Internet and mobile ratings panels[1]

Philip M. Napoli[a], Paul J. Lavrakas[b], and Mario Callegaro[c]
[a]*Fordham University, USA*
[b]*Independent Research Psychologist/Research Methodologist, USA*
[c]*Google UK*

17.1   Introduction

Internet ratings panels represent a distinctive type of online panel. Internet ratings panels are designed to provide systematic data on audience size, demographics, and exposure patterns for individual websites and mobile applications. These ratings data play a central role in the transactions that take place between online content providers and advertisers, in addition to serving a variety of other analytical purposes.

These panels are typically constructed and managed by third-party measurement firms. These firms provide data on a subscription basis to website operators, advertisers, media buyers, academic researchers, policy analysts, and any other individuals or organizations with an interest in understanding the size, demographic composition, and behavioral patterns of online audiences.

Ratings panels are used to provide key website performance criteria such as the number of unique audience members visiting a website in a given month; the active reach (which is

---

[1] We thank Sherrill Mane from the Interactive Advertising Bureau (IAB) for her support in writing this chapter. Alexei Zverovitch (Google) helped with the Java code and parsing of the XML output from Alexa Top Sites in order to create Figure 17.1.

Online Panel Research: A Data Quality Perspective, First Edition.
Edited by Mario Callegaro, Reg Baker, Jelke Bethlehem, Anja S. Göritz, Jon A. Krosnick and Paul J. Lavrakas.
© 2014 John Wiley & Sons, Ltd. Published 2014 by John Wiley & Sons, Ltd.
Companion website: www.wiley.com/go/online_panel

the percentage of web users who visited a site in a given month); and the time per person spent on the site (see Nielsen, 2012b). Such performance metrics can also be segmented via demographic categories.

This chapter examines how Internet (PC and mobile) ratings panels are constructed, managed, and utilized. We provide an overview of the history and evolution of Internet/ mobile ratings panels. It examines the methodological challenges associated with creating and maintaining accurate and reliable Internet/mobile ratings panels. Research that has assessed the accuracy and validity of online panel data is critically examined as well as research that illustrates the type of scholarly and applied research questions that can be investigated using online ratings panel data. The chapter concludes with a discussion of the future of online ratings panels within the rapidly evolving field of Internet audience measurement.

17.2   History and development of Internet ratings panels

The commercialization of the Internet – and (equally important) the development of the World Wide Web – began in the early to mid-1990s. The first banner ad appeared on the website of *HotWired* magazine in 1995 (Bermejo, 2009). It is around this time that the first systems for providing audience ratings for the fledgling medium emerged (see Flynn, 1995). Many of these systems relied on online panels to produce their audience ratings (see Coffey, 2001).

However, even in the earliest days of the web, online panels were just one of a variety of possible ways of producing Internet audience ratings. Other approaches, such as relying on the analysis of website server logs, or analyzing web traffic data aggregated by ISPs, have also been – and continue to be – employed (see, e.g., Bennett, et al., 2009; Bermejo, 2007).

The earliest online ratings panels began in 1995 with a service called PC-Meter. Created by the NPD Group, the PC-Meter service operated from a 500-household panel (Bejmejo, 2007). This earliest panel highlights one of the persistent challenges in the provision of Internet audience ratings – creating and maintaining a representative panel of sufficient size to account for the tremendous number of website vying for audience attention. Even in the early days of the web in the mid-1990s, a 500-household panel would be capable of providing meaningful audience size and demographic estimates for a very small fraction of the websites in operation.

By 1997, additional ratings providers had entered the marketplace – Relevant Knowledge and NetRatings; followed in 1998 by NetValue, and in 1999 by a measurement service provided by computer sales and marketing firm PC Data (Bermejo, 2007). As often happens in the field of audience ratings, consolidation took place fairly quickly (Napoli, 2003). Many of the early providers of panel-based audience ratings data quickly merged; and, perhaps most significant, television ratings giant Nielsen moved into the field, partnering with NetRatings in 1998 (Bermejo, 2007). Nielsen would later acquire the firm, with Nielsen NetRatings continuing to serve as one of the leading providers of Internet audience ratings today (now under the name Nielsen Online), with the panel-centric measurement service now operating under the name Nielsen Netview.

As might be expected in the nascent stages of a new field of audience measurement, these early online ratings providers employed a number of different methodological approaches in the construction and management of their panels. Early Internet audience measurement firm Media Metrix, for instance, employed an approach involving installing software in the operating systems of the computers of participating panelists (Bermejo, 2007). Net-Value, in contrast, employed an approach that involved installing measurement software at the level of the Internet protocol suite (TCP/IP). And finally, the measurement system developed and employed by Nielsen NetRatings (now Nielsen Online) was installed at the browser level.

The implications of these different approaches to installing the measurement software were significant. Software installed in the operating system, for instance, could gather information on all computer activities (including non-Internet-related activities), but could not detect, for instance, the distinct elements within individual web pages (such as banner ads). Measurement software installed at the TCP/IP level could monitor any form of online communication (including instant messaging, email, etc.), and could detect all of the distinct elements within individual web pages. However, this approach could not measure the time spent on the last page that was viewed in the browser, since the end of time spent on a web page was determined via the next page request (Bermejo, 2007). Measurement software installed at the browser level was also capable of gathering information on individual web page elements (such as banner ads), but could not gather information about other forms of online activity (such as instant messaging, emailing, etc.). This approach also was incapable of measuring the duration of time spent on the last page requested.

What ultimately emerged from this competitive environment were two firms that, for some time, championed somewhat different (though now converging) approaches to the construction and management of online ratings panels, and the production of Internet audience ratings estimates. Nielsen NetRatings began with (and until recently focused exclusively on) a system based upon a traditional random sample of Internet users, recruited via traditional random digit dialing recruitment procedures (see Goosey, 2003), albeit with extremely low response rates. This panel includes both home- and work-based participants, and has peaked in size at around 30000 individuals (Bennett, Douglas, Rogers, & Broussard, 2009).

In contrast to Nielsen and its traditional sampling approach, leading competitor comScore pioneered "Global Macro-Panels" (Bermejo, 2007, p. 161). By 2007, comScore had managed an online panel of over 1.5 million Internet users (Bermejo, 2007). Today, that number exceeds two million (comScore, 2012), though gained with extremely low participation rates. Moreover, smartphone users are now added to the panel together with tablet owners and game consoles and Smart TVs (comScore, 2013). Panelists are recruited online, via banner ads placed on websites that collaborate with comScore. Incentives for participation include faster browsing speeds, antivirus protection, participation in a sweepstakes, and even offers by comScore to plant trees in exchange for participation (Abraham, 2012; Fulgoni, 2003). Nielsen has since adopted a similar approach, employing a larger non-random sample alongside its traditional random sample (Bennett et al., 2009; Nielsen, 2012a).

It is important to emphasize that this sampling approach, involving very large, non-random samples, has proved controversial. Participants are not always clear that they are agreeing to have their online behaviors measured. And in some instances measurement firms have gathered additional user information (passwords, financial data) without the participants' consent. Occurrences such as these have created a somewhat blurred line between online audience measurement and spyware (see, e.g., Levine & Finkle, 2011).

Obviously, such a recruitment approach lacks the traditional sampling rigor associated with panel-based research. Data from the smaller panel, along with RDD enumeration surveys are employed to weight the data obtained from the larger sample. This strategy of employing various types of "calibration" panels to weight data obtained by other means that may either lack the ability to be representative, or may simply lack the ability to obtain demographic information about the audience, is now somewhat standard.

comScore, for instance, like Nielsen, maintains a smaller, 120000-person "calibration" panel, recruited via traditional random digit dialing methods. This calibration panel is comprised of 50000 home-based participants, 50000 work-based participants, and 20000 university-based participants (Bennett et al., 2009). This panel is used to weight the results obtained from the larger panel of participants recruited online (Cook & Pettit, 2009).

A more recent entrant into the Internet audience measurement arena, Compete, maintains a panel with a sample size of roughly two million participants. Compete engages in what has been described as "panelist multisourcing," which involves gathering data from the company's own proprietary panels, which is combined with licensed clickstream data obtained from third party partners such as ISPs (Bennett et al., 2009, p. 169). Via this approach, Compete (2011) claims to be able to provide detailed audience information for approximately one million websites (see Bennett et al., 2009).

Some tech companies such as Facebook and Google have started building panels in order to be able to meet their advertising clients' needs. Facebook partnered with Nielsen to measure their effectiveness of internet advertising (Nielsen, 2009) and also more recently with Datalogix (Reynolds, 2012) to measure offline advertising. In the Facebook case, the pool of Facebook users is considered their "panel." In other words, the concept of a panel is qualitatively different than what else has been discussed in this chapter. Google is experimenting with online panel partners such as Gfk-Knowledge Networks in the United States (Arini, 2012b; (http://www.google.com/landing/screenwisepanel/) and with Kantar in the United Kingdom (Connected Vision 2011) in order to provide advertisers with an additional way to measure online advertising. For example, a panel was launched in conjunction with the 2012 Olympics and in collaboration with the NBC network in the United States. This panel of approximately 3000 households measured Olympics media consumption from PCs, tablets, TVs and smartphones (Spangler, 2012; Steinberg, 2012).

Finally, it is important to emphasize that the Internet audience has been, in its relatively short history, something of a moving target, as technological, demographic, and behavioral changes have consistently posed new challenges to effectively capturing, aggregating, and reporting audience exposure to online content options. This is why, for instance, panel-based measurement services must continually engage in enumeration surveys in order to try to maintain accurate and up-to-date information on the size and demographic characteristics of the universe of Internet users (something that changes from month to month as Internet penetration across various access technologies continues to advance; see Cook & Pettit, 2009).

17.3   Recruitment and panel cooperation

By recruitment, we mean what happens during the initial stages of creating an audience measurement panel when the online panel researchers try to gain *cooperation* from potential panel members by getting them to "join" the panel. During this stage the panel member (1) must be found to be eligible to join the panel; (2) must agree to do so, including agreeing to all privacy and confidentiality policies; and (3) must engage in the set-up (aka "installation") behaviors that allow for the start of the measurement of her/his Internet usage. Exactly what "join" and "cooperation" and "installation" mean vary by panel and are discussed below.

Of note, we differentiate the recruitment (cooperation-gaining) stage from the notion of maintaining *compliance* among panel members, which refers to the on-going behavior of those who have qualified to participate, and who have agreed to cooperate with what is asked of them, and who actually have started to do so. Longer-term compliance in the panel often presents severe and separate problems for the panel vendor. These are addressed in the next section on panel attrition.

Recruitment of the general public into an audience measurement panel is a very challenging endeavor for several major reasons:

- the time burden on potential panel members to go through the "eligibility screener" process;

- the considerable cognitive burden on potential panel members to read, understand, and make the decision about whether to agree to what can be very convoluted and lengthy confidentiality/privacy agreements the panel vendor has in place,

- the time, intellectual, and/or physical burden on the panel members to install (or to have installed) the measurement system onto their computers and/or smartphones;

- the time and cognitive burden on panel members to use the system properly;

- the invasion of privacy burden on a panel member to allow a research company to learn a great deal about their Internet usage.

Because of these factors, and despite the considerable efforts which some audience measurement companies engage in to motivate cooperation, only a small portion of the public is willing to participate in such a panel – usually less than 5%, and quite often less than 1%. Such participation rates call into question the representativeness of any of these unweighted audience measurement panels.[2]

Recruitment sampling for Internet measurement panels occurs in one of three basic ways:

1. A new panel may be built from scratch using a probability sampling method.

2. A new panel may be built from scratch using a nonprobability sampling method.

3. A new panel may be built from members of a pre-existing panel.

17.3.1 Probability sampling for building a new online Internet measurement panel

When a probability sampling approach is used to start a new online audience measurement panel, it often signals that the representativeness of the unweighted panel is a key concern for the panel researchers.

Currently, the two primary probability sampling designs used for recruiting these online panels are: (1) a form of dual frame (both landline and mobile/cell) random-digit dialing (RDD) sampling with telephone as the mode of contact (AAPOR, 2010); or (2) a form of multistage area sampling using an address-based frame with mail as the first mode of contact (Iannacchione, 2011). In the case of the former sampling technique, RDD surveys in the United States have undergone considerable modification during the past decade due to the considerable growth in the proportion of the population that can only be reached via mobile phone numbers; estimated at approximately 40% as of the end of 2012 (Blumberg & Luke, 2013). Thus, when an RDD sampling design is used, mobile numbers must make up a significant proportion of those who are in the final recruited sample. In the case of address-based

---

[2] These estimates are drawn from the extensive professional experience of one of the authors. The reader may be surprised to learn that these serious scientific concerns do not appear to harm the business prospects of most panel vendors. That is because most clients appear not to understand or care about the threats these problems cause for the reliability and validity of the data they are purchasing.

probability sampling, some research companies supplement the use of mail for the recruitment mode with telephone contact to those nonresponding addresses that can be matched to a phone number. To our knowledge, no panel vendor is as yet building an online audience measurement panel by relying on in-person recruitment as part of a probability sampling design.

17.3.2 Nonprobability sampling for a new online Internet measurement panel

When a nonprobability sampling approach is used to start a new online audience measurement panel, it often signals that the representativeness of the unweighted panel is not of paramount concern for the researchers. Rather, the paramount concern often seems to be to get as many "bodies" (almost regardless of who they are) to join the panel and to keep the costs of doing so as low as possible. Researchers appear to do this because of their apparent belief that the lack of representation in their unweighted panel can be adequately corrected by various statistical adjustments (or so many tell their clients).

Some of the nonprobability sampling approaches use systematic methods (e.g., filling pre-determined demographic quotas) to try to recruit members into the panel, but most use various uncontrolled "sampling" methods that essentially amount to no more than what is correctly described as a haphazard self-selected convenience sample. The methods used to make potential panel members aware of the opportunity to join the new panel vary but are most often made up of Internet-placed "invitations" to do so. Thus whoever happens to notice such invitations has a chance of actually reading it and if they do, then may decide to look into what s/he is being invited to do. Recruitment invitations also may come in the form of an email or via traditional mail. In some cases, a form of telephone recruitment may also be used, but this is more likely to utilize nonrandom lists of telephone numbers rather than a technique like RDD to generate representative telephone numbers. These nonprobability sampling methods most often achieve fewer than 1% of those exposed to the invitations agreeing to join the new panel.[3]

17.3.3 Creating a new panel from an existing Internet measurement panel

For online panel companies that already have general population online panels in place, it is very cost effective to use their existing panel members as the sampling frame to create a new panel. Sampling from this frame can be done in numerous ways, but in almost all cases it would be foolish not to use a probability sampling design, in particular one that stratifies selection along whatever variables are judged to be key to what will be measured (e.g., geography, age, gender, race, Hispanic ethnicity, presence of children in the household, household size, etc.).

In the case of some audience measurement companies (e.g., Arbitron and Nielsen), they already have scientifically sampled panels to measure the audience for media other than the Internet. For them, it is a simple sampling task to try convert some or all of these households so that they agree to have their online behavior measured in addition to what other media exposure/consumption is measured. These "cross-media" or multimedia panels have the great advantage of measuring media behaviors on more than one medium from the same people. Their disadvantages are that it is not a random subset of existing panel members who will agree to allow their Internet behaviors to be measured in addition to the media behavior they

---

[3] This estimate comes from the extensive professional experience of one of the authors. Such response data are typically kept proprietary by online audience measurement firms.

previously agreed to have measured. But, as a counter to this disadvantage, these companies are in the excellent position to study the presence of nonresponse bias (its size and nature) in the new audience measurement panel because they already know so much about each panel member/household in the entire frame from which the new panel is being sampled.

When these existing panels have been formed via probability sampling methods, the original response rates can be relatively high (e.g., in the 20%–40% range for Nielsen's People Meter panels). But when panel households are asked to also allow their Internet usage to be measured, the subset of exiting panel members who agree to cooperate is not a random subset of the original panel sample and typically is well less than 50% of those who are asked to do so.[4]

17.3.4   Screening for eligibility, privacy and confidentiality agreements, gaining cooperation, and installing the measurement system

Regardless of which sampling design is used to get people into an Internet measurement panel, the first step in the actual recruitment of these individuals (and possibly the recruitment of others in their households) is to determine whether they meet the panel's eligibility requirements. Such requirements typically are technology-based, but other than being a resident of the geo-political area being measured and age, they are not often demographic-based.

To accomplish this, a person is asked to complete an "intake" questionnaire that first determines one's eligibility. For those who are found to be eligible, they typically are asked to read and agree to whatever privacy requirements and confidentiality pledges the panel vendor (and possibly its client) has in place. It is at this point where a number of people refuse to continue. This is a difficult tradeoff for Internet measurement companies, because they need and want to protect themselves from various privacy-related liabilities and yet by disclosing in great detail all the data that will be gathered and what they will do with the data, they scare away a good proportion of people from agreeing to join the panel. And there is no good reason to assume that the people who drop out of the recruitment process at this point are a random subset of those who began the process.

For those who agree to the privacy/confidentiality stipulations, they then are informed in more detail about what "joining" the panel entails and then are asked to agree to do so. The burden of actually serving in such a panel typically is not an onerous one, and in some cases is essentially a one-time shot with little or no continued effort required from the panel member.

After they agree to join, there are two basic means by which Internet usage is measured. One system is entirely software-based, in which an app or other type of software is downloaded or otherwise installed onto their computers, tablets, and/or smartphones. The other system involves the installation and use of a piece of hardware in the household – an Internet "meter" – such as a wireless router, in addition to software that may need to be installed onto the devices someone uses to reach the Internet. For the most part, the operation of these systems is invisible to the panel member, with the occasional need to update demographic data and answer some topic-specific questionnaires, all of which is done online.

17.3.5   Motivating cooperation

As noted above, there are a great many hurdles to overcome between initially contacting a potential panel member and gaining a final commitment for cooperation from that person.

---

[4] These data come from the extensive professional experience of one of the authors. As was noted in the previous footnote, such response rates are typically kept confidential by online audience measurement firms.

Lavrakas and colleagues (2012) have advised that the entire process is best viewed as a multi-step challenge and that recruitment procedures should be devised and implemented for achieving success at each of the steps. Thus, the notion of addressing only the end-goal – gaining final cooperation – is not prudent.

Instead the strategies that are used to eventually gain cooperation are dependent on successfully advancing a potential panel member along the entire continuum from sampling to usage of the Internet measurement system. So, for example, offering a contingent incentive only to those persons who are found to be qualified and eventually agree to cooperate is not enough. Ideally sampled individuals will be motivated to accomplish each step and then move onto and through the next step: starting with visiting the website to begin the intake questionnaire, completing the eligibility section, agreeing to the confidentiality/privacy stipulations, agreeing to join the panel, installing the measurement systems, and then starting to use them. Experience shows that achieving cooperation will depend directly on:

- how interested the person is in what is being measured;

- how important s/he thinks the measurement of this behavior is and who s/he perceives will benefit, and by how much, from the existence of the panel;

- how much s/he trusts that no harm will come to her/him or their household from participating;

- how much initial and on-going burden s/he will experience to be a panel member;

- the "value" of the material and nonmaterial incentives s/he and her/his household will receive for joining and staying active in the panel.

Internet usage measurement companies that strive to get only "a lot of bodies" into their panels will use the least costly and therefore the least effective techniques for recruitment, thinking that the cooperation rate does not matter. But the science of building such online audience panels is still in its infancy and no one knows for certain whether it is more cost-effective to build a panel by spending more or less money on each successfully recruited member.

17.4 Compliance and panel attrition

As noted above, compliance is the term that defines how well Internet audience measurement panel members "behave" after joining the panel, including how long they remain as members compared to the length of time for which they originally agreed to stay in the panel. The latter speaks to the issue of *panel attrition*, i.e., whether, and why, someone drops out or is forced out before their original commitment has ended.

From an on-going compliance standpoint, experience shows that within a general population panel, there will be three groups of panelists: (1) those who comply well with what they agreed to do and need little or no encouragement from the panel managers; (2) those who comply well enough, but need occasional encouragement (intervention) from the panel managers, and (3) those who do not comply well at all and need a lot of intervention by the panel managers.

In the extreme case, some among this last group will have such high noncompliance and will require so much attention from the panel managers that a point will be reached where they are likely to be forced to drop out of the panel. This process has been euphemistically called by some "unforced attrition," because it happens before the panelists reach the point (if there is one in a particular panel) when they would be forced to leave the panel due to their tenure reaching the maximum time permitted for the panel.

There are two broad types on noncompliance – often termed *faulting* – within audience measurement panels. One form of faulting is associated with technical problems with the system that is used to measure a panelist's Internet usage. These technical faults may occur because the system hardware and/or software itself fails to work reliably. In these cases, the system needs to be repaired or new hardware/software needs to be installed to bring the panelist back into compliance (i.e., eliminate the faulting). However, in some instances these technical faults occur because of some behavior in which the panelist has or has not engaged. Examples include problems such as buying a new computer or smartphone and not informing the panel company so that the measurement system can be properly installed on these new devices; or moving a desktop computer or the router to another room and not reconfiguring it to work properly with the measurement system. In these cases it is not the equipment's "fault" that the panelist is no longer in compliance, but rather the panel member's fault. Although these latter types of technical faults may occur with just about any panelist, experience shows that they occur much more frequently with certain types of panelists (see below).

A second category of noncompliance occurs because of behavioral problems associated with the panelist. These social or behavioral faults are associated with panelists who simply forget or refuse to continue doing what they originally agreed to do as part of their panel membership agreement. For example, some online Internet usage measurement systems may require that a panelist login to the measurement software to indicate which household member is presently using the computer. In some measurement systems the user may be prompted occasionally to re-login, in case the user has changed since the last login and the previous user failed to properly log out. In the case of smartphone software, the login–logout problem is moot as most panel measurement companies tend to assume there is only one user for a given smartphone. For some people, this requirement to login/logout is an onerous burden, and thus they will be noncompliant for various times when they are not properly logged in as themselves.

A related, but different type of behavioral faulting occurs in some measurement systems in which Internet users can log themselves in as a "guest" without disclosing anything else about themselves. Were legitimate panel members to do this because they did not want the Internet usage in which they were about to engage to be directly associated with them, they would technically be noncompliant for that time of Internet usage. Were they to do this over and over again selectively (and there is good reason to believe that some people do this quite often, e.g., those panel members who do not want anyone else to know every time they visit porn sites), then this would constitute a social/behavioral form of noncompliance.

Unfortunately for audience measurement companies and their clients, it is not a random subgroup of the panel membership that consistently has technical and/or behavioral problems with noncompliance (Holden, Heng, Lavrakas, Bell, & Flizik, 2006). Experience has shown that low compliance in audience measurement panels is likely to be correlated with panel members who are younger (< 35 years), African Americans, Spanish-Dominant Hispanics, have lower income, have lower educational attainment, have at least one child (< 18 years) living in their household, and/or are members of large-sized households. To the extent this happens, the data being gathered each day about the panel's Internet usage will be biased, as there will be missing data that are "not missing at random."

To address these issues, Internet audience measurement companies can implement various strategies to try to reduce or possibly eliminate a panel member's faulting. These efforts can be as simple as sending the noncompliant person/household a reminder email. At the other end of the spectrum is the deployment of a "personal coaching" program, whereby a member of the panel management staff telephones or even may visit in person the problem panelists and attempt to motivate them to improve compliance (Holden et al., 2006). Depending on the approaches used, this can be an extremely expensive enterprise. To our knowledge, very few business models for Internet audience measurement companies support doing more than sending email reminders alerting panelists of their noncompliance problems, if even that. Instead, the *modus operandi* for most of these panels for reasons of costs seems to be to simply drop those panelists who are consistently noncompliant, and replace their number, if not their "type," with new panelists.

Another form of attrition is linked to those who drop out of an online Internet measurement panel prematurely (before their originally agreed to tenure ends) because of a growing concern about the invasion of privacy to which they are subjecting themselves by allowing their Internet behavior to be measured. These are likely to be individuals who originally joined the panel with some reluctance because of their privacy concerns, and these concerns begin to "eat away at them" over time. They finally reach a point where they think "enough is enough," even if they have no tangible evidence that they have been harmed by their participation in the panel. There is no valid reason to believe that those who leave Internet measurement panels for these reasons are a random subset of the entire panel membership, though we are unaware of any definitive empirical evidence on this issue.

17.5   Measurement issues

There are many measurement issues to consider when measuring audience behavior with online panels (Media Rating Council, 2007). Measurement issues can be classified into the following three broad groups: (1) coverage of Internet access points; (2) confounding who is measured; and (3) differences in metering technologies.

17.5.1   Coverage of Internet access points

With the proliferation of devices that can access the Internet (tablets, smartphones, laptop and desktop computers at both home and work), the challenge is to measure all of them. For example, Internet browsing at work is not easily captured and linked back to the same household panel member. Tablets are becoming more and more prevalent in households and many studies are converging on the finding that smartphones and tablets are used as a second screen when watching TV (Smith & Boyles, 2012). In the United States, for example, half of the adult mobile phone owners have used their phone for engagement, diversion (e.g., during commercial breaks), and fact checking when watching TV (Smith & Boyles, 2012). In another report (Nielsen, 2011), it was estimated that "36% of people 35–54 and 44% of people 55–64 use their tablets to dive deeper into the TV program they are currently watching" (p. 4). These data points are important to understanding the need to measure every device in order to have a complete picture of Internet behaviors.

According to recent estimates, about 60% of the US adult population owns a smartphone, while 40% own a tablet and 32% an ebook reader (Zickur & Rainie, 2014). In the United Kingdom, for example, 59% of adults 16 and older use a smartphone while tablet ownership is at 35% at a household level (Ofcom, 2014). These technologies are becoming an increasingly prominent means by which individuals go online. And, in some countries, these devices are far and away the *primary* means by which the bulk of the population goes online (Napoli & Obar, 2013).

According to current assessments, however, the measurement of online audiences via mobile platforms has not kept pace with the rate at which audiences are using these devices to go online (see, e.g., Gluck, 2011). For instance, measurement in this area has yet to fully move to the third party provider model that characterizes other media, and that is generally considered essential for generating the necessary levels of advertiser trust and confidence in the audience estimates. Further, existing measurement systems remain, at this point, largely unable to describe the mobile audience in terms of unique visitors (Gluck, 2011), which, of course, undermines the audience targeting models traditionally employed by advertisers, as well as the ability to conduct comparative analyses across media platforms. As research by Gluck (2011) has indicated, "Publishers and advertisers feel frustrated that mobile, to date, does not offer the same metrics and accountability found in PC-based digital media" (p. 6).

This situation is due, in part, to the greater complexity associated with measuring audience behavior via mobile devices. This complexity arises from a range of factors, including substantial variability in terms of the devices, operating systems, and carriers that need to be accounted for in the measurement process; as well as more complex privacy issues that arise within the context of mobile-based Internet audience measurement; and the prominence of mobile applications (i.e., apps) rather than web pages as the mechanism via which individuals engage with online content (Gluck, 2011).

Panel-based measurement of Internet audiences via mobile devices is very much in its infancy. Nielsen, for instance, released its first mobile rankings of websites and mobile applications in late 2011 (see BusinessWire, 2011). These rankings were derived in part from a panel of 5000 smartphones installed with metering software (BusinessWire, 2011). comScore launched a panel-based mobile measurement system, Mobile Metrix 2.0, in mid-2012 (Opidee, 2012). Through these panels, data are gathered on a wide range of mobile-based activities, including application usage, website visits, geographic location, and even date and duration of phone calls (Nielsen, 2012c). US radio ratings firm Arbitron (recently acquired by Nielsen) recently launched its Mobile Trend Panel in the United States, the United Kingdom, Germany, France, Denmark and Finland. It is a consumer opt-in panel in which panel members install an app on their smartphones and tablets that tracks their Internet and phone behavior. A questionnaire can also be triggered from the app based on context-specific situations (Arbitron 2012a, 2012b).

17.5.2   Confounding who is measured

Internet rating panels rely on panel members and on technology to identify who is using what device. In industry language this is called "Attribution of user sessions to specific users" (Cook & Pettit, 2009). This is the case of shared devices such as laptops, desktops, and tablets. For example, Nielsen (2011) estimated that 43% of tablet users share their device with household members. One way to assess who is using what device is to require panel members to identify themselves on the device used at certain intervals. This strategy is prone to compliance issues (see above). Another method relies on technology to identify the device user. For example, comScore (Cook & Pettit, 2009) uses a patented procedure called User

Demographic Reporting (UDR) that analyzes mouse and keystroke patterns in order to build a profile for each panel member that is then cross-checked with a self-identification survey.

When talking about mobile rating panels, the measurement issues are slightly different. First, there is very little sharing of smartphones among family members. Measuring Internet behaviors on smartphone generally entails panel members installing an app that tracks their Internet traffic. As with Internet panel members, compliance is an issue. Another way is to measure smartphone Internet behaviors only at home when the panel members use their smartphone connected to their home Wi-Fi and the Wi-Fi router sends data to the panel company. This methodology has the advantage of not requiring panel members to install an app and the disadvantage of not measuring smartphone Internet behavior outside the home (for a discussion of additional measurement issues, see Gluck, 2011).

17.6   Long tail and panel size

Perhaps the single greatest challenge to effectively utilizing panels to calculate online audience ratings involves the construction and maintenance of sufficiently large panels to account for the unprecedented levels of media and audience fragmentation that characterize the online space. Audiences are fragmented across the array of websites available to them (see FAST/ARF, 2000; Fulgoni, 2003; Napoli, 2011), which creates the persistent problem of effectively capturing the distribution of audience attention across these content options. The pattern that we typically see, in terms of the distribution of audience attention, is one in which a large proportion of the audience is clustered around relatively few popular sites, with the remaining audience attention very widely dispersed across an incredibly large array of relatively unpopular content options. This pattern has been termed the "long tail" by media analyst Chris Anderson (2008). This pattern is represented in Figure 17.1 with Alexa data as of June 2013 for US traffic.

In Figure 17.1, we can see that to be among the top 1000 websites, very few sites capture more than 5% of overall traffic in the United States, very quickly the majority of sites capture 1% or fewer visitors per month; and perhaps more importantly, once outside the top 100 websites, we are well into the Internet's exceptionally long tail. For example 0.005% of traffic is captured by site rank number 877. This tail is so long that, according to one recent estimate, companies such as Nielsen and comScore can "provide detailed audience measurement for [only] fifteen to thirty thousand of the estimated 187 million available websites" (Napoli, 2011, p. 74). The audiences for the remaining 186 million plus websites remain essentially unmeasurable via purely panel-based measurement systems. Granted, the audiences for each of these individual websites are incredibly small; however, in the aggregate they represent a substantial portion of online audience attention. This dynamic creates a powerful incentive for online audience measurement systems to try to reach as far down the long tail as is practically possible.

The unprecedented fragmentation of the online space means, however, that it is for all practical purposes impossible for panel-based ratings systems to provide accurate and reliable measurements of even a fraction of the totality of available websites. The costs associated with recruiting and maintaining panels impose significant limits on the extent to which panels are able to provide detailed audience data on the websites that populate the long tail of audience attention (Fulgoni, 2003).

Other measurement approaches, such as the analysis of server logs, offer ways around the long tail problem (Bennett et al., 2009). However, these analytical approaches are beset with their own distinctive methodological shortcomings. In particular, these

Figure 17.1 June 2013 percentage of US traffic for the top 1,000 U.S. websites. Reproduced by permission of Amazon Web Services. *Source:* Data from Alexa Top Sites http://aws.amazon.com/alexatopsites/

measurement approaches lack the ability to accurately and reliably provide the kind of audience demographic information that advertisers have come to demand. As the Interactive Advertising Bureau (2009) notes in their Audience Reach Measurement Guidelines,

> While it is possible for census-based measures to produce counts of Unique Users . . . the threshold of measurement difficulty for achieving this measure in a census-based environment is quite high (generally because of the difficulty of being able to identify a cookie as a unique person persistently during the measurement period).

(p. 4)

Census-based approaches are used for measuring smartphone traffic as well. For example, comScore in collaboration with some UK mobile operators carriers collects anonymized census-level data traffic (comScore, 2013).

17.7   Accuracy and validation studies

The history of online audience ratings has, from its very beginnings, been characterized by persistent evidence of substantial discrepancies across different measurement services (see, e.g., Fitzgerald, 2004). Anecdotal examples of extreme discrepancies in audience estimates across measurement services – even among those employing methodologically similar approaches (e.g., across panel-based measurement services) are commonplace (Story, 2007; Graves, Kelly, & Gluck, 2010). It is not uncommon, for instance, for the two leading panel-centric online ratings firms to fail to even be in agreement as to what is the second most popular online destination in a given month (Graves et al., 2010). Some analyses have contended that these discrepancies, and the uncertainty that they engender, have been the primary impediment to the growth of online advertising (see, e.g., Graves et al., 2010; Story, 2007).

Detailed audits of the accuracy and reliability of audience measurement systems are seldom made publicly available in the United States, where such audits typically are handled on a confidential basis by the Media Rating Council (MRC). While the MRC will make public the outcome of such an audit, in terms of whether the audit resulted in the MRC accrediting the measurement service, the underlying analyses typically are kept confidential. As of this writing, the MRC is engaged in audits of the leading panel-centric Internet audience measurement systems in the United States. Such audits can take into consideration each of the various stages in the process of generating audience ratings, from the enumeration survey, to the metering process, to sample recruitment and weighting (see Haering, 2002, for an example of an audit conducted on three panel-based online audience ratings services operating in France at the time).

In 2009, the Advertising Research Foundation released a fairly detailed assessment of the comScore methodology (Cook & Pettit, 2009). This assessment relied heavily on the assertions and representations of comScore and thus was explicitly described as "not an . . . audit" (Cook & Pettit, 2009, p. 2). Nonetheless, the report contains more methodological detail than can typically be found in publicly accessible documents; as well as a brief validation study in which the authors compare comScore data with other available data sources. Specifically, the authors compare comScore data on ad impressions with impression data obtained directly by the advertisers (finding variations of between 3–5% between the two data sources). In the end, the authors conclude with a very positive assessment of the comScore methodology and data (Cook & Pettit, 2009).

Of course, one of the greatest challenges in determining the accuracy and validity of any audience ratings system is finding any kind of meaningful baseline for comparison. When it comes to online ratings panels, there is one approach that is commonly relied upon. Specifically, as comScore's Gian M. Fulgoni (2003) has noted, "Unlike offline media, there is a reliable way to confirm the validity of sample-based online audience data: namely a comparison to site server logs, which measure machine-based visits" (p. 9). However, as Fulgoni (2003) also notes, this approach has, "historically, been a source of much heated debate between syndicated audience measurement suppliers and site operators, with syndicated suppliers generally showing lower audience estimates" (p. 9). Research has found discrepancies across the two approaches ranging from 35%– 350% (McDonald & Collins, 2007).

Reasons for the significant discrepancies include the fact that panel-based systems have tended to underrepresent work- and school-based Internet usage; the fact that website visits can come from around the globe, though most panels are not global in nature; and the fact that server logs measure visits by individual computers/devices rather than individual people (Fulgoni, 2003). From this standpoint, the closer that panel-based estimates correlate with server-log estimates can be seen as reflective of the greater comprehensiveness of the panel being utilized (see Fulgoni, 2003, for an example of a comScore-conducted validation analysis comparing comScore's audience estimates for individual websites with those of competing providers).

Of course, such approaches to assessing the validity of panel-based audience ratings services are somewhat limited in that, while they can facilitate comparative analyses of the size of the audiences measured by different analytical approaches, they cannot provide meaningful insights into the accuracy or validity of the demographic data obtained by panel-based ratings systems.

As should be clear, the limited analyses that are publicly accessible suggest persistent disagreements between the various methodological approaches to measuring audiences online. What is perhaps most striking about this state of affairs is the fairly recent observation by industry research professionals that "we know of no endeavors to attempt detailed reconciliations between the various methods, which might lead to an understanding and quantification of various areas of discrepancy" (Bennett et al., 2009, p. 165), and that, over time confusion, contradictions, and uncertainty, both within and across audience ratings systems have become worse, rather than better (Graves et al., 2010).

17.8   Statistical adjustment and modeling

Because most unweighted Internet audience panels are unrepresentative of the populations they purport to measure (as explained above), panel vendors use a variety of statistical adjustments and modeling techniques to try to correct these serious limitations. These techniques all amount to some form of statistical weighting and in some cases utilize a so-called "calibration panel" to assist weighting.

Calibration panels are ones that are formed by a probability sampling design (e.g., RDD or address-based sampling) and utilize more muscular and expensive recruitment protocols to try to yield a truly representative sample of the target population. These calibration panels are made up of members who agree to engage in the same Internet measurement systems in which the main measurement panel members engage. In theory, if the calibration panel is in fact representative of the target population, then the main panel can be modeled on the calibration panel, thereby making the main panel representative of the target population. However, we are unaware of the existence of any publicly available detailed evidence that this approach is reliable and valid.

Another major problem in this area is that no detailed evidence has been reported to allow any outsiders to know with confidence whether any of the weighting that is used to adjust Internet measurement panels actually is successful in fixing them to be representative of their target population. However, it is unlikely that they are because of the nature of the factors that lead such panels to be unrepresentative in the first place. For all Internet measurement panels, there is a considerable amount of noncoverage or nonresponse that takes place when they first are built. Although the panel vendors can statistically adjust the panel to the demographic factors that are measured from the panel members and are available on the population, there is no reliable evidence of which we are aware that demonstrates that these demographic factors are in fact the characteristics that cause the lack of representation in these panels. Instead, it is likely that there are a host of unmeasured psychographic and lifestyle variables that account for why someone does or does not get sampled for an Internet measurement panel and for why someone who is sampled does or does not join the panel. Until panel vendors address these issues and release detailed evidence that (1) they have identified the key characteristics that cause their unweighted panels to be unrepresentative; and (2) they have used valid statistical means to take these key characteristics into account, it is our position that one cannot have confidence in the effect of any of the weighting schemes being deployed.

17.9   Representative research

The audience behavior and demographic data obtained via online ratings panels can, of course, be applied to a variety of both applied and academic research purposes. Advertisers and media buyers use the data to identify websites that attract audiences that meet their desired audience size and demographic criteria. Website operators use the data to assess the performance of their sites, and of individual pages within their sites. Such analyses can involve assessing perfor-mance patterns over time; and engaging in comparative performance analyses with competing websites. Such analyses can be used to formulate and adjust strategy. Policy-makers will, on occasion, utilize such data to inform both economic and social policy concerns, such as assess-ing the competitive environment and determining which demographic groups do, and do not, have access to the Internet. Government agencies have utilized ratings data to assess usage patterns and audience demographics of government-operated websites, as one of a variety of mechanisms employed to evaluate whether these sites are effectively meeting the population's information needs (Wood, 2005).

Online ratings panel data have also been utilized in academic research across a variety of fields, including communications, information science, and marketing. However, given the costs associated with obtaining access to such data, the use of such data for academic research purposes has been infrequent.

One important way that online ratings panel data have been utilized in academic research has been to develop and test theories of audience behavior. The bulk of the extant audience behavior research has been conducted within the context of traditional media such as televi-sion; and so one key question that has engaged researchers is whether patterns of audience behavior that have characterized traditional media are replicating themselves online.

Television audience researcher, James Webster, for instance, has been engaged in a line of research examining the extent to which patterns of audience duplication, fragmenta-tion, and polarization that have come to characterize television viewing also characterize online behaviors. Webster and Lin (2002), for instance, used Nielsen NetRatings data to examine whether patterns of audience distribution and audience duplication across websites exhibited patterns found in the television context. In both cases, the authors found significant commonalities across television and online audience behaviors, leading them to conclude that "when users of the World Wide Web are seen as a mass audience, they exhibit straightforward, law-like behaviors similar to those of older media" (Webster & Lin, 2002, p. 8).

This line of inquiry is extended in Webster and Ksiazek's (2012) analysis of audience duplication *across* television channels and websites; this time utilizing a Nielsen "convergence panel" of individuals who simultaneously had their web browsing and television viewing monitored. Their results indicate levels of audience duplication across television channels and websites that would seem to call into question commonly-raised concerns about media fragmentation leading to audiences segmenting themselves into narrowly defined, homogeneous enclaves (Webster & Ksiazek, 2012).

Similar concerns underlie Tewksbury's (2005) analysis of online news audiences. Utilizing Nielsen NetRatings data, Tewksbury (2005) addresses the political dimension of online media consumption, in terms of whether individual news sites are extremely specialized in terms of the types of audiences (in terms of demographics) they attract, and in terms of the type of content that is produced and consumed. The results indicate high levels of specialization across both audience demographics and content production and consumption (Tewksbury, 2005).

Chiang, Huang, and Huang (2010) look specifically at the well-known "double jeopardy" effect (a pattern in which there is a strong correlation between audience size and audience loyalty), in an effort to determine whether this pattern is prevalent in patterns of online audience behavior. Utilizing ratings panel data provided by the Taiwanese audience measurement firm InsightXplorer Limited, the authors find strong evidence that the double jeopardy effect is as prominent in online audience behavior as it has been in the behavior of traditional media audiences (Chiang et al., 2010).

Marketing researchers Danaher, Mullarkey, and Essegaier (2006) utilized online panel data to explain patterns in the *duration* of audiences' visits to individual websites. Given the widely acknowledged importance of the duration of audience visits to maximizing audiences' exposure – and potential response – to online advertisements, the authors sought to identify those factors that affect visit duration. Utilizing Nielsen NetRatings data, the authors found a wide range of significant predictors of visit duration, including: demographic characteristics such as age (older audiences = longer visit duration) and gender (female = longer visit duration); as well as content characteristics (auction, entertainment, and news sites = longer visit duration). Importantly, the authors also found that the amount of advertising content is significantly negatively related to visit duration (Danaher et al., 2006).

17.10   The future of Internet audience measurement

It is important to emphasize that today, almost 20 years after the birth of Internet audience ratings systems, most of the stakeholders involved remain highly dissatisfied with the nature of the information available to them. As one recent trade press account began, "Online measurement is broken. That, few would dispute" (Neff, 2011, p. 8). This same account details recent efforts by major advertisers, trade organizations, and media companies to establish acceptable standards for the measurement of online audiences and the metrics to be reported. Called "Making Measurement Make Sense," this initiative in many ways reflects how little progress has been made over the past 20 years in terms of measurement firms being able to satisfy the demands of their client base. It is also a reflection of the enormous complexity and challenges associated with the task of producing online audience ratings. It seems reasonable to question whether a system that genuinely satisfies all of the relevant stakeholder groups is even achievable under the challenging conditions of the online media environment.

The key trend going forward appears to be one in which online panels no longer function as a stand-alone source of ratings data. The current state of the art in Internet ratings involves a hybrid approach that integrates panel-centric and site-centric data. In 2009, for instance, comScore (2009) launched Media Metric 360, a service that derives website ratings from what it terms a "panel-centric hybrid" that combines the ability of site-centric data to capture audience exposure in the long tail with the demographic data that can best be obtained via traditional panels (for more detail, see Pellegrini, 2009). Such an approach provides a means of extending the reach of an Internet ratings system further into the tail than can be achieved via a pure reliance on panels. However, such an approach does not completely "solve" the problem of the long tail (discussed above).

An additional methodological approach, therefore, is likely to become more fully integrated into the production of Internet audience ratings as these measurement services continue to evolve. Specifically, "direct" measurement of Internet traffic data obtained from ISPs is also increasingly being integrated into the production of online audience ratings (by firms such as Quantcast). Thus, according to some industry professionals, for "the immediate future and beyond, we'll be living in a three-sourced world for Internet audience estimates: Panel, server and direct" (Bennett et al., 2009, p. 166).

Finally, probably the biggest challenge in Internet audience estimates is the proliferation of the number of devices per user. It is very common that in developed countries the same people access the web from their desktop computer at work, a laptop computer, a tablet, and a smartphone. Measuring the "whole picture" is extremely complex and technically challenging, not to mention intrusive. Advertisers are also interested in learning about the interaction of traditional media such as TV and print media with Internet advertisements. Measuring such exposure within the same households, for example, adds another layer of complexity to an already fragmented media environment. This is, however, the direction that Internet rating companies are pursuing with different measurement solutions.

References

Abraham, M. M. (2012, April 23). comScore plants 3 millionth tree as part of "trees for knowledge" program. Retrieved July 1, 2013, from: http://www.comscore.com/Insights/Blog /comScore_Plants_3_Millionth_Tree_as_Part_of_Trees_for_Knowledge_Program.

Anderson, C. (2008). *The long tail: How the future of business is selling more of less* (Updated ed). New York: Hyperion.

Arbitron. (2012a). Media tracker. Retrieved August 24, 2012, from: http://www.zokem.com /products/media-tracker/.

Arbitron. (2012b). Arbitron mobile infographics. Retrieved August 24, 2012, from: http://www .zokem.com/arbitron-mobile-infographics/.

Arini, N. (2012a, March 19). Measuring modern media consumption. Retrieved August 24, 2012, from: http://adwordsagency.blogspot.co.uk/2012/03/measuring-modern-media-consumption.html.

Arini, N. (2012b) http://www.google.com/landing/screenwisepanel/.

Bennett, D., Douglas, S., Rogers, B., & Broussard, G. (2009). Digital research – online measurement too many numbers – too many different relationships. Paper presented at the Worldwide Readership Research Symposium, Valencia, Spain. Retrieved July 2, 2012, from: http://www.printanddigitalresearchforum.com/online-measurement-%E2%80%93-too-many-numbersrelationships.

Bermejo, F. (2007). *The Internet audience: Constitution & measurement.* New York: Peter Lang.

Bermejo, F. (2009). Audience manufacture in historical perspective: From broadcasting to Google. *New Media & Society, 11,* 133–154.

Blumberg S. J., & Luke J. V. (2013) *Wireless substitution: Early release of estimates from the National Health Interview Survey, July–December 2012.* National Center for Health Statistics. June 2013. Retrieved June 29, 2013 from:http://www.cdc.gov/nchs/data/nhis/earlyrelease/wireless201306.pdf.

BusinessWire (2011, September 13). Nielsen releases first mobile media rankings based on audience measurement data from Android smartphone meters. Retrieved August 22, 2012, from: http://www.businesswire.com/news/home/20110913005752/en/Nielsen-Releases-Mobile-Media-Rankings-Based-Audience.

Chiang, I., Huang, C., & Huang, C. (2010). Traffic metrics and web 2.0-ness. *Online Information Review, 34,* 115–126.

Coffey, S. (2001). Internet audience measurement: A practitioner's view. *Journal of Interactive Advertising, 1,* 10–17.

Compete (2011). Overview of data methodology and practices. Retrieved August 19, 2012, from: http://media.compete.com/site_media/upl/img/Compete_Data_Methodology_3.pdf.

comScore (2009). comScore announces Media Metric 360: The next generation of global digital audience measurement. Retrieved August 28, 2012, from: http://www.comscore.com/Press_Events/Press_Releases/2009/5/comScore_Announced_Media_Metrix_360.

comScore (2012). Investor FAQs. Retrieved August 28, 2012, from: http://ir.comscore.com/faq.cfm.

comScore (2013). comScore brings big data to mobile and multi-platform analytics. Retrieved February 22, 2013, from: http://www.comscore.com/Insights/Press_Releases/2013/2/comScore_Brings_Big_Data_to_Mobile_and_Multi-Platform_Media_Analytics.

comScore (n.d.). GSMA mobile media metrics. Retrieved February 22, 2013, from: http://www.comscore.com/Products/Audience_Analytics/GSMA_Mobile_Media_Metrics_MMM.

Connected Vision. (2011, July 8). Google and Kantar develop measurement panel. Retrieved August 24, 2012, from: http://informitv.com/news/2011/07/08/googleandkantar/.

Cook, W. A., & Pettit, R.C. (2009, April 6). comScore Media Metrix U.S. Methodology. An ARF research review. Retrieved July 1, 2013, from: http://thearf-org-aux-assets.s3.amazonaws.com/downloads/research/comScore-RReview.pdf.

Danaher, P. J., Mullarkey, G. W., & Essegaier, S. (2006). Factors affecting website visit duration: A cross-domain analysis. *Journal of Marketing Research, 43,* 182–194.

Earle, B., Shaw, K., & Moy, C. (2011). The GfK NOP Media Efficiency Panel. Retrieved August 24, 2012, from: http://www.mrs.org.uk/article/item/356.

FAST/ARF (2000). Principles for online media audience measurement. Retrieved August 2, 2012, from: http://s3.amazonaws.com/thearf-org-aux-assets/downloads/research/ARF-FAST-Principles-2000.pdf.

Fitzgerald, K. (2004, March 15). Debate grows over Net data. *Advertising Age,* pp. 4, 78.

Flynn, L. (1995, May 29). In search of Nielsens for the Internet. *New York Times.* Retrieved July 25, 2012, from: http://www.nytimes.com/1995/05/29/business/in-search-of-nielsens-for-the-internet.html?pagewanted=all&src=pm.

Fulgoni, G. M. (2003, June). Measuring consumers' Internet behavior. Paper presented at the ESOMAR/ARF Worldwide Audience Measurement Conference, Los Angeles, CA.

Gluck, M. (2011). State of mobile measurement. Paper prepared for the Interactive Advertising Bureau. Retrieved August 2, 2012, from: http://www.iab.net/media/file/Mobile-measurement-paper-final-draft2.pdf.

Goosey, R. W. (2003, June). Defining the gold standard for user centric global online audience measurement. Paper presented at the ESOMAR/ARF Worldwide Audience Measurement Conference, Los Angeles.

Haering, H. (2002, June). Monitoring Internet audience measurement panels. Paper presented at the ESOMAR/ARF Worldwide Audience Measurement Conference, Cannes, France.

Holden, R., Heng, L. M., Lavrakas, P. J., Bell, S., & Flizik, A. (2006). Personal contact and performance-based incentives to raise long-term panel compliance and to reduce missing data. Paper presented at the 61st annual conference of the American Association for Public Opinion Research Conference, Montreal.

Iannacchione, V. G. (2011). The changing role of address-based sampling in survey research. *Public Opinion Quarterly, 75*, 556–575.

Interactive Advertising Bureau (2009). Audience reach measurement guidelines version 1.0. Retrieved July 24, 2012, from: http://www.iab.net/media/file/audience_reach_022009.pdf.

Kok, F., Appel, M., Verhulst, E., & Pellegrini, P. (2011). Internet audience measurement now – sites, content and campaigns – anywhere, anyhow. Paper presented at the Print and Digital Research Forum, San Francisco. Retrieved August 19, 2012, from: http://www.printanddigitalresearchforum.com/internet-audience-measurement-now-sites-content-and-campaigns-anywhere-anyhow.

Lavrakas, P. J., Dennis, M. J., Peugh, J., Shand-Lubbers, J., Lee, E., Charlebois, O., & Murakami, M. (2012). An experimental investigation of the effects of noncontingent and contingent incentives in recruiting a long-term panel: Testing a Leverage Salience Theory hypothesis. Paper presented at the 37th Conference of the Midwest Association for Public Opinion Research. Chicago, IL. Retrieved from: http://www.mapor.org/confdocs/absandpaps/2012/2012_papers/5C4_Lavrakas_paper.pdf.

Levine, D., & Finkle, J. (2011, August 23). Exclusive: Privacy lawsuit targets comScore. *Reuters*. Retrieved February 22, 2013, from: http://www.reuters.com/article/2011/08/23/us-comscore-lawsuit-idUSTRE77M76O20110823.

McDonald, S. & Collins, J. (2007). Internet site metrics and print media measurement – from dissonance to consonance? Paper presented at the Worldwide Readership Symposium, Vienna. Retrieved August 19, 2012, from: http://www.gfkmri.com/PDF/WP%20Internet%20Site%20Metrics%20and%20Print%20Media%20Measurement%20From%20Dissonance%20to%20Consonance.pdf.

Media Rating Council. (2007). A guide to understanding Internet measurement alternatives. Retrieved July 1, 2013, from: http://mediaratingcouncil.org/MRC%20POV%20General%20Internet%20080307.pdf.

Napoli, P. M. (2003). *Audience economics: Media institutions and the audience marketplace.* New York: Columbia University Press.

Napoli, P. M. (2011). *Audience evolution: New technologies and the transformation of media audiences.* New York: Columbia University Press.

Napoli, P. M., & Obar, J. (2013). Mobile leapfrogging and digital divide policy: Assessing the limitations of mobile Internet access. New America Foundation White Paper. Retrieved July 1, 2013, from: http://newamerica.net/sites/newamerica.net/files/policydocs/MobileLeapfrogging_Final.pdf.

Neff, J. (2011, Sept. 19). Coca-Cola, AT&T and others out to reinvent web measurement. *Advertising Age*, p. 8.

Nielsen. (2009, September 22). Nielsen in a relationship with Facebook. Retrieved from: http://www.nielsen.com/us/en/newswire/2009/nielsen-in-a-relationship-with-facebook.html.

Nielsen. (2011, May 5). Connected devices: how we use tablets in the US. Retrieved from: http://blog.nielsen.com/nielsenwire/online_mobile/connected-devices-how-we-use-tablets-in-the-u-s/.

Nielsen (2012a). The industry benchmark for Internet audience measurement. Retrieved July 5, 2012, from: http://nielsen.com/content/dam/nielsen/en_us/documents/pdf/Fact%20Sheets/NetView_US.pdf.

Nielsen (2012b). Press center: Press FAQs. Retrieved August 22, 2012, from: http://www.nielsen-online.com/press.jsp?section=pr_faq.

Nielsen (2012c). Smartphone study: FAQs & contacts. Retrieved August 22, 2012, from: https://mobilepanel.nielsen.com/enroll/help.d0?l=en_us&pid=1.

Nielsen. (2012d). *The cross platform report: Q2 -2012 - US*. Retrieved July 1, 2013, from http://www.nielsen.com/content/dam/corporate/us/en/reports-downloads/2012-Reports/Nielsen-Cross-Platform-Report-Q2-2012-final.pdf.

Ofcom (2014). Ofcom technology tracker Wave 3, 2013. Retrieved from: http://stakeholders.ofcom.org.uk/binaries/research/statistics/2014Jan/Ofcom_Technology_Tracker_data_tables_for_publication_Wave_3_2013.pdf.

Opidee, I. (2012, May 8). comScore launches smartphone metrics services. *FolioMag*. Retrieved August 22, 2012, from: http://www.foliomag.com/2012/comscore-launches-smartphone-metrics-service.

Pellegrini, P. (2009). Towards panel-centric hybrid measurement: Enhancing online audience measurement. Paper presented at the Worldwide Readership Research Symposium, Valencia, Spain. Retrieved August 12, 2012, from: http://www.printanddigitalresearchforum.com/papers/921.pdf.

Pew Internet. (2012). Adult gadget ownership over time (2006–2012). Retrieved August 26, 2012, from: http://pewinternet.org/Trend-Data-%28Adults%29/Device-Ownership.aspx.

Reynolds, J. (2012, September 24). Facebook partners with Datalogix for ad tracking project. Retrieved from: http://mediaweek.co.uk/news/1151339/Facebook-partners-Datalogix-ad-tracking-project/.

Smith, A., & Boyles, J. L. (2012, July 17). The rise of the "connected viewer". Retrieved July 1, 2013, from: http://www.pewinternet.org/~/media//Files/Reports/2012/PIP_Connected_Viewers.pdf.

Spangler, T. (2012, February 13). NBCU enlists Google, comScore to track multiscreen Olympics viewing. Retrieved August 24, 2012, from: http://www.multichannel.com/article/480514-NBCU_Enlists_Google_ComScore_To_Track_Multiscreen_Olympics_Viewing.php.

Steinberg, B. (2012, February 13). NBC Universal's new Olympics challenge: Screen-jumping. Retrieved August 24, 2012, from: http://adage.com/article/media/nbc-universal-s-olympics-challenge-screen-jumping/232676/?page=1.

Story, L. (2007, October 22). How many site hits? Depends who's counting. *New York Times*. Retrieved July 22, 2012, from: http://www.nytimes.com/2007/10/22/technology/22click.html?_r=1&pagewanted=all.

Tewksbury, D. (2005). The seeds of audience fragmentation: Specialization in the use of online news sites. *Journal of Broadcasting & Electronic Media, 49*, 332–348.

Webster, J. G., & Ksiazek, T. B. (2012). The dynamics of audience fragmentation: Public attention in an age of digital media. *Journal of Communication, 62*, 39–56.

Webster, J. G., & Lin, S. (2002). The Internet audience: Web use as mass behavior. *Journal of Broadcasting & Electronic Media, 46*, 1–12.

Webster, J. G., Phalen, P. F., & Lichty, L. W. (2005). *Ratings analysis: The theory and practice of audience research*. Mahwah, NJ: Erlbaum.

Wood, F.B. (2005). Use of Internet audience measurement data to gauge market share for online health information services. *Journal of Medical Internet Research, 7*. Retrieved August 28, 2012, from: http://www.jmir.org/2005/3/e31/.

Zickuhr, K and Raine, L. (2014). E-Reading rises and device ownership jumps. Retrieved January 26 from http://pewinternet.org/Reports/2014/E-Reading-Update.aspx.

Part VII

OPERATIONAL ISSUES IN ONLINE PANELS

Introduction to Part VII

Paul J. Lavrakas[a] and Anja S. Göritz[b]

[a]*Independent Research Psychologist/Research Methodologist, USA*
[b]*University of Freiburg, Germany*

VII.1 Management of sampling and data quality processes in online panels

As with any computer-assisted survey mode, there are myriad operational issues that will affect the quality of the sampling that is achieved in an online survey and the data quality produced. From a Total Survey Error perspective, these operational issues are principally related to sampling error, nonresponse error, questionnaire-related measurement error, respondent-related measurement error, and data processing error.

From a sampling standpoint, those who are managing an online survey need a systematic way to sample the correct panel members (i.e., those eligible for a specific survey) to be invited to participate in a given client's survey. This may require complex sampling designs and the management procedures must be able to support drawing such samples from the panel membership without error. For example, panel managers know a great deal about their panel members in advance of sampling for a specific survey, thus clients may want to take advantage of this and have complex stratified samples drawn.

From a nonresponse standpoint, despite panel members voluntarily joining panels in order to complete surveys, many do not comply with the survey invitations they receive. Depending how important the response rate for a given survey is for the client, the panel managers may need to deploy strategies during the survey's field period to gain compliance from initial non-responders. For example, a reminder email with special persuading text may be sent and/or a different incentive may be offered. Thus, panel managers must have a system in place where they can assess the status of every panelist who has been sampled for a given survey in real time. In addition, online panels are ripe for nonresponse bias analyses, given all that is known

---

Online Panel Research: A Data Quality Perspective, First Edition.
Edited by Mario Callegaro, Reg Baker, Jelke Bethlehem, Anja S. Göritz, Jon A. Krosnick and Paul J. Lavrakas.
© 2014 John Wiley & Sons, Ltd. Published 2014 by John Wiley & Sons, Ltd.
Companion website: www.wiley.com/go/online_panel

about every panelist, and panel managers need to be able to accurately append such auxiliary data onto each record in a given survey.

From a questionnaire standpoint, the software that is used to present the questionnaire to respondents must have capabilities to accurately control the flow of questions based on the contingency of the answers the respondent provides. The software also must accurately execute randomization in question order and content that the researchers may build into the questionnaire. The software also must be able to present the questionnaire in a standardized format across the many browsers that respondents will use.

From a respondent standpoint, it is well known that data quality can be a special problem among online panel members. Many online panel members are not committed to take on the burdens required to provide full and accurate answers to survey questions. Thus, panel managers need to be able to monitor the data quality that panel members are generating. To this end, they should have software that is capable of assessing data quality metrics, such as the amount of missing data for a given respondent, the prevalence of straight-lining, cross-item response inconsistencies, completing questionnaires too quickly, etc. Using such data quality metrics will allow panel managers to coach their panelists as needed, and to warn certain panelists that unless their data quality improves they will not earn their incentives and eventually will be terminated from the panel. Panel managers also need to assure their clients that the members of the panel do have legitimate identities and are not masquerading as someone other than who they are.

From a data processing standpoint, panel managers must have foolproof systems to capture all the data (including open-ended text verbatims and the audio and video input the respondent may provide in a given survey) that is produced by panel members. The managers also must have systems that accurately and fully capture a host of paradata that is generated for each survey. And, all these data must be accurately compiled into the datasets that are delivered to clients.

VII.2   The chapters in Part VII

Chapter 18 by Tim Macer addresses the issues of online panel software. As Macer correctly observes, "The software behind [online panels] has evolved to make them more useful and more easily achievable. Technology is not only what drives these methods, but is also what has shaped them." The software that online panel vendors use to support the surveys their clients purchase must control: (1) the database of panel members; and (2) the questionnaires that panel members are invited to complete.

To his considerable credit and to the benefit of the chapter readers, Macer conducted a comprehensive survey of panel software vendors to provide the substance of what the chapter reports about the current state of online panel software. Using the information from his survey, Macer discusses a broad array of issues including: panel software typologies, database platforms, database limitations, support for panel recruitment, profile data and verification, incentive management, panelist portals, sample selection, data capture and linkage, panel health, and diagnostics.

This chapter will be of interest to anyone who wants to purchase panel software, to those who may want to develop or revise their own panel management software, and to clients and other users of online panel data who would like to better understand where the data come from.

Chapter 19 by Reg Baker, Chuck Miller, Dinaz Kachhi, Keith Lange, Lisa Wilding-Brown, and Jacob Turner addresses the issue of validating respondent identity in online access panels.

As Baker et al. note, "The focus of this chapter is the challenge posed by people who create false identities as a way to maximize their survey opportunities." This problem is a very serious one for online panels, especially given the size of its suspected prevalence; some research suggests it is as high as one-third of panel members. As such, there now are several companies that provide services to online panel vendors to help "clean" their sample of dubious members.

To address the issues of how prevelant fake panel members are and how well these validation services are able to identify them, Baker et al. conducted an orginal study for their chapter. The study involved testing four validation services applied to the same panel member sample. The study began by conducting a brief online panel survey (N = 7435) about Internet usage in early July and August of 2012. As part of the questionnaire the respondents were asked to share basic personal contact information such as name, address, etc. Then, each validation service identified those cases where the respondent's identify could be verified, sometimes coding by level of certainty.

Baker et al. found nearly one-fourth (24%) of the panelists who completed the questionnaire had refused to provide the personal identifying information asked of them in the questionnaire and they present findings on factors that differentiated those who did provide such information from those who did not. Among those 5640 panelists whose personal data were sent to the validation companies, the average rate of validation was 83%. Although that is a high rate, it also signals that approximately one in six panel members who did provide personal identifying information could not be validated. Adding this to the large proportion of the original sample who did not provide personal identifying information and thus could not be processed by the validation companies, the study shows that falsification of personal identify among online panel members is a substantial problem.

The authors go on to show how their survey data would be different if the unverified panelists were excluded from the final dataset. They conclude their chapter with several recommendations for how the problem of false identities might be addressed in the future.

18

Online panel software

Tim Macer
Meaning Ltd, UK

18.1 Introduction

It could be argued that online research, and online panel research in particular, are opportunistic research methods that have arisen because technology provided the opportunity to do research in a new way. They have survived and developed because they are useful – and because the software behind them has evolved to make them more useful and more easily achievable. Technology is not only what drives these methods, but is also what has shaped them.

Put at its simplest, the software one needs to do online panel research comes in two parts: a database for the panel, and a survey tool to deliver the online survey. Because the requirements for the survey tool are no different whether the sample is drawn from a panel or any other source, this chapter will focus on the software capabilities that are unique to panel-based online research, specifically, the administration of the panel database, and the activities that bridge between the panel database and the survey. There are other texts and sources of information that provide useful material on online survey software (e.g., Čehovin & Vehovar, 2012; Kaczmirek, 2008; Poynter, 2010).

As there is little empirical information on online panel software to refer to, this chapter is based upon original research carried out among specialist software developers into the capabilities offered within the panel software they provide. The findings of this research are presented in this chapter. The chapter should therefore be of interest to:

- anyone wishing to select a suitable panel software product;
- researchers or technology specialists wishing to develop their own panel software;

- researchers who wish to gain a better understanding of how recent technological developments can improve their panel-based online research.

Good tools will not necessarily result in good research, but tools that are deficient make it harder to do research well. The aim of this chapter is to provide an understanding of what panel management software needs to do to support good panel research, in light of a growing body of literature concerned with limitations, deficiencies, and even bad practices in online panel research, which are discussed in the other chapters in this text. Furthermore, this chapter aims to identify innovations that can improve the quality of panel-based research, reduce the administrative burden, and ensure compliance with professional and ethical standards as they continue to evolve.

The questions from the original survey conducted for this chapter also may provide a useful checklist for anyone evaluating panel software tools for their own use, and are available in the online appendix at www.wiley.com/go/online_panel_research.

18.2   What does online panel software do?

There are many different panel management systems in use today, with significant variations in what they provide and how they interpret that provision. These variations in the systems used to manage panels can have an impact on the performance of the panel and the research arising from it (Vonk, van Ossenbruggen, & Willems, 2006, p. 55), so it is important to be aware of the differences between software packages, as well as their similarities.

Nevertheless, all panel solutions tend to provide support in these six general areas of functionality:

1. A database to store the panel members and related management data.

2. Administrative tools to assist with recruiting, capturing and entering new members into the panel.

3. Tools to support routine and on-going panel administration tasks, including self-administration.

4. Tools and processes to allow the extraction of samples by selection criteria from the panel for survey activities.

5. Tools and processes to feed data back into the panel database after research interactions with panel members.

6. Diagnostic reports and processes to assist in maintaining a balanced, healthy panel.

These six areas of functionality were used to provide a framework for the survey of panel software developers, with a series of more detailed questions asked in each of these categories, in order to differentiate their products.

Within each section of the questionnaire, questions were constructed with particular consideration of guidance published by ESOMAR (2012) for buyers of online research to ensure that any panel management software would be able to satisfy these requirements. Although frequent reference to these questions is made below, they describe the same recommendations to be found in codes of conduct and recommended practice guidelines issued by many other professional research associations including AAPOR (Baker, Blumberg, Brick, et al., 2010) and CASRO (2011).

18.3   Survey of software providers

Research for this chapter involved a survey of panel software providers concerning some of the capabilities they support. Eligible products were identified from online software directories such as Quirk's Market Research Portal,[1] Web Survey Methods,[2] together with the directory that I maintain.[3] Companies that operated access panels but did not allow others to use their software were excluded from the survey. A questionnaire comprising 37 questions set out as a form in Microsoft Excel, was distributed to 17 identified providers, 14 of whom responded. Responses included 11 from commercial survey software providers and three from open source, not-for-profit developers, as shown in Table 18.1.

All contributors except one company agreed for their data to be published in detail, with their names and responses identifiable. Therefore, tables in this chapter that present aggregated data include all 14 responses; those that identify product by name show 13 products.

The majority of the questions asked whether specific functionality was supported, was planned to be offered in the near future, or was not supported. As there is too much data to be presented within this single chapter, this dataset is available on the companion website at www.wiley.com/go/online_panel_research. I am most grateful to these 14 organizations for so willingly providing this information.

The survey revealed that all the tools examined do provide extensive functional support within each of these categories, though with considerable variation in how this is delivered.

Table 18.1   Companies and products surveyed.

Commercial providers	Open-source developers
Cint: Engage	LimeSurvey
Confirmit: Confirmit Professional Panel	OpenPool (Leiner, 2012)
Kinesis Survey Technologies: Kinesis Panel	phpPanelAdmin (Göritz, 2009)
MARSC: MARSC.net	
Nebu: Dub Knowledge	
QuestBack: EFS Panel	
Verint: Vovici	
Vision Critical: Vision Critical Community Panels	
Voxco: Command Center Panel Management	
WorldAPP: Key Survey	
plus, one undisclosed provider	

---

[1] Quirk's Worldwide Market Research Portal, www.quirks.com.
[2] Web Survey Methods, www.websm.org.
[3] Research Software Central, www.meaning.uk.com/your-resources/software-database/.

18.4   A typology of panel research software

Based on the survey findings, panel systems can be considered to fall into three broad but overlapping categories:

- standalone panel software programs
- integrated panel research software
- online research community software.

18.4.1   Standalone panel software

Standalone panel software products typically will consist of a database and associated panel management tools that can be used in association with external online survey tools. The external survey tools with which they communicate could be commercially developed, open source or homegrown in origin. For any standalone system, communication with the survey software needs to be bi-directional and highly automated, in order to eliminate the scope for error that arises when manual intervention is involved.

Standalone panel management databases typically tend to be used to store data that are relevant to sampling and panel management, but they usually do not store all the survey-specific case data completed by panel members. They may contain some data collected during survey encounters with members, which are deemed sufficiently useful for panel management purposes for these data to be added retrospectively to the panel member's record.

18.4.2   Integrated panel research software

With an integrated panel research software program, the panel database and panel management functionalities are built into a larger software product that is also used for survey design, survey management, and possibly data analysis.

Most integrated panel and survey tools still segregate the panel data from the survey data, either in a separate database or discrete tables within the database. This separation often is justified by panel software designers on the grounds that it simplifies the database structure, and also because panel profile data usually are maintained to a higher standard of completeness and verification than survey data.

18.4.3   Online research community software

Online research communities, also known as Market Research Online Communities or MROCs, share many characteristics of panel-based online research. What makes them distinctive is their tendency to engage with panel or community members on a continual basis, with semi-structured or unstructured qualitative research activities on a related topic, rather than through a series of discrete quantitative survey interventions and with direct interactions between respondents (Heist & Sanders, 2008). From a technical standpoint, Comley (2008, p. 680) notes that the software used to manage an online community will need both conventional survey and polling tools as well as a content management system with capabilities to support forums, blogging, videos and email contact mechanisms.

From a technological standpoint, this means a much closer relationship between panel data and case data than is found in the more conventional, quantitative survey-oriented online panel. This provides an operational reason for holding both panel data and response data in the same database. It also results in a much more complex database layout or schema, and a more eclectic mix of research tools in the resulting software.

Next, I will consider some of the specific tools required, and examine the extent to which providers are offering these more advanced tools as an evolution of their online panel platforms.

18.5   Support for the different panel software typologies

Most panel operators require flexibility in their panel software, as the demands they will face for their panel cannot be predicted. Having the ability to use an integrated panel tool in standalone mode, or the possibility of community capabilities, keeps these options open, even if there is no immediate need. Fortunately, among the 14 tools surveyed, the majority offered standalone capabilities, with some exceptions as seen in Table 18.2. There are still not many tools that support communities, though this is likely to grow over time.

18.5.1   Mobile research

The most recent development in online research is the arrival of mobile research – where respondents use either a web browser or a specific survey app on their smartphones to access surveys (Hairong & Townsend, 2008). Panel databases have particular relevance to mobile research, due to the difficulty of accessing respondents for mobile research via conventional email invitations, which can be overcome by using a mobile-enabled panel (Funke & Wachenfeld, 2012; Vicente, Reis, & Santos, 2009). Mobile research can improve coverage and response among some populations, and deliver much faster responses too (Friedrich-Freksa & Metzger, 2009). Some of these impacts are also discussed in Chapters 16 and 17 of this book.

Table 18.2   Panel software by type.

	Standalone	Integrated	Communities
Cint Engage	■	■	■
Confirmit Professional Panel	■	■	
Dub Knowledge	■	■	■
Kinesis Panel		■	
LimeSurvey°	■		■
MARSC.net	■	■	□
OpenPool°	■		
phpPanelAdmin°	■		
QuestBack EFS Panel	■	■	■
Verint Vovici		■	■
Vision Critical Community Panels	□	■	■
Voxco Command Center Panel Management		■	
WorldAPP Key Survey		■	

Key: ° Open source software ■ Supported □ Planned.

The ideal panel software today is multi-modal, and should offer mobile support capabilities for recruitment and routine self-administration by panel members. It can be operationally convenient if it also includes some mobile-friendly survey tools. I will present some of the desirable features throughout this chapter, and show how providers are rising to this latest challenge.

18.6   The panel database

The primary function of a panel database is to store panel member records. The simplest panel may be no more than a spreadsheet, listing panel members, their email addresses and some key demographics. However, panel software products typically use a relational database such as Microsoft SQL Server, or MySQL, to organize the database into different tables for different record types, such as sample frames, incentive payments and records of many other aspects of panel administration.

Survey researchers, used to working with datasets of hundreds or thousands of respondents, have to raise their sights by several orders of magnitude when thinking about the potential size of a panel. One may wish to retain members' records in the database long after they have left the panel, so it is conceivable that panels can have millions of records within them.

The use of modern high-performance database technology is therefore essential for anyone operating a panel. Excel and Access, which are useful office desktop tools, are wholly unsuited to the capacity and performance demands of a panel.

Beyond this, for maximum flexibility, it is important that the database should be "open," allowing its data to be examined using a variety of statistical or reporting tools, or shared with other applications. This is another capability that relational databases provide with little or no additional development effort required.

18.6.1   Deployment models

The two most common used commercially-provided relational databases today are Microsoft SQL Server and Oracle. MySQL, which is an open source database, is an equally capable relational database for the purpose of most panel management applications. All of the software providers I contacted use one of these databases, and several offer the panel manager a choice (see Table 18.3, Section A). All three open source tools used MySQL, though one, LimeSurvey, will also work with Microsoft's proprietary SQL Server database. Interestingly, several of the commercial providers have also based their panel systems on the open source MySQL database system.

18.6.2   Database architecture

Database technology also permits linkage to take place across panels – so that one panel may be logically divided into subpanels, which may be operated independently, but share records or characteristics across the panels. They may even share members without needing to duplicate that record. The best database designs are those that avoid duplication of data, or *redundancy*, as it is termed in database design. It can be highly convenient to create a sub-panel or virtual panel for a specific task – such as a mobile panel, or when forming a research community drawn from a more universal panel.

Table 18.3   Database platforms and design.

	A. Database platform				B. Database design to allow for multiple panels
	SQL Server	My SQL	Oracle	Other	
Cint Engage	■				Virtual
Confirmit Professional Panel	■				Separate
Kinesis Panel		■			Virtual
LimeSurvey°	■	■		■[1]	Virtual
MARSC.net	■				Separate
Nebu Dub Knowledge	■				Virtual
OpenPool°		■			Other [3]
phpPanelAdmin°		■			Separate
QuestBack EFS Panel		■			Separate
Verint Vovici	■				Virtual
Vision Critical Community Panels	■		■	■[2]	Separate
Voxco CC Panel Management	■				Separate
WorldAPP Key Survey	■	■	■		Other [4]

Key: °Open source software ■ Supported.
[1] Postgres.
[2] MongoDB and SOLR.
[3] Only one panel is supported.
[4] Virtual database simulated in memory.

Duplication can be undesirable in other ways too. When panels involve some kind of incentive, it has been observed that "professional respondents"[4] will register multiple times for different panels, which has a highly damaging effect on data quality (Baker & Downes-LeGuin, 2007; Knapton & Garlick, 2007, p. 58; Rivers, 2008). Some overlap is inevitable and may not always be harmful, but it is good practice to be alert to this phenomenon, and this is recommended in the ESOMAR (2012) guidelines. A database design which allows all one's panels to be interlinked and reported on together will make it easier to detect hyperactive panel members in one's own panels and curtail their activities.

Providers were asked if their respective database design used (1) a separate database for each panel; (2) a single database organized into any number of virtual, functionally independent databases; or (3) something else. At present, there are only a few products on the market that allow one to define multiple panel databases or create virtual panels from within the one database (see Table 18.3, Section B).

18.6.3   Database limitations

Using modern database technology, one would not anticipate constraints on the number of panel members or records the panel could contain. Nine of the providers contacted stated there was no effective limit beyond the rational limits of the database or hardware being used.

---

[4] Respondents who join many different panels with the aim of earning an income from survey participation.

Table 18.4 System architecture and deployment characteristics.

	A. Software deployment		B. Data protection	
	Locally installed	SaaS	European servers	Safe harbor
Cint Engage		■	■	■
Confirmit Professional Panel	■	■	■	■
Kinesis Panel	■	■	■	■
LimeSurvey°	■	■	■	
MARSC.net	■	■	■	
Nebu Dub Knowledge	■	■	■	■
OpenPool°	■		n/a	
phpPanelAdmin°	■		n/a	
QuestBack EFS Panel	■	■	■	
Verint Vovici	■	■	■	■
Vision Critical Community Panels		■	■	
Voxco CC Panel Management	■	■	■	■
WorldAPP Key Survey	■	■		■

Key: ° Open source software ■ Supported n/a Not applicable.

Two of the commercial providers declared practical limits of several million members, which still is likely to exceed most panel owners' demands.

18.6.4 Software deployment and data protection

Databases, particularly those offering Internet access, are technically demanding to install and run. If a panel operator does not have access to in-house IT resources, finding a panel system which is provided on a cloud computing or "Software as a Service" (SaaS) basis means that one passes the responsibility for maintaining hardware and round-the-clock Internet access to one's software provider, and accesses their database and software securely across the Internet (Macer, 2008), a model frequently applied to online survey software (J. D. Baker, 2013). There is a good choice among providers for either deployment method (see Table 18.4, Section A), and all the commercial products are available in SaaS mode.

Panels, by their nature, will contain a large amount of confidential data. In Europe, Data Protection legislation will limit where researchers may store any personally identifiable data. Compliance is normally achieved by working with an SaaS provider using servers located somewhere within the European Union. If a researcher finds he or she is subject to EU data protection laws, it is still possible to work with a provider in the United States if that provider has entered into a *Safe Harbor agreement* with the US Department of Commerce (USDoC, 2000), which is accepted under EU law as providing an equivalent level of protection. Data protection is also the target of Question 25 within the ESOMAR 28 Questions (ESOMAR, 2012) and it is something that all panel providers need to be concerned about. Any contract or service level agreement with an SaaS provider should make clear their duty of care in maintaining the confidentiality of the data.

Panel software providers appear to be well aware of these requirements, and there is a very high level of compliance among those surveyed (see Table 18.4, Section B).

18.7   Panel recruitment and profile data

The second major functional area of the panel system is in administering panel members: a process that starts with recruitment. Although it may be surprising to some observers, panel databases tend not to be configured to contain survey case data – but it is a requirement that they contain *profile data* (Duffy, Smith, Terhanian, & Bremer, 2005). Profile data comprise demographic and other personal data collected from the panel member at the time they are recruited, in a special profiling questionnaire or *profiler*, which may be added to or enriched later with more information from supplementary profile surveys (Tress, 2011).

There is no technical reason why a panel database using a modern relational database system cannot be configured to hold an unlimited amount of case data – but there is usually no operational benefit from doing so, and the effort involved can be considerable.

An exception arises in the case of panels that are used in association with online communities, where extensive data collection takes place within the panel environment, rather than within the context of discrete (and usually detached) surveys.

Panel software therefore needs to contain within it the means to set up profile surveys, and process new members through the registration stage, and offer flexibility over what data are collected. Ideally, it should also allow for this information to be added to, over time, and periodically re-validated or updated.

18.7.1   Panel recruitment methods

New panel members need to be found from somewhere – and panel owners report that this is a constant process, as *attrition* (sometimes called "churn") means a proportion of existing members become inactive or ask to leave. Thus, panel software needs to offer a choice of ways of reaching new respondents, assisting panel managers in that process, with a wide choice of recruitment methods. Attrition has been discussed in Chapters 4 and 5 of this volume, and is the over-arching subject of coverage in Chapter 6.

The standard recruitment methods usually deployed are:

- by a targeted email invitation;
- through websites in various ways – a static link on a website;
- web-based advertising;
- website pop-up pages.

For mobile panels, a more appropriate method than a standard web URL or hyperlink is to provide a QR Code. It is essential that the would-be panelist is subsequently presented with a mobile-friendly experience when following the QR Code link, and is able to view and complete any subsequent profile surveys in full on his or her mobile device.

Those providers surveyed offer reasonably broad coverage of these methods (see Table 18.5, Section A). Most, but not all, offer integrated email recruitment. Without this, panel managers would need to extract lists and use separate bulk emailing software, and then manage the response – which is likely to be a burdensome and error-prone manual task.

Table 18.5   Recruitment support capabilities among providers contacted.

	Yes	Planned	No
A. Invitation method			
Email invitation	12	1	1
Website link	12	1	1
Website advertising	10	1	3
Website pop-up	9	2	3
QR Code or advertised link for mobile devices	9	2	3
B. Import and load external data			
Load data directly into database	12	1	1
Import and hold samples for a single use	12	1	1
C. Original recruitment method or source identified in database	12	1	1

Note: N = 14 panel software providers.

For the mobile panel user, support for QR Codes is far from universal at present. Panels may also need to import records from other sources, on a permanent, or on a temporary basis. A temporary import would allow a customer list to be used to recruit members to the panel, offering them the choice to opt in or not, or to use the panel database for an ad hoc project. All of the commercial products and one of the open source tools offered these capabilities (Table 18.5, Section B).

ESOMAR (2012) requires that panel owners be able to trace the origin of all samples back to their original source in order to satisfy several of its 28 Questions (specifically Q2–Q6, Q21 and Q22). Most but not all providers support this automatically (see Table 18.5, Section C).

18.7.2   Double opt-in

Question 23 of ESOMAR's 28 Questions and the guidelines issued by many other professional research bodies require that panel members should be subject to a double opt-in during the recruitment process.[5] Some 11 of the 14 providers confirmed they currently support this; three did not, but two of these were planning to introduce double opt-in soon.

18.7.3   Verification

Applying some form of independent verification to establish that a newly recruited panelist is genuine is a practical and easily achievable quality control measure (see also Chapters 10, 11, and 19). It is something which Question 22 of the ESOMAR 28 Questions looks at.

Providers were asked, in an open question, what verification methods they supported. Ten of the 14 providers described verification measures they support, which range from using the double opt-in method to verify email address (which is a very basic form of validation) through

---

[5] A double opt-in is achieved by the panel recruit confirming at the time of invitation that they wish to join the panel, and then a second time, usually at the time they complete their panel profile.

Table 18.6 Profile data capture methods.

Built-in standard profile form [a]	4
Custom profile form [b]	11
Survey import [c]	11
Either a standard or custom profile [a or b]	13
Choice of standard or custom profile [a and b]	2
Choice of built-in form or survey import [(a or b) and c]	10

Note: N = 14 panel software providers.

to using digital fingerprinting[6] and commercial third-party verification services, where identity can be checked against independently maintained verification databases. Four providers specifically mentioned using digital fingerprinting services, including TrueSample, the Verity and Relevant ID services from Imperium and CropDuster from Mktg. Inc.

It therefore appears that only a minority of panel software providers surveyed are offering tools to identify and weed-out professional or fraudulent respondents from the panel recruitment stage.

18.7.4 Profile data capture

Profile data are usually built up over time. Some essential data will be captured at recruitment, but other data may be captured by follow-up profilers, or brought in from surveys deployed to panel members. Panel software needs to offer the flexibility to enhance or enrich panel profile data over time, and also to keep it up-to-date by periodically refreshing the data.

Some panel software designers choose not to create a full-blown survey tool in their products merely to capture profile data, and instead offer simple forms to collect essential demographics, which can then be supplemented by fielding a profile survey using an external survey tool. Others have gone further in developing their profile tools to offer a higher level of customizability – even to the extent that it becomes a survey or polling tool in its own right.

Mindful of this, providers were asked if they offered either "a pre-set or limited profile form", or "a fully customizable profile survey" in their tools, and if they allowed profile data to be imported from an external survey. Summarized responses, showing the different combinations of responses, appear in Table 18.6.

It appears this is an area where the choices available to the panel manager may be constrained by the software design and methods available, as many solutions do not offer a choice between internal and external survey tools.

Providers also were asked if they limited the number of *profile variables* a panel may contain. One stated a specific numeric limit, another stated there was a predefined set of profile variables, but for all others, there was no effective limit.

18.8 Panel administration

The administrative burden in running a panel can be considerable. Quite apart from the requests that arise from researchers and panel clients, there are many routine tasks in taking

---

[6] Digital fingerprinting relies on analyzing information passively gathered from the user's device, along with other information such as IP address which, when combined, can be considered unique, or at least, highly characteristic of that one device, to the exclusion of all others.

care of panel members that require a fast and efficient response from the panel administrator, such as:

- Responding to panel members' queries and accessing individual member's records to verify or update information.

- Terminating membership on request and preventing further contact.

- Managing incentive fulfillment and reward redemption, including running sweepstakes, if these are used.

Having an effective panel portal that allows panel members to deal with some of these matters for themselves can eliminate a lot of administrative work. Some key capabilities for this are considered in Section 18.8.1.

Ideally, any panel management software tool today needs to have a simple graphical interface that brings together reports and tools for the panel administrator to use. All of the products that were surveyed offered these capabilities.

18.8.1   Member administration and opt-out requests

Compliance with opt-out requirements only makes sense if there is a speedy remedy to requests at any time to opt out. Software providers were specifically asked:

> *Q: Can you immediately suspend further contact and reminders, e.g., upon learning the panel member has died?*

The above example applies equally to any routine opt-out request.

All products surveyed recognized the importance of this capability by supporting it in full.

18.8.2   Incentive management

Incentive management concerns many market research and some social research panel operators. Whereas online panels may operate successfully without any reward being offered to the respondent, one may wish to offer the chance of winning a prize, or even offer points which can be redeemed for cash rewards or gifts, with the intention of improving sample coverage and response rates. Prize draws and points are the most commonly used incentive methods (Brüggen, Wetzels, de Ruyter, & Schillewaert, 2011; Göritz, 2004) in commercial panels, and prize draws and other types of incentives also may be used in panels used in social research.

Research communities very often operate successfully without incentives, but an additional reward mechanism has created the context where participation is visible to other members of the panel, which is to award "kudos" points which enhance the status of the member for regular or high quality contributions (Comley, 2008). These three methods were explored in a question in the survey.

None of the open source panels provided support for any kind of reward management, but the ability to control the awarding of points was universally found among the commercial products. Although prize draws are probably even more widely used than points, built-in support for prize draws is less common than points that can be converted into rewards, in the tools that were researched (see Table 18.7).

Table 18.7 Incentive administration.

Q: What incentive management capabilities do you provide?	
Points for conversion into rewards at variable rates	●●●●●●●●●●●○●●●
Entry and administration of prize draws	●●●●●●●●○●●●●●●
Kudos points and awards, such as badges or enhanced status	●●●●○●●●●●●●●●

Notes: N = 14 panel software providers. Key: ● Supported ○ Planned ● Not supported.

18.9   Member portal

Self-administration has brought the cost of research down in the case of online panels, and it is a useful principle to follow in keeping the cost of online panel administration to a minimum. Panel software ideally should make it easy for panel members to be aware of their survey assignments, update their profiles and deal with routine administrative inquires on a self-service basis. Well-designed software tools therefore will help to improve the quality of the panel by reducing the burden of keeping panel profile data up to date.

18.9.1   Custom portal page

Ideally, each panel should have its own web page or "portal" – so called because it functions as the gateway to all the members' panel interactions. Many commercial operators choose to apply branding, or even to brand each panel they operate in a distinctive way which is appropriate for the panel's target subpopulation (e.g., teenagers, physicians, frequent flyers). Those operating a research community are very likely to want give it a distinctive identity and presence on the web.

The act of creating such pages can easily involve many hours of custom development work by web programmers if not supported within the software. Some panel management tools can provide the kinds of capabilities you can find in content management tools such as WordPress or Google Sites, to create and "skin" a site with an appropriate theme, and populate it with content, with no technical skill.

One question in the survey asked providers if such themed portals or community pages could be defined, and if so, how technical the task was (see Table 18.8).

Table 18.8 Capabilities for creating themed panel or community pages.

Q. Can each panel have its own themed portal or community page?	
a) Yes, this is a task for a software consultant/web designer	5
b) Yes, can be created by a panel administrator	6
c) No, the respondent interface is standard	2
d) No, but we plan to develop user-defined themed panels	1

Note: N = 14 panel software providers.

Although all tools offered a web interface, and most allowed this to be themed, only a minority of those contacted had achieved the goal of allowing panel managers to do all of this, unassisted, using a web-based interface.

18.9.2   Profile updating

Question 12 of the ESOMAR 28 Questions asks how profile data are kept up-to-date, and this is an important factor in the quality of the panel and its effectiveness as a source of sample. There are many different methods in which profile information can be kept up-to-date, ranging from prompting members to update the information in various ways, to selectively refreshing it from surveys. However, one is likely to need a combination of these strategies to keep on top of the quality of the profile data, with the more, the better.

Since panel members may use many different channels of communications, an update strategy should allow as many channels as possible, both for prompting and obtaining updated information from members. Some channels will be more appropriate to some populations than others – so once again, this is an area where panel software should offer some flexibility.

In the survey, providers were asked which methods panelists could use to update their profile data; which methods panel managers could use to solicit updates, and whether survey data could be used to update panel data.

Though many of these different channels are offered in the products available, only a few offer a full spectrum of easy update routes, as shown in Table 18.9. All but two allow survey data to be used to update profile data. However, since most surveys are on a small sample drawn from a larger panel, this is a limitation. Instead, the panel manager will need to administer a profile update survey periodically: another method with broad but not universal support.

18.9.3   Mobile apps

Anyone operating a mobile panel would want to have the capability for members to access the panel portal via a mobile device, which could provide access to survey activities, allow profile updates and other self-service administrative activities.

One benefit of using a mobile application is that it can work in a more convenient and natural way for the panel member, with greater control that is easily achievable using mobile-friendly web pages. A major advantage to the panel operator is that a mobile app can use the device's built-in notification system to alert panel members that there is a new survey assignment, rather than rely on email (Macer, 2011).

According to the survey (see Table 18.9 and also the last row of Table 18.10), only four providers have developed mobile apps for panel members to use, but as many are planning to introduce them. Furthermore, three of the four providing apps also use mobile notifications to provide a prompt to members, and two more are developing this feature. This indicates this is an area that is relatively undeveloped in panel software products at the moment, but one that appears to be developing rapidly.

18.9.4   Panel and community research tools

Clearly, panel members need to do more than update their profiles, and some developers are building more tools into the panel portal, to provide support for running online research communities in addition to operating a more conventional online panel. Some are highly relevant

Table 18.9   Methods of keeping profile data up-to-date.

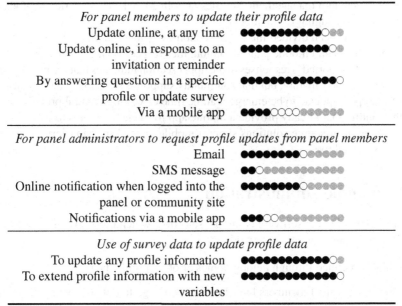

For panel members to update their profile data	
Update online, at any time	●●●●●●●●●●●○◉◉
Update online, in response to an invitation or reminder	●●●●●●●●●●●○◉◉
By answering questions in a specific profile or update survey	●●●●●●●●●●●●●○
Via a mobile app	●●●●○○○○◉◉◉◉◉◉
For panel administrators to request profile updates from panel members	
Email	●●●●●●●●○◉◉◉◉◉
SMS message	●●○◉◉◉◉◉◉◉◉◉◉◉
Online notification when logged into the panel or community site	●●●●●●●●○◉◉◉◉◉
Notifications via a mobile app	●●●○○◉◉◉◉◉◉◉◉◉
Use of survey data to update profile data	
To update any profile information	●●●●●●●●●●●●○◉
To extend profile information with new variables	●●●●●●●●●●●●●○

Notes: N = 14 panel software providers Key: ● Supported ○ Planned ◉ Not supported.

Table 18.10   Panel and community member tools.

Panel-oriented tools offered	
Surveys	●●●●●●●●●●●○◉◉
Reward history	●●●●●●●●○◉◉◉◉
Reward redemption	●●●●●●●○◉◉◉◉◉
Direct messages with panel admin	●●●●●●●●●○○◉◉
Community-oriented tools offered	
Polls	●●●●●●●●●●○◉◉◉
Discussion forums	●●●●●○○◉◉◉◉◉◉
Blogs	●●●●●○○◉◉◉◉◉◉
Co-creation tools	●○○○◉◉◉◉◉◉◉◉◉
Public profiles	●●●●○○◉◉◉◉◉◉◉
Direct messages with other members	●●○○◉◉◉◉◉◉◉◉◉
Integrated mobile application offered	●●●●●○○○◉◉◉◉◉

Notes: N = 14 panel software providers Key: ● Supported ○ Planned ◉ Not supported.

to conventional online panels, such as viewing current survey assignments, checking reward balances and redeeming rewards or sending direct messages to panel administrators. Others are more relevant if one wishes to run an online research community, such as polls, discussion forums, blogs, displaying your member profile (which also is where any enhanced status or kudos would be visible), and even allowing members to exchange messages directly.

If one is operating a panel, these may be of little interest, though it may be convenient to use some of the content management facilities or polls to make the experience of visiting the panel portal more engaging for members.

Table 18.10 has been divided into three sections, to show functionality relevant to online panels, communities and mobile panels/communities respectively. The greater concentration of positive (i.e., "supported") responses in the upper section indicates that the panel portal tools surveyed are now fairly mature, but that community support is still largely a work-in-progress.

Mobile support appears to be emergent at the time of writing, several providers are about to join the handful that already provide a mobile app for panel members to use.

None of the open-source solutions support mobile apps, discussion forums or any other community-oriented tools.

18.10   Sample administration

The *raison d'être* of panel software is sample selection, which must take into account a wide range of variables and conditions, including:

- eligibility, defined through demographics or other profile data;

- how recently panel members have been invited, e.g., to define "resting rules" to avoid over-frequent invitations;

- the panel member's participation history.

Several of ESOMAR's 28 Questions relate to some best practice aspects of sample selection, notably over-frequent solicitation (Q19) and frequency of participation (Q20).

In market research, where sample frames are typically defined as quota samples (Cape, 2011; Stoop, 2006, pp. 7–8), panel management tools will typically create a multi-dimensional quota matrix based on the profile variables selected. Into the cells of this matrix, the panel administrator or panel researcher enters the targets for each quota cell, which the system then will try to retrieve from the database.

After specifying quota variables, there are often several other criteria to be set in defining or limiting eligibility:

- Whether the panel member has been selected for other samples at present or is in a "resting period" as defined in a resting rule, e.g., no more than one survey completed per week or per month.

- How frequently the member has been invited or has participated, in order to balance invitations over a longer period by giving preference to those less frequently contacted.

After other conditions have been applied, the system will calculate and present the number of eligible samples within each quota cell, and identify any shortfalls. Where a shortfall is identified, the user will either need to relax the targets to allow oversampling in other areas, relax some of the other eligibility criteria, or supplement the sample from other sources, such as other access panels.

Interviewing more subjects than necessary can incur unnecessary cost, especially when costs are associated with sample invitations, or completed interviews achieved. More sophisticated panel systems will use some kind of heuristic model based on past participation

Table 18.11   Supported sampling capabilities.

Sample selection criteria supported	
Any profile data	●●●●●●●●●●●●●●
Previous survey responses	●●●●●●●●●●●○◐◐
Original recruitment source or method	●●●●●●●●●●●●○◐
Participation history and level of engagement	●●●●●●●●●●●○◐◐
History that sample selection criteria can take into account	
Frequency of invitation	●●●●●●●●●●●●○◐
Frequency of participation	●●●●●●●●●●○●◐◐
Recency of invitation	●●●●●●●●●●●●●○
Recency of participation	●●●●●●●●●●●○◐◐
Sample selection may also take into account samples being sourced from other panels	
If the sample records are imported	●●●●●●●○○○○○○○
Without importing the records	○○○○○○○●●○○○○
Not supported	○○○○○○○○○○●●●
Heuristic or predictive model for sample selection	●●●●●●○○○○○○○○

Notes: N = 14 panel software providers Key: ● Supported ○ Planned ◐ Not supported.

to predict the number of invitations that need to be issued, in order to achieve the target response; simpler ones allow for a factor to be applied to take into account incidence or "strike rate" – that is, the ratio of invites to completed responses likely to be needed. All systems should then select respondents randomly.

To test the providers' capabilities in sampling, I selected a small number of proxy questions that help to identify the level of sophistication or maturity of the products on offer. One question set out the different selection stages permissible, and the second focused on using frequency and recency of both invitation and participation. These questions and the aggregated responses appear in Table 18.11.

For the most part, sample selection is well supported within the products and seems to have reached a state of maturity: no "planned" features were reported in this area.

One specific question asked if any heuristic algorithm was used to select samples of the appropriate size to meet sample targets, based on participation history.

The ability to use any profile data for selections is universally met. A separate question asked providers if there was any limit on how many profile variables could be used in making selections. All stated there was no effective limit.

A small point of difference arises with respect to frequency and recency of invitation, which relies on having the actual response history available in the panel database. Most, though not all providers, store participation history.

However, few providers take advantage of any heuristics or predictive modeling, based on participation history, in determining how many samples to draw to meet targets. One specific question asked whether this was supported or not. Only six of the commercial providers and none of the open source providers included any predictive models for sample selection.

18.11   Data capture, data linkage and interoperability

A flexible panel database needs to be able to acquire and share data with other sources of data and other applications, especially from the survey applications used to field surveys using samples drawn from the panel database. At the very least, it must be able to acquire information about participation for all surveys to which samples were issued, in order to build up a complete participation history for each panel member. It must also allow panel profile data to be enriched over time, not simply to keep it up-to-date, such as to add new variables to the profile.

18.11.1   Updating the panel history: Response data and survey paradata

Each completed survey response provides a valuable opportunity to update the panel database with a variety of information, and then, in some cases, to act upon that information. For example, to suspend a member if multiple invitations have failed to result in any response and the latest nonresponse exceeds the pre-determined limit.

From the panel administrator's perspective, nonresponse is as valuable an outcome as a response, so that a complete picture of a panel member's participation can be built up. The most convenient method by which survey and panel software tools communicate in real time is to use a "redirect" at the end of the survey, to take the respondent back to the panel portal. Code added to the redirect query string allows the panel software to identify the respondent, record the survey outcome and credit them with any reward for which they are eligible, and/or issue a thank you message for their participation.

Survey redirects, however, do not provide sufficient information to distinguish a nonresponse from a partial response (those who started the questionnaire but did not complete it). For this level of detail, a supplementary data feed is needed with some form of batch update, at a time when the survey is deemed to be closed and sampled panel members may no longer participate. If the sampling is based on resting rules, which will prevent panel members from being selected if they are already "booked out" on another project, then a batch update is essential in order to release these panel members and make them available for other survey selections.

I already have examined why it is unusual to load survey data into a panel database, but it may be useful to collect survey paradata (process data), such as the characteristics of the web browser or device used by the respondent, timing information, or even location, imputed from the IP address of the device used to participate (Callegaro, 2013; Jeavons, 2001). Such data need to be used selectively, and with caution, but they can be useful in identifying panel members using mobile devices or a particular kind of hardware. By capturing a combination of this data for each panel member, it is possible to detect suspicious behavior and identify fraudulent panel members. It is the basis of the digital fingerprinting methods discussed earlier.

Most of the software providers in the survey had developed update capabilities to deal with all the routine situations outlined above (see Update Methods section in Table 18.12). Real-time connections, which are operationally the most convenient, are only supported in half of the products surveyed. All of the commercial products allowed direct updates when used in the survey tool provided by the panel software developer, and most allowed update runs to be performed in batches, when other survey software is being used. The ability to update profile data also benefits from these automated batch updates – usually profile data and activity data will be updated at the same time (see Table 18.9).

Table 18.12 Feedback from survey activities: capabilities and methods.

Update capabilities	
To update panelists' history with the survey outcome	●●●●●●●●●●●●○○●
To suspend a panelist, if necessary	●●●●●●●●●●●●○○●
Update methods	
Automatic update when used with an integrated survey tool	●●●●●●●●●●●○○●
Batch (or user-initiated) import from an external survey tool	●●●●●●●●●●●●●●
Real-time connection with the other survey tool via a Web services link or similar	●●●●●●○●●●●●●●
Import from email	●●●●●●●●●●●●●

Notes: N = 14 panel software providers Key: ● Supported ○ Planned ◉ Not supported.

Only one product (Kinesis) offered updates directly from received email messages. Two of the open source tools offered no built-in support for updates: users would need to create their own processes to update the database with new information collected in any survey.

18.11.2   Email bounce-backs

Where survey invitations are issued by email, another valuable source of process data is the information received in email bounce-backs. Some messages may be an auto-response, such as an "out of office" message. Others may be from the mail server denoting that the mail was undelivered for various reasons. Some transmission errors may be temporary while others may be permanent, signifying the email address is dead. Some responses will be genuine messages sent by panelists, including removal requests. Without automated processes in place to deal with bounced messages, the panel administrator will be overwhelmed by incoming email. Manual processing will inevitably result in errors being made.

Providers were asked what support they offered for handling email bounce-backs in an open-ended question. The responses have been categorized in Table 18.13. This shows a variety of approaches was offered among the commercial providers. The open source solutions did not offer any automated way to deal with bounce-backs. As this was an open question, it is possible that providers do support more than is reported here. For example, if the information is recorded in the panelist's history, then providing a summary report should not be difficult to achieve.

18.11.3   Panel enrichment

Panel managers may wish to capture certain data from a survey to extend the profile data, as well as to update existing profile data, as discussed earlier. Ideally, one should be able to import supplementary panel data from surveys, or add new profile variables and request panel members to provide these data, or merge in data from other datasets, such as a customer file. Almost all the tools surveyed provide good coverage for these requirements (Table 18.14).

Table 18.13   Feedback from survey activities: capabilities and methods.

Automated support for handling email bounce-backs	
Provide summary report of undelivered invitations	●●●●●
Record undelivered in panelist history	●●●●●●
Suspend panelists when "hard" bounce-back/dead email detected	●●●●●●
Attempt retries when "soft" bounce-back detected	●●
Any support provided (net total of the above)	●●●●●●●●●●●●
Not supported	●●

Notes: N = 14 panel software providers Key: ● Supported.

Table 18.14   Methods supported for enriching panel data.

By importing them from a survey	●●●●●●●●●●●○◉
By defining new profile variables and inviting panelists to provide the data	●●●●●●●●●●●●●○
By uploading them from an external dataset	●●●●●●●●●●●○◉

Notes: N = 14 panel software providers Key: ● Supported ○ Planned ◉ Not supported.

18.11.4   Interoperability

It is highly desirable for panel management systems to be able to communicate and exchange data with other systems, and to do so in a highly automated way. Batch file updates, for example, unless they have been automated, require manual intervention to apply. Delays can affect the efficiency and performance of the panel. As already noted, an update to release a sample from one project, if delayed, can diminish the pool of those eligible for selection for other projects. Those in hard-to-reach groups always will be in the highest demand, and therefore most prone to being locked out in this way.

Data exchange or interoperability methods can be viewed along a continuum, from simple exchange of "flat" files (such as comma separated value or CSV files), through to the exchange of data and metadata in a highly structured way, e.g., using an agreed XML description, or through real-time system calls via an application programming interface (API) or the modern, Internet-based equivalent, a web services interface.

Triple-S is a commonly used standard (though not an ISO-recognized standard) for the exchange of survey data and metadata which many survey software tools have adopted, particularly in the commercial sector (Hughes, Jenkins, & Wright, 1999). It can provide a convenient way to export samples and profile data in a way that is intelligible to other survey tools, and also to import survey data, such as additional profile variables, into the panel database. It can also be convenient as a means to automate tasks and reduce data transcription or transmission errors in situations where an API or web services interface is not viable.

The responses from providers are shown in Table 18.15, which presents the capabilities in the order of increasing level of sophistication. Although the majority supports the simplest exchange mechanisms, support is increasingly limited among the more technically advanced methods. However, it appears to be an improving situation, with several planned enhancements in all of the more advanced methods. Triple-S data is more widely supported for export (i.e., sending samples) than it is for import (i.e., enriching panel data).

Table 18.15 Interoperability protocols for exchanging data with other panel or survey platforms.

Simple CSV or Excel data import and export	●●●●●●●●●●●●●○
Triple S data import	●●●●○○◐◐◐◐◐◐
Triple S data export	●●●●●●●○○◐◐◐◐
Exchange of structured data and metadata	●●●●●●○○◐◐◐◐◐
API or web services interface	●●●●●●●●●○○○◐◐

Notes: N = 14 panel software providers Key: ● Supported ○ Planned ◐ Not supported.

18.12   Diagnostics and active panel management

Panel quality is an area where every professional association involved with research standards and ethics has recently expressed concern or set up committees of enquiry (e.g., R. Baker, et al., 2010). Panel attrition is a phenomenon that pre-dates the advent of online research, but which applies equally to the online panel (Fitzgerald, Gottschalk, & Moffitt, 1998). It is also discussed elsewhere in this book, in Chapters 4, 5 and 6.

It is the responsibility of any panel manager to ensure that new members are continually recruited to replace those lost to attrition for a variety of reasons. Indeed, Bethlehem and Stoop (2007) assert that the actual point at which sampling occurs in a panel is at the time members are recruited, not at the time respondent lists are created. This demands great care over who is recruited to replace those who are lost over time. Gittleman and Trimarchi (2010) argue that membership of multiple panels and hyperactive survey-taking can introduce measurable response biases in panels over time, and that what they term "long tenure" among panel members also weakens a panel's performance as the population in the panel becomes habituated to research questions and behaves less and less like the wider population that the panel is intended to represent.

All panels, and especially panels where incentives are offered, are prone to survey cheating or satisficing, where respondents take the easy route through a survey in order simply to claim their incentive. This common source of measurement error is discussed in Chapters 8, 9, 10 and 11.

It is relatively simple to incorporate tests or "traps" to detect response patterns that betray satisficing behavior, such as speeding, straight-lining (selecting identical response options to different questions) and giving answers that contradict previous answers in that survey or earlier surveys (Kaminska, McCutcheon, & Billiet, 2010; Levine & Holmes, 2007). The length of time taken to complete the survey may be present in the paradata provided by the survey tool, though not all survey tools provide this automatically. Other measures will need to be collected systematically during each survey encounter, then imported into the panel database and associated with the panelist's history record.

A healthy panel is the aim of every panel manager, but as with issues of public health, keeping an entire population healthy means being constantly alert to issues of hygiene and early diagnosis of recognized diseases. Online panels therefore require very careful and considered active management by ensuring that:

- inactive members are purged;

- hyperactive or satisficing members are identified, removed and prevented from re-joining;

- new members are selected and screened with care, to ensure that the panel remains true to its sample frame at all times.

An advantage of modern database technology and the statistical and reporting tools that complement it is that panel management software can offer excellent diagnostic tools to the panel manager. These can help in maintaining panel health by identifying problems or shortfalls early, and in monitoring the effect of recent recruitment, in order to fine tune on-going recruitment efforts and achieve long-term consistency in the panel's composition.

I will consider this problem in two parts:

- whether panel software systems routinely store the data required for monitoring panel health;

- whether these systems provide the tools to measure and diagnose panel health.

The first is a prerequisite of the second. The absence of either or both does not render the task impossible for the panel manager, but it will make the task more onerous and probably more costly, as ad hoc workarounds will be required.

18.12.1   Data required for monitoring panel health

To maintain good panel health, panel management software should allow panel managers to understand the composition of their panel by a very wide range of indicative variables, including:

- profile variables to understand the demographic composition of the sample and how this relates to original sample design criteria;

- the recruitment method or source originally used to acquire the panelist;

- panel history: invitation issues and the history of the member's response those invitations;

- other variables imported from surveys, including survey paradata, such as interview duration, or variables built into surveys to detect satisficing behavior and allow a repeated pattern of such behavior to be detectable from the panelist's history.

These were articulated as six specific capabilities in the review of panel products (see Table 18.16) which revealed that these capabilities were only present in one of the open source products reviewed (LimeSurvey), but were widely supported within the commercial products. As can be seen in Table 18.17, 11 products supported either all or five of the six measures. If these capabilities are not supported by built-in functionality, the panel manager will need to make a much greater effort when measuring panel health, by explicitly importing data – provided the panel software permits such imports.

18.12.2   Tools required for monitoring panel health

Provided that the necessary data are available, the panel management system ideally should also provide the researcher with a range of diagnostic tools for measuring panel health and actively supervising quality.

Table 18.16 Types of information stored in the panel database that is available to researchers or panel or community managers.

Profile data	●●●●●●●●●●●●○◉
Original recruitment method or source	●●●●●●●●●●●○◉◉
Sample selection history	●●●●●●●●●○○◉◉◉
Response history	●●●●●●●●●●●○◉◉
Survey responses	●●●●●●●●●○◉◉◉◉
User-defined data	●●●●●●●●●●○◉◉◉

N = 14 panel software providers Key: ● Supported ○ Planned ◉ Not supported.

Table 18.17 Types of information stored: number of different types supported.

Types supported	All 6	5	4-3	2	1	None
N (providers)	5	6	0	1	0	2

Table 18.18 Panel health and active quality supervision methods.

Across the panel, at a macro level	
Activity and responsiveness	●●●●●●●●●●●●●●○
Quality of response	●●●●○○◉◉◉◉◉◉◉◉
Inactive or dormant panelists	●●●●●●●●●○○◉◉◉
Hyperactivity or multiple panel membership	●●●○○◉◉◉◉◉◉◉◉◉
Suspicious or undesirable activity	●●●○○◉◉◉◉◉◉◉◉◉
At the individual panelist level	
Level of activity	●●●○○◉◉◉◉◉◉◉◉◉
Quality of response	●●●○○◉◉◉◉◉◉◉◉◉
Suspected inappropriate or abusive behavior	●●●○○◉◉◉◉◉◉◉◉◉
Panel health dashboard provided	●●●●●○○○○○◉◉◉◉

N = 14 panel software providers Key: ● Supported ○ Planned ◉ Not supported.

In the survey of providers, I identified five useful quality measures at a macro level, and three measures which could be applied at an individual level, in order to identify panel members who should be retired or removed. In addition, in order to evaluate the extent to which software developers are taking a holistic approach to quality control measures, I asked whether the software provided a simple overview of the different panel health measures in the form of a "dashboard," or single summary screen that highlighted issues of panel health.

As found previously with panel health data, there also is wide variation in the range of capabilities offered for diagnostic tools (see Table 18.18). Overall, however, diagnostic support tools are even less well-developed (Table 18.19). Those selecting panel software products need to examine their requirements in the area of panel health reporting and diagnostics with care.

Some areas are well supported, though. Among the different measures of panel health I identified, activity and responsiveness, assessed from invitation and response history, was very

Table 18.19   Range of panel health and quality methods supported.

Methods supported	All 9	8	7-6	5-4	3-2	1	None
N (providers)	1	0	3	4	2	3	1

N = 14 panel software providers.

widely supported (all of the commercial products support this, and two of the three open source products report on activity), and inactivity or dormant panelists were also readily identifiable.

Yet detecting hyperactivity, multiple panel membership and other means to identify suspicious or undesirable activity – the subject of ESOMAR's Question 22 – was addressed by only three of the commercial providers contacted (with two more planning to) and none of the open source tools.

Similarly, at a micro-level, few of the products reviewed provided tools to assess the quality of response or identify satisficing or fraudulent behavior, in order to identify panel members who need to be asked to leave. Without the tools in place, the burden falls on the panel manager to run her/his own queries or create reports, if the panel database allows this, in order to operate the panel in a way that will satisfy good practice as described by many professional bodies. The data discussed here is the kind of data that the ESOMAR 28 Questions expect panel providers to be able to provide (notably Questions 18–22). Without built-in support, the panel manager will be left to create her/his ad hoc solutions for satisficing and fraud detection, which is an unsatisfactory situation, given how universal the problem is.

Software developers need to be mindful of the pressure coming from the industry and its professional bodies to improve the quality of panels and the accountability of panel providers to their research customers. There is currently a shortfall in what most of the software products provide, with respect to the tools needed to assess panel health and maintain panel quality. It is incumbent on those procuring such systems that they request software developers to provide these capabilities.

18.13   Conclusion and further work

The panel manager is well supported when it comes to choosing a panel management system, though the level of sophistication offered by these products varies considerably. Software, by its definition, continues to develop, and by the time you are reading this, it is certain that some functions identified here as missing will no longer be so.

Many of the lacks identified here do not prevent the panel manager or operator from running a panel with due attention to quality, but this will affect the efficiency and cost-economies achievable, and make the task harder to achieve.

It should also be borne in mind that the survey described in this chapter has only examined publicly available applications. In previous research (Macer & Wilson, 2009), I have identified that a large proportion of the software in use by market research companies has been developed in-house, and the same is true in many academic research centers. It would be surprising if the capabilities in own-developed panel tools consistently exceeded those examined, which have benefited from considerable investment and specialist development effort.

Therefore, in concluding this chapter, it is worth emphasizing those areas where more development appears to be required, as either "further work" for developers, or as areas for those choosing or using panel software to scrutinize.

18.13.1   Recent developments: Communities and mobiles

The practice of online research is also continuing to develop rapidly in some areas, with the emergence of *online communities*. Panel software will need to accommodate more of these qualitative interactions with members – and currently very few of the tools surveyed offer these capabilities.

Smartphone usage has an impact on online research in two different ways, which can also be characterized as an opportunity and a threat. The opportunity is to embrace mobile research as a primary research channel and to do more "moment of truth" and in-location research. This means panel members must be able to do *all* of their panel and survey interactions from their smartphone or tablet. Currently, it is an area only partially supported by products on the market. The threat is that the very short time-limit that a mobile survey permits (Macer & Wilson, 2013), and the high drop-off rate experienced as survey duration increases (Bosnjak, Poggio, & Funke, 2013) will make many kinds of research nonviable.

Many researchers are inclined to dismiss mobile research as a niche application and not relevant to their work. However, this fails to consider the increasing predisposition for survey participants to complete surveys on their smartphones, regardless or not of whether the survey was designed for mobile participation. Panel tools therefore need to provide capabilities to support mobile and tablet access, so as to avoid introducing coverage error because they exclude important sub-populations.

18.13.2   Demands for interoperability and data exchange

Beyond the horizons of research, an increased awareness and interest in big data, in the sense of a large external database of passive data or data collected for other purposes, mean that all data users, including those conducting surveys, are seeking ways to combine data from different sources, and expecting this to be the norm. The impact of this on the survey will mean an increased demand for data linkage and interoperability between the panel system, survey databases and external databases, panel management tools, and external systems.

This too is an area that is currently under-developed in many of the panel tools surveyed. Even though all of the systems are built on industry-standard database platforms, the tools provided to facilitate the flow of data from one tool to another were largely absent from many of the products analyzed here. In particular, developers need to consider implementing web services interfaces, to facilitate better connectivity with other data sources and also other survey tools for data collection or for analysis and reporting.

18.13.3   Panel health

Against a backdrop of universal concern among research professional bodies around the world about the quality of online panels and the research derived from them, active support for the means to measure and maintain quality and "health" in any online panel would seem to be an imperative – in particular, to identify and weed out inactive members and ensuring the replacements achieve a demographically balanced panel. This too is an area where support among providers is patchy and limited. The relevant information that can be accumulated easily in a panel from profile data and from each interaction with panel members is a valuable resource. However, the tools surveyed often appear to be incomplete in this area, and place the responsibility on the panel manager to create and apply their own means of measuring panel health.

18.13.4   Respondent quality

Related to this is general concern in the wider research community over *fraudulent behavior*, particularly on any incentivized research. Widespread abuse has been reported. Panel management software uniquely provides an opportunity to detect such fraud, yet at present, only a small number of the products available support the methods that coordinate the efforts effectively. Verifying emails and checking duplicates is not sufficient. Using digital fingerprinting, within a panel and preferably across panels (including other organizations' panels), and providing the means to build up a pattern of incidents of suspected cheating (speeding, satisficing, and so on) over time is what is needed, provided this is managed ethically. It is not what is being offered at present, except in a minority of solutions.

In summary, it is essential for developers, and for acquirers or users of panel software to recognize that an effective panel management tool must not only support administrative functions, but must also actively support and encourage quality management, which is both a constant and an evolving issue.

References

Baker, J. D. (2013). Online survey software. In M. C. Bocarnea, R. A., Reynolds, & J. D. Baker (Eds.), *Online instruments, data collection, and electronic measurements: Organizational advancements.* (pp. 328–334). Hershey, PA: Information Science Reference.

Baker, R., Blumberg, S. J., Brick, J. M., and others (2010). *AAPOR report on online panels.* Deersville, IL: American Association for Public Opinion Research.

Baker, R., & Downes-LeGuin, T. (2007). Separating the wheat from the chaff: ensuring data quality in internet panel samples. Paper presented at the Fifth International Conference of the Association of Survey Computing: The Challenges of a Changing World, Southampton.

Bethlehem, J. G., & Stoop, I. (2007). Online panels: A paradigm theft? Paper presented at the Fifth International Conference of the Association of Survey Computing: The Challenges of a Changing World, Southampton.

Bosnjak, M., Poggio, T., & Funke, F. (2013). Online survey participation via mobile devices: Findings from seven access panel studies. Paper presented at the The American Association for Public Opinion Research (AAPOR) 68th Annual Conference.

Brüggen, E., Wetzels, M., de Ruyter, K., & Schillewaert, N. (2011). Individual differences in motivation to participate in online panels. *International Journal of Market Research, 53*, 369–390.

Callegaro, M. (2013). Paradata in web surveys. In F. Kreuter (Ed.), *Improving surveys with paradata: Analytic use of process information.* (pp. 261–279). Hoboken, NJ: John Wiley & Sons, Inc.

Cape, P. J. (2011). Quota controls: science or merely sciencey? Paper presented at the General Online Research (GOR), Düsseldorf, Germany.

CASRO. (2011). *Code of standards and ethics for survey research.* Port Jefferson, NY: Council of American Survey Research Organizations (CASRO).

Čehovin, G., & Vehovar, V. (2012). WebSM study: overview of features of software packages. Retrieved from: http://www.websm.org/db/12/15753/.

Comley, P. (2008). Online research communities. *International Journal of Market Research, 50*, 679–684.

Duffy, B., Smith, K., Terhanian, G., & Bremer, J. (2005). Comparing data from online and face-to-face surveys. *International Journal of Market Research, 47*, 615–639.

ESOMAR. (2012). 28 questions to help buyers of online samples. Retrieved from: http://www.esomar.org/knowledge-and-standards/research-resources/28-questions-on-online-sampling.php.

Fitzgerald, J., Gottschalk, P., & Moffitt, R. (1998). An analysis of sample attrition in panel data. *Journal of Human Resources, 33*, 251–299.

Friedrich-Freksa, M., & Metzger, G. (2009). Evaluating two different mobile survey approaches: Personal mobile panel research and ad-hoc mobile portal research. Paper presented at the Mobile Research Conference 2009 (MRC 2009), London.

Funke, F., & Wachenfeld, A. (2012). High potential for mobile web surveys: Findings from a survey representative for German Internet users. Paper presented at the General Online Research (GOR). Retrieved from: http://frederikfunke.net/papers/2012_gor_c.php.

Gittleman, S., & Trimarchi, E. (2010). How attrition/conditioning effects impact response bias in online panels. *Quirk's Marketing Research Review, November*, 48.

Göritz, A. S. (2004). The impact of material incentives on response quantity, response quality, sample composition, survey outcome, and cost in online access panels. *International Journal of Market Research, 46*, 327–345.

Göritz, A. S. (2009). Building and managing an online panel with phpPanelAdmin. *Behavior Research Methods, 41*, 1177–1182.

Hairong, L., & Townsend, L. (2008). Mobile research in marketing: design and implementation issues. *International Journal of Mobile Marketing, 3*, 32–40.

Heist, G. S., & Sanders, M. S. (2008). How do online communities compare to online panels? *Quirk's Marketing Research Review, 38*.

Hughes, K., Jenkins, S., & Wright, G. (1999). Triple-S: A standard within a standard. Paper presented at the Third ASC International Conference, Edinburgh.

Jeavons, A. (2001). Paradata. concepts and applications. Paper presented at the Net Effects 4: Worldwide Internet Conference and Exhibition, 2001, Barcelona, Spain.

Kaczmirek, L. (2008). Internet survey software tools *The SAGE handbook of online research methods*. London: Sage.

Kaminska, O., McCutcheon, A. L., & Billiet, J. (2010). Satisficing among reluctant respondents in a cross-national context. *Public Opinion Quarterly, 74*, 956–984.

Knapton, K., & Garlick, R. (2007). Catch me if you can. *Quirk's Marketing Research Review, November*, 58.

Leiner, D. (2012). *Super convenience samples*. Munich: Institut für Kommunikationswissenschaft und Medienforschung Ludwig-Maximilians-Universität München.

Levine, A., & Holmes, M. G. (2007). Speed traps: A test of seven panels measured the impact of three problem respondent types. *Quirk's Marketing Research Review, July*, 30.

Macer, T. (2008). Technology review of the year. *Research, December*.

Macer, T. (2011). Making it fit: how survey technology providers are responding to the challenges of handling web surveys on mobile devices. Paper presented at the Sixth international conference of the Association of Survey Computing: Shifting the Boundaries of Research, Bristol, UK.

Macer, T., & Wilson, S. (2009). Confirmit annual market research software survey 2008: report and key findings. Retrieved from: http://www.meaning.uk.com/resources/reports/2008-confirmit-mr-software-survey.pdf.

Macer, T., & Wilson, S. (2013). A report on the Confirmit market research software survey. *Quirk's Marketing Research Review, June, 50*.

Poynter, R. (2010). *The handbook of online and social media research*. Chichester: John Wiley & Sons Ltd.

Rivers, D. (2008). The "professional respondent" problem in web surveys. Paper presented at the American Association for Public Opinion Research 63rd Annual Conference, New Orleans.

Stoop, I. (2006). Access panels and online surveys: mystifications and misunderstandings. Paper presented at the DANS symposium: Access Panels And Online Research, Panacea Or Pitfall?, Amsterdam, Netherlands.

Tress, F. (2011). Rich profiles – or: what's the problem with self-disclosure data? Paper presented at the General Online Research (GOR), Düsseldorf.

USDoC. (2000). *Safe Harbor privacy principles*. Retrieved from: http://export.gov/safeharbor/eu /eg_main_018475.asp.

Vicente, P., Reis, E., & Santos, M. (2009). Using mobile phones for survey research. *International Journal of Market Research, 51*, 613–633.

Vonk, E., van Ossenbruggen, R., & Willems, P. (2006). A comparison study across 19 online panels (NOPVO 2006). Paper presented at the Access panels and Online Research, Panacea Or Pitfall? DANS symposium, Amsterdam, Netherlands.

19

Validating respondents' identity in online samples

The impact of efforts to eliminate fraudulent respondents

Reg Baker[a], Chuck Miller[b], Dinaz Kachhi[c], Keith Lange[c], Lisa Wilding-Brown[c] and Jacob Tucker[c]

[a]*Market Strategies International, USA*

[b]*Digital Marketing and Measurement, USA*

[c]*uSamp, Encino, CA, USA*

19.1 Introduction

In the past 15 years we have seen truly dramatic growth in research over the Internet. In 2012, *Inside Research* estimated that online research, a method that barely existed in the mid-1990s, accounted for almost $6 billion in market research revenues worldwide. In the US market research sector where telephone research once dominated, online has edged ahead of telephone as the leading method for quantitative research. ESOMAR (2012a) estimated that in 2011 the global spend on online research was twice that of telephone.

The vast majority of this research relies on samples from online access panels, which, despite their rapid growth, continue to be dogged by questions about data quality (*Inside Research*, 2013). Those questions fall into two broad categories. In the first category are questions about sampling methods. Most US panels rely on volunteers and therefore are susceptible

Online Panel Research: A Data Quality Perspective, First Edition.
Edited by Mario Callegaro, Reg Baker, Jelke Bethlehem, Anja S. Göritz, Jon A. Krosnick and Paul J. Lavrakas.
© 2014 John Wiley & Sons, Ltd. Published 2014 by John Wiley & Sons, Ltd.
Companion website: www.wiley.com/go/online_panel

to selection bias, raising legitimate concerns about the accuracy of inferences made from these samples (Baker et al., 2010). In the second category are questions about the panelists themselves. Many of these questions revolve around concerns that the incentivized nature of panels may encourage some people to go to elaborate lengths to maximize participation to their financial advantage (See, for example, Fulgoni, 2005; Downes-LeGuin, Mechling, & Baker, 2006; Kuwahara-Elrod & Weber 2006).

This concern about what we might call, "respondent quality," manifests itself in several different ways, and so ensuring data quality is a multi-layered task. The focus of this chapter is the challenge posed by people who create false identities as a way to maximize their survey opportunities. Chapter 10 considers the threat to data quality posed by professional respondents, that is, individuals who regularly complete a large number of surveys. Chapter 11 looks at the problem of respondents who rush through questionnaires, seemingly putting forth little or no cognitive effort toward answering the survey's questions.

In an influential 2009 paper, Conklin argued that online data quality assurance practices should include methods to deal with three major problems: (1) respondents using false identities; (2) respondents who complete the same survey more than once; and (3) respondents who rush through the survey without providing honest and thoughtful responses. Conklin found that a sample drawn from a single online panel included 24% of respondents whose identity could not be validated, 3% who took the survey more than once, and 2% who exhibited satisficing behaviors. When two panels were used to draw the sample, Conklin found the proportion of duplicate respondents jumped to 22%. On the specific issue of respondent validation, Conklin examined the mean responses to a set of attribute questions and found that respondents whose identities could not be validated consistently produced higher means than those who could be validated.

Conklin's results reflected an emerging consensus among market researchers that some method of sample hygiene (i.e., cleaning) is essential when using online panels. Indeed, it already was common practice in some companies to clean out so-called "bad respondents" in post survey processing (Baker & Downes-LeGuin, 2007). The international standard, "ISO 26362 – Access Panels in Market, Opinion, and Social Research," (ISO 2009) recognizes the use of such processes and requires that the specific methods used and the number of excluded respondents be documented in reports to clients. ESOMAR (2012b) recommends that buyers of online sample first ask the prospective supplier whether it has in place "a confirmation of respondent identity procedure" (p. 15.).

Most large panel companies have developed some processes to protect against false registrations. Several companies provide this service along with a number of other data cleansing operations such as identifying duplicate respondents or those whose response patterns indicate lack of engagement (see, e.g., Imperium, 2010; MarketTools, 2011). All of these services compare the basic personally identifying information (hereafter referred to as PII) of potential panelists or respondents in the United States to third party databases purporting to have data on a very high percentage of US households. In the simplest terms, if a prospective respondent's name, address match, and date of birth can be matched to one or more of these sources, then he or she is assumed to be a real person.

But how effective are these validation processes and what impact, if any, do they have on the validity of survey results? In 2011, Courtright and Miller reported on a study designed to answer these questions. A significant number of respondents in their study refused to provide basic PII (name, address, and date of birth) so that their identity could be validated. Among those who provided their PII, 15%–20% failed validation, depending on the validation service used. Further, Courtright and Miller found that the validation process had a demographic bias and that respondents who were validated sometimes answered attitudinal and behavioral questions differently from those who did not. Their findings suggest that the presence of non-validated respondents in a sample can lead to bias.

The net result was the exclusion of large numbers of people from Courtright and Miller's survey, as much as a third of the sample, either because they refused to provide the information needed to validate their identity or because the information provided could not be successfully matched to a valid name and address. Given the high demand for online survey respondents and especially less-experienced respondents, this shrinking of the pool of available respondents is troubling. It becomes more troubling still when one considers that these excluded respondents were sometimes demographically, behaviorally, and attitudinally different from those completing the survey, suggesting the possibility that the validation process may be increasing bias.

In 2012, we designed a similar study to take a second and more detailed look at this problem. Results from this second study are the primary focus of this chapter. But before moving to a discussion of the research, we feel it important to clarify our use of the term *validation*. Those familiar with the historical lexicon of survey research may be confused by our use of the term here. In the context of data collection, validation often refers to a set of procedures designed to "ensure that data collection was conducted according to an agreed upon specification" (ISO 20252, 2012, p. 7). For example, telephone interviews are sometimes validated by a second interviewer calling a previously interviewed respondent to verify that the person was interviewed and the answers to selected questions recorded correctly. In this chapter we use the term differently, as shorthand for validation of identity, meaning a set of procedures designed to verify that a person registering for an online panel or otherwise responding to an online survey request is a real person who can be tied to a specific name, postal address, and perhaps other identifiers. When we refer to a *validated respondent*, we mean an individual whose identity has been confirmed by reference to third party records and databases.

19.2   The 2011 study

The design of the Courtright and Miller's (2011) study was very straightforward. A short questionnaire comprised of demographic, lifestyle, attitudinal, and behavioral items was administered online to a sample of 7200 US adults. The study was fielded in the first two weeks of January 2011 and the average interview length was 3.6 minutes. Roughly 60% of the sample came from the DMS Insights panel with the remainder from third party sources. Quotas were set for age, gender, income, and ethnicity based on the 2010 US Census. Prior to beginning the survey, respondents were asked to share their name, postal address, and date of birth. Post data collection the PII for those who agreed to supply it was sent to four validation services. Each service flagged those records where the respondent's identify could be verified, sometimes coding by level of certainty.

Results from the study were intriguing:

- Of the 7200 people who responded to Courtright and Miller's survey invitation, 16% refused to provide their name, mailing address and date of birth.

- The average rate of validation across the four services for the 6000 respondents who supplied their PII was 83%.

- There were important demographic differences between those who could be validated and those who could not. Respondents under 25 years old (especially males) and those claiming to be of Asian or Hispanic descent validated at a much lower rate than others. People under 25 validated at about half the rate of the sample as a whole. Validation rates for Asians and Hispanics were 20–30 points lower than whites and African-Americans.

- There also were important differences in behavioral characteristics relative to technology use, buying behavior, and home ownership.

- Finally, even in this very short questionnaire, Courtright and Miller found evidence of classic poor respondent behaviors including straight-lining and speeding. Respondents whose identity could not be validated were about 50% more likely to exhibit a response quality problem than those who were validated.

Overall, this study offered both good and bad news. On the one hand, validation seemed to find and eliminate a number of poor respondents – people who did not take the survey seriously. This was encouraging. Unfortunately, validation also eliminated people disproportionally from some of the otherwise hardest to reach demographic groups – those under 25 years old, Asians, and Hispanics. Further, differences in consuming behavior among those who could be validated and those who could not signaled a potential bias in consumption patterns that might be very troubling, especially to market researchers.

19.3   The 2012 study

In 2012 we replicated the 2011 study to see how, if at all, the dynamics and impact of the validation process might have changed and whether the impact on sample composition was as significant as it had appeared in the original study. We hypothesized that the relatively high rate of noncompliance with the PII request in 2011 might have receded somewhat as such requests have become more commonplace in online surveys. In addition, we wanted to see how improvements that validation service providers claim to have made in their processes were affecting the rate of validation.

We also planned to extend the analysis beyond differences between those who validated and those who did not and to assess the overall impact on the characteristics of the achieved sample. Here we hypothesized that while differences reported in the 2011 study might persist, they also might wash out in the overall sample because the proportion of respondents eliminated due either to refusing to supply PII or failing to validate was relatively small. Finally, we wanted some new data with which to speculate about reasons why some people refused the request for PII and the implications for online research going forward.

The questionnaire (Appendix 19.A) for the 2012 study was essentially the same as that used in 2011, with only a few modest changes. We added questions to assess respondents' views of online privacy and their comfort with purchasing online. We also asked about the type and frequency of their Internet use.

As in 2011, respondents were asked prior to beginning the survey to provide their name, postal address, and date of birth. Even respondents who refused to supply their PII were allowed to continue, so that we could compare the questionnaire responses of those who did not cooperate with the request for PII to those who did. At the end of the questionnaire we added a question that asked respondents to supply their email address because some validation services use this information in their validation process.

Table 19.1 Distribution of completed interviewers by age, gender, ethnicity, and income compared to targets.

		Completed interviews (%)	From uSamp (%)	From river (%)	Target (%)
Age	18–34	26	26	27	33
	35–54	39	35	43	34
	55+	35	40	30	33
Gender	Male	45	50	41	50
	Female	55	50	59	50
Ethnicity	Caucasian	73	62	83	67
	African-American	12	17	8	12
	Asian	4	3	4	4
	Other	2	2	3	2
	Hispanic	16	25	7	15
N			7435	3558	3877

The study was fielded from July 5, 2012, through August 6, 2012, to an online sample comprised of approximately 50% members of uSamp's panel and 50% from other panels, including river samples. Age, gender, income, and race/ethnicity quotas were used to achieve a sample that approximated the distributions in the US population as reported by the 2010 US Census. We collected a total of 7435 completed interviews and the mean completion time was 5.1 minutes. Table 19.1 shows the distribution by age, gender, and ethnicity for all completed surveys.

The PII (name, address, date of birth, and email address) for respondents providing it was sent to four different validation services. Three of the services were the same as those used in the 2011 study and a fourth service, not previously used, was added. Each of these services uses somewhat different methods and sources for determining whether a respondent is indeed a real person, and they report their results in somewhat different ways. As a condition of their participation the services asked that they not be identified by name, but each provided a brief description of their sources and methods.

- One service uses public records such as driver's license and property tax records to find respondents and match address and date of birth. It sends back multiple validation indicators such as address does not match, date of birth does not match, multiple records found, newer record found, or no date of birth available, among other indicators. Based on these indicators, it also sends a decision flag of match, partial match, or no match. For this research we considered only respondents who fully matched to be valid.

- The second service uses a proprietary and undisclosed method. It uses both address and email address to determine whether person is real. There is no partial match category.

- The third service uses proprietary data from telephone companies to assign a confidence score for each person (name, address, email address). The score is from 0–99 and then grouped into five different tiers with a higher score equaling a stronger association. For this analysis, those in tier A (70+ score) were considered fully validated.

- The final service uses multiple third party data sources to return a six point (0–5) scoring system that reflects whether it is appropriate to accept or reject individuals. It can match based on name, address, email address, and/or phone number. In addition to this scoring system, normalized data are provided. We only considered respondents in the top category to be validated.

At some future date we plan to analyze how these different validation approaches impact sample compositions, within and across providers. However, in the analysis that follows those differences are ignored.

19.4   Results

As in 2011, a significant proportion of the sample (24%) refused to provide their PII. Of the 1790 respondents who did not provide all of the requested PII, 19% provided name and address but not email; 42% provided email address but nothing else; and 39% provided no information at all. Put another way, roughly one in four respondents refused to comply with our PII request, a higher rate of noncompliance than the 17% observed in the 2011 study. However, in 2012, we added a request for email address because some services use email address in their validation process. We adjusted the noncompliance rate for those who provided name, address, and data of birth but refused on email address. This adjusted rate was 19%, slightly higher than the 17% reported by Courtright and Miller in 2011. Thus, despite the trend toward more surveys requesting PII, potential respondents seemed no more willing to provide their PII in 2012 than in 2011.

There were a number of statistically significant differences between those willing to provide PII versus those who refused, but in many instances those differences were quite small. For example, 77% of respondents who said that they read the Sunday newspaper regularly provided their PII while the comparable figure for those saying they didn't regularly read the Sunday newspaper was 74%. We put questions with these significant but small differences aside in our analysis and focused instead on those where differences were sufficiently large to be potentially important.

Table 19.2 shows six questionnaire items where there were significant differences between those providing PII versus those refusing. Younger people, especially those under the age of 25, were less likely to provide full PII than older people, with the exception of those aged 60 and over. Respondents identifying as either Asian or Other refused PII at a somewhat higher rate than other groups. Respondents who said they spent less than 10 hours a week online were also less likely to provide full PII than those who were online more often.

There also were potentially important differences in willingness to provide PII among those worried about online privacy compared to those who were less worried. Among respondents who said they were "very worried" about their privacy, 70% provided their PII compared to 85% of those who said they were "not very worried." In the attitudinal battery, we asked 11 agree/disagree items using a seven-point scale. Two of those questions related to privacy. We recoded the scale to two values: 1–5 at the bottom and 6–7 at the top (Top 2 Box) for respondents indicating a high level of agreement. Eighty percent of respondents who fell in the Top 2 Box for the item, "I am comfortable making purchases online," provided their full PII, while 71% of those answering 1–5 complied with the PII request. For the item, "I am very cautious about sharing my personal information online," 69% of those in the Top 2 Box provided their PII versus 81% of those choosing 1–5.

Table 19.2 Items with significant differences between those providing PII and those refusing.

		Provided PII (%)	Refused PII (%)
Age	18–24	6.2	8.5
	25–34	19.3	20.5
	35–44	17.2	15.5
	45–59	34.6	30.3
	60 and over	22.7	25.2
Ethnicity	Caucasian	73.6	69.3
	African-American	12.2	13/3
	American Indian	1.1	0.9
	Asian	3.3	5.0
	Other	0.8	1.3
	Hispanic	15.8	14.9
Hours online (weekly)	Less than 10	37.1	47.9
	10 or more	62.9	52.1
Privacy online	Very worried	18.6	25.3
	Somewhat worried	50.4	50.4
	Neutral	20.1	17.0
	Not very worried	9.1	5.2
	Not worried at all	1.9	2.1
Agree/Disagree	Comfortable making purchases online (Top 2 Box)	63.4	50.9
	Very cautious about sharing personal information online (Top 2 Box)	42.1	58.4
N		5640	1790

19.4.1   Outcomes from the validation process

We sent all 5640 respondents who provided their full PII to four validation services. As noted above, three of those services were used in the 2011 study and an additional service was added. Different services report their results differently. One service returns a simple yes/no. Another service reports two levels of validation: full and partial. Still another service reports a range of certainty on a scale of 1–5. Within those limitations the percentage validated by clearing the highest bar for all services varied from a low of 66% to a high of 95% with an average of 85%. This average was essentially the same as the 83% average reported by Courtright and Miller for the 2011 study. Looking just at the three services used in both 2011 and 2012, the means diverge more substantially, from 85% in 2011 to 91.3% in 2012.

Overall, there were few differences between those who validated and those who did not. As with PII cooperation, younger respondents validated at a lower rate than older respondents. For example, 88% of those aged 18–25 validated compared to 98% for those aged 35–44 and 98% for the 45–59 groups. Ninety percent of Asians were validated compared to 96% of Hispanics, 97% of African-Americans, and 98% of whites.

Somewhat surprisingly, there were no differences in home ownership, having a bank account, or owning a credit card. There also were no significant differences in responses to the three privacy questions.

Finally, there was one especially important difference that links directly to data quality. Respondents who failed verification were nearly three times as likely to fail at least one quality check. The most frequent failure was answering inconsistently on two of the agree/disagree items in the attitudinal battery. One item said, "Price is more important to me than brand names." The other said, "Brand names are more important to me than price." Among those who did not validate, 18% chose conflicting answers on those items versus 6% for those who validated.

19.4.2   The impact of excluded respondents

The last step in our analysis was to assess the impact of the validation process on the final sample. We started with 7435 interviews. We first dropped 24% of the total because they refused to provide PII. We dropped another 2% because they could not be validated by at least one provider. That left us with a sample of 5475 validated respondents. We then compared this achieved sample to the original sample of 7435. We wanted to know what impact, if any, exclusion of almost one in four respondents would have on the demographic makeup of the sample, the behavioral items we included, and attitudes on a number of issues.

The only area where we saw changes of any magnitude was in the demographic distributions (Table 19.3) and even those were small, generally less than 2 percentage points. There were no meaningful differences in ownership of technology products or most of the attitudinal measures. However, there were two exceptions. The percentage of respondents in the final sample who said they worried about sharing personal information online (Top 2 Box) dropped from 46% in the original sample to 42% after validation. The percentage saying they were comfortable buying online (Top2 Box again) increased from 60% to 64%. Finally, the

Table 19.3   Impact of exclusions on demographic distributions.

		Original sample (%)	Sample after exclusions (%)	Excluded sample (%)
Age	18–24	7	6	10
	25–34	20	19	22
	35–44	17	17	16
	45–59	34	35	29
	60 and over	23	23	24
Gender	Male	45	44	49
	Female	55	56	52
Ethnicity	Caucasian	73	74	68
	African-American	12	12	13
	American Indian	1	1	1
	Asian	4	3	6
	Other	1	1	1
	Hispanic	16	16	16
N		7435	5475	1960

removal of almost 25% of the sample due to the validation process had no impact on the percentage of respondents failing one or more response quality checks.

19.5   Discussion

We undertook this research to replicate Courtright and Miller's (2011) study with some augmentations that we hoped would help us understand the dynamics of respondent validation in more detail. We focused on two specific issues.

The first area of focus was the refusal by a large proportion of the sample to provide name, postal address, birth date, and email address so that we could validate their identity. We had hypothesized that respondents might be more willing to provide their PII given that doing so has become an increasingly common part of online research. Such was not the case. An even greater percentage of respondents refused to provide PII in 2012 than did in 2011. Twenty-four percent refused in 2012 as compared with 17% in 2011. Even after adjusting for having asked for an additional bit of PII in 2012 (email address), the rate for people refusing to provide PII was higher in 2012 than in 2011.

We speculated that the somewhat higher refusal rate in 2012 might be due to sample source. The sample was comprised partly of members of the uSamp panel (who also hosted the survey) and part from other panels. It seemed reasonable that a respondent who already had a relationship with uSamp might be more willing to give us their PII than a respondent from some other panel with no prior relationship. Such was not the case. There were no differences in cooperation by sample source. It seems that, at least for now, researchers must accept a high rate of refusal to questions asking for PII.

Our data were more helpful in understanding why people may be reticent to provide their PII. We saw elevated concerns about privacy and sharing of information online among those who refused PII versus those who provided it. The refusal group also spent less time online in general and made fewer online purchases than those who provided their PII. All of which is not to say that not everyone who refused did so because of online privacy concerns or a general reticence about online in general. Some may well have refused because they planned to exploit the survey opportunity solely for the incentive, though here we note that those refusing to provide PII were no more likely to show a poor quality score than those who provided it.

A second concern was the rate at which those providing PII validated. Recall that when we looked just at the three validation services used in both 2011 and 2012 that the average validation rate across the surveys rose from 85% in 2011 to 91% in 2012. The validation rate for one service increased from 86% in 2011 to 95% in 2012; another increased from 81% to 84%; and a third increased from 87% to 95%. In addition, all three of these services substantially increased their validation rates of 18–24-year-olds, Hispanics, and Asians.

There are two possible explanations. One is that the validation services have improved their methods. Using a broader set of sources and improved matching algorithms may be resulting in fewer false negatives. A second possibility is that the widespread use of validation is discouraging fraudulent respondents from attempting to subvert the survey process. We cannot tell which of these forces might be at work here. However, we note that respondents who did not validate scored poorly on our response quality measurement compared those who validated. So validation techniques are finding at least some poor respondents and disqualifying them.

Finally, and perhaps most importantly, there is the issue of the impact of the validation process on sample composition and the potential for bias. The combination of refusal to provide

PII and the validation process itself resulted in about one in four prospective respondents being removed from the sample. And as we saw, those removed respondents were different in a number of ways – younger; more likely to be Asian; worried about privacy (especially online); and generally spending less time online. As it turns out, these differences, while significant, are not of sufficient magnitude to seriously alter the targeted demographic distribution of the sample, nor were their major changes in the behavioral or attitudinal measures in our short questionnaire. Despite losing people from some of the hardest groups to get online (young people, non-white ethnics, people who spend less time online, etc.), the overall character of the sample was little changed. While we found evidence of poor survey-taking behavior among respondents who did not validate, that group was so small that their overall impact on the sample also was small and possibly ignorable.

19.6   Conclusion

Validation of respondent identify is one of several responses to perceived shortcomings in the quality of data coming from online samples. While published sources are few, there is a strongly held belief among researchers that some unknown proportion of online respondents are gaming the system and deliberately producing bad data. Whether the magnitude of that behavior is sufficient to produce poor results and, as Conklin claims, "bad business decisions" is debatable.

Our findings from this study and the earlier study by Courtright and Miller suggest that validation is something of a mixed bag. It seems to root out some bad actors, which is the main reason for doing it. But these bad actors are easily caught by other quality checks. Even if they are not, it is not clear that they exist in such numbers that they might influence a survey's results. Online research has not cornered the market on poor performing survey respondents and it may well be that the problem is no worse with online than with other offline methods.

On the other hand, the validation process appears to be excluding a significant number of willing respondents at a time when the demand for respondents is high. Most are being excluded only because they are reticent about sharing their personal information online. Further, a significant number of these excluded respondents have characteristics that can be difficult to get in an online sample. It seems fair to ask whether the benefits of validation are sufficiently great to balance the loss of as much as a quarter of the sample for a study – and a higher percentage among certain demographic groups.

Respondent validation also faces some very significant challenges going forward. From the outset there have been those who have wondered about its legality (Feinberg, 2009). While that issue has been largely settled in the United States, it is unclear whether validation will ever be possible in countries where privacy protections are more stringent or where the data available for matching are less complete and robust. In addition, as online sample suppliers increasingly move to dynamic sourcing models under which samples are drawn more or less in real-time from traditional panels, social networks, river samples and web intercepts, one wonders whether the proportion of people willing to provide their PII will increase, decrease, or remain the same.

All of which raises the question of whether validation should be done at all. On the one hand, the results of this study suggest that the problems described by Conklin may not be as widespread or as detrimental to data quality as he claimed. Validation also may be excluding a potentially significant number of qualified respondents from online surveys. On the other hand, there continue to be significant concerns about the quality of online data and a strong respondent validation process presumably offers some assurances to data users, lending greater face validity to survey results.

Despite our findings in this study and the earlier study by Courtright and Miller, we do not recommend that validation no longer be used. However, we believe that more such studies are needed to understand the process dynamics more clearly so that the risks and benefits can be more fully assessed.

References

Baker, R., Blumberg, S. J., Brick, J. M., Couper, M. P., Courtright, M., Dennis, J. M., Dillman, D. A., et al. (2010). Research synthesis: AAPOR report on online panels. *Public Opinion Quarterly, 74,* 711–781.

Baker, R., & Downes-LeGuin, T. (2007). Separating the wheat from the chaff: Ensuring data quality in Internet samples In M. Trotman, T. Burrell, L. Gerrard, K. Anderton, G. Basi, M. Couper, K. Moris, et al. (Eds.), in *Proceedings of the Fifth International Conference of the Association for Survey Computing: The Challenges of a Changing World* (pp. 157–166). Berkeley, UK: ASC.

Conklin, M. (2009). What impact do "bad" respondents' have on business decisions? Retrieved from: http://www.markettools.com/downloads/WP_BadRespondents.pdf?mkt_tok=3RkMMJWWfF9.

Courtright, M., & Miller, C. (2011). Respondent validation: So many choices! Paper presented at the CASRO Online Research Conference, Las Vegas, NV.

Downes-Le Guin, T. (2005). Satisficing behavior in online panelists: What's a research buyer to do? Paper presented at the MRA Annual Conference, Chicago, IL.

Downes-LeGuin, T., Mechling, J., & Baker, R. (2006). Great results from ambiguous sources: Cleaning Internet panel data. Paper presented at the ESOMAR Panel Research Conference, Barcelona, Spain.

ESOMAR (2012a). *Global Market Research: An ESOMAR Industry Report*. Amsterdam: ESOMAR.

ESOMAR (2012b). 28 questions to ask buyers of online samples. Retrieved from: http://www.esomar.org/knowledge-and-standards/research-resources/28-questions-on-online-sampling.php.

Feinberg, H. (2009). Is respondent validation legal? Weeding out cheater, repeaters and professional respondents could bring legal and regulatory problems. *MRA's Alert Magazine*. October, pp. 39–40.

Fulgoni, G. M. (2005). The professional respondent problem in online survey panels today. Paper presented at the MRA Annual Conference, Chicago, IL.

Imperium (2010) World's top panel companies join forces with Imperium to advance data quality across the market research industry. Retrieved from: http://www.imperium.com/press/pdfs/ICE.pdf.

Inside Research. (2012). Strong growth in WW online spend. March, 4–5.

Inside Research. (2013). Special report: U.S. online MR spend stuck. January, 15–17.

International Organization for Standardization. (2009). ISO 26362:2009 – Access panels in market, opinion and social research – Vocabulary and service requirements. Geneva, Switzerland.

International Organization for Standardization. (2012). ISO 20252:2006 – Market, opinion and social research – Vocabulary and service requirements. Geneva, Switzerland.

Kuwahara-Elrod, M., & Weber, M. B. (2006). Ensuring respondent quality in online research. Retrieved from: http://www.sigmavalidation.com/tips/06_02_07_Online_Research_White_Paper.pdf.

Markettools (2011). Online research quality: The next frontier. Retrieved from: https://www.markettools.com/sites/default/files/resources/white_paper/TrueSample_Online_Research_Quality-The_Next_Frontier_whitepaper.pdf.

Appendix 19.A

19.A.1 Questionnaire: Respondent Validation Research: Phase II

MASTER SCREENER QUESTIONS (OP screener – use same questions for all samples)

Overall Respondent Qualifications:

	N = 6000
Gender Quota	
Male	50%
Female	50%
Age Quota	
Adults 18–34	33%
Adults 35–54	34%
Adults 55+	33%
Income Quota	
Under $50K	30%
$50K– $99K	40%
$100K+	30%
Ethnicity Quota	
White	67%
Hispanic/Latino	15%
African American	12%
Asian	4%
Other	2%
Sourcing Quota	
Returning Member	50%
uSamp	25%
Third Party	20%
Other	5%

COLLECT NAME, ADDRESS, AND BIRTHDATE (INCLUDE PRIVACY POLICY STATEMENT)

BASE: ALL RESPONDENTS

Q1. **Which of the following do you own? (select all that apply) [ROTATE except last option]**

01 Wifi/wireless internet system (in your home)

02 Laptop

03 Portable MP-3 player / IPOD

04 Tablet /e-Reader (such as iPad, Kindle Fire, Samsung Galaxy, etc.)

05 DVR/PVR (such as TiVO)

06 Smartphone (such as iPhone, Droid, or Blackerry)

07 Videogame system that attaches to a TV set (Wii, Playstation, X-Box, Nintendo GameCube, etc.)

08 Hand-held videogame system (such as PSP, Game Boy, DS Lite, etc.)

09 Hybrid car

10 HDTV

11 None of these [ANCHOR]

BASE: ALL RESPONDENTS

Q2. In general, how worried would you say you are about your personal privacy?

01 Very worried

02 Somewhat worried

03 Neutral

04 Not very worried

05 Not worried at all

BASE: ALL RESPONDENTS

Q3. How much do you agree or disagree with the following statements? [GRID; ROTATE]

7 Strongly Agree … … … … … 1 Strongly Disagree

01 Buying American products is important to me.

02 I like to shop around before making a purchase.

03 I tend to be an impulse buyer, making purchase decisions very quickly.

04 If I really want something I will buy it on credit card rather than wait.

05 Price is more important to me than brand names.

06 Brand names are more important to me than price.

07 I'm not all that worried about environmental issues.

08 I would be willing to accept a lower standard of living to conserve energy.

09 I am very concerned about global warming.

10 I am very cautious about sharing any personal information online, even my name and address.

11 I am comfortable making purchases online.

BASE: ALL RESPONDENTS

> **Q4.** Please indicate the approximate number of hours spent on the following in a given week?

	Less than an hour	1–3 hours	3–6 hours	6–9 hours	More than 9 hours
Accessing Internet					
Accessing Social Media Network(s)					

BASE: ALL RESPONDENTS

> **Q5.** Do you … ? [YES/NO GRID; ROTATE]
>
> 01 Read the Sunday newspaper?   Yes / No
>
> 02 Hold a valid US passport?   Yes / No
>
> 03 Own your own home (Not rent)?   Yes / No
>
> 04 Have a bank account?   Yes / No
>
> 05 Smoke cigarettes at least occasionally?   Yes / No

BASE: ALL RESPONDENTS

> **Q6.** Please indicate the number of personal email accounts you actively use.
>
> 01 One
>
> 02 Two
>
> 03 Three
>
> 04 Four or more
>
> 05 I don't have a personal email account.

BASE: ALL RESPONDENTS

> **Q7.** How many credit cards do you own?
>
> 01 0
>
> 02 01
>
> 03 02
>
> 04 03
>
> 05 More than three

BASE: ALL RESPONDENTS

Q8. Have you owned a home within the last 3 years?

01 Yes

02 No

03 Not sure

BASE: ALL RESPONDENTS

Q9. Do you consider yourself to be left-handed, right-handed, or use both hands equally?

01 Left-handed

02 Right-handed

03 Use both hands equally

BASE: ALL RESPONDENTS

Q10. Which of those brands of mouthwash have you ever purchased? [MULTI SELECT GRID; ROTATE]

01 Biotene	Heard Of / Purchased/ Neither
02 Crest Pro Health Rinse	Heard Of / Purchased/ Neither
03 Lavoris	Heard Of / Purchased/ Neither
04 Listerine	Heard Of / Purchased/ Neither
05 Reach ACT	Heard Of / Purchased/ Neither
06 Scope	Heard Of / Purchased/ Neither
07 Tom's of Maine	Heard Of / Purchased/ Neither
08 Targon	Heard Of / Purchased/ Neither

BASE: ALL RESPONDENTS

Q11. Which one of the following categories best describes your educational status?

01 Less than high school graduate

02 High school graduate

03 Some college (includes two-year Associate degrees)

04 Trade, technical or vocational school

05 College graduate

06 Post-graduate work or degree

07 Prefer not to answer

BASE: ALL RESPONDENTS

Q12. What is your employment status?

01 Employed full time

02 Employed part time

03 Retired

04 A homemaker

05 Not working right now

06 A student

07 Other

08 Don't know

09 Prefer not to answer

Lastly, we would like to collect your email address. This information will be used entirely for research purposes, and will be kept entirely confidential by the party conducting the research. It will not be used for any form of solicitation. By providing this information you will be entered into a sweepstakes to win an iPad. NO PURCHASE NECESSARY TO ENTER OR WIN. To review our official rules, eligibility requirements and alternative methods of entry, please review our Official Sweepstakes Rules at www.ipadsweepstakesgiveaway.com. Are you willing to provide your email address?

1. Yes

2. No
   [IF NO IS SELECTED GO TO END]

Email address: [   ]

Verify email address: [   ]

[MAKE SURE BOTH EMAIL ADDRESSES ARE THE SAME]

Index

Page numbers in *italics* refer to Figures, those in **bold** refer to Tables.

Online Panel Research: A Data Quality Perspective, First Edition.
Edited by Mario Callegaro, Reg Baker, Jelke Bethlehem, Anja S. Göritz, Jon A. Krosnick and Paul J. Lavrakas.
© 2014 John Wiley & Sons, Ltd. Published 2014 by John Wiley & Sons, Ltd.
Companion website: www.wiley.com/go/online_panel

WILEY SERIES IN SURVEY METHODOLOGY
Established in Part by WALTER A. SHEWHART AND SAMUEL S. WILKS

Editors: *Mick P. Couper, Graham Kalton, J. N. K. Rao, Norbert Schwarz, Christopher Skinner*
Editor Emeritus: *Robert M. Groves*

The *Wiley Series in Survey Methodology* covers topics of current research and practical interests in survey methodology and sampling. While the emphasis is on application, theoretical discussion is encouraged when it supports a broader understanding of the subject matter.

The authors are leading academics and researchers in survey methodology and sampling. The readership includes professionals in, and students of, the fields of applied statistics, bio-statistics, public policy, and government and corporate enterprises.

*Now available in a lower priced paperback edition in the Wiley Classics Library.

*HANSEN, HURWITZ, and MADOW · Sample Survey Methods and Theory, Volume II: Theory

HARKNESS, BRAUN, EDWARDS, JOHNSON, LYBERG, MOHLER, PENNELL, and SMITH (editors) · Survey Methods in Multinational, Multiregional, and Multicultural Contexts

HARKNESS, VAN DE VIJVER, and MOHLER (editors) · Cross-Cultural Survey Methods

HUNDEPOOL, DOMINGO-FERRER, FRANCONI, GIESSING, NORDHOLT, SPICER, and DE WOLF · Statistical Disclosure Control

KALTON and HEERINGA · Leslie Kish Selected Papers

KISH · Statistical Design for Research

*KISH · Survey Sampling

KORN and GRAUBARD · Analysis of Health Surveys

KREUTER (editor) · Improving Surveys with Paradata: Analytic Uses of Process Information

LEPKOWSKI, TUCKER, BRICK, DE LEEUW, JAPEC, LAVRAKAS, LINK, and SANGSTER (editors) · Advances in Telephone Survey Methodology

LESSLER and KALSBEEK · Nonsampling Error in Surveys

LEVY and LEMESHOW · Sampling of Populations: Methods and Applications, *Fourth Edition*

LUMLEY · Complex Surveys: A Guide to Analysis Using R

LYBERG, BIEMER, COLLINS, de LEEUW, DIPPO, SCHWARZ, TREWIN (editors) · Survey Measurement and Process Quality

LYNN · Methodology of Longitudinal Surveys

MADANS, MILLER, and MAITLAND (editors) · Question Evaluation Methods: Contributing to the Science of Data Quality

MAYNARD, HOUTKOOP-STEENSTRA, SCHAEFFER, and VAN DER ZOUWEN · Standardization and Tacit Knowledge: Interaction and Practice in the Survey Interview

PORTER (editor) · Overcoming Survey Research Problems: New Directions for Institutional Research, No. 121

PRESSER, ROTHGEB, COUPER, LESSLER, MARTIN, MARTIN, and SINGER (editors) · Methods for Testing and Evaluating Survey Questionnaires

RAO · Small Area Estimation

REA and PARKER · Designing and Conducting Survey Research: A Comprehensive Guide, *Third Edition*

SARIS and GALLHOFER · Design, Evaluation, and Analysis of Questionnaires for Survey Research, *Second Edition*

SÄRNDAL and LUNDSTRÖM · Estimation in Surveys with Nonresponse

SCHWARZ and SUDMAN (editors) · Answering Questions: Methodology for Determining Cognitive and Communicative Processes in Survey Research

SIRKEN, HERRMANN, SCHECHTER, SCHWARZ, TANUR, and TOURANGEAU (editors) · Cognition and Survey Research

SNIJKERS, HARALDSEN, JONES, and WILLIMACK · Designing and Conducting Business Surveys

STOOP, BILLIET, KOCH and FITZGERALD · Improving Survey Response: Lessons Learned from the European Social Survey

SUDMAN, BRADBURN, and SCHWARZ · Thinking about Answers: The Application of Cognitive Processes to Survey Methodology

*Now available in a lower priced paperback edition in the Wiley Classics Library.

Printed in the United States
By Bookmasters